Springer Series in Information Sciences 20

Editor: Thomas S. Huang

Springer Series in Information Sciences

1 **Content-Addressable Memories**
By T. Kohonen 2nd Edition

2 **Fast Fourier Transform and Convolution Algorithms**
By H. J. Nussbaumer 2nd Edition

3 **Pitch Determination of Speech Signals**
Algorithms and Devices By W. Hess

4 **Pattern Analysis** By H. Niemann

5 **Image Sequence Analysis**
Editor: T.S. Huang

6 **Picture Engineering**
Editors: King-sun Fu and T.L. Kunii

7 **Number Theory in Science and Communication** With Applications in Cryptography, Physics, Digital Information, Computing, and Self-Similarity By M.R. Schroeder
2nd Edition

8 **Self-Organization and Associative Memory** By T. Kohonen 3rd Edition

9 **Digital Picture Processing**
An Introduction By L.P. Yaroslavsky

10 **Probability, Statistical Optics and Data Testing** A Problem Solving Approach By B.R. Frieden

11 **Physical and Biological Processing of Images** Editors: O.J. Braddick and A.C. Sleigh

12 **Multiresolution Image Processing and Analysis** Editor: A. Rosenfeld

13 **VLSI for Pattern Recognition and Image Processing** Editor: King-sun Fu

14 **Mathematics of Kalman-Bucy Filtering**
By P.A. Ruymgaart and T.T. Soong
2nd Edition

15 **Fundamentals of Electronic Imaging Systems** Some Aspects of Image Processing By W.F. Schreiber

16 **Radon and Projection Transform-Based Computer Vision** Algorithms, A Pipeline Architecture, and Industrial Applications By J.L.C. Sanz, E.B. Hinkle, and A.K. Jain

17 **Kalman Filtering** with Real-Time Applications By C.K. Chui and G. Chen

18 **Linear Systems and Optimal Control**
By C.K. Chui and G. Chen

19 **Harmony: A Psychoacoustical Approach** By R. Parncutt

20 **Group-Theoretical Methods in Image Understanding** By Kenichi Kanatani

21 **Linear Prediction Theory**
A Mathematical Basis for Adaptive Systems By P. Strobach

22 **Optical Signal Processing**
By P.K. Das

Kenichi Kanatani

Group-Theoretical Methods in Image Understanding

With 138 Figures

Springer-Verlag Berlin Heidelberg New York
London Paris Tokyo Hong Kong

Professor Kenichi Kanatani Ph.D.
Department of Computer Science, Gunma University
Kiryu, Gunma 376, Japan

ISBN 3-540-51253-5 Springer-Verlag Berlin Heidelberg New York
ISBN 0-387-51253-5 Springer-Verlag New York Berlin Heidelberg

Library of Congress Cataloging in Publication Data.
Kanatani, Kenichi, 1947– Group theoretical methods in image understanding/Kenichi Kanatani. p. cm. — (Springer series in information sciences; vol. 20) Includes bibliographical references.
ISBN 0-387-51253-5 (U.S.)
1. Image processing—Digital techniques. 2. Computer graphics. I. Title. II. Series: Springer series in information sciences: 20.

Typesetting: Macmillan India Ltd., Bangalore 25
Offset printing: Colordruck, Berlin. Bookbinding: Lüderitz & Bauer-GmbH, Berlin.
2154/3020-543210—Printed on acid-free paper.

Preface

Image understanding is an attempt to extract knowledge about a 3D scene from 2D images. The recent development of computers has made it possible to automate a wide range of systems and operations, not only in the industry, military, and special environments (space, sea, atomic plants, etc.), but also in daily life. As we now try to build ever more intelligent systems, the need for "visual" control has been strongly recognized, and the interest in image understanding has grown rapidly. Already, there exists a vast body of literature—ranging from general philosophical discourses to processing techniques. Compared with other works, however, this book may be unique in that its central focus is on "mathematical" principles—Lie groups and group representation theory, in particular.

In the study of the relationship between the 3D scene and the 2D image, "geometry" naturally plays a central role. Today, so many branches are interwoven in geometry that we cannot truly regard it as a single subject. Nevertheless, as Felix Klein declared in his *Erlangen Program*, the central principle of geometry is *group theory*, because geometrical concepts are abstractions of properties that are "invariant" with respect to some group of transformations.

In this text, we specifically focus on two groups of transformations. One is 2D rotations of the image coordinate system around the image origin. Such coordinate rotations are indeed irrelevant when we look for intrinsic image properties. The other is 3D rotations of the "camera" around the center of its lens. We will study the 2D image transformation induced by such a 3D camera rotation.

Since we are specifically dealing with 3D and 2D spaces, it seems that we do not need sophisticated mathematics, which is usually concerned with general n-dimensional spaces. Some people even suspect that general results, say for Lie groups, will all reduce to triviality if we put $n = 2$ or $n = 3$. They believe that elementary geometry, trigonometry, and vector and matrix calculus can take care of all practical problems. There is some truth in this belief: if a vision problem can be solved by using advanced mathematics, it can, in most cases, also be solved by some elementary means. After all, advanced mathematics is built on the basis of elementary components, to which everything can be reduced "eventually".

However, this reduction often results in complication. There are many papers filled with pages of equations that could be expressed tersely in abstract terms. The merit of advanced mathematics is not limited to making expressions compact. Abstract concepts are constructed from elementary ideas through some rational reasoning, which allows us to find new applications: if one problem is solved, other problems which seem entirely different in appearance but have the same mathematical structure can also be solved by the same means.

The trouble is that image understanding is now regarded as a branch of *computer science*, and students of computer science are usually trained in modern "discrete" and "algebraic" mathematics—computability, automata, languages, graphs, combinatorics, algorithms, coding, and the like; they seldom have opportunities to study classical "continuous" and "geometric" mathematics—Lie groups, manifolds, and differential geometry, for instance. As a result, they sometimes write Ph.D. theses filled with tedious equations, "discover" the same thing in different forms again and again, and believe that they have established state-of-the-art mathematical achievements.

Although mathematicians often mock the "mathematical achievements" of engineering students, part of the blame lies with mathematicians themselves. They are accustomed to write textbooks intended solely for students in mathematics (and sometimes theoretical physics). In pursuit of generality, their books are packed with abstract concepts, most of which are irrelevant to engineering applications. Thus, engineering scientists have enormous difficulty in skimming through them to search for knowledge useful to their specific subjects of interest.

Unfortunately, many mathematicians do not know, or are not interested in, engineering applications. There is a need for people who have a good knowledge of both mathematics and engineering applications to write textbooks supplying mathematical tools for application problems. This book is such an attempt although the author's mathematical knowledge is very limited.

The text consists of two parts. In Part I, we discuss invariance of image characteristics with respect to 2D image coordinate rotation and 3D camera rotation in relation to representations of $SO(2)$ and $SO(3)$. We incorporate various mathematical tools such as Lie groups, Lie algebras, projective geometry, differential geometry, tensor calculus, harmonic analysis, and topology—in such a way that readers with nonmathematical backgrounds can understand such abstract notions in terms of real problems. In Part II, we discuss various 3D recovery problems. The mathematical principles introduced in Part I underlie all the problems. At the same time, much attention is paid to technical aspects such as measurement, image processing, and computation, although we do not go into the details of technicalities. Thus, readers whose background is in mathematics can understand such practical aspects in mathematical terms.

This work is based on a series of lectures given by the author at the Center for Automation Research, University of Maryland in 1985–1986. Almost all the materials are taken from the author's own work. The author thanks Azriel Rosenfeld and Larry Davis of the University of Maryland for their constant encouragement. He also thanks many of his colleagues and friends for valuable discussions: Yannis Aloimonos and Tsai-Chia Chou of the University of Maryland, Chung-Nim Lee of Pohang Institute of Science and Technology,

Korea, Jan-Olof Eklundh of the Royal Institute of Technology, Sweden, Shun-ichi Amari, Kokichi Sugihara and Kazuo Murota of the University of Tokyo, and Saburo Saitoh of Gunma University. Special thanks are due to Michael Plastow of NHK for thorough advice on English writing.

This work was supported in part by the Defense Advanced Research Projects Agency and the US Army Night Vision and Electro-Optics Laboratory under Contract DAAB07-86-K-F073 and by the Casio Science Promotion Foundation, the Yazaki Memorial Foundation for Science and Technology, and the Inamori Foundation, Japan.

Gunma, Japan
September 1989

Kenichi Kanatani

Contents

1. Introduction 3
1.1 What is Image Understanding? 3
1.2 Imaging Geometry of Perspective Projection 5
1.3 Geometry of Camera Rotation 9
1.3.1 Projective Transformation 12
1.4 The 3D Euclidean Approach 13
1.5 The 2D Non-Euclidean Approach 16
1.5.1 Relative Geometry and Absolute Geometry 18
1.6 Organization of This Book 19

2. Coordinate Rotation Invariance of Image Characteristics 21
2.1 Image Characteristics and 3D Recovery Equations 21
2.2 Parameter Changes and Representations 22
2.2.1 Rotation of the Coordinate System 25
2.3 Invariants and Weights 25
2.3.1 Signs of the Weights and Degeneracy 27
2.4 Representation of a Scalar and a Vector 28
2.4.1 Invariant Meaning of a Position 29
2.5 Representation of a Tensor 29
2.5.1 Parity 34
2.5.2 Weyl's Theorem 35
2.6 Analysis of Optical Flow for a Planar Surface 35
2.6.1 Optical Flow 45
2.6.2 Translation Velocity, Rotation Velocity, and Egomotion . 46
2.6.3 Equations in Terms of Invariants 48
2.7 Shape from Texture for Curved Surfaces 48
2.7.1 Curvatures of a Surface 56
Exercises 58

3. 3D Rotation and Its Irreducible Representations 61
3.1 Invariance for the Camera Rotation Transformation 61
3.1.1 *SO*(3) Is Three Dimensional and Not Abelian 62
3.2 Infinitesimal Generators and Lie Algebra 63
3.3 Lie Algebra and Casimir Operator of the 3D Rotation Group . 69
3.3.1 Infinitesimal Rotations Commute 71
3.3.2 Reciprocal Representations 73
3.3.3 Spinors 74

3.4 Representation of a Scalar and a Vector 74
3.5 Irreducible Reduction of a Tensor 76
3.5.1 Canonical Form of Infinitesimal Generators 80
3.5.2 Alibi vs Alias 82
3.6 Restriction of $SO(3)$ to $SO(2)$ 83
3.6.1 Broken Symmetry 85
3.7 Transformation of Optical Flow 85
3.7.1 Invariant Decomposition of Optical Flow 98
Exercises 99

4. Algebraic Invariance of Image Characteristics 103
4.1 Algebraic Invariants and Irreducibility 103
4.2 Scalars, Points, and Lines 105
4.3 Irreducible Decomposition of a Vector 109
4.4 Irreducible Decomposition of a Tensor 112
4.5 Invariants of Vectors 115
4.5.1 Interpretation of Invariants 119
4.6 Invariants of Points and Lines 121
4.6.1 Spherical Geometry 126
4.7 Invariants of Tensors 129
4.7.1 Symmetric Polynomials 134
4.7.2 Classical Theory of Invariants 136
4.8 Reconstruction of Camera Rotation 137
Exercises 142

5. Characterization of Scenes and Images 147
5.1 Parametrization of Scenes and Images 147
5.2 Scenes, Images, and the Projection Operator 148
5.3 Invariant Subspaces of the Scene Space 152
5.3.1 Tensor Calculus 154
5.4 Spherical Harmonics 157
5.5 Tensor Expressions of Spherical Harmonics 161
5.6 Irreducibility of Spherical Harmonics 165
5.6.1 Laplace Spherical Harmonics 170
5.7 Camera Rotation Transformation of the Image Space 172
5.7.1 Parity of Scenes 176
5.8 Invariant Measure 176
5.8.1 First Fundamental Form 178
5.8.2 Fluid Dynamics Analogy 179
5.9 Transformation of Features 181
5.10 Invariant Characterization of a Shape 186
5.10.1 Invariance on the Image Sphere 191
5.10.2 Further Applications 192
Exercises 194

6. Representation of 3D Rotations 197
6.1 Representation of Object Orientations 197
6.2 Rotation Matrix 197
6.2.1 The nD Rotation Group $SO(n)$ 201
6.3 Rotation Axis and Rotation Angle 202
6.4 Euler Angles 205
6.5 Cayley–Klein Parameters 207
6.6 Representation of $SO(3)$ by $SU(2)$ 210
6.6.1 Spinors 215
6.7 Adjoint Representation of $SU(2)$ 215
6.7.1 Differential Representation 218
6.8 Quaternions 220
6.8.1 Quaternion Field 224
6.9 Topology of $SO(3)$ 225
6.9.1 Universal Covering Group 229
6.10 Invariant Measure of 3D Rotations 230
Exercises 233

7. Shape from Motion 239
7.1 3D Recovery from Optical Flow for a Planar Surface 239
7.2 Flow Parameters and 3D Recovery Equations 240
7.2.1 Least Squares Method 243
7.3 Invariants of Optical Flow 245
7.4 Analytical Solution of the 3D Recovery Equations 249
7.5 Pseudo-orthographic Approximation 253
7.5.1 Robustness of Computation 256
7.6 Adjacency Condition of Optical Flow 256
7.7 3D Recovery of a Polyhedron 260
7.7.1 Noise Sensitivity of Computation 262
7.8 Motion Detection Without Correspondence 263
7.8.1 Stereo Without Correspondence 271
Exercises 274

8. Shape from Angle 278
8.1 Rectangularity Hypothesis 278
8.2 Spatial Orientation of a Rectangular Corner 280
8.2.1 Corners with Two Right Angles 288
8.3 Interpretation of a Rectangular Polyhedron 289
8.3.1 Huffman–Clowes Edge Labeling 299
8.4 Standard Transformation of Corner Images 301
8.5 Vanishing Points and Vanishing Lines 309
8.5.1 Spherical Geometry and Projective Geometry 322
Exercises 323

9. Shape from Texture 327
9.1 Shape from Texture from Homogeneity 327
9.2 Texture Density and Homogeneity 328
9.2.1 Distributions 333
9.3 Perspective Projection and the First Fundamental Form 334
9.4 Surface Shape Recovery from Texture 337
9.5 Recovery of Planar Surfaces 340
9.5.1 Error Due to Randomness of the Texture 341
9.6 Numerical Scheme of Planar Surface Recovery 343
9.6.1 Technical Aspects of Implementation 353
9.6.2 Newton Iterations 353
Exercises 354

10. Shape from Surface 356
10.1 What Does 3D Shape Recovery Mean? 356
10.2 Constraints on a $2\frac{1}{2}$D Sketch 359
10.2.1 Singularity of the Incidence Structure 364
10.2.2 Integrability Condition 367
10.3 Optimization of a $2\frac{1}{2}$D Sketch 368
10.3.1 Regularization 374
10.4 Optimization for Shape from Motion 377
10.5 Optimization of Rectangularity Heuristics 379
10.6 Optimization of Parallelism Heuristics 384
10.6.1 Parallelogram Test and Computational Geometry 392
Exercises 394

Appendix. Fundamentals of Group Theory 398
A.1 Sets, Mappings, and Transformations 398
A.2 Groups 402
A.3 Linear Spaces 405
A.4 Metric Spaces 408
A.5 Linear Operators 412
A.6 Group Representation 415
A.7 Schur's Lemma 418
A.8 Topology, Manifolds, and Lie Groups 421
A.9 Lie Algebras and Lie Groups 426
A.10 Spherical Harmonics 431

Bibliography 436

Subject Index 448

Part I
Group Theoretical Analysis of Image Characteristics

1. Introduction

This chapter describes what is meant by *image understanding*. We begin with the imaging geometry of *perspective projection*, and introduce the *camera rotation transformation*, which plays a central role in subsequent chapters. Then, we discuss two typical mathematical approaches to the three-dimensional (3D) recovery problem—the *3D Euclidean approach* and the *two-dimensional* (*2D*) *non-Euclidean approach*. Finally, the organization of this book is described.

1.1 What is Image Understanding?

Image understanding is the study of images. In this text, a function $F(x, y)$ of two variables (taking binary, real, or vector values) is called an *image* if it is regarded as describing a *three-dimensional scene*. If we do not consider any three-dimensional meanings, it is simply called a *pattern*.

Pattern recognition is the study of the analysis of patterns (e.g., characters, symbols, figures), treating them as purely two-dimensional entities. Based on the extracted numerical *features* (Chap. 5), the input pattern is classified ("Which category does it belong to?") and identified ("Which prototype does it correspond to?"). Image understanding, by contrast, attempts to extract the *three-dimensional* meanings of two-dimensional images.

Pattern recognition has by now a long history, and many techniques have been developed and used in various industrial applications—in character reading machines, the automatic inspection of circuits and machine parts, and the computer analysis of photographs in material science, biology and medicine, to name a few. As the scope of pattern recognition has grown wider and wider, the necessity of dealing with three-dimensional objects has become ever more strongly recognized. This desire is especially strong in such fields as industrial automation, mobile robot control, autonomous land vehicles, and automatic reconnaissance.

First, attempts were made to apply the techniques of pattern recognition directly, without regard to the three-dimensionality of objects. One example was the attempt to recognize and identify ships and airplanes from their silhouettes. However, the limitations of such approaches were soon recognized: the same object can look very different when viewed from different angles. The consensus today is that we must first establish *intermediate descriptions* from original images before we start the classification and identification.

In order to classify or identify three-dimensional objects, we need two types of information. One is the three-dimensional *description* of objects (size, shape, position, orientation, etc.) extracted from the intermediate description, and the other is particular *knowledge* about the scene (or a *model* of the domain). For example, in an "office scene", there should be a desk with a large planar surface, on which a telephone is placed with a high probability, etc. The stage of extracting the intermediate description is called *low-level* (or *early*) *vision*, while the stage of classification and identification is called *high-level* (or *expert*) *vision*.

Many researchers agree that, for low-level vision at the signal level, images should be processed *bottom-up* without using particular knowledge about the scene, as opposed to high-level vision at the symbolic level, where knowledge about the scene is used *top-down*. (However, there are also many researchers who have different philosophies, and various strategies have been proposed to combine the image data and the a priori knowledge.) Our aim in this book is in a sense *intermediate vision* at the *analytical level*, analyzing intermediate descriptions supplied by the low-level vision, and extracting three-dimensional interpretations to send to high-level vision.

Several levels exist for intermediate description. The most fundamental is the *line drawing* obtained by image processing techniques. Gray-level images have too much information, most of which is irrelevant for object recognition. This observation is based on the fact that humans can easily recognize three-dimensionality by merely seeing line drawings.

Once a line drawing is obtained, we can interpret it as a three-dimensional scene. There exist many schemes of line drawing interpretation, among which the *Huffman–Clowes edge labeling* (Chap. 8) is best known. We can also estimate the gradient of the object surface from various clues: motion (Chaps. 2–4, 7, 10), texture (Chaps. 1, 9), and shading, for instance. The problem is often referred to as *shape from* . . . depending on the clues used ("shape from motion", for example). Sometimes, we invoke simple assumptions about the object, such as rectangularity of corners (Chaps. 8, 10), parallelism of edges (Chaps. 8, 10), and symmetry of shapes. Such fundamental assumptions are called *constraints*, as opposed to *knowledge* about the scene. If estimates of surface gradients or 3D edge orientations are obtained, the intermediate description is called a *$2\frac{1}{2}$D sketch* (Chap. 10).

Often, the constraints on objects are chosen in relation to the psychology of human perception, since humans can easily perceive three-dimensionality from two-dimensional images. In this respect, psychologists have greatly contributed to this area. At the same time, studies of computer vision have also exerted a great influence on psychologists.

Since we focus on *analytical* aspects of 3D recovery problems, we do not go into the details of signal-level image processing techniques, which include *edge detection*, *image segmentation*, *stereo matching*, and *optical flow detection* (Chaps. 2, 7, 10). Although such techniques are vital and are currently being studied extensively, we simply assume that they are available whenever necessary. Also, we do not go into the symbolic treatment of high-level vision, for which objects must be modeled (*wire-frame model, surface model, solid model, generalized cone* (or *cylinder*) *representation, extended Gaussian sphere representation, etc.*), properties and interrelationships of objects must be represented (*first-order predicate logic, production rules, frames, semantic network*, etc.), and reasoning strategies must be established (*backward, forward, hierarchical, fuzzy*, etc.). The study of high-level vision is now one of the central subjects of *artificial intelligence.*

1.2 Imaging Geometry of Perspective Projection

We start with the mechanism by which 3D objects are projected onto 2D images. For our purpose, we need not go into the details of the optics and photo-electronics of the camera. All we need is the fact that the *camera lens catches the ray of light that passes through the center of the lens.* Figure 1.1 gives a schematic illustration of an ideal camera: l is the distance of the object from the center of the lens, and f is the distance between the center of the lens and the surface of the film. These two distances are related by the *lens equation*

$$\frac{1}{f}+\frac{1}{l}=\frac{1}{F}, \tag{1.2.1}$$

where F is the *focal length* of the lens. The lens equation (1.2.1) does not necessarily hold for real cameras; image distortion—called *aberration*—inevitably occurs to some extent, producing an effect as if the ray of light has been deflected or the film's surface is curved. Therefore, appropriate correction is necessary, and *camera calibration* is currently an important issue in computer vision.

The imaging mechanism of Fig. 1.1 is conveniently represented by the coordinate system of Fig. 1.2. An XYZ coordinate system is placed in the scene, and the plane $Z = f$ is regarded as the *image plane.* A point P in the scene is

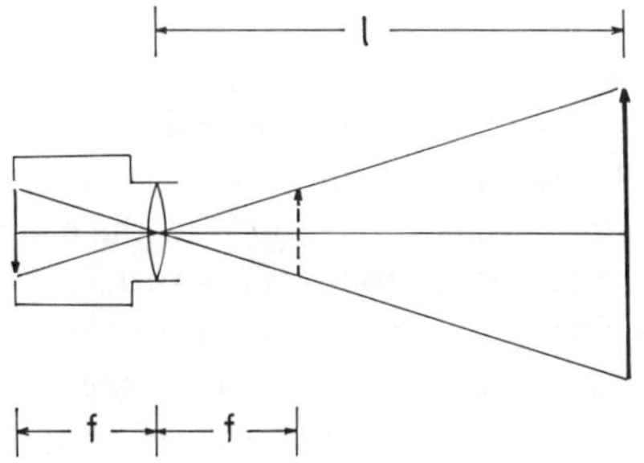

Fig. 1.1. Imaging geometry of an ideal camera. l: distance of the object from the center of the lens; f: distance between the center of the lens and the surface of the film; F: focal length of the lens

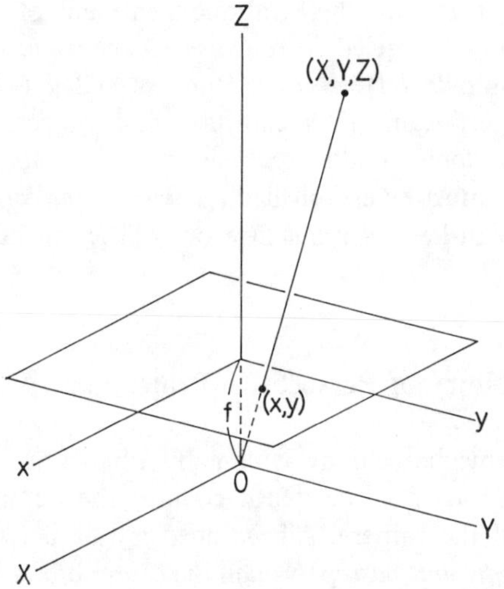

Fig. 1.2. The viewer-centered coordinate system. A point (X, Y, Z) is projected onto point (x, y) on the image plane $Z = f$ by perspective projection from viewpoint O

projected onto the intersection of the image plane with the ray connecting the point P and the coordinate origin O. This process is called *perspective* (or *central*) *projection* from the *viewpoint* (or *center of projection*) O.

Let us take a Cartesian xy coordinate system on the image plane in such a way that the x- and the y-axes are parallel to the X- and the Y-axes, respectively. The origin of the image plane is $(0, 0, f)$, the intersection of the Z-axis with the image plane. We call this xy coordinate system the *image coordinate system.* As we can see from Fig. 1.2, a point in the scene whose coordinates are (X, Y, Z)—called the *scene coordinates*—is projected onto a point on the image plane whose coordinates are (x, y)—called the *image coordinates*—given by

$$x = \frac{fX}{Z}, \qquad y = \frac{fY}{Z}. \tag{1.2.2}$$

We call these two equations the *projection equations.*

Comparing Figs. 1.1 and 1.2, we notice that the viewpoint O corresponds to the center of the lens, the Z-axis corresponds to the camera's *optical axis*, and the image plane $Z = f$ corresponds to the film's surface, except that the relationship is symmetrically reversed with respect to the viewpoint O. The distance f between the viewpoint O and the image plane depends, as shown by the lens equation (1.2.1), on the distance currently in focus. For simplicity, we do not explicitly state the current focus, assuming that f has some fixed value. If the distance l is much larger than the lens focal length F, as is usually the case in most computer vision applications, the lens equation (1.2.1) implies $f \approx F$, and for this reason, the distance f is often referred to simply as the *focal length*. It is also called the *camera constant*.

The coordinate system of Fig. 1.2 is not the only possibility. An alternative is shown in Fig. 1.3, where the XY-plane plays the role of the image plane, and the X- and Y-axes are respectively identified with the image x- and y-axes. The viewpoint $(0, 0, -f)$ is a distance f away from the image plane on the negative side of the Z-axis. The projection equations (1.2.2) are now replaced by

$$x = \frac{fX}{f+Z}, \qquad y = \frac{fY}{f+Z}. \tag{1.2.3}$$

We call this coordinate system the *image-centered coordinate system*, as opposed to that of Fig. 1.2, which we call the *viewer-centered coordinate system*.

Both the image-centered and the viewer-centered coordinate systems are used in the computer vision literature, depending on authors and books. Both are useful, and the choice often depends on the type of analysis. For example, the viewer-centered coordinate system is very convenient when we consider camera rotations, which we will introduce in the next section. Also, this coordinate system is suitable when we apply *projective geometry*, which identifies a "ray" starting from the origin O as a "point", as we will show in the next section.

On the other hand, the image-centered coordinate system is very useful when we analyze 3D recovery problems. For one thing, this configuration naturally includes *orthographic* (or *parallel*) *projection* in the limit $f \to \infty$ (Fig. 1.4). Orthographic projection is the projection by parallel rays orthogonal to the image plane. Then the projection equations are given simply by $x = X$, $y = Y$.

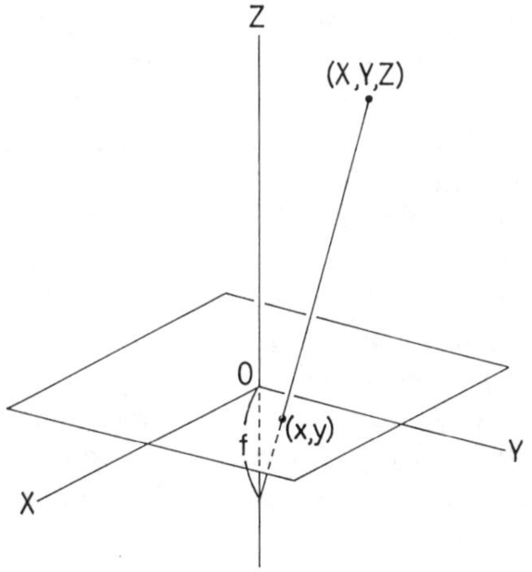

Fig. 1.3. The image-centered coordinate system. A point (X, Y, Z) is projected onto point (x, y) on the image plane $Z = 0$ by perspective projection from viewpoint $(0, 0, -f)$

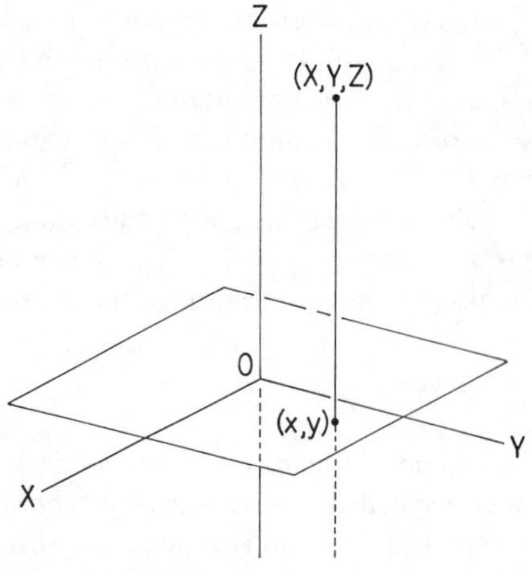

Fig. 1.4. Orthographic projection. A point (X, Y, Z) is projected onto point (x, y) on the image plane such that $x = X$, $y = Y$. This relation results in the limit $f \to \infty$ if the image-centered coordinate system of Fig. 1.3 is used

This is a good approximation when the object is, compared with the focal length f, very small or located very far away from the viewer, or when a telephoto lens of large focal length is used. When we obtain some results with reference to the image-centered coordinate system, the corresponding results for orthographic projection are automatically obtained by taking the limit $f \to \infty$. The use of the image-centered coordinate system also makes it easy to see the transition from perspective to orthographic as the focal length f become larger and larger (Chapters 7, 10). In this text, we use both systems, depending on the type of analysis. We use the viewer-centered coordinate system whenever camera rotation is involved (Chaps. 3–5, 8, 9), and the image-centered coordinate system to analyze 3D recovery equations (Chaps. 2, 7, 10).

There is one convention we adopt that requires some care from readers. Whenever we show images together with the image coordinate system, we take the x-axis *vertically* and the y-axis *horizontally* (Fig. 1.5). For description of geometrical relationship in a scene, it is convenient to use a right-hand system. Since we take the Z-axis in the direction away from the camera, it is natural to regard the Z-axis as passing through the image away from the viewer. If we take the x-axis vertically and the y-axis horizontally, we can obtain a right-hand system.[1] Mathematically, nothing is changed by our convention, but one curcial

[1] Note that in mathematics the Z-axis is customarily assumed to point toward the viewer whenever we consider an XY-plane. Accustomed to graphic displays, some researchers in computer science today take the x-axis horizontally as usual, but the y-axis *downward*.

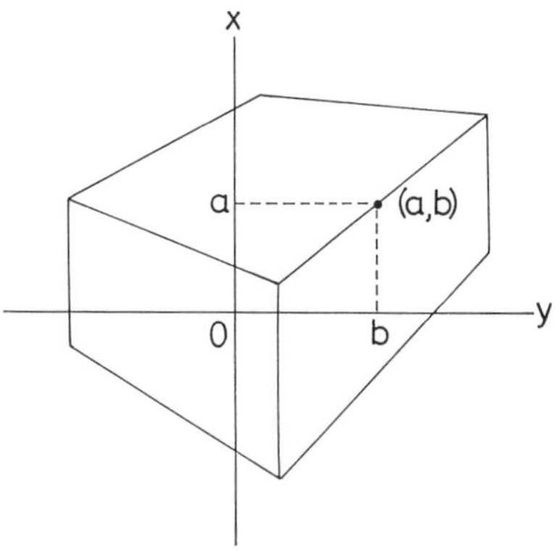

Fig. 1.5. The image xy coordinate system. The x-axis is taken vertically, and the y-axis horizontally. A positive rotation is a *clockwise* rotation

difference arises when we mention rotations: We say that a rotation is *positive* if the positive part of the x-axis is rotated toward the positive part of the y-axis, in conformity with the literature in mathematics. Hence, a positive rotation is a *clockwise* rotation on the image plane.

1.3 Geometry of Camera Rotation

In this section, we consider rotation of the camera around the center of the lens, which plays a crucial role in subsequent chapters (Chaps. 3–5, 8, 9). Its significance lies in the fact that *the information contained in the image is preserved by such camera rotations.* As shown in Fig. 1.6, the rays of light entering the camera are the same if the camera is rotated around the center of the lens. (For the moment, let us forget about the boundary of the film.) As a result, the incoming information is also the same, since the input to a camera is the *rays of light that pass through the center of the lens.*

The viewer-centered coordinate system is convenient for dealing with camera rotations. Consider a rotation of the camera by a 3D orthogonal matrix

$$\boldsymbol{R} = \begin{pmatrix} r_{11} & r_{12} & r_{13} \\ r_{21} & r_{22} & r_{23} \\ r_{31} & r_{32} & r_{33} \end{pmatrix} \tag{1.3.1}$$

whose determinant is 1 (We call such a matrix a 3D *rotation matrix*). Consider a point (X, Y, Z) in the scene. Let (x, y) be its image coordinates for the initial camera position, and (x', y') those after the rotation (Fig. 1.7). The transformation from (x, y) to (x', y') is given as follows.

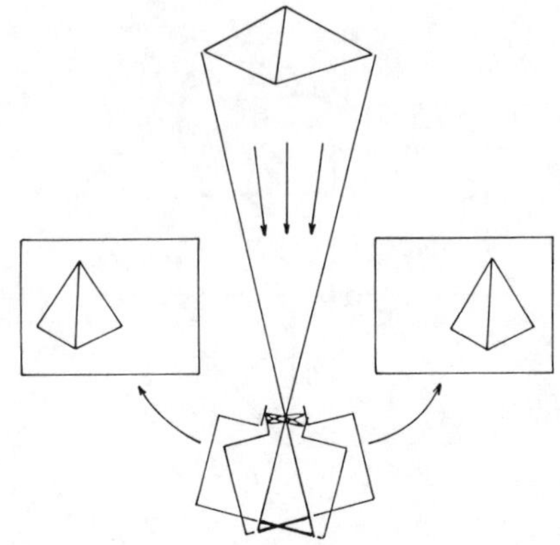

Fig. 1.6. Incoming rays of light are identical if the camera is rotated around the center of the lens. The object image retains the same information as it had before the camera rotation

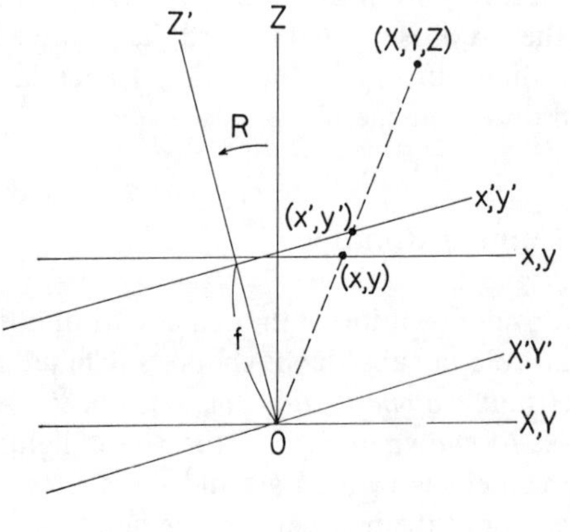

Fig. 1.7. If the camera is rotated relative to a stationary scene, a point in the scene is moved to the intersection of the corresponding ray with the new image plane $Z' = f$. This can also be viewed as a rotation of the scene in the opposite sense relative to the camera

Theorem 1.1. *The image transformation due to camera rotation* $\boldsymbol{R} = (r_{ij})$ *is given by*

$$x' = f\frac{r_{11}x + r_{21}y + r_{31}f}{r_{13}x + r_{23}y + r_{33}f}, \qquad y' = f\frac{r_{12}x + r_{22}y + r_{32}f}{r_{13}x + r_{23}y + r_{33}f}. \tag{1.3.2}$$

Proof. A rotation of the camera by $\boldsymbol{R}$ is equivalent to the rotation of the scene in the opposite sense. If point (X, Y, Z) is rotated in the scene around the origin O by $\boldsymbol{R}^{-1}$, it moves to

$$\begin{pmatrix} X' \\ Y' \\ Z' \end{pmatrix} = \begin{pmatrix} r_{11} & r_{21} & r_{31} \\ r_{12} & r_{22} & r_{32} \\ r_{13} & r_{23} & r_{33} \end{pmatrix} \begin{pmatrix} X \\ Y \\ Z \end{pmatrix} . \tag{1.3.3}$$

(Note that $\boldsymbol{R}$ is an orthogonal matrix, so $\boldsymbol{R}^{-1} = \boldsymbol{R}^{\mathrm{T}}$, where T denotes matrix transpose.) Equations (1.3.2) are obtained by combining (1.3.3) and the projection equations

$$x = \frac{fX}{Z} , \qquad y = \frac{fY}{Z} , \qquad x' = \frac{fX'}{Z'} , \qquad y' = \frac{fY'}{Z'} . \tag{1.3.4}$$

Replacing $\boldsymbol{R}$ by $\boldsymbol{R}^{\mathrm{T}}(= \boldsymbol{R}^{-1})$, we obtain:

Corollary 1.1. *The inverse of the transformation* (1.3.2) *is given by*

$$x = f\frac{r_{11}x' + r_{12}y' + r_{13}f}{r_{31}x' + r_{32}y' + r_{33}f} , \qquad y' = f\frac{r_{21}x' + r_{22}y' + r_{23}f}{r_{31}x' + r_{32}y' + r_{33}f} . \tag{1.3.5}$$

Equations (1.3.2) define a nonlinear transformation from (x, y) to (x', y'), which we call the *camera rotation transformation.* The existence of its inverse assures that no information is lost, since the original image is recovered by applying the inverse.

Under matrix multiplication, the set of 3D rotation matrices forms a *group* called the *three-dimensional special orthogonal group* and denoted by $SO(3)$. For simplicity, we also call it the *3D rotation group.* If we denote the transformation (1.3.2) symbolically as

$$(x', y') = M_R(x, y) , \tag{1.3.6}$$

we can easily see that $\{M_R | \boldsymbol{R} \in SO(3)\}$ is a *group of transformations* (See the Appendix for the definition of a group and a group of transformations). Moreover, we see that

$$M_{R'} \circ M_R = M_{RR'} , \qquad \boldsymbol{R}', \boldsymbol{R} \in SO(3) , \tag{1.3.7}$$

$$M_I = I , \qquad (M_R)^{-1} = M_{R^{-1}} , \tag{1.3.8}$$

where $\boldsymbol{I}$ is the unit matrix. This means that the group of transformations $\{M_R\}$ is a *representation* of $SO(3)$. (Strictly speaking, $\{M_R\}$ is a *reciprocal representation* in the sense that the order of $\boldsymbol{R}'$ and $\boldsymbol{R}$ is reversed on the two sides of (1.3.7). This occurs because the camera is rotated by $\boldsymbol{R}$. We would obtain $M_{R'} \circ M_R = M_{R'R}$ if the scene were to be rotated relative to a fixed camera.)

The 3D rotations of the camera that form the 3D rotation group $SO(3)$ also contain rotations around the Z-axis. These rotations form a subgroup of $SO(3)$ denoted by $SO(2)$. If the camera rotation is restricted to $SO(2)$, the induced image transformation is equivalent to the 2D rotation of the image coordinate system.

In Chap. 2, we first study the invariance properties of the image coordinate rotation, and show how this study yields a powerful tool for solving 3D recovery problems. In Chap. 3, we will go on to the study of the (full) camera rotation transformation. The problem of how the invariance properties change if $SO(3)$ is restricted to $SO(2)$ is also discussed in Chap. 3. Then, we will study how the camera rotation transformation preserves algebraic expressions in image characteristics (Chap. 4). We will also study properties of images viewed as continuous functions (Chap. 5). The camera rotation transformation also plays an important role in numerical computation of various 3D recovery problems (Chaps. 8, 9).

1.3.1 Projective Transformation

A transformation of the form of (1.3.2) is called a 2D *projective transformation* if the matrix (1.3.1) is an arbitrary three-dimensional nonsingular matrix and if the image plane is extended to include *points at infinity* and the *line at infinity*. This extended xy-plane is called the 2D *projective space*.

A convenient way to deal with the projective space is to use *homogeneous coordinates* (u, v, w): if $w \neq 0$, a "point" (u, v, w) of the projective space is identified with the point $(u/w, v/w)$ on the xy-plane; $x = u/w$ and $y = v/w$ are called the *inhomogeneous coordinates*. If $w = 0$, the "point" $(u, v, 0)$ is regarded as being located *at infinity*. The set of all such points is called the *line at infinity*; its equation is $w = 0$.

A "line" in a 2D projective space is the set of points constrained by a linear equation $Au + Bv + Cw = 0$, where A, B, C are not all zero. The three coefficients A, B, C are called the *homogeneous coordinates* of the line. If A or B is not zero, the line appears on the xy-plane as $Ax + By + C = 0$. If $A = B = 0$, the line is thought of as located at infinity.

The homogeneous coordinates can be multiplied by an arbitrary nonzero number; the "point" represented by them is still the same. In terms of the homogeneous coordinates, a projective transformation is written as

$$\begin{pmatrix} u' \\ v' \\ w' \end{pmatrix} = c\boldsymbol{P}\begin{pmatrix} u \\ v \\ w \end{pmatrix}, \tag{1.3.9}$$

where c is an arbitrary nonzero number, and $\boldsymbol{P}$ is a nonsingular matrix.

The most important fact about the projective transformation is that *a line is mapped onto a line.* A line whose homogeneous coordinates are A, B, C is mapped by the transformation (1.3.9) onto a line whose homogeneous coordinates A', B', C' are given by

$$\begin{pmatrix} A' \\ B' \\ C' \end{pmatrix} = c'(\boldsymbol{P}^{\mathrm{T}})^{-1}\begin{pmatrix} A \\ B \\ C \end{pmatrix}, \tag{1.3.10}$$

where c' is an arbitrary nonzero number. Points in a projective space are said to be *collinear* if there exists a line passing through them. Thus, a projective transformation maps collinear points onto collinear points. Conversely, we can prove that a one-to-one mapping from a projective space onto itself which maps collinear points onto collinear points—which is called a *collineation*—must be a projective transformation. (This is not necessarily true for a *complex* projective space. Another important quantity preserved by a projective transformation is the *cross* (or *anharmonic*) *ratio*.)

Since the camera rotation transformations constitutes a subgroup of the group of 2D projective transformations, all properties of the 2D projective transformation, such as the preservation of collinearity, also hold for the camera rotation. However, there are many aspects of the camera rotation which do not necessarily hold for the full 2D projective transformation; the projective transformation is too general. In subsequent chapters, we mainly focus on the camera rotation alone. Also, we do not use "homogeneous coordinates"; "inhomogeneous coordinates" are very convenient for treating image data.[2]

1.4 The 3D Euclidean Approach[3]

"Understanding" implies *modeling* of the objects. Reconstructions of 3D raw data—called *3D images*—by direct measurements such as *laser* or *ultrasonic ranging*, *stereo*, and *computer tomography* (*CT*) cannot be called understanding; a 3D image is only a collection of data in a three-dimensional array, just as a 2D image is a collection of data in a two-dimensional array. It is not until a model is fitted that we can say something about the object.

A model is specified by a small number of parameters. For example, an object can be modeled by a line, a plane, a quadric surface, a sphere, a cylinder, a cone, or a combination of these—a polyhedron (Chaps. 8, 10). A line is specified by a point and the unit vector along it; a plane by a point and the unit surface normal; a quadric surface by the coefficients of the defining equation; a sphere by its center and radius. The 3D position of an object is specified by the position vector. The 3D orientation of an object is specified by three mutually orthogonal vectors starting from a fixed point on the object—or rotation matrix, Euler angles, quaternion, etc. (Chap. 6). If the object is in motion, its 3D motion is specified by a velocity at a reference point (*translation velocity*) and a *rotation*

---

[2] However, we will introduce many formulations which "implicitly" use homogeneous coordinates. For example, the image coordinates $x = fX/Z$, $y = fY/Z$ can be regarded as defining the inhomogeneous coordinates of a point (X, Y, Z) in the scene identified with a 2D projective space. (The constant f is not essential.) In other words, the scene coordinates (X, Y, Z) are essentially the homogeneous coordinates of the point of the image plane.

[3] Sections 1.4 and 1.5 may be too abstract for readers without any knowledge of 3D recovery problems. These two sections can be skipped for the first reading; the discussion here will be understood more clearly after subsequent chapters have been read.

velocity, which is represented by an axis and an angular velocity around it (Chaps. 2, 3, 7). Let us call these parameters that specify the object model *object parameters*. (They are called *structure and motion parameters* in some literature.)

Let us call the numerical data that characterize the observed image *image characteristics*. (Some authors call such data *image features* or *image properties*. We will use the term "feature" in Chap. 5 in a slightly different sense. We will also introduce the term "observable" in Chaps. 7, 9.) They may reflect the gray levels of the image, the texture of the object surface (Chaps. 2, 9), the object contour, the intensity of light reflectance, or the optical flow if the object is in motion (Chaps. 2, 7). Stated in this way, the 3D recovery problem is viewed as *estimation of the object parameters from the image characteristics*. This problem is called *shape from* . . . according to the source of the image characteristics—*shape from texture, shape from shading, shape from motion*, etc. Therefore, the 3D recovery problem can be solved if equations relating the object parameters and the image characteristics are obtained. We call such equations *3D recovery equations*. It is desirable that the 3D recovery equations have simple forms, hopefully yielding analytical solutions. Otherwise, solutions must be computed numerically, which often results in various difficulties: multiple solutions may exist, iterations may not converge, and computation time may be too long. Therefore, it is very important to obtain "good" 3D recovery equations. All subsequent processes are affected by the choice of the 3D recovery equations.

One approach to obtaining 3D recovery equations is to treat all quantities "in 3D space". First, an image itself is regarded as a 3D object by setting the image plane in the scene according to the camera model. Next, the image is *backprojected* into the scene by introducing unknown parameters (Fig. 1.8). For example, a point P on the image plane is backprojected onto the point whose position vector is $r\overrightarrow{OP}$, where O is the viewpoint and r is an unknown

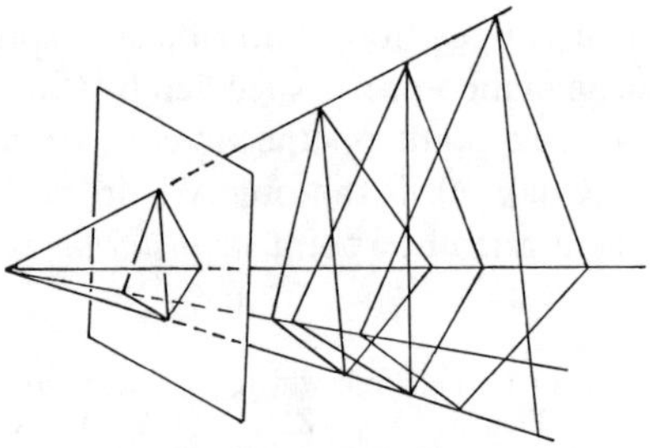

Fig. 1.8. Backprojection. A family of infinitely many object shapes are constructed such that they all yield the same image after projection. From among them, the one that satisfies the appropriate constraints is selected

parameter. Similarly, a line l on the image plane is backprojected onto a line lying somewhere on the plane defined by the viewpoint O and the line l. In this way, starting from observed data, we can construct a *family of infinitely many candidates* of the object shape in such a way that all of them yield the same image data after perspective projection. Then, one solution is selected. The selection is dictated by the *constraints* on the object: we select, from among the backprojected candidates, the one that possesses properties that the object model is required to have.

Object constraints are usually expressed in terms of the 3D Euclidean metric. For example, a particular line segment of object must have a fixed length, two line segments must make a fixed angle or intersect perpendicularly, particular line segments must be coplanar, and all lengths and angles are preserved if the object is in rigid motion. This means that the object constraints themselves serve the 3D recovery equations. We call this approach the *3D Euclidean approach*, because 3D recovery equations have geometrical meanings in "3D Euclidean space"; they may specify conditions concerning lengths, angles, parallelism, orthogonality, planarity, rigidity, etc. They are usually expressed as 3D vector and matrix equations, so subsequent analyses are done in terms of vector calculus and matrix algebra.

The great majority of past research on 3D recovery has been done with this approach, since it is very natural to ask what family of objects can be projected onto the observed image and then choose the one that best fits our prior knowledge about the object. In fact, this viewpoint has long been adopted by psychologists in the study of human visual perception, and although psychologists are not so very interested in the computational aspects, their approaches have exerted a great influence on studies of computer vision.

A major disadvantage of this approach is that the resulting 3D recovery equations become very complicated. Often, analytical solutions are difficult to obtain. One reason is that image characteristics and object parameters are chosen so as to make the backprojection easy. For example, backprojection is easy if the image data have straightforward meanings like positions, orientations, lengths, and angles. However, such straightforward data may not necessarily have meanings inherent to the image. The object parameters, on the other hand, also tend to be quantities describing the object *relative to the image plane*, such as its distance from the image plane. Hence, they may not necessarily indicate properties inherent to the object.

Another point to note is that this approach results in 3D vector and matrix equations, which are easy to analyze *if the X-, Y-, and Z-axes play interchangeable roles*. In the 3D recovery problem, however, the Z-axis, which corresponds to the camera optical axis, plays a special role. This means that the three components do not all have the same geometrical meaning; it only has symmetry with respect to the x- and y-components. Is there any way to exploit this fact explicitly?

1.5 The 2D Non-Euclidean Approach[4]

The 3D Euclidean approach starts with an observed 2D image and ends with constraints on the 3D object model. An alternative approach is to start with a *parameterized* 3D *object model.* Since the imaging geometry is simple, it is easy to compute the expected projection image. This process, called *projection* as opposed to *backprojection*, defines a *parametrized family of infinitely many images* (Fig. 1.9). These images are regarded as "2D quantities"; their 3D origins can be ignored.

Once a family of 2D images is defined, the next step is to *define* image characteristics. Since the images are parametrized, the image characteristics thus defined are functions of object parameters. Then, we turn to the observed image. The 3D recovery equations are obtained by measuring these image characteristics on the observed image and equating their values with the theoretical expressions for them.

Since we start with an object model, we can choose object parameters in such a way that they have desirable properties. It is desirable that the parameters have geometrical meanings inherent to the object itself and independent of the choice of the coordinate system. The ease of subsequent analysis is affected by the choice of these parameters. Thus, this approach allows more freedom and flexibility than the 3D Euclidean approach, where the choice of object parameters is virtually dictated by the convenience of backprojection.

After the object parameters are chosen, there is still another choice to make. We must define "good" image characteristics that are capable of distinguishing

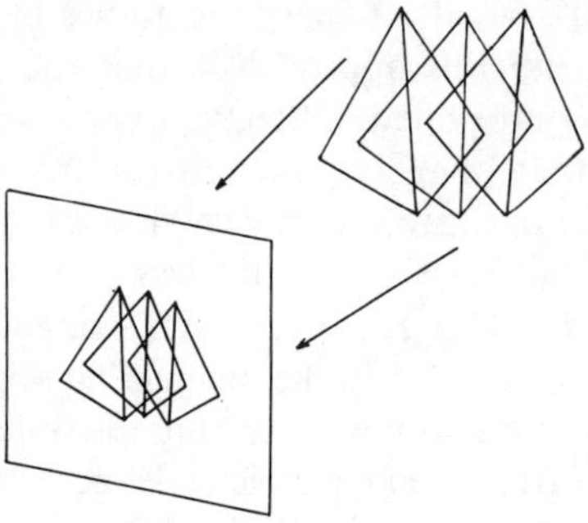

Fig. 1.9. A parametrized 3D object model is projected onto the image plane, resulting in a parametrized family of infinitely many images. From among them, the one which best matches the observed image is selected. The matching is done by comparing a small number of image characteristics

[4] See footnote 3, page 13.

images with different object parameter values. Since no coordinate system is inherent to the images, the x- and y-axes must play interchangeable roles. Furthermore, if another $x'y'$ coordinate system is taken by rotating the original xy coordinate system around the image origin, these two must play equivalent roles. Hence, the image characteristics should be symmetric with respect to the x- and y-coordinates and have *the same geometrical meaning* if the coordinate system is rotated (Chap. 2). In this sense, 3D *recovery is possible if and only if the image characteristics have coordinate rotation invariance*. The fact that the image characteristics are *defined*, as opposed to *given* as for the 3D Euclidean approach, provides great freedom and flexibility in analyzing 3D recovery problems. In particular, image characteristics defined as *linear functionals* (e.g., weighted sums or averages of image data) play important roles (Chaps. 5, 7, 9).

A line in the scene is projected onto a line on the image plane, but projections of parallel or orthogonal lines are generally no longer parallel or orthogonal on the image plane if measured in the 2D Euclidean metric of the image plane. Also, the length of a line segment in the scene is not preserved by projection if measured in the 2D Euclidean metric of the image plane. The distortion due to projection differs from position to position if measured in the 2D Euclidean metric of the image plane. However, if we adopt an appropriate model of the object, we can define parallelism or orthogonality on the image plane which reflect parallelism or orthogonality in the scene. In other words, if we introduce an object model, we can define a *2D non-Euclidean metric* on the image plane in such a way that the lengths and angles measured in that metric have a geometrical meaning in the original 3D space (Chap. 9). For example, if the object is a plane, the Euclidean geometry on it is regarded as 2D *projective geometry* on the image plane (Chap. 8). From these observations, let us call the approach described in this section the *2D non-Euclidean approach.*

The 3D Euclidean approach first observes image characteristics on the image plane, then *backprojects* them into the scene, and applies *object constraints* expressed in terms of 3D Euclidean geometry. The mathematical tools are 3D vector calculus and matrix algebra. In contrast, the 2D non-Euclidean approach first *models* the object, then *projects* it onto the image plane, and *defines* image characteristics in terms of an appropriate 2D non-Euclidean geometry. As a result, various mathematical tools become available, *differential geometry* (Chaps. 2, 5, 9) and *tensor calculus* (Chap. 5), for example. Moreover, the 2D non-Euclidean approach can exploit various *invariance properties* over some *groups of transformations*: the geometrical interpretation of image characteristics should be invariant under coordinate rotation (Chap. 2), and the information contained in an image is preserved by camera rotation (Chaps. 3–5).

If the object satisfies some constraints (e.g., collinearity, coplanarity, parallelism and orthogonality), they give rise to *consistency conditions* that the projected object image must satisfy (Chaps. 8, 10). If a given image does not satisfy the consistency conditions, that image is a *false image*; it cannot be obtained by projecting a real object. Due to the existence of noise, real images often do not exactly satisfy the consistency conditions even if they are projections of real objects. As a result, if we try to backproject them, the sets of candidates constructed in the 3D scene may be empty or may not contain the true solution. This inconsistency can be easily overcome by the 2D non-Euclidean approach, for essentially what it does is *matching* on the image plane. We need not necessarily seek an "exact match"; we need only seek the "best match".

The 2D non-Euclidean approach does not directly match images; matching is done at the level of the image characteristics, and other image properties are ignored. In other words, our choice of image characteristics *defines* the matching, and the 3D recovery equations are viewed as the *matching conditions.* Robustness to noise can be increased if we choose, as image characteristics, sums or averages of a large number of measured values.

Another advantage of the 2D non-Euclidean approach is that we need not identify the *three-dimensional meanings* of the image characteristics; the image characteristics are defined as purely 2D properties of the image (e.g., the average intensity, the area inside the object contour, etc.). If the object is in motion, its 3D motion can be determined without detecting *point-to-point correspondences*, i.e., without requiring knowledge of which point corresponds to which between two image frames (Chap. 7).

The distinction between the 3D Euclidean and the 2D non-Euclidean approaches is sometimes not so clear cut. It depends on our *interpretation* of whether the 3D recovery equations are regarded as "3D constraints" or "2D matching conditions".

1.5.1 Relative Geometry and Absolute Geometry

An analogy from a different viewpoint is also found in the differential geometry of surfaces (or hypersurfaces). One approach is *relative geometry*, so to speak, which regards the surface (or hypersurface) as a subset of a 3D (or higher-dimensional) Euclidean space. The quantities that describe the surface characteristics, such as the *curvature*, are defined with reference to the coordinate system of the *outside Euclidean space.* The other is *absolute geometry*, which does not assume the existence of an outside Euclidean space. The (hyper)surface itself is the only existing entity equipped with a *distorted non-Euclidean geometry.* All geometrical characteristics, such as the curvature, are expressed in terms of *internal quantities* defined with reference to the coordinate system of the (hyper)surface. Such internal quantities include the *metric* and the *affine connection.* Einstein's general theory of relativity describes 4D space-time in terms of an absolute geometry called *Riemannian geometry.*

1.6 Organization of This Book

The rest of this volume consists of two separate parts. In Part I, we study fundamental mathematical aspects of image understanding. All topics in Part I are related to *Lie groups* and their *representations*. This part is designed to give the reader, in the course of studying problems of image understanding, a clear idea of what Lie group and group representation theory is. An abstract theory is best understood by seeing how the theory works in actual applications. Hence, mathematical aspects are emphasized, while many technical aspects (e.g., computer implementation, computation time and memory storage, noise sensitivity) are omitted for the sake of consistency.

In Chap. 2, we study the transformations and *invariants* of image characteristics when the image coordinate system is rotated. There, the role of group representation theory becomes clear. We introduce a philosophy called *Weyl's thesis*, which enables us to understand the geometrical meanings of quantities involved in the problem. The use of invariants also provides a powerful tool for solving 3D recovery equations. We apply our technique to the analysis of *optical flow* and the problem of *shape from texture*.

In Chap. 3, we study the transformations and invariants of image characteristics when the camera is rotated three dimensionally. The analysis involves *irreducible representations* of $SO(3)$, their *Lie algebras*, *infinitesimal generators*, *commutation relations*, and the *Casimir operator*. The results are applied to the camera rotation transformation of optical flow, and the *invariant decomposition* of optical flows are derived.

In Chap. 4, we study *algebraic invariants*, i.e., invariants which are algebraic expressions in image characteristics. Some of the results are stated in terms of *projective geometry*, but many new concepts specific to the camera rotation transformation are also introduced. They include the *invariant length*, the *invariant angle*, and the *invariant area*. Then, we present a scheme for reconstructing the 3D camera rotation from the values of image characteristics. We also give a brief summary of the *classical theory of invariants*.

In Chap. 5, we view images as functions of two variables, and study how to extract invariant *features* of images. Here, we introduce *tensor calculus* and describe the irreducible reduction of the *symmetric tensor representation* of $SO(3)$ in terms of *spherical harmonics*. The beauty of classical tensor analysis culminates here.

In Chap. 6, we study various representations of $SO(3)$. We start with the *rotation matrix*, the *rotation axis* and *angle*, and *Euler angles*. Then, we introduce the *Cayley–Klein parameters*, *adjoint representations*, *quaternions*, and study the relationship between $SU(2)$ and $SO(3)$, incorporating the *topology* and *invariant measures* of $SO(3)$.

In Part II, we study typical image understanding problems in detail. While the mathematical principles introduced in Part I underlie all the problems, much attention is also paid to "technical" aspects such as measurement, image processing, and computation, although we do not go into the details of technicalities.

In Chap. 7, we analyze *optical flows* and derive analytical solutions. Using these results, we discuss various aspects of the *shape-from-motion* problem, including the *adjacency condition* of optical flows. Then, by defining *observables* of images, we present a scheme of motion detection without using the point-to-point correspondence.

In Chap. 8, we study 3D recovery from "angle clues" observed on the image plane. First, we present an analysis based on the *rectangularity hypothesis*. Then, we construct an interpretation scheme of images of *rectangular polyhedra*. The camera rotation transformation plays a crucial role when perspectively projected images are analyzed; the analysis is done by means of the *standard transformation* of images and computation of *canonical angles*. We also describe typical effects of perspective projection, such as *vanishing points* and *vanishing lines*, in terms of the *duality theorem* of projective geometry. Much consideration is given to the technical issue of confining the computation within the finite domain.

In Chap. 9, we study the *shape-from-texture* problem, invoking the *theory of distributions* and *differential geometry*. We will show how our definition of texture *homogeneity* leads to practical 3D recovery algorithms. The camera rotation transformation also plays an important role.

In Chap. 10, we study the problem of reconstructing a consistent polyhedral shape from inconsistent and inaccurate data. This is a very important issue, since error is unavoidable in real problems. We present an *optimization* scheme to provide a consistent solution which conforms to the observed data *on the average*. We first apply this scheme to the *shape-from-motion* problem. Then, the optimization scheme is coupled with such heuristics as the *rectangularity hypothesis* and the *parallelism hypothesis*. We also show some examples based on real images. Much consideration is given to technical issues such as noise and overflow of computation.

In the Appendix, we give a brief summary of fundamental mathematical concepts such as *sets*, *mappings*, *transformations*, *groups*, *linear spaces*, *linear operators*, *metric spaces*, *group representation*, and prove a theorem known as *Schur's lemma*, which plays a fundamental role in group representation theory. We also summarize basic notions of *topology*, *manifolds*, *Lie groups*, *Lie algebras*, and *spherical harmonics*.

At the end of each chapter are given Exercises. Most of the problems in the Exercises are easy confirmations of the statements in the text. Problems which require knowledge not mentioned in the text are marked by $*$.

The Bibliography, to be found at the end of the volume, is not intended as an exhaustive list of all related literature; only books to be recommended and directly related papers are listed.

2. Coordinate Rotation Invariance of Image Characteristics

In this chapter, we present a powerful mathematical tool for analyzing 3D recovery equations. The underlying principle is that the image coordinate system plays only an auxiliary role because no coordinate system is inherent to the image; any coordinate systems obtained by rotations around the image origin can be used equivalently. Based on this observation, we rearrange observed image characteristics into "invariants". Invoking a general discipline which we call "Weyl's thesis", we show that the geometrical meanings of involved quantities become clear if the 3D recovery equations are written in terms of invariants. We also see that analytical solutions often emerge themselves. To illustrate this, we apply our method to optical flow analysis and the shape-from-texture problem.

2.1 Image Characteristics and 3D Recovery Equations

Suppose an object model is specified by parameters $\alpha_1, \ldots, \alpha_n$. They may be the scene coordinates of particular points, or coefficients of the equations defining surfaces, edges and faces. If the object is in motion, they may also include the translation and rotation velocity components (Sect. 2.6). Let $c_1, \ldots, c_m$ be the image characteristics that characterize the projection image. They may reflect the gray levels, the surface texture (Sect. 2.7), the object contour, the light reflectance, or shading. If the object is in motion, they may also be parameters of the optical flow (Sect. 2.6). Our aim is to estimate the object parameters $\alpha_1, \ldots, \alpha_n$ from the image characteristics $c_1, \ldots, c_m$. The problem is often referred to as *shape from* ..., where ... indicates the source of the image characteristics (texture, shading, motion, etc.). Since the object model is parametrized, the image characteristics to be observed can be derived theoretically from the imaging geometry of perspective projection (Sect. 1.2), resulting in equations like

$$c_i = F_i(\alpha_1, \ldots, \alpha_n) , \qquad i = 1, \ldots, m . \tag{2.1.1}$$

Let us call these *3D recovery equations*. (Namely, we take the 2D non-Euclidean approach discussed in Sect. 1.5.) The values of the object parameters $\alpha_1, \ldots, \alpha_n$ are determined by solving (2.1.1) after substituting the observed values of $c_1, \ldots, c_m$.

Unfortunately, the 3D recovery equations for most problems are nonlinear and difficult to solve analytically. A brute force method may be the exhaustive search, trying all possible values of $\alpha_1, \ldots, \alpha_m$ in some heuristic order, each time computing $F_i(\alpha_1, \ldots, \alpha_n)$, and checking if the resulting $c_1, \ldots, c_n$ coincide with the observed values. Obviously, this process is very inefficient. One alternative is the *table lookup*: the values of functions $F_i(\alpha_1, \ldots, \alpha_n)$, $i = 1, \ldots, n$, are stored in a lookup table beforehand, and we search this table for the observed values. This method may be computationally efficient but often impractical because of the large memory space it requires. Another alternative is to apply a numerical scheme such as the Newton iterations (Sect. 9.6). This is probably the most widely used approach in computer vision. However, numerical iterations usually require a good initial guess, which is often very difficult to obtain. Also, in many cases, there is no guarantee of convergence. Furthermore, we cannot tell whether there exist unique or multiple solutions.

In view of these observations, it would be desirable to obtain an explicit solution in an analytically closed form. In this chapter, we present a powerful tool for finding analytical solutions by exploiting the fact that the 3D recovery equations have a "structure" that reflects the imaging geometry of perspective projection. In general, there is no systematic way to solve arbitrarily given nonlinear equations, but we may be able to solve nonlinear equations that have some internal structure.

What we are going to do is to focus on the *invariance properties* of the object parameters and image characteristics. We know that the image coordinate system plays only an auxiliary role, since no coordinate system is inherent to the image; any coordinate system obtained by rotations around the image origin can be used equivalently. We will show that if the 3D recovery equations are expressed in terms of *invariants*, the solutions often emerge by themselves. The merit of using invariants is not limited to finding analytical solutions. The geometrical "meaning" of the parameters becomes very clear if they are expressed in terms of invariants. We refer to this fact as *Weyl's thesis.* We illustrate our procedure by analyzing optical flow (Sect. 2.6) and the shape-from-texture problem (Sect. 2.7).

2.2 Parameter Changes and Representations

Suppose a set of image characteristics $c_1, \ldots, c_m$ is measured with reference to a given image xy coordinate system. Since the image itself does not have any inherent coordinate system, the orientation of the x- or y-axis is completely arbitrary. Therefore, we may use any other $x'y'$ coordinate system obtained by rotating the original xy coordinate system by an arbitrary angle θ clockwise. (Recall that a clockwise rotation is a positive rotation (Sect. 1.2).) Let $c'_1, \ldots, c'_m$ be the new image characteristics obtained by the same measurement but with reference to a new $x'y'$ coordinate system. Suppose the new values $c'_1, \ldots, c'_m$

are related to the original values $c_1, \ldots, c_m$ *linearly* in the form

$$\begin{pmatrix} c'_1 \\ \vdots \\ c'_m \end{pmatrix} = \boldsymbol{T}(\theta) \begin{pmatrix} c_1 \\ \vdots \\ c_m \end{pmatrix} . \tag{2.2.1}$$

In the following, we consider image characteristics that undergo, under coordinate rotation, nonsingular linear transformations in the form (2.2.1). This means that the matrix $\boldsymbol{T}(\theta)$ defines a *representation* of the 2D rotation group $SO(2)$. Namely, the correspondence from a coordinate rotation by angle θ to matrix $\boldsymbol{T}(\theta)$ is a *homomorphism* from $SO(2)$ into the group of matrices. (See the Appendix, Sect. A.9, for the precise definition.)

Proposition 2.1. *The correspondence between a coordinate rotation by angle θ and matrix $\boldsymbol{T}(\theta)$ is a homomorphism, and* (2.2.1) *defines a representation of the 2D rotation group $SO(2)$.*

Proof. Consider another $x''y''$ coordinate system obtained by rotating the $x'y'$ coordinate system by angle θ' clockwise. From (2.2.1), we obtain

$$\begin{pmatrix} c''_1 \\ \vdots \\ c''_m \end{pmatrix} = \boldsymbol{T}(\theta')\,\boldsymbol{T}(\theta) \begin{pmatrix} c_1 \\ \vdots \\ c_m \end{pmatrix} . \tag{2.2.2}$$

The $x''y''$ coordinate system is also obtained by rotating the xy coordinate system by angle $\theta' + \theta$. Hence, we have

$$\begin{pmatrix} c''_1 \\ \vdots \\ c''_m \end{pmatrix} = \boldsymbol{T}(\theta' + \theta) \begin{pmatrix} c_1 \\ \vdots \\ c_m \end{pmatrix} . \tag{2.2.3}$$

Since (2.2.2, 3) must hold for arbitrary values of $c_1, \ldots, c_m$, we conclude that

$$\boldsymbol{T}(\theta')\,\boldsymbol{T}(\theta) = \boldsymbol{T}(\theta' + \theta) . \tag{2.2.4}$$

Thus, a composition of rotations corresponds to matrix multiplication, and $\boldsymbol{T}(\theta)$ defines a homomorphism from rotations into matrices.

Now, we must notice an important fact: image characteristics are obtained by particular measurements, and *each of them does not necessarily have its own meaning separate from the others.* Indeed, there exist infinitely many ways of choosing equivalent parameters. For example, instead of using parameters c_1

and c_2, we can equivalently use new parameters $C_1 = c_1 - c_2$ and $C_2 = c_1 + c_2$. The two sets $\{c_1, c_2\}$, $\{C_1, C_2\}$ indicate one and the same property.

Suppose we use new image characteristics $C_1, \ldots, C_m$ obtained by taking *linear combinations* of $c_1, \ldots, c_m$. (We assume that the original image characteristics $c_1, \ldots, c_m$ are real numbers, but we allow the coefficients of the linear combinations to be complex numbers, so the new image characteristics $C_1, \ldots, C_m$ can be complex numbers.) Suppose the transformation of the new parameters $C_1, \ldots, C_m$ takes the form

$$\begin{pmatrix} C_1' \\ \vdots \\ C_l' \\ C_{l+1}' \\ \vdots \\ C_m' \end{pmatrix} = \left(\begin{array}{c|c} * & O \\ \hline O & * \end{array}\right) \begin{pmatrix} C_1 \\ \vdots \\ C_l \\ C_{l+1} \\ \vdots \\ C_m \end{pmatrix} \tag{2.2.5}$$

for any θ. This means that the two sets $\{C_1, \ldots, C_l\}$, $\{C_{l+1}, \ldots, C_m\}$ are transformed independently of each other. Hence, it may be reasonable to say that *the two sets indicate different properties.*

The process of decoupling is called the *reduction* of the representation: the representation $\boldsymbol{T}(\theta)$ is *reduced* to the *direct sum* of two representations. If a representation can be reduced, it is said to be *reducible.* In the above case, we may be able to apply the same process to the two sets $\{C_1, \ldots, C_l\}$ and $\{C_{l+1}, \ldots, C_m\}$, and then to each of the resulting sets, and so on until no further reduction is possible. Finally, we may end up with the form

$$\begin{pmatrix} C_1' \\ \vdots \\ C_m' \end{pmatrix} = \begin{pmatrix} * & & & \\ & * & & \\ & & \ddots & \\ & & & * \end{pmatrix} \begin{pmatrix} C_1 \\ \vdots \\ C_m \end{pmatrix} \tag{2.2.6}$$

We then say that the representation has been reduced to the direct sum of *irreducible representations.* If a representation can be reduced to the direct sum of irreducible representations, the representation is said to be *fully reducible* (see the Appendix, Sect. A.6, for a general account of group representation).

If a representation is irreducible, the parameters cannot be separated in any way into independently transforming subsets. Hence, it is natural to regard such a set of parameters as indicating a *single* property, whereas a set defining a reducible representation indicates two or more different properties simultaneously. Thus, the vague notion of "separating into individual properties" can be given a definite meaning as the *irreducible reduction* of a representation with respect to some group of transformations. This view was emphasized by Hermann Weyl (1885–1955) who asserted that a set of measurement data, or *observables*, can be regarded as indicating a "single" physical property if and only if it defines an irreducible representation of a group of transformations that does not change the meaning of the phenomenon. He illustrated this viewpoint by describing quantum mechanics in terms of group representation theory. In the following, we call this view *Weyl's thesis*.

2.2.1 Rotation of the Coordinate System

The coordinate rotation can be viewed as a special case of the camera rotation transformation (Sect. 1.3). As shown in Sect. 1.2, the image xy coordinate system is determined uniquely with reference to the scene XYZ coordinate system. If the XYZ coordinate system (i.e., the camera) is rotated around the Z-axis, the xy coordinate system is rotated around the image origin accordingly. Conversely, if we define an image xy coordinate system, the XYZ coordinate system is determined accordingly. It follows that if the image xy coordinate system is rotated around the image origin, the scene XYZ coordinate system is also rotated around the Z-axis in such a way that the projection equations (1.2.2), or (1.2.3), remain unchanged. Consequently, the object parameters also undergo corresponding transformations (Sects. 2.6, 7).

2.3 Invariants and Weights

It can be shown that for coordinate rotations *all representations are fully reducible* and *all irreducible representations are one-dimensional*. In other words, given image characteristics $c_1, \ldots, c_m$ that define a representation, we can always find, by taking appropriate linear combinations, a new set of parameters $C_1, \ldots, C_m$ such that each is transformed separately:

$$C'_k = T_k(\theta)C_i \ , \qquad k = 1, \ldots, m \ . \tag{2.3.1}$$

(We will discuss why this is possible later.)

Since a representation is a homomorphism from rotations, the coefficient $T_k(\theta)$ must satisfy

$$T_k(\theta')T_k(\theta) = T_k(\theta' + \theta) \ . \tag{2.3.2}$$

Since a rotation by 2π is the same as no rotation, $T_k(\theta)$ must be a periodic function in θ:

$$T_k(0) = 1 \ , \qquad T_k(\theta + 2\pi) = T_k(\theta) \ . \tag{2.3.3}$$

Since $T_k(\theta)$ must be a continuous function, it must be $e^{-in\theta}$ for some integer n, where i is the imaginary unit (Exercise 2.1). Let us call the integer n the *weight* of the irreducible representation. We adopt the convention of using $e^{-in\theta}$ to define the weight, instead of $e^{in\theta}$. This choice is convenient for treating rotations of the coordinate system. If we were to consider rotations of images relative to a fixed coordinate system, it would be convenient to use $e^{in\theta}$. We call an image characteristic of weight 0 an *absolute invariant* and that of nonzero weight n a *relative invariant* of weight n. We also call both absolute and relative invariants simply *invariants.*[1]

What we have shown above is that any representation (2.2.1) defined by image characteristics $c_1, \ldots, c_m$ is reduced, by taking appropriate linear combinations, to the direct sum of one-dimensional irreducible representations such that the new image characteristics $C_1, \ldots, C_m$ are transformed under coordinate rotation in the form

$$\begin{pmatrix} C_1' \\ \vdots \\ C_m' \end{pmatrix} = \begin{pmatrix} e^{-in_1\theta} & & \\ & \ddots & \\ & & e^{-in_m\theta} \end{pmatrix} \begin{pmatrix} C_1 \\ \vdots \\ C_m \end{pmatrix}. \tag{2.3.4}$$

The reduction of a representation is formally stated as follows. Let the new parameters $C_1, \ldots, C_m$ be defined by linear combinations of the original image characteristics $c_1, \ldots, c_m$:[2]

$$\begin{pmatrix} C_1 \\ \vdots \\ C_m \end{pmatrix} = \boldsymbol{P} \begin{pmatrix} c_1 \\ \vdots \\ c_m \end{pmatrix}. \tag{2.3.5}$$

Then, we have

$$\begin{pmatrix} C_1' \\ \vdots \\ C_m' \end{pmatrix} = \boldsymbol{P} \begin{pmatrix} c_1' \\ \vdots \\ c_m' \end{pmatrix}$$

$$= \boldsymbol{P}\boldsymbol{T}(\theta) \begin{pmatrix} c_1 \\ \vdots \\ c_m \end{pmatrix}$$

$$= \boldsymbol{P}\boldsymbol{T}(\theta)\boldsymbol{P}^{-1} \begin{pmatrix} C_1 \\ \vdots \\ C_m \end{pmatrix}. \tag{2.3.6}$$

[1] Care is required: Many authors use the term "invariant" to mean "having a constant value", but in this book an "invariant" may not necessarily have a constant value; we mean that it always has the "same meaning".

[2] The requirement that the original and the new image characteristics should be equivalent demands that $\boldsymbol{P}$ should be a nonsingular (in general, complex) matrix. Hence, its inverse $\boldsymbol{P}^{-1}$ exists.

Thus, we find that

$$PT(\theta)P^{-1} = \begin{pmatrix} e^{-in_1\theta} & & \\ & \ddots & \\ & & e^{-in_m\theta} \end{pmatrix} . \qquad (2.3.7)$$

The left-hand side is called the *similarity transformation* of matrix $T(\theta)$ by a nonsingular matrix P. Application of a similarity transformation to make a matrix diagonal is called *diagonalization*, and the resulting diagonal matrix is called the *canonical form*. The important fact to note here is that a *single* matrix P, *which does not depend on* θ, can diagonalize the matrix $T(\theta)$ for *all* values of θ simultaneously. This is due to the fact that the 2D rotation group $SO(2)$ is a *compact* group[3] and consequently all representations are fully reducible (Appendix, Sect. A.6). Since it is also an *Abelian* group,[4] all representations are reduced into one-dimensional irreducible representations. (This is proved from *Schur's lemma*—see the Appendix, Sect. A.7, for details.)

2.3.1 Signs of the Weights and Degeneracy

If the original image characteristics $c_1, \ldots, c_m$ are always real for any choice of the coordinate system, the transformation matrix $T(\theta)$ is also a real matrix. It can be shown that the new image characteristics $C_1, \ldots, C_m$ which define irreducible representations consist of real quantities and complex conjugate pairs (we omit the proof). A quantity, say C_1, is real if and only if it is an absolute invariant:

$$C_1' = C_1 , \qquad (2.3.8)$$

If one quantity, say C_2, is a relative invariant of weight n ($\neq 0$), it must be a complex quantity (its imaginary part may vanish at particular angles of coordinate rotation), and there must exist another, say C_3, which is the complex conjugate of C_2 and has weight $-n$:

$$C_3 = C_2^* , \qquad C_2' = e^{-in\theta}C_2 , \qquad C_3' = e^{in\theta}C_3 . \qquad (2.3.9)$$

If there appear, after irreducible reduction, more than one invariant of the same weight, arbitrary linear combinations of them are also invariants of the same weight, and hence the decomposition is not unique. This situation is called *degeneracy*. It also arises in quantum mechanics—the degeneracy of quantum states. In such a case, let us interpret Weyl's thesis as follows: we cannot

---

[3] The domain of the parameter θ is confined in the interval $[0, 2\pi]$ modulo 2π. (See the Appendix, Sect. A.8, for the precise definition of a compact group.)

[4] The composite of rotation by angle θ' and rotation by angle θ is rotation by angle $\theta' + \theta$, irrespective of the order of application. (See the Appendix, Sect. A.2, for more details.)

differentiate invariants that have the same weight simply by applying coordinate rotations. In order to differentiate them any further, another group of transformations must be applied: *If a property really has some intrinsic meaning, there should exist a group of transformations that can single it out.*

2.4 Representation of a Scalar and a Vector

If a parameter c does not change its value when the coordinate system is rotated, i.e.,

$$c' = c \ , \tag{2.4.1}$$

the parameter c is called a *scalar* with respect to the 2D coordinate rotation.[5] For example, the image intensity of a particular point of the object image is a scalar. Equation (2.4.1) trivially defines a representation called the *identity representation.* Hence, a scalar is an absolute invariant.

A set of two parameters $\{a, b\}$ is caled a *vector* with respect to the coordinate rotation if it is transformed by coordinate rotation of angle θ clockwise as follows:[6]

$$\begin{pmatrix} a' \\ b' \end{pmatrix} = \begin{pmatrix} \cos\theta & \sin\theta \\ -\sin\theta & \cos\theta \end{pmatrix} \begin{pmatrix} a \\ b \end{pmatrix} . \tag{2.4.2}$$

For example, the image coordinates (x, y) of a particular point of the object image are a vector. If we put

$$\boldsymbol{R}(\theta) = \begin{pmatrix} \cos\theta & \sin\theta \\ -\sin\theta & \cos\theta \end{pmatrix} , \tag{2.4.3}$$

we can easily confirm that

$$\boldsymbol{R}(\theta')\boldsymbol{R}(\theta) = \boldsymbol{R}(\theta' + \theta) \ . \tag{2.4.4}$$

Hence, the matrix $\boldsymbol{R}(\theta)$ defines a *faithful* representation of $SO(2)$—called the *vector representation.* (A representation is faithful if different group elements correspond to different matrices. See the Appendix, Sect. A.6, for more details.) This representation is not irreducible; if we take linear combinations $a + ib$, $a - ib$, we obtain

---

[5] Whenever we say that a parameter is a "scalar", we must specify the underlying group of transformations. However, we often do not mention the group when it can be understood. If we need to distinguish a scalar with respect to the 2D coordinate rotation from a scalar with respect to the 3D camera rotation, we refer to them as a *2D scalar* and *3D scalar. A 2D scalar is not necessarily a 3D scalar* (Chap. 3).

[6] Equation (2.4.2) is a consequence of our convention that the coordinate system is rotated. If the image is rotated, the matrix in (2.4.2) is replaced by its transpose. As in the case of scalars, we often do not mention the underlying group of transformations. If we must distinguish $SO(2)$ from $SO(3)$, we use the terms *2D vectors* and *3D vectors.*

$$\begin{pmatrix} a' + ib' \\ a' - ib' \end{pmatrix} = \begin{pmatrix} e^{-i\theta} & \\ & e^{i\theta} \end{pmatrix} \begin{pmatrix} a + ib \\ a - ib \end{pmatrix} . \tag{2.4.5}$$

In other words, by taking the linear combination

$$\begin{pmatrix} z \\ z^* \end{pmatrix} = \begin{pmatrix} 1 & i \\ 1 & -i \end{pmatrix} \begin{pmatrix} a \\ b \end{pmatrix} , \tag{2.4.6}$$

the matrix $\boldsymbol{R}(\theta)$ is diagonalized as

$$\begin{pmatrix} 1 & i \\ 1 & -i \end{pmatrix} \begin{pmatrix} \cos\theta & \sin\theta \\ -\sin\theta & \cos\theta \end{pmatrix} \begin{pmatrix} 1 & i \\ 1 & -i \end{pmatrix}^{-1} = \begin{pmatrix} e^{-i\theta} & \\ & e^{i\theta} \end{pmatrix} , \tag{2.4.7}$$

where *Euler's formula* $e^{i\theta} = \cos\theta + i\sin\theta$ is used.

From this, we conclude as follows:

Theorem 2.1. *If image characteristics* $\{a, b\}$ *are transformed as a vector, their linear combinations*

$$z = a + ib , \qquad z^* = a - ib \tag{2.4.8}$$

are relative invariants of weight 1 *and* -1 *respectively:*

$$z' = e^{-i\theta} z , \qquad z^{*\prime} = e^{i\theta} z^* . \tag{2.4.9}$$

2.4.1 Invariant Meaning of a Position

The xy coordinates (a, b) define a single complex number $z = a + ib$. Conversely, a complex number $z = a + ib$ can be identified with a position on the image plane regarded as the complex number (or Gaussian) plane, the x- and the y-axes being respectively identified with the real and imaginary axes. However, the position indicated by a pair of real numbers, or equivalently by a single complex number, does not necessarily have an *invariant meaning* independent of the choice of the coordinate system; it merely indicates a location on the image plane. If a position has an invariant meaning—say, the brightest spot or the centroid of some region—it must be transformed as a vector, and the converse is also true according to Weyl's thesis. *A position has an invariant meaning if and only if it is transformed as a vector.*

2.5 Representation of a Tensor

A set of parameters $\{A, B, C, D\}$ is called a *tensor* (of *degree* 2) with respect to the coordinate rotation if it is transformed by rotation of angle θ clockwise in the form[7]

---

[7] In matrix notion, (2.5.1) is written as $\boldsymbol{M}' = \boldsymbol{R}(\theta)\boldsymbol{M}\boldsymbol{R}(\theta)^{\mathrm{T}}$, where $\boldsymbol{R}(\theta)$ is defined by (2.4.3). If we rotate the image, instead of the coordinate system, the matrices $\boldsymbol{R}(\theta)$ and $\boldsymbol{R}(\theta)^{\mathrm{T}}$ must be interchanged. As in the case of scalars and vectors, we often do not mention the underlying group of transformations. We use the terms *2D tensors* and *3D tensors* if distinction is necessary between $SO(2)$ and $SO(3)$.

$$\begin{pmatrix} A' & B' \\ C' & D' \end{pmatrix} = \begin{pmatrix} \cos\theta & \sin\theta \\ -\sin\theta & \cos\theta \end{pmatrix} \begin{pmatrix} A & B \\ C & D \end{pmatrix} \begin{pmatrix} \cos\theta & -\sin\theta \\ \sin\theta & \cos\theta \end{pmatrix} . \tag{2.5.1}$$

This equation defines a linear mapping from A, B, C, D onto A', B', C', D', and hence a representation of $SO(2)$. This representation is called the *tensor representation*. If we pick out the matrix elements, (2.5.1) is rearranged into the following form (Exercise 2.3):

$$\begin{pmatrix} A' \\ B' \\ C' \\ D' \end{pmatrix} = \begin{pmatrix} \cos^2\theta & \cos\theta\sin\theta & \cos\theta\sin\theta & \sin^2\theta \\ -\cos\theta\sin\theta & \cos^2\theta & -\sin^2\theta & \cos\theta\sin\theta \\ -\cos\theta\sin\theta & -\sin^2\theta & \cos^2\theta & \cos\theta\sin\theta \\ \sin^2\theta & -\cos\theta\sin\theta & -\cos\theta\sin\theta & \cos^2\theta \end{pmatrix} \begin{pmatrix} A \\ B \\ C \\ D \end{pmatrix} . \tag{2.5.2}$$

This representation is not irreducible. Reduction into irreducible representations is done systematically as follows.

First, note that a square matrix is uniquely decomposed into its *symmetric* and *antisymmetric* (or *skew-symmetric*) parts:

$$\begin{pmatrix} A & B \\ C & D \end{pmatrix} = \begin{pmatrix} A & (B+C)/2 \\ (B+C)/2 & D \end{pmatrix} + \begin{pmatrix} 0 & -(C-B)/2 \\ (C-B)/2 & 0 \end{pmatrix} . \tag{2.5.3}$$

The proof given below does not depend on the dimensionality of the matrix, and the uniqueness of decomposition holds for any dimensionality.

Lemma 2.1. *The decomposition of a square matrix into its symmetric and antisymmetric parts is unique.*

Proof. Suppose a matrix $\boldsymbol{M}$ has two decompositions

$$\boldsymbol{M} = \boldsymbol{S}_1 + \boldsymbol{A}_1 , \qquad \boldsymbol{M} = \boldsymbol{S}_2 + \boldsymbol{A}_2 , \tag{2.5.4}$$

where $\boldsymbol{S}_1$ and $\boldsymbol{S}_2$ are symmetric matrices

$$\boldsymbol{S}_1^{\mathrm{T}} = \boldsymbol{S}_1 , \qquad \boldsymbol{S}_2^{\mathrm{T}} = \boldsymbol{S}_2 , \tag{2.5.5}$$

and $\boldsymbol{A}_1$ and $\boldsymbol{A}_2$ are antisymmetric matrices

$$\boldsymbol{A}_1^{\mathrm{T}} = -\boldsymbol{A}_1 , \qquad \boldsymbol{A}_2^{\mathrm{T}} = -\boldsymbol{A}_2 . \tag{2.5.6}$$

By subtraction, we obtain from (2.5.4)

$$(\boldsymbol{S}_1 - \boldsymbol{S}_2) + (\boldsymbol{A}_1 - \boldsymbol{A}_2) = \boldsymbol{O} , \tag{2.5.7}$$

where $\boldsymbol{O}$ is the zero matrix. Hence, we have

$$\boldsymbol{S}_1 - \boldsymbol{S}_2 = -(\boldsymbol{A}_1 - \boldsymbol{A}_2) . \tag{2.5.8}$$

The left-hand side is a symmetric matrix, while the right-hand side is an anti-symmetric matrix, which is a contradiction unless both sides are zero matrices. Thus, we conclude

$$\boldsymbol{S}_1 = \boldsymbol{S}_2 , \qquad \boldsymbol{A}_1 = \boldsymbol{A}_2 . \tag{2.5.9}$$

The following result is essential.

Proposition 2.2. *The decomposition of a tensor into its symmetric and anti-symmetric parts is invariant; the symmetric and antisymmetric parts are transformed independently as tensors.*

Proof. Suppose a matrix $\boldsymbol{M}$ is transformed as a tensor:

$$\boldsymbol{M}' = \boldsymbol{R}(\theta)\boldsymbol{M}\boldsymbol{R}(\theta)^{\mathrm{T}} , \tag{2.5.10}$$

where the matrix $\boldsymbol{R}(\theta)$ is defined by (2.4.3). The matrix $\boldsymbol{M}$ is decomposed into its symmetric part $\boldsymbol{S}$ and antisymmetric part $\boldsymbol{A}$:

$$\boldsymbol{M} = \boldsymbol{S} + \boldsymbol{A} . \tag{2.5.11}$$

Substituting this into (2.5.10), we obtain

$$\boldsymbol{M}' = \boldsymbol{R}(\theta)\boldsymbol{S}\boldsymbol{R}(\theta)^{\mathrm{T}} + \boldsymbol{R}(\theta)\boldsymbol{A}\boldsymbol{R}(\theta)^{\mathrm{T}} . \tag{2.5.12}$$

The matrix $\boldsymbol{M}'$ is also decomposed into its symmetric and antisymmetric parts:

$$\boldsymbol{M}' = \boldsymbol{S}' + \boldsymbol{A}' . \tag{2.5.13}$$

Now, $\boldsymbol{R}(\theta)\boldsymbol{S}\boldsymbol{R}(\theta)^{\mathrm{T}}$ is a symmetric matrix, while $\boldsymbol{R}(\theta)\boldsymbol{A}\boldsymbol{R}(\theta)^{\mathrm{T}}$ is antisymmetric:

$$(\boldsymbol{R}(\theta)\boldsymbol{S}\boldsymbol{R}(\theta)^{\mathrm{T}})^{\mathrm{T}} = \boldsymbol{R}(\theta)\boldsymbol{S}\boldsymbol{R}(\theta)^{\mathrm{T}} , \qquad (\boldsymbol{R}(\theta)\boldsymbol{A}\boldsymbol{R}(\theta)^{\mathrm{T}})^{\mathrm{T}} = -\boldsymbol{R}(\theta)\boldsymbol{A}\boldsymbol{R}(\theta)^{\mathrm{T}} . \tag{2.5.14}$$

By Lemma 2.1, the decomposition of a square matrix into its symmetric and antisymmetric parts is unique. Hence, (2.5.12) and (2.5.13) must coincide term by term:

$$\boldsymbol{S}' = \boldsymbol{R}(\theta)\boldsymbol{S}\boldsymbol{R}(\theta)^{\mathrm{T}} , \qquad \boldsymbol{A}' = \boldsymbol{R}(\theta)\boldsymbol{A}\boldsymbol{R}(\theta)^{\mathrm{T}} . \tag{2.5.15}$$

Proposition 2.2 tells us that the two sets of parameters $\{A, D, B + D\}$, $\{B - C\}$ are transformed independently of each other. In fact, the transformation (2.5.2) is reduced as follows (Exercise 2.6):

$$\begin{pmatrix} A' \\ D' \\ B'+C' \\ B'-C' \end{pmatrix} = \left(\begin{array}{ccc|c} \cos^2\theta & \sin^2\theta & \cos\theta\sin\theta & \\ \sin^2\theta & \cos^2\theta & -\cos\theta\sin\theta & \\ -\sin 2\theta & \sin 2\theta & \cos 2\theta & \\ \hline & & & 1 \end{array}\right) \begin{pmatrix} A \\ D \\ B+C \\ B-C \end{pmatrix} . \tag{2.5.16}$$

The symmetric part is further decomposed into its *scalar part* (multiple of the unit matrix) and *deviator part* (symmetric matrix of trace 0) uniquely:

$$\begin{pmatrix} A & (B+C)/2 \\ (B+C)/2 & D \end{pmatrix} = \frac{A+D}{2}\begin{pmatrix} 1 & 0 \\ 0 & 1 \end{pmatrix} + \begin{pmatrix} (A-D)/2 & (B+C)/2 \\ (B+C)/2 & -(A-D)/2 \end{pmatrix} . \tag{2.5.17}$$

The proof for the following proposition does not depend on the dimensionality of the matrix, and the uniqueness of decomposition holds for any dimensionality.

Lemma 2.2. *The decomposition of a symmetric matrix into its scalar and deviator parts is unique.*

Proof. Suppose a symmetric matrix $\boldsymbol{M}$ has two decompositions

$$\boldsymbol{M} = \boldsymbol{T}_1 + \boldsymbol{D}_1 , \qquad \boldsymbol{M} = \boldsymbol{T}_2 + \boldsymbol{D}_2 , \tag{2.5.18}$$

where $\boldsymbol{T}_1$ and $\boldsymbol{T}_2$ are multiples of the unit matrix

$$\boldsymbol{T}_1 \propto \boldsymbol{I} , \qquad \boldsymbol{T}_2 \propto \boldsymbol{I} , \tag{2.5.19}$$

and $\boldsymbol{D}_1$ and $\boldsymbol{D}_2$ are traceless symmetric matrices

$$\mathrm{Tr}\{\boldsymbol{D}_1\} = 0 , \qquad \mathrm{Tr}\{\boldsymbol{D}_2\} = 0 . \tag{2.5.20}$$

(Tr denotes the trace.) By subtraction, we obtain from (2.5.18)

$$(\boldsymbol{T}_1 - \boldsymbol{T}_2) + (\boldsymbol{D}_1 - \boldsymbol{D}_2) = \boldsymbol{O} , \tag{2.5.21}$$

and hence, we have

$$\boldsymbol{T}_1 - \boldsymbol{T}_2 = -(\boldsymbol{D}_1 - \boldsymbol{D}_2) . \tag{2.5.22}$$

The left-hand side is a multiple of the unit matrix, while the right-hand side is a traceless symmetric matrix, which is a contradiction unless both sides are zero matrices. Thus, we conclude

$$T_1 = T_2 , \qquad D_1 = D_2 .. \tag{2.5.23}$$

The following result is also essential.

Proposition 2.3. *The decomposition of a symmetric tensor into its scalar and deviator parts is invariant; the scalar and deviator parts are transformed independently as tensors.*

Proof. Suppose a symmetric matrix $\boldsymbol{M}$ is transformed as a tensor:

$$\boldsymbol{M}' = \boldsymbol{R}(\theta)\boldsymbol{M}\boldsymbol{R}(\theta)^{\mathrm{T}} . \tag{2.5.24}$$

The matrix $\boldsymbol{M}$ is decomposed into its scalar part $\boldsymbol{T}$ and deviator part $\boldsymbol{D}$:

$$\mathrm{M} = \boldsymbol{T} + \boldsymbol{D} . \tag{2.5.25}$$

Substituting this into (2.5.24), we obtain

$$\boldsymbol{M}' = \boldsymbol{R}(\theta)\boldsymbol{T}\boldsymbol{R}(\theta)^{\mathrm{T}} + \boldsymbol{R}(\theta)\boldsymbol{D}\boldsymbol{R}(\theta)^{\mathrm{T}} . \tag{2.5.26}$$

The matrix $\boldsymbol{M}'$ is also symmetric, cf. the first equation of (2.5.15), and hence can be decomposed into its scalar and deviator parts:

$$\boldsymbol{M}' = \boldsymbol{T}' + \boldsymbol{D}' . \tag{2.5.27}$$

Now, $\boldsymbol{R}(\theta)\boldsymbol{T}\boldsymbol{R}(\theta)^{\mathrm{T}}$ is equal to $\boldsymbol{T}$ itself, while $\boldsymbol{R}(\theta)\boldsymbol{D}\boldsymbol{R}(\theta)^{\mathrm{T}}$ is a traceless matrix:

$$\begin{aligned} \boldsymbol{R}(\theta)\boldsymbol{T}\boldsymbol{R}(\theta)^{\mathrm{T}} = \boldsymbol{T}\boldsymbol{R}(\theta)\boldsymbol{R}(\theta)^{\mathrm{T}} = \boldsymbol{T} , \quad \mathrm{Tr}\{\boldsymbol{R}(\theta)\boldsymbol{D}\boldsymbol{R}(\theta)^{\mathrm{T}}\} &= \mathrm{Tr}\{\boldsymbol{D}\boldsymbol{R}(\theta)^{\mathrm{T}}\boldsymbol{R}(\theta)\} \\ &= \mathrm{Tr}\{\boldsymbol{D}\} = 0 , \end{aligned} \tag{2.5.28}$$

where use has been made of the fact that a multiple of the unit matrix $\boldsymbol{I}$ commutes with any square matrix of the same dimension, and that $\mathrm{Tr}\,\{\boldsymbol{AB}\} = \mathrm{Tr}\,\{\boldsymbol{BA}\}$ holds for all square matrices of the same dimension. By Lemma 2.2, the decomposition of a symmetric matrix into its scalar and deviator parts is unique. Hence, (2.5.26) and (2.5.27) must coincide term by term:

$$\boldsymbol{T}' = \boldsymbol{R}(\theta)\boldsymbol{T}\boldsymbol{R}(\theta)^{\mathrm{T}} , \qquad \boldsymbol{D}' = \boldsymbol{R}(\theta)\boldsymbol{D}\boldsymbol{R}(\theta)^{\mathrm{T}} . \tag{2.5.29}$$

Proposition 2.3 tells us that the two sets of parameters $\{A + D\}$, $\{A - D, B + C\}$ are transformed independently of each other. In fact, the transformation (2.5.16) is further reduced as

$$\begin{pmatrix} A'+D' \\ A'-D' \\ B'+C' \\ B'-C' \end{pmatrix} = \begin{pmatrix} 1 & & & \\ & \cos 2\theta & \sin 2\theta & \\ & -\sin 2\theta & \cos 2\theta & \\ & & & 1 \end{pmatrix} \begin{pmatrix} A+D \\ A-D \\ B+C \\ B-C \end{pmatrix} . \tag{2.5.30}$$

Now, the pair $\{A-D, B+C\}$ is transformed just like a vector but the arguments of the sines and cosines are 2θ, instead of θ. It is easy to show that the transformation is reduced as follows:

$$\begin{pmatrix} (A'-D')+\mathrm{i}(B'+C') \\ (A'-D')-\mathrm{i}(B'+C') \end{pmatrix} = \begin{pmatrix} \mathrm{e}^{-2\mathrm{i}\theta} & \\ & \mathrm{e}^{2\mathrm{i}\theta} \end{pmatrix} \begin{pmatrix} (A-D)+\mathrm{i}(B+C) \\ (A-D)-\mathrm{i}(B+C) \end{pmatrix} . \tag{2.5.31}$$

Applying (2.5.31) to (2.5.30), the transformation of (2.5.2) is finally reduced into

$$\begin{pmatrix} A'+D' \\ B'-C' \\ (A'-D')+\mathrm{i}(B'+C') \\ (A'-D')-\mathrm{i}(B'+C') \end{pmatrix} = \begin{pmatrix} 1 & & & \\ & 1 & & \\ & & \mathrm{e}^{-2\mathrm{i}\theta} & \\ & & & \mathrm{e}^{2\mathrm{i}\theta} \end{pmatrix} \begin{pmatrix} A+D \\ B-C \\ (A-D)+\mathrm{i}(B+C) \\ (A-D)-\mathrm{i}(B+C) \end{pmatrix} . \tag{2.5.32}$$

Thus, we conclude as follows:

Theorem 2.2. *If image characteristics $\{A, B, C, D\}$ are transformed as a tensor, the following quantities are invariants and define irreducible representations of $SO(2)$:*

$$T = A + D \ , \quad R = B - C \ ,$$
$$S = (A-D) + \mathrm{i}(B+C) \ , \quad S^* = (A-D) - \mathrm{i}(B+C) \ . \tag{2.5.33}$$

Here, T and R are absolute invariants, and S and S^* are relative invariants of weights 2 and -2:

$$T' = T \ , \quad R' = R \ , \quad S' = \mathrm{e}^{-2\mathrm{i}\theta} S \ , \quad S^{*\prime} = \mathrm{e}^{2\mathrm{i}\theta} S^* \ . \tag{2.5.34}$$

2.5.1 Parity

Since invariants T and R have the same weight 0, the decomposition (2.5.32) has degeneracy (Sect. 2.3). Hence, T and R could be replaced by their arbitrary (independent) linear combinations as far as coordinate rotations are concerned. However, *they should have different geometrical meanings*, since they have different origins—the scalar and the deviator parts of a tensor. Hence, *there should exist another group of transformations which differentiate between them.*

Indeed, they can be differentiated by the coordinate reflections. Consider the reflection with respect to the y-axis:

$$x' = -x , \qquad y' = y . \tag{2.5.35}$$

It is easy to show that T and R undergo different transformations:

$$T' = T , \qquad R' = -R . \tag{2.5.36}$$

We say that T and R have *different parities*: T has *even parity* while R has *odd parity*.

2.5.2 Weyl's Theorem

In Sect. 2.5 we attained irreducible representations by decomposing a tensor according to its symmetry and trace. According to a theorem due to Hermann Weyl, this is true of *tensor representations* of any dimensions and degrees. A *tensor* of dimension n and degree r is a set of indexed numbers $A_{i_1 \ldots i_r}$, $i_1, \ldots, i_r = 1, \ldots, n$, which are transformed by n-dimensional coordinate rotation in the form

$$A'_{i_1 \ldots i_r} = \sum_{j_1=1}^{n} \cdots \sum_{j_r=1}^{n} r_{i_1 j_1} \cdots r_{i_r j_r} A_{j_1 \ldots j_r} , \tag{2.5.37}$$

where r_{ij} is the ij element of the n-dimensional rotation matrix (Chap. 5).

Equation (2.5.37) defines a tensor representation of $SO(n)$. For $n > 2$, *Weyl's theorem* asserts that *all irreducible representations of any tensor representation can be found systematically by applying permutations and summations over appropriate indices.*[8] However, the actual procedure is very complicated. It requires knowledge of the *group of permutations* (or the *symmetric group*) and the *Young diagram.* A special case of a tensor of dimension 2 and degree 3 is given in Exercise 2.16. In general, the procedure becomes very simple for *symmetric tensors* (Chap. 5).

2.6 Analysis of Optical Flow for a Planar Surface

Suppose a planar surface is moving in the scene and we are looking at its image orthographically projected along the Z-axis (Fig. 2.1). (In chap. 7, we will give a detailed analysis under perspective projection.) Let

$$Z = pX + qY + r \tag{2.6.1}$$

---

[8] Hence, the reduction is done in the *real* domain for $n > 2$. The two-dimensional rotation group $SO(2)$ is an exception in that it is Abelian; all $SO(n)$, $n > 2$, are non-Abelian. For Abelian groups, all representations are reduced to one-dimensional irreducible representations in the complex domain. See the Appendix, Sect. A.7, for the proof.

be the equation of the surface. The pair (p, q) designates the *surface gradient*, while r designates the distance of the surface from the image plane along the Z-axis, which we call the *absolute depth*.

An instantaneous rigid motion is specified by the velocity (a, b, c)—called the *translation velocity*—at an arbitrarily fixed reference point and the *rotation velocity* $(\omega_1, \omega_2, \omega_3)$ around it. The rotation velocity $(\omega_1, \omega_2, \omega_3)$ defines an instantaneous rotational motion around an axis passing through the reference point. The axis has orientation $(\omega_1, \omega_2, \omega_3)$, and the angular velocity around it is $\sqrt{\omega_1^2 + \omega_2^2 + \omega_3^2}$ (rad/s) screw-wise. We choose as our reference point the point $(0, 0, r)$—the intersection of the Z-axis with the surface (Fig. 2.1).

Thus, the configuration of the problem is specified by the surface gradient (p, q), the absolute depth r, the translation velocity (a, b, c), and the rotation velocity $(\omega_1, \omega_2, \omega_3)$. These nine parameters are the "object parameters" for this problem. (The values of the velocity component c along the Z-axis and the absolute depth r cannot be recovered as long as the projection is orthographic.)

If the motion is as described above, a velocity field $\dot{x} = u(x, y)$, $\dot{y} = v(x, y)$, which we call *optical flow*, is induced on the image plane.

Proposition 2.4 (*Flow equations*). *The optical flow induced by orthographic projection of planar surface motion has the form*

$$\begin{aligned} u(x, y) &= a + p\omega_2 x + (q\omega_2 - \omega_3)y \ , \\ v(x, y) &= b - (p\omega_1 - \omega_3)x - q\omega_1 y \ . \end{aligned} \tag{2.6.2}$$

Proof. The velocity of point (X, Y, Z) in the scene is given by

$$\begin{pmatrix} \dot{X} \\ \dot{Y} \\ \dot{Z} \end{pmatrix} = \begin{pmatrix} a \\ b \\ c \end{pmatrix} + \begin{pmatrix} \omega_1 \\ \omega_2 \\ \omega_3 \end{pmatrix} \times \begin{pmatrix} X \\ Y \\ Z - r \end{pmatrix} . \tag{2.6.3}$$

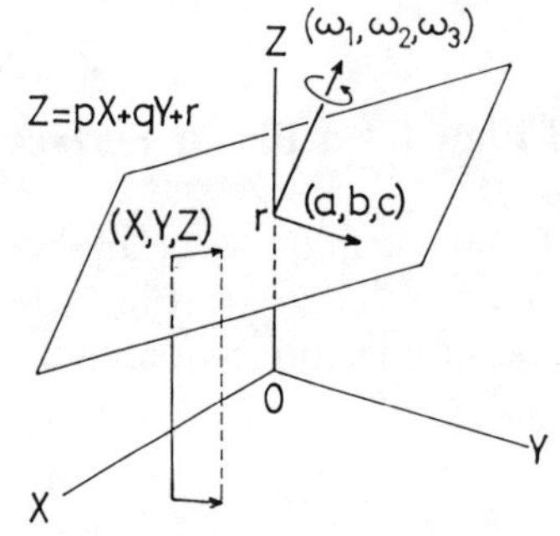

Fig. 2.1. A plane having equation $Z = pX + qY + r$ is moving with translation velocity (a, b, c) at $(0, 0, r)$ and rotation velocity $(\omega_1, \omega_2, \omega_3)$ around it. An optical flow is induced on the image plane by orthographic projection along the Z-axis

If we substitute the surface equation (2.6.1) into this and note $\dot{x} = \dot{X}$, $\dot{y} = \dot{Y}$, we obtain the flow equations (2.6.2).

The flow equations (2.6.2) are rewritten in matrix form as

$$\begin{pmatrix} u \\ v \end{pmatrix} = \begin{pmatrix} a \\ b \end{pmatrix} + \begin{pmatrix} A & B \\ C & D \end{pmatrix} \begin{pmatrix} x \\ y \end{pmatrix} . \tag{2.6.4}$$

where

$$A = p\omega_2 , \quad B = q\omega_2 - \omega_3 , \quad C = -p\omega_1 + \omega_3 , \quad D = -q\omega_1 . \tag{2.6.5}$$

We call the six parameters a, b, A, B, C, D the *flow parameters*. Suppose the optical flow is detected at points (x_i, y_i), $i = 1, 2, \ldots$. Let (u_i, v_i) be the velocities at these points. Then, the flow parameters a, b, A, B, C, D can be determined by fitting the flow equations (2.6.4). For example, we can use the least squares method (Chap. 7), minimizing

$$M = \sum_i [(a + Ax_i + By_i - u_i)^2 + (b + Cx_i + Dy_i - v_i)^2] , \tag{2.6.6}$$

where summation is taken over all the points at which the velocity is observed. Since the flow parameters can be determined from a given optical flow, they can be regarded as "image characteristics". After a, b, A, B, C, D are computed, the translation velocity components a, b are available at hand, and the rest of the parameters are determined by solving (2.6.5) (c and r cannot be determined as long as the projection is orthographic). Hence, (2.6.5) are the 3D recovery equations for this problem.

Since (2.6.5) provides four equations for five unknowns p, q, ω_1, ω_2, ω_3, one degree of freedom remains indeterminate. It seems, at first sight, that we can solve (2.6.5) by choosing one unknown, say p, and expressing the rest in terms of it. In this way, however, we may not be able to "understand" the geometrical implications of the indeterminacy; we cannot tell, for example, which quantities are invariant and what their geometrical meanings are. This is a flaw inherent to algorithmic approaches in general.

Here, we note the following fact: An optical flow is described with reference to an image xy coordinate system, but *the choice of the image coordinate system can be arbitrary* as long as the image origin O corresponds to the camera optical axis. Suppose we use an $x'y'$ coordinate system obtained by rotating the original xy coordinate system around the image origin O by angle θ clockwise. Since we are also observing the rigid motion of a planar surface with respect to the new coordinate system, the flow equations must have the same form

$$\begin{pmatrix} u' \\ v' \end{pmatrix} = \begin{pmatrix} a' \\ b' \end{pmatrix} + \begin{pmatrix} A' & B' \\ C' & D' \end{pmatrix} \begin{pmatrix} x' \\ y' \end{pmatrix} . \tag{2.6.7}$$

The values of the coefficients may be different, but the form of the equation must be the same. From this fact, we can observe the following transformation rules of the flow parameters.

Proposition 2.5. *The pair* $\{a, b\}$ *is a vector, while the set* $\{A, B, C, D\}$ *is a tensor*:

$$\begin{pmatrix} a' \\ b' \end{pmatrix} = \begin{pmatrix} \cos\theta & \sin\theta \\ -\sin\theta & \cos\theta \end{pmatrix} \begin{pmatrix} a \\ b \end{pmatrix} , \tag{2.6.8}$$

$$\begin{pmatrix} A' & B' \\ C' & D' \end{pmatrix} = \begin{pmatrix} \cos\theta & \sin\theta \\ -\sin\theta & \cos\theta \end{pmatrix} \begin{pmatrix} A & B \\ C & D \end{pmatrix} \begin{pmatrix} \cos\theta & -\sin\theta \\ \sin\theta & \cos\theta \end{pmatrix} . \tag{2.6.9}$$

Proof. The original coordinates (x, y) and the new coordinates (x', y') are related by

$$\begin{pmatrix} x' \\ y' \end{pmatrix} = \begin{pmatrix} \cos\theta & \sin\theta \\ -\sin\theta & \cos\theta \end{pmatrix} \begin{pmatrix} x \\ y \end{pmatrix} , \quad \begin{pmatrix} x \\ y \end{pmatrix} = \begin{pmatrix} \cos\theta & -\sin\theta \\ \sin\theta & \cos\theta \end{pmatrix} \begin{pmatrix} x' \\ y' \end{pmatrix} . \tag{2.6.10}$$

The original flow (u, v) and the new flow (u', v') are also related by

$$\begin{pmatrix} u' \\ v' \end{pmatrix} = \begin{pmatrix} \cos\theta & \sin\theta \\ -\sin\theta & \cos\theta \end{pmatrix} \begin{pmatrix} u \\ v \end{pmatrix} . \tag{2.6.11}$$

Substituting the flow equations (2.6.4) into (2.6.11) and using (2.6.10), we obtain

$$\begin{pmatrix} u' \\ v' \end{pmatrix} = \begin{pmatrix} \cos\theta & \sin\theta \\ -\sin\theta & \cos\theta \end{pmatrix} \begin{pmatrix} a \\ b \end{pmatrix} + \begin{pmatrix} \cos\theta & \sin\theta \\ -\sin\theta & \cos\theta \end{pmatrix} \begin{pmatrix} A & B \\ C & D \end{pmatrix} \begin{pmatrix} \cos\theta & -\sin\theta \\ \sin\theta & \cos\theta \end{pmatrix} \begin{pmatrix} x' \\ y' \end{pmatrix} . \tag{2.6.12}$$

Comparing this with the new flow equations (2.6.7), we obtain (2.6.8, 9).

Thus, we obtain the invariants

$$V = a + \mathrm{i}b , \quad T = A + D . \quad R = C - B ,$$
$$S = (A - D) + \mathrm{i}(B + C) . \tag{2.6.13}$$

The complex number V is a relative invariant of weight 1, T and R are absolute invariants, and S is a relative invariant of weight 2:

$$V' = \mathrm{e}^{-\mathrm{i}\theta} V , \quad T' = T , \quad R' = R , \quad S' = \mathrm{e}^{-2\mathrm{i}\theta} S . \tag{2.6.14}$$

Here and in the following, we omit S^* from our analysis, since, as far as real-valued images are concerned, a complex image characteristic and its complex conjugate have the same information. Since these invariants define irreducible representations, each one should have a distinct meaning (Weyl's thesis). In order to see their meanings, let us consider what flow will result if all the parameters but one are zero.

The meaning of invariant V is easy to see. If all other parameters are zero, the flow takes the form

$$\begin{pmatrix} u \\ v \end{pmatrix} = \begin{pmatrix} a \\ b \end{pmatrix}, \tag{2.6.15}$$

which is *translational flow* (Fig. 2.2).

If all the parameters but T are zero, the flow takes the form

$$\begin{pmatrix} u \\ v \end{pmatrix} = \frac{T}{2}\begin{pmatrix} 1 & 0 \\ 0 & 1 \end{pmatrix}\begin{pmatrix} x \\ y \end{pmatrix}, \tag{2.6.16}$$

which is *divergent flow* (Fig. 2.3). The value T is called the *divergence*, since we can write[9]

$$T = \frac{\partial u}{\partial x} + \frac{\partial v}{\partial y}. \tag{2.6.17}$$

Similarly, if all the parameters but R are zero, the flow takes the form

$$\begin{pmatrix} u \\ v \end{pmatrix} = \frac{R}{2}\begin{pmatrix} 0 & -1 \\ 1 & 0 \end{pmatrix}\begin{pmatrix} x \\ y \end{pmatrix}, \tag{2.6.18}$$

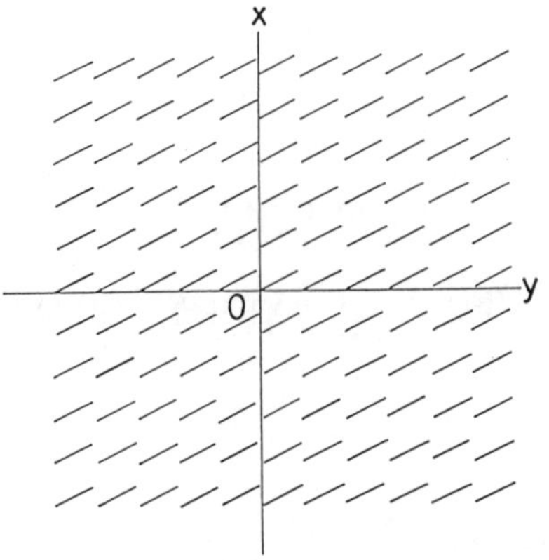

Fig. 2.2. Translational flow

[9] Since the flow equations (2.6.2) are linear in x and y, the divergence (2.6.17) and the rotation (2.6.19) take the same values everywhere. For a general flow field, they may take different values from position to position.

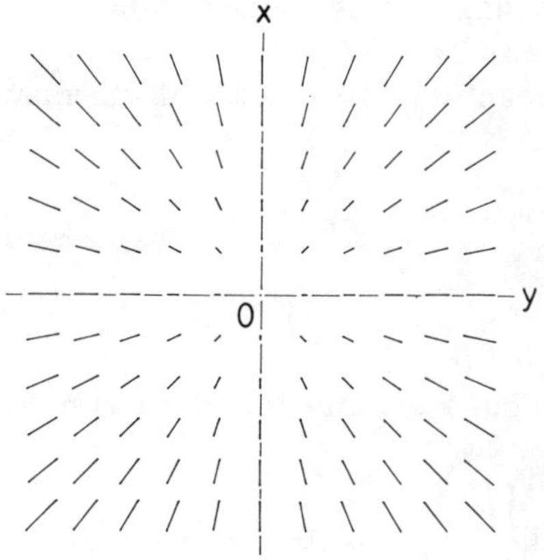

Fig. 2.3. Divergent flow

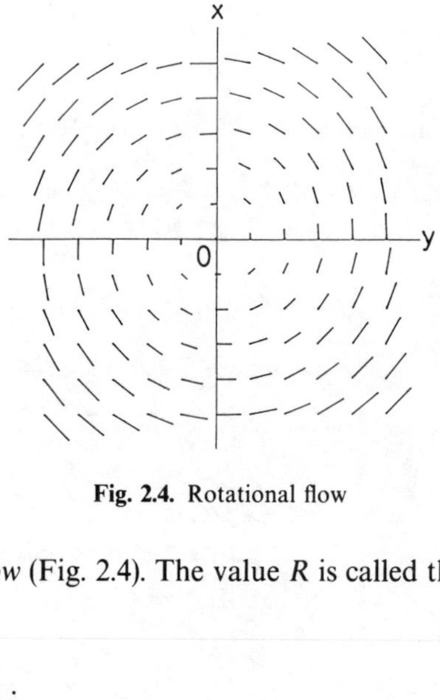

Fig. 2.4. Rotational flow

which is *rotational flow* (Fig. 2.4). The value R is called the *rotation* or *vorticity*, since we can write[9]

$$R = \frac{\partial v}{\partial x} - \frac{\partial u}{\partial y} \ . \tag{2.6.19}$$

Now, consider S, which is a complex number. Let Q_1 be its complex square root normalized to unit modulus $\sqrt{S}/\sqrt{|S|}$, and let $Q_2 = \mathrm{i}Q_1$:

$$Q_1 = \mathrm{e}^{\mathrm{i}\arg(S)/2} \ , \qquad Q_2 = \mathrm{i}\mathrm{e}^{\mathrm{i}\arg(S)/2} \ . \tag{2.6.20}$$

Since S is transformed as a relative invariant of weight 2, Q_1 and Q_2 are both transformed as relative invariants of weight 1:

$$Q_1' = \mathrm{e}^{-\mathrm{i}\theta} Q_1 \ , \qquad Q_2' = \mathrm{e}^{-\mathrm{i}\theta} Q_2 \ . \tag{2.6.21}$$

This means that the complex numbers Q_1 and Q_2 both indicate orientations that have invariant meanings if the image plane is identified with the complex number plane (Sect. 2.4). In other words, *they should represent orientations on the image plane independent of the choice of the coordinate system.* In fact, we obtain the following observation:

Proposition 2.6. *Invariants Q_1 and Q_2 indicate the orientations of the maximum extension and maximum contraction of the optical flow, respectively.*

Proof. If all the invariants but S are zero, the flow takes the form

$$\begin{pmatrix} u \\ v \end{pmatrix} = \frac{1}{2}\begin{pmatrix} S_1 & S_2 \\ S_2 & -S_1 \end{pmatrix}\begin{pmatrix} x \\ y \end{pmatrix}, \tag{2.6.22}$$

where S_1 and S_2 are the real and imaginary parts of $S\ (= S_1 + \mathrm{i}S_2)$. It is easy to prove that the matrix in (2.6.22) has two eigenvalues $\pm |S|$. Hence, if we take a new $x'y'$ coordinate system in such a way that the x'- and y'-axes coincide with the orientations of its eigenvectors, the flow must have the canonical form

$$\begin{pmatrix} u' \\ v' \end{pmatrix} = \frac{|S|}{2}\begin{pmatrix} 1 & 0 \\ 0 & -1 \end{pmatrix}\begin{pmatrix} x' \\ y' \end{pmatrix}. \tag{2.6.23}$$

This flow is called *shear flow* (Fig. 2.5), and $|S| = \sqrt{S_1^2 + S_2^2}$ is called the *shear strength.* We can easily see that the flow extends along the x'-axis and contracts along the y'-axis. We can also see that the complex numbers $Q_1'\ (= 1)$, $Q_2'\ (= \mathrm{i})$ respectively indicate the orientations of the *maximum extension* and *maximum contraction* in "this" canonical form. However, this must be true for *any* coordinate system, *because Q_1 and Q_2 are relative invariants of weight* 1.

Thus, invariants V, T, R, S indicate properties inherent to the flow itself, while the original flow parameters a, b, A, B, C, D specify the flow with reference to the current image coordinate system. Therefore, one cannot say, for instance,

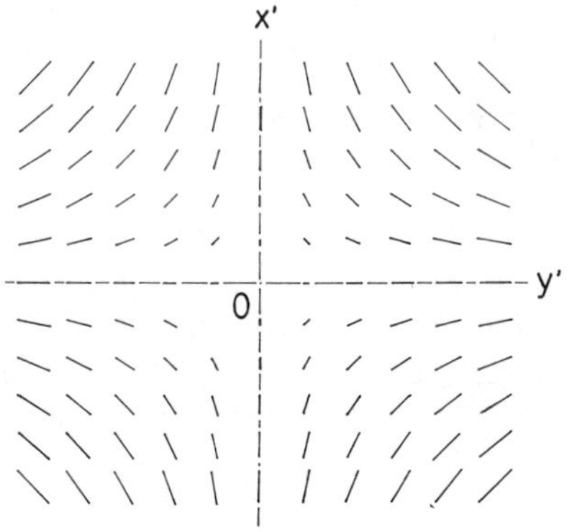

Fig. 2.5. Shear flow

"parameter A indicates such and such a property of the flow." Such a statement makes sense only for invariants. Indeed, this is what we mean by Weyl's thesis.

Next, we consider the transformation rules of the object parameters $p, q, r, \omega_1, \omega_2, \omega_3$. As discussed in Sect. 2.2, whenever the image xy coordinate system is rotated around the image origin, the corresponding scene XYZ coordinate system is also rotated around the Z-axis accordingly. We observe the following fact:

Proposition 2.7. (i) *The surface gradient* (p, q) *is a vector, while the absolute depth* r *is a scalar:*

$$\begin{pmatrix} p' \\ q' \end{pmatrix} = \begin{pmatrix} \cos\theta & \sin\theta \\ -\sin\theta & \cos\theta \end{pmatrix} \begin{pmatrix} p \\ q \end{pmatrix}, \qquad r' = r . \tag{2.6.24}$$

(ii) *The rotation velocity components* $\{\omega_1, \omega_2\}$ *are a vector, while* ω_3 *is a scalar:*

$$\begin{pmatrix} \omega_1' \\ \omega_2' \end{pmatrix} = \begin{pmatrix} \cos\theta & \sin\theta \\ -\sin\theta & \cos\theta \end{pmatrix} \begin{pmatrix} \omega_1 \\ \omega_2 \end{pmatrix}, \qquad \omega_3' = \omega_3 . \tag{2.6.25}$$

Proof. (i) The surface equation (2.6.1) is written in vector form as

$$Z = r + (p \quad q) \begin{pmatrix} X \\ Y \end{pmatrix} . \tag{2.6.26}$$

After the XYZ coordinate system is rotated around the Z-axis, the equation must still have the same form:

$$Z' = r' + (p' \quad q') \begin{pmatrix} X' \\ Y' \end{pmatrix} . \tag{2.6.27}$$

(Any planar surface must always have this form.) The XYZ and the $X'Y'Z'$ coordinate systems are related by

$$\begin{pmatrix} X \\ Y \end{pmatrix} = \begin{pmatrix} \cos\theta & -\sin\theta \\ \sin\theta & \cos\theta \end{pmatrix} \begin{pmatrix} X' \\ Y' \end{pmatrix}, \qquad Z = Z' . \tag{2.6.28}$$

On substitution of these, the surface equation (2.6.26) becomes

$$Z' = r + (p \quad q) \begin{pmatrix} \cos\theta & -\sin\theta \\ \sin\theta & \cos\theta \end{pmatrix} \begin{pmatrix} X' \\ Y' \end{pmatrix} . \tag{2.6.29}$$

Comparing this with the new surface equation (2.6.27), we obtain (2.6.24).

(ii) The rotation velocity components $\{\omega_1, \omega_2, \omega_3\}$ are transformed as a 3D vector under 3D coordinate rotations (Chap. 3). As a result, $\{\omega_1, \omega_2\}$ are transformed as a 2D vector and ω_3 as a 2D scalar under 2D coordinate rotations. (See Sect. 3.6 for a more detailed discussion.)

From this result, we find that r and ω_3 are absolute invariants in themselves, while

$$P = p + iq , \qquad W = \omega_1 + i\omega_2 , \tag{2.6.30}$$

are relative invariants of weight 1:

$$P' = e^{-i\theta}P \ , \qquad W' = e^{-i\theta}W \ . \tag{2.6.31}$$

Proposition 2.8 (*3D recovery equations*).

$$PW^* = (2\omega_3 - R) - iT \ , \qquad PW = iS \ . \tag{2.6.32}$$

Proof. From (2.6.5), invariants T and R are written as

$$T = p\omega_2 - q\omega_1 \ , \qquad R = 2\omega_3 - p\omega_1 - q\omega_2 \ , \tag{2.6.33}$$

which are combined into one complex expression

$$R + iT = 2\omega_3 - p\omega_1 - q\omega_2 + i(p\omega_2 - q\omega_1) = 2\omega_3 - PW^* \ . \tag{2.6.34}$$

The relative invariant S is written as

$$S = p\omega_2 + q\omega_1 + i(q\omega_2 - p\omega_1) = -iPW \ . \tag{2.6.35}$$

Hence, the 3D recovery equations (2.6.5) are rewritten as (2.6.32).

Thus, four real equations (2.6.5) in five unknowns p, q, ω_1, ω_2, ω_3 reduce to two complex equations in three unknowns P, W, ω_3. The solution of (2.6.32) is given as follows.

Theorem 2.3 (*Solution of the 3D recovery equations*).

$$\omega_3 = \frac{1}{2}(R \pm \sqrt{SS^* - T^2}) \ , \tag{2.6.36}$$

$$W = k \exp\left[i\left(\frac{\pi}{4} + \frac{1}{2}\arg(S) - \frac{1}{2}\arg((2\omega_3 - R) - iT)\right)\right] \ , \tag{2.6.37}$$

$$P = \frac{S}{k}\exp\left[i\left(\frac{\pi}{4} - \frac{1}{2}\arg(S) + \frac{1}{2}\arg((2\omega_3 - R) - iT)\right)\right] \ , \tag{2.6.38}$$

where k is an indeterminate scale parameter. For each k, two sets of solutions exist for p, q, ω_1, ω_2, ω_3.

Proof. Since $|PW^*| = |PW|$, the right-hand sides of (2.6.35) have the same modulus. Hence,

$$[(2\omega_3 - R) - iT][(2\omega_3 - R) + iT] = SS^* \ , \tag{2.6.39}$$

from which we obtain (2.6.36). Now, we can immediately see that if W and P satisfy (2.6.35), so do cW and P/c, where c is a nonzero real constant. Hence, the magnitude $k = |W|$ of W can be taken as an indeterminate scale factor. Elimination of P from (2.6.35) by taking the ratio yields

$$\frac{W}{W^*} = e^{2\arg(W)} = \frac{iS}{(2\omega_3 - R) - iT} \ . \tag{2.6.40}$$

Taking the argument of both sides, we obtain

$$2\arg(W) = \frac{\pi}{2} + \arg(S) - \arg((2\omega_3 - R) - \mathrm{i}T) \qquad (\mathrm{mod}\ 2\pi)\ , \quad (2.6.41)$$

hence

$$\arg(W) = \frac{\pi}{4} + \frac{1}{2}\arg(S) - \frac{1}{2}\arg((2\omega_3 - R) - \mathrm{i}T) \qquad (\mathrm{mod}\ \pi)\ . \quad (2.6.42)$$

However, "mod π" can be ignored by allowing the scale factor k to be negative. Then, we can always write $W = k\exp[\mathrm{i}\arg(W)]$, from which follows (2.6.37). Equation (2.6.38) is obtained from $P = \mathrm{i}S/W$.

From this analytical solution, we can deduce several properties that the solution must possess. First, from (2.6.33), we obtain the following *rigidity condition*:

Corollary 2.1 (*Rigidity condition*). *The magnitude of divergence should not be greater than the shear strength*:

$$|T| \leqq |S|\ . \qquad (2.6.43)$$

In other words, if inequality (2.6.43) is not satisfied, the flow cannot be regarded as induced by rigid planar motion.

As we discussed earlier, a relative invariant of weight 1 indicates a 2D orientation that has an invariant meaning on the image plane. Consider P and W, which are both relative invariants of weight 1. For simplicity, let us call the orientations indicated by P and W simply the orientations of P and W. We find the following:

Corollary 2.2. *The orientations of P and W are symmetric with respect to the orientation of maximum shearing.*

Proof. From (2.6.37, 38), we find that

$$\frac{1}{2}[\arg(P) + \arg(W)] = \frac{1}{2}\arg(S) \pm \frac{\pi}{4}\ , \qquad (2.6.44)$$

where $\pm$ corresponds to the sign of the scale factor k. Since $\arg(S)/2$ indicates the orientation of maximum extension (Proposition 2.6), the right-hand side of (2.6.44) indicates the orientation that bisects the orientations of maximum extension and maximum contraction. This orientation is known in fluid mechanics as the orientation of *maximum shearing*. (Viscosity is maximum along this orientation.)

Theorem 2.3 shows that there are two solutions—one is true and the other is spurious. We also find the following.

Corollary 2.3. *The orientations of true P and spurious W are mutually orthogonal, and so are the orientations of true W and spurious P. The orientations of true and spurious W are always symmetric with respect to the principal axes of S, and so are the orientations of true and spurious P.*

By *principal axes* of S, we mean the two principal axes (the orientations of the two mutually orthogonal eigenvectors) of the coefficient matrix in (2.6.22). The proof is left as an exercise (Exercise 2.10).

Example 2.1. Consider the flow of Fig. 2.6. The flow parameters are $a = 0.1$, $b = 0.1$, $A = 0.087$, $B = -0.227$, $C = 0.087$, $D = 0.0524$. Hence, the divergence is $T = 0.140$, the rotation is $R = 0.314$, and the shearing is $S = 0.035 - 0.140i$. Since the shear strength is $|S| = 0.144$ ($> |T|$), the condition (2.6.43) is satisfied. Equation (2.6.36) yields $\omega_3 = 10, 8$ deg/s, and (2.6.37, 38) yield two solutions $\{W_1 = (0.706 + 0.708i)k$ rad/s, $P_1 = (0.123 - 0.074i)/k\}$ and $\{W_2 = (0.516 + 0.857i)k$ rad/s, $P_2 = (0.102 - 0.102i)/k\}$, where k is an indeterminate scale factor. Figure 2.7 shows the solution for $k = 0.5$ on the complex number plane, where the orientations of maximum extension, maximum contraction and maximum shearing are also indicated. We can easily confirm that Corollaries 2.2 and 2.3 are indeed satisfied.

2.6.1 Optical Flow

The term *optical flow* was first used by American psychologist James Jerome Gibson (1904–1979) in his study of human visual perception. This term originally designated motions of visual pattern perceived on the human retina, but many researchers of computer vision have realized that it is essential to motion analysis to establish techniques for detecting "optical flow" from images.

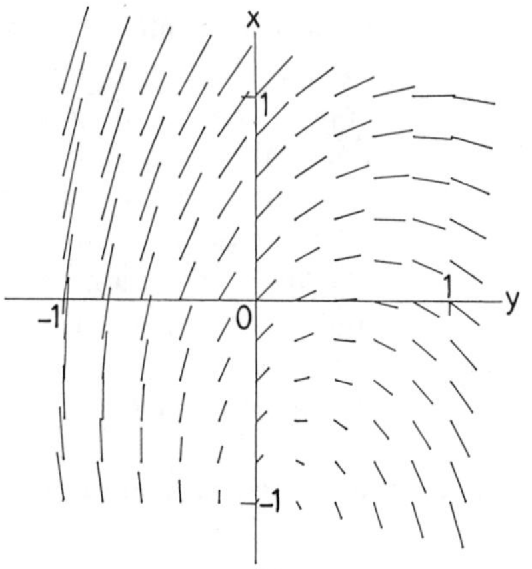

Fig. 2.6. An example of optical flow

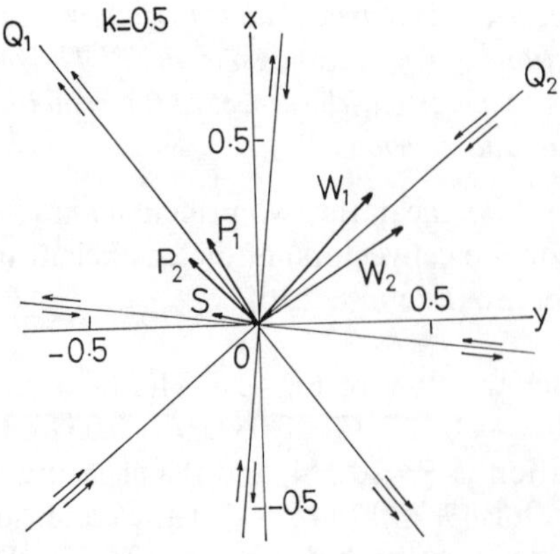

Fig. 2.7. Result of the analysis of the flow of Fig. 2.6. Two solutions exist, one true and the other spurious. The orientations of maximum extension, maximum contraction and maximum shearing are also indicated

These techniques are supposed to detect small (but finite) *displacements of corresponding points* between two image frames. Many approaches have been proposed and tested. They include: (i) the *gradient-based approach*, which computes the time and space derivatives of the gray level of the images; (ii) the *correlation-based approach*, which searches one image for the part that is most correlated to the specified part of the other image—the matching is based on either gray-level patterns or specially identifiable markings (*tokens* or *feature points*); and (iii) the *spatio-temporal approach*, which computes phase shifts of Fourier components of the images.

However, in many cases (typically in the gradient-based approach), the correspondence is established between "points of the same light intensity" that do not necessarily correspond to the same point of the object. This difficulty is easily understood if we consider a motion of a vertical edge with one side completely white and the other side completely black (Fig. 2.8). For such a pattern, we cannot say anything about the motion along the edge—the *tangential velocity*; we can only compute its perpendicular component—the *vertical* (or *transversal*) *velocity*. This difficulty is known as the *aperture problem*.

There exist many terms to describe the motion on the image plane. They include *image flow*, *motion field*, and *vector field*, but their precise meanings differ from author to author. In this book, the term *optical flow* means the *instantaneous velocity field* on the image plane.

2.6.2 Translation Velocity, Rotation Velocity, and Egomotion

As is well known in physics, the motion of a rigid object can be specified by the *translation velocity* at a fixed reference point and the *rotation velocity* around

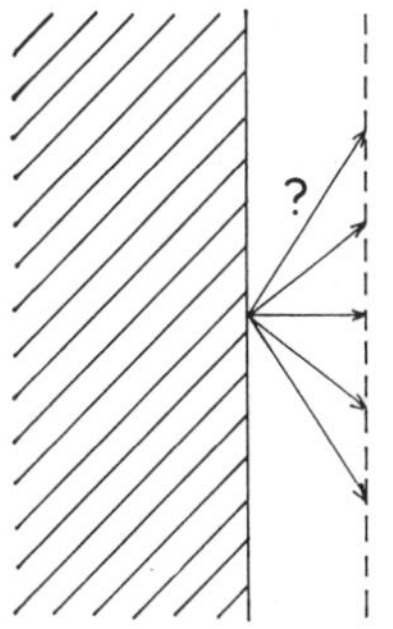

Fig. 2.8. The aperture problem. If an edge dividing the image into two regions of different gray levels moves, the velocity component along the edge cannot be determined; only the component perpendicular to it is determined

that reference point. However, the choice of the reference point is completely arbitrary. Suppose a rigid motion is prescribed by translation velocity (a, b, c) and rotation velocity $(\omega_1, \omega_2, \omega_3)$ with respect to a reference point (X_0, Y_0, Z_0). If we take another reference point (X_0', Y_0', Z_0'), the same motion is now prescribed by translation velocity (a', b', c') and rotation velocity $(\omega_1', \omega_2', \omega_3')$ given by

$$\begin{pmatrix} a' \\ b' \\ c' \end{pmatrix} = \begin{pmatrix} a \\ b \\ c \end{pmatrix} + \begin{pmatrix} \omega_1 \\ \omega_2 \\ \omega_3 \end{pmatrix} \times \begin{pmatrix} X_0' - X_0 \\ Y_0' - Y_0 \\ Z_0' - Z_0 \end{pmatrix}, \tag{2.6.45}$$

$$\omega_1' = \omega_1 , \qquad \omega_2' = \omega_2 , \qquad \omega_3' = \omega_3 . \tag{2.6.46}$$

The reference point need not be inside or on the surface of the object; it can be taken far away from the object. Also, it can be fixed in the scene or moved together with or independently of the object. Therefore, if one encounters a statement like "the translation velocity cannot be determined uniquely" or "computation of the translation velocity involves a large amount of error", one must be careful about where the reference point is taken, because such a statement does not have an invariant meaning; it makes sense only when the choice of the reference point is specified.

In our analysis, the reference point is at the intersection of the planar surface with the Z-axis. An alternative is the origin O of the scene XYZ coordinate system. Then, the rotation velocity components $\omega_1, \omega_2, \omega_3$ can be respectively regarded as angular velocities around the X-, Y-, Z-axes. This choice is equivalent to interpreting the object motion as caused not by the motion of the object but by the motion of the viewer—often referred to as *egomotion*. If the object is moving with translation velocity (a, b, c) and rotation velocity $(\omega_1, \omega_2, \omega_3)$ with respect to the coordinate origin O, the coordinate system is moving relative to the object with translation velocity $(-a, \ -b, \ -c)$ and rotation velocity $(-\omega_1, \ -\omega_2, \ -\omega_3)$. Hence, *any rigid motion can be interpreted as caused by egomotion.* However, the best choice of the reference point depends on the problem to be analyzed, because analysis becomes often easier with one choice than with another.

2.6.3 Equations in Terms of Invariants

As we saw in Sect. 2.6, if we write 3D recovery equations in terms of invariants, the number of equations and the number of unknowns become almost half those in their original forms. As a result, the expressions become succinct, and subsequent analysis becomes very easy. However, the merit of using invariants is not limited to this. Note that the complex conjugate of a relative invariant of weight n is a relative invariant of weight $-n$, while the product of relative invariants of weight n and n' is a relative invariant of weight $n + n'$. The important fact to note is that *only terms of the same weight can be added or subtracted.* Hence, in an equation written in invariants, *all terms must have the same weight*, and *both sides must also have the same weight*. This fact enables us to check equations very easily and sometimes guess the final solution.

2.7 Shape from Texture for Curved Surfaces

Consider a planar surface in the scene, and let $\boldsymbol{n} = (n_1, n_2, n_3)$ be its unit surface normal. The three components are called *directional cosines*, since if α, β, γ are the angles that the vector $\boldsymbol{n}$ makes with respectively the X-, Y-, and Z-axes, we have

$$n_1 = \cos\alpha \ , \qquad n_2 = \cos\beta \ , \qquad n_3 = \cos\gamma \ . \tag{2.7.1}$$

Let $Z = pX + qY + r$ be the equation of the surface. It is easy to see that the directional cosines $\cos\alpha$, $\cos\beta$, $\cos\gamma$ are given as (Exercise 2.11)

$$\cos\alpha = -\frac{p}{\sqrt{p^2 + q^2 + 1}} \ , \qquad \cos\beta = -\frac{q}{\sqrt{p^2 + q^2 + 1}} \ ,$$

$$\cos\gamma = \frac{1}{\sqrt{p^2 + q^2 + 1}} \ . \tag{2.7.2}$$

If the planar surface is projected orthographically onto the image plane, the projected region becomes $\cos\gamma$ times as large as the original region of the surface. The same relationship also holds for a curved surface if p and q are identified with $\partial Z/\partial X$ and $\partial Z/\partial Y$, respectively. Hence, if we consider an infinitesimal square on the image plane defined by four points (x, y), $(x + dx, y)$, $(x, y + dy)$, $(x + dx, y + dy)$, the area of the corresponding portion of the surface is given by $\sqrt{(\partial Z/\partial X)^2 + (\partial Z/\partial Y)^2 + 1}\, dx\, dy$ (Fig. 2.9).

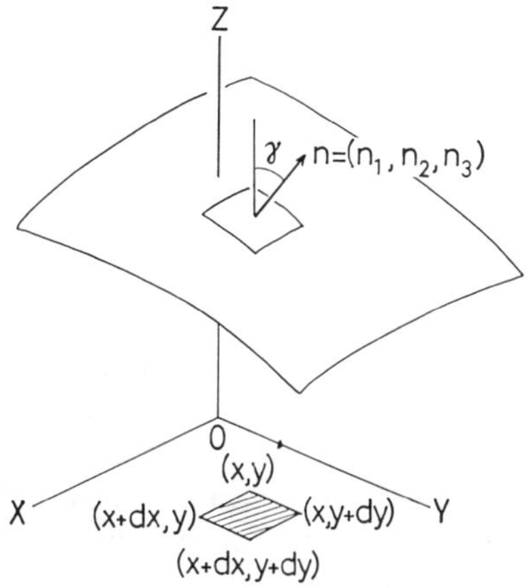

Fig. 2.9. Under orthographic projection, the area of the region on the surface which corresponds to the infinitesimal square on the image plane defined by four points (x, y), $(x + dx, y)$, $(x + dx, y + dy)$, $(x, y + dy)$ is $dx\,dy/\cos\gamma$, where γ is the angle the unit surface normal $\boldsymbol{n}$ makes with the Z-axis

Suppose a surface $Z = Z(X, Y)$ is *homogeneously textured.*[10] Consider its orthographically projected image, and let $\Gamma(x, y)$ be the *texture density* on the image plane. It follows from the above consideration that the texture density $\Gamma(x, y)$ is given by

$$\Gamma(x, y) = \rho\sqrt{\left(\frac{\partial Z}{\partial X}\right)^2 + \left(\frac{\partial Z}{\partial Y}\right)^2 + 1}\ , \tag{2.7.3}$$

where ρ is the true texture density on the surface.

Suppose the surface is quadric:

$$Z = r + pX + qY + aX^2 + 2bXY + cY^2\ . \tag{2.7.4}$$

Hence, r, p, q, a, b, c plus the true texture density ρ are the object parameters for this textured surface. Noting that $x = X$ and $y = Y$ under orthographic projection, we obtain from (2.7.3) and the surface equation (2.7.4)

$$\Gamma(x, y) = A_0\sqrt{1 + A_1x + A_2y + A_3x^2 + 2A_4xy + A_5y^2}\ , \tag{2.7.5}$$

[10] *Texture* is a fine structured pattern characteristic of the material surface. It usually consists of somewhat regularly distributed *texture elements* or *texels.* The texture elements need not have precisely the same shape or size. Also, their distribution need not be strictly regular. The only requirement here is that the distribution be *homogeneous.* For the moment, let us be content to regard the term "homogeneity" with common sense intuition to mean, say, that the number of texture elements per unit area is everywhere the same. The precise definition will be given in Chap. 9. There, we will also present an indirect method for measuring the texture density.

where

$$A_0 = \rho\sqrt{1+p^2+q^2}\,, \quad A_1 = \frac{4(ap+bq)}{1+p^2+q^2}\,, \quad A_2 = \frac{4(bp+cq)}{1+p^2+q^2}\,,$$

$$A_3 = \frac{4(a^2+b^2)}{1+p^2+q^2}\,, \quad A_4 = \frac{4b(a+c)}{1+p^2+q^2}\,, \quad A_5 = \frac{4(b^2+c^2)}{1+p^2+q^2}\,. \tag{2.7.6}$$

By fitting the form of (2.7.5) to the observed $\Gamma(x, y)$, we can estimate the parameters A_i, $i = 0, \ldots, 5$, which are the "image characteristics" for this problem. (In Chap. 9, we will show an indirect method for determining these parameters without measuring the texture density.) If the values of A_i, $i = 0, \ldots, 5$, are given, the values of the "object parameters" ρ, r, p, q, a, b, c are determined by solving (2.7.6). Hence, (2.7.6) are the "3D recovery equations" for this problem. First, let us study the invariance properties under coordinate rotations.

Proposition 2.9. *Parameter A_0 is a scalar, the pair $\{A_1, A_2\}$ is a vector, and the set $\{A_3, A_4, A_4, A_5\}$ is a tensor.*

Proof. Equation (2.7.5) is written in matrix form as

$$\Gamma(x, y) = A_0\left[1 + (A_1 \;\; A_2)\begin{pmatrix} x \\ y \end{pmatrix} + (x \;\; y)\begin{pmatrix} A_3 & A_4 \\ A_4 & A_5 \end{pmatrix}\begin{pmatrix} x \\ y \end{pmatrix}\right]^{1/2}. \tag{2.7.7}$$

If we rotate the image xy coordinate system, the *form* of the texture density $\Gamma(x, y)$ must be the same, since the surface is still quadric with respct to the new coordinate system. Hence, we must have

$$\Gamma(x', y') = A'_0\left[1 + (A'_1 \;\; A'_2)\begin{pmatrix} x' \\ y' \end{pmatrix} + (x' \;\; y')\begin{pmatrix} A'_3 & A'_4 \\ A'_4 & A'_5 \end{pmatrix}\begin{pmatrix} x' \\ y' \end{pmatrix}\right]^{1/2} \tag{2.7.8}$$

with respect to the new x' y' coordinate system. Noting that the texture density $\Gamma(x, y)$ is a scalar function, and substituting (2.6.10b) into (2.7.7), we obtain

$$\Gamma(x', y') = A_0\left[1 + (A_1 \, A_2)\begin{pmatrix} \cos\theta & -\sin\theta \\ \sin\theta & \cos\theta \end{pmatrix}\begin{pmatrix} x' \\ y' \end{pmatrix}\right.$$

$$\left. + (x' \, y')\begin{pmatrix} \cos\theta & \sin\theta \\ -\sin\theta & \cos\theta \end{pmatrix}\begin{pmatrix} A_3 & A_4 \\ A_4 & A_5 \end{pmatrix}\begin{pmatrix} \cos\theta & -\sin\theta \\ \sin\theta & \cos\theta \end{pmatrix}\begin{pmatrix} x' \\ y' \end{pmatrix}\right]^{1/2}. \tag{2.7.9}$$

Comparing this with (2.7.8), we see that A_0 is transformed as a scalar, $\{A_1, A_2\}$

are transformed as a vector, and $\{A_3, A_4, A_4, A_5\}$ are transformed as a tensor:

$$A_0' = A_0 \ , \quad \begin{pmatrix} A_1' \\ A_2' \end{pmatrix} = \begin{pmatrix} \cos\theta & \sin\theta \\ -\sin\theta & \cos\theta \end{pmatrix} \begin{pmatrix} A_1 \\ A_2 \end{pmatrix} ,$$

$$\begin{pmatrix} A_3' & A_4' \\ A_4' & A_5' \end{pmatrix} = \begin{pmatrix} \cos\theta & \sin\theta \\ -\sin\theta & \cos\theta \end{pmatrix} \begin{pmatrix} A_3 & A_4 \\ A_4 & A_5 \end{pmatrix} \begin{pmatrix} \cos\theta & -\sin\theta \\ \sin\theta & \cos\theta \end{pmatrix} . \tag{2.7.10}$$

From Proposition 2.9, we can define invariants

$$V = \frac{A_1 + iA_2}{4} , \qquad T = \frac{A_3 + A_5}{8} , \qquad S = \frac{A_3 - A_5}{8} + i\frac{A_4}{4} . \tag{2.7.11}$$

The complex number V is a relative invariant of weight 1, T is an absolute invariant, S is a relative invariant of weight 2, and A_0 is an absolute invariant:

$$A_0' = A_0 \ , \qquad T' = T \ , \qquad V' = e^{-i\theta} V \ , \qquad S' = e^{-2i\theta} S \ . \tag{2.7.12}$$

Next, consider the object parameters. As discussed in Sect. 2.2, if the image xy coordinate system is rotated around the image origin, the scene XYZ coordinate system also undergoes a corresponding rotation around the Z-axis. We observe as follows.

Proposition 2.10. *Parameter r is a scalar, the pair $\{p, q\}$ is a vector, and the set $\{a, b, b, c\}$ is a tensor.*

Proof. The surface equation (2.7.4) is written in matrix form as

$$Z = r + (p \ q) \begin{pmatrix} X \\ Y \end{pmatrix} + (X \ Y) \begin{pmatrix} a & b \\ b & c \end{pmatrix} \begin{pmatrix} X \\ Y \end{pmatrix} . \tag{2.7.13}$$

After the XYZ coordinate system is rotated around the Z-axis, a quadric surface is still a quadric surface, and hence the *form* of the equation must still be the same:

$$Z' = r' + (p' \ q') \begin{pmatrix} X' \\ Y' \end{pmatrix} + (X' \ Y') \begin{pmatrix} a' & b' \\ b' & c' \end{pmatrix} \begin{pmatrix} X' \\ Y' \end{pmatrix} . \tag{2.7.14}$$

If we substitute (2.6.28) into the surface equation (2.7.13), we have

$$\begin{aligned} Z' = r + (p \, q) &\begin{pmatrix} \cos\theta & -\sin\theta \\ \sin\theta & \cos\theta \end{pmatrix} \begin{pmatrix} X' \\ Y' \end{pmatrix} \\ + (X' \ Y') &\begin{pmatrix} \cos\theta & \sin\theta \\ -\sin\theta & \cos\theta \end{pmatrix} \begin{pmatrix} a & b \\ b & c \end{pmatrix} \begin{pmatrix} \cos\theta & -\sin\theta \\ \sin\theta & \cos\theta \end{pmatrix} \begin{pmatrix} X' \\ Y' \end{pmatrix} . \end{aligned} \tag{2.7.15}$$

Comparing this with the new surface equation (2.7.14), we see that r is transformed as a scalar, $\{p, q\}$ are transformed as a vector, and $\{a, b, b, c\}$ are transformed as a tensor:

$$r' = r \ , \qquad \begin{pmatrix} p' \\ q' \end{pmatrix} = \begin{pmatrix} \cos\theta & \sin\theta \\ -\sin\theta & \cos\theta \end{pmatrix} \begin{pmatrix} p \\ q \end{pmatrix} ,$$

$$\begin{pmatrix} a' & b' \\ b' & c' \end{pmatrix} = \begin{pmatrix} \cos\theta & \sin\theta \\ -\sin\theta & \cos\theta \end{pmatrix} \begin{pmatrix} a & b \\ b & c \end{pmatrix} \begin{pmatrix} \cos\theta & -\sin\theta \\ \sin\theta & \cos\theta \end{pmatrix} . \tag{2.7.16}$$

From Proposition 2.10, we can define the following invariants:

$$k = \sqrt{1 + p^2 + q^2} \ , \quad v = \frac{p + \mathrm{i}q}{k} \ , \quad t = \frac{a + c}{2k} \ , \quad s = \frac{a - c}{2k} + \mathrm{i}\frac{b}{k} \ . \tag{2.7.17}$$

(Here, k is not a linear combination of the original parameters. We will discuss nonlinear expressions of invariants in Chap. 4, but the reader can easily check that k is indeed a scalar.) Parameters k and t are absolute invariants, v is a relative invariant of weight 1, s is a relative invariant of weight 2, and both ρ and r are absolute invariants:

$$\rho' = \rho \ , \quad r' = r \ , \quad k' = k \ , \quad t' = t \ , \quad v' = \mathrm{e}^{-\mathrm{i}\theta} v \ , \quad s' = \mathrm{e}^{-2\mathrm{i}\theta} s \ . \tag{2.7.18}$$

In terms of these invariants, the 3D recovery equations (2.7.6) are now rewritten as follows (Exercise 2.13):

Proposition 2.11 (3D *recovery equations*).

$$\rho k = A_0 \ , \quad tv + sv^* = V \ , \quad t^2 + ss^* = T \ , \quad ts = \frac{1}{2}S \ . \tag{2.7.19}$$

The solution is given as follows.

Theorem 2.4 (*Solution of the* 3D *recovery equations*). *If* $t \neq 0$ *and* $t^2 - ss^* \neq 0$, *the solution of* (2.7.19) *is given by*

$$t = \pm\sqrt{\frac{T \pm \sqrt{T^2 - SS^*}}{2}} \ , \qquad s = \frac{S}{2t} \ ,$$

$$v = \frac{tV - sV^*}{t^2 - ss^*} \ , \qquad k = \frac{1}{\sqrt{1 - vv^*}} \ , \qquad \rho = \frac{A_0}{k} \ , \tag{2.7.20}$$

where the two $\pm$ *signs in the first equation are independent. The object parameters are*

$$\rho = A_0/k \ , \qquad p = k\mathrm{Re}\{v\} \ , \qquad q = k\mathrm{Im}\{v\} \ ,$$

$$a = k(t + \mathrm{Re}\{s\}) \ , \qquad b = k\mathrm{Im}\{s\} \ , \qquad c = k(t - \mathrm{Re}\{s\}) \ . \tag{2.7.21}$$

Proof. Equation of (2.7.20b), is obtained from (2.7.19d). Substituting it into (2.7.19c), we obtain

$$t^4 - Tt^2 + \frac{1}{4}SS^* = 0 \ , \tag{2.7.22}$$

hence (2.7.20a). The complex conjugate of (2.7.19b) yields $tv^* + s^*v = V^*$. Eliminating v^*, we obtain (2.7.20c). Noting

$$1 - vv^* = 1 - \frac{p^2}{1+p^2+q^2} - \frac{q^2}{1+p^2+q^2} = \frac{1}{1+p^2+q^2} = \frac{1}{k^2} \ , \tag{2.7.23}$$

we obtain (2.7.20d). Equation of (2.7.20e), is obtained from (2.7.19a). Equations (2.7.21) are easily obtained from (2.7.17).

We thus find that the solution is not unique; in other words, the 3D interpretation is *ambiguous*. This means that there exist some transformations of the surface *that do not affect the 3D recovery equations*. First, we observe:

Corollary 2.4. *The distance r of the surface along the Z-axis from the image plane cannot be determined.*

This is because parameter r is not included in the 3D recovery equations (2.7.6) (or (2.7.19)). Next, we observe that the first $\pm$ in (2.7.20) corresponds to the following obvious fact (Fig. 2.10):

Corollary 2.5. *Two surfaces that are mirror images for each other with respect to a mirror perpendicular to the Z-axis cannot be distinguished.*

This is an immediate consequence of the fact that the 3D recovery equations contain the object parameter in quadratic forms; equations (2.7.6a–f) contain only quadratic terms in p, q, a, b, c. Hence, if p, q, a, b, c are a solution, so are

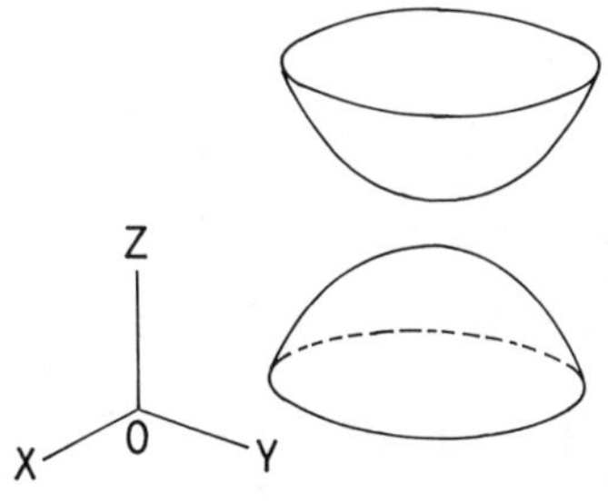

Fig. 2.10. Two surfaces that are mirror images of each other with respect to a mirror perpendicular to the Z-axis cannot be distinguished under orthographic projection

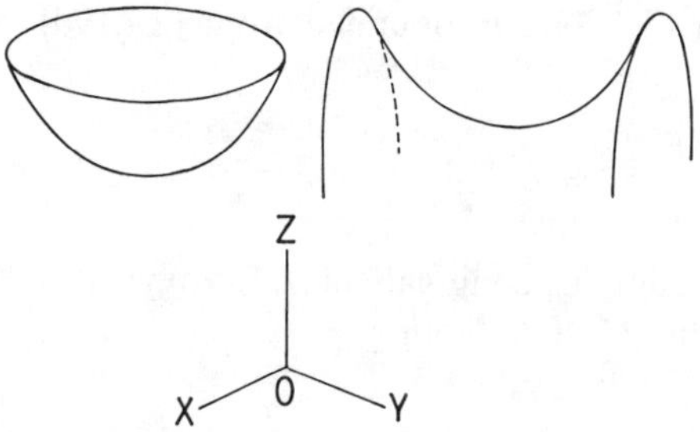

Fig. 2.11. An elliptic surface cannot be distinguished from a hyperbolic surface if the two principal curvatures have the same absolute values

$-p$, $-q$, $-a$, $-b$, $-c$ (Fig. 2.10). The other ambiguity resulting from the second $\pm$ in (2.7.20) is described as follows (Fig. 2.11):

Corollary 2.6. *An elliptic surface cannot be distinguished from a hyperbolic surface if the two principal curvatures have the same absolute values.*

Proof. Since the projection is orthographic, the relationship between the surface and the image plane is the same if the surface is translated parallel to the image plane or rotated around an axis perpendicular to the image plane. Suppose the surface is not a plane, i.e., a, b, c are not all zero. Then, as is well known in linear algebra, the equation of the surface is reduced, by appropriately translating and rotating the surface, to either of the following forms (Exercise 2.17): (i)

$$Z = r + aX^2 + cY^2 \ , \qquad a, c \neq 0 \ , \tag{2.7.24}$$

for which the 3D recovery equations are

$$A_0 = \rho \ , \quad A_1 = 0 \ , \quad A_2 = 0 \ , \quad A_3 = 4a^2 \ ,$$
$$A_4 = 0 \ , \quad A_5 = 4c^2 \ , \tag{2.7.25}$$

or (ii)

$$Z = r + pX + cY^2 \ , \qquad c \neq 0 \ , \tag{2.7.26}$$

for which the 3D recovery equations are

$$A_0 = \rho\sqrt{1 + p^2} \ , \quad A_1 = 0 \ , \quad A_2 = 0 \ ,$$
$$A_3 = 0 \ , \quad A_4 = 0 \ , \quad A_5 = \frac{4c^2}{1 + p^2} \ . \tag{2.7.27}$$

Hence, changing the sign of a or c in (2.7.25) (or c in (2.7.27)) does not affect the 3D recovery equation.

An intuitive interpretation is that the texture density can provide information only about the angle of the surface orientation (the *slant*) and nothing about the orientation of inclination (the *tilt*).

Combination of Corollaries 2.5 and 2.6 results in four interpretations in general. However, there are two exceptional cases in which infinitely many interpretations exist. These two cases occur when the assumptions in Theorem 2.4 are not satisfied. First, we assumed in Theorem 2.4 that $t \neq 0$. If $t = 0$, there exist infinitely many interpretations. Namely:

Corollary 2.7. *If the surface is hyperbolic with mean curvature zero, the principal directions of the surface are indeterminate.*

Proof. As we have already discussed, we do not lose generality by assuming that the equation of the surface is given by either (2.7.24) or (2.7.26). In the former case, we have from (2.7.25)

$$T = \frac{a^2 + b^2}{2}, \qquad S = \frac{a^2 - b^2}{2}, \tag{2.7.28}$$

In the latter case, we have

$$T = \frac{1}{2}\frac{c^2}{1 + p^2}, \qquad S = \frac{1}{2}\frac{c^2}{1 + p^2}. \tag{2.7.29}$$

In the former case, (2.7.20) gives $t = \pm(a \pm b)/2$. In the latter case, (2.7.20a) gives $t = \pm c/2\sqrt{1 + p^2}$, which is not zero. Hence, $t = 0$ is possible only when $a = \pm b$, i.e., when the surface has either of the following two forms:

$$Z = r + a(X^2 + Y^2), \qquad Z = r + a(X^2 - Y^2). \tag{2.7.30}$$

The first one is a *circular paraboloid*, which is symmetric around the Z-axis. According to Corollary 2.7, the surfaces of the first and the second look the same when projected. Hence, the latter hyperbolic surface also looks the same when projected after rotation by an arbitrary angle around the Z-axis. Thus, infinitely many interpretations arise.

We also assumed $t^2 - ss^* \neq 0$ in Theorem 2.4. If $t^2 - ss^* = 0$, there exist infinitely many interpretations (Fig. 2.12).

Corollary 2.8. *If the surface is parabolic, the depth gradient of the asymptotic direction is indeterminate.*

Proof. As before, we can assume without losing generality that the surface is given by either (2.7.24) or (2.7.26). In the former case, if we substitute (2.7.28) into (2.7.20a, b), we find $t^2 - ss^* = \pm ac$, which is not zero. In the latter case, we find $t^2 - ss^* = 0$. Thus, $t^2 - ss^* = 0$ is possible only in the latter case. From (2.7.27), we see that p is indeterminate, and infinitely many interpretations arise.

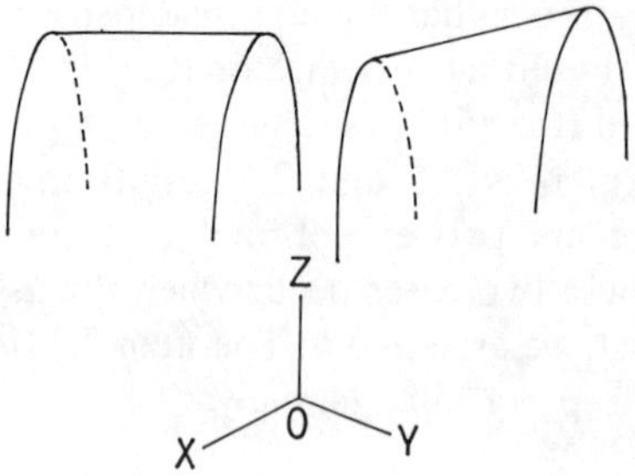

Fig. 2.12. If the surface is parabolic, the depth gradient along the asymptotic direction is indeterminate

An intuitive interpretation is as follows. A parabolic surface does not have a local maximum or minimum; it has one *asymptotic direction*, or a ridge, along which the depth changes linearly (Fig. 2.12). It follows that the gradient of the depth along the asymptotic direction cannot be determined from the texture density alone. This is easily understood if we recall the fact that the gradient of a planar surface, for which the depth changes linearly in all directions, is indeterminate under orthographic projection.

2.7.1 Curvatures of a Surface

Consider a smooth surface in space. Let P be a point on it. Consider a plane tangent to the surface at point P. Take a Cartesian XYZ coordinate system with origin P such that the X- and Y-axes are on the tangent plane (Fig. 2.13). By

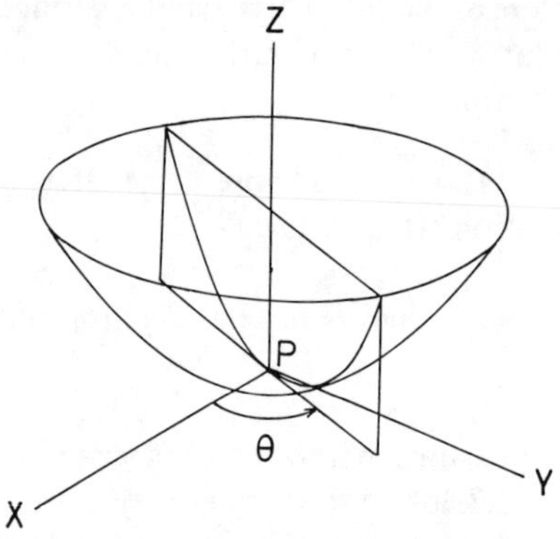

Fig. 2.13. The normal curvature is the curvature at P of the curve of intersection of the surface and the plane containing the Z-axis

Taylor expansion, the equation of the surface with respect to this coordinate system is given by

$$Z = \frac{1}{2}(aX^2 + 2bXY + c^2Y^2) + \dots = \frac{1}{2}(X \quad Y)\begin{pmatrix} a & b \\ b & c \end{pmatrix}\begin{pmatrix} X \\ Y \end{pmatrix} + \dots, \tag{2.7.31}$$

where ... denotes higher-order terms in X and Y. The determinant Δ of the coefficient matrix is called the *Hessian*:

$$\Delta = ac - b^2 . \tag{2.7.32}$$

Suppose the surface is cut by a plane that contains the Z-axis (Fig. 2.13). Let θ be the angle that the plane makes with the X-axis. The intersection of the surface with this plane is a space curve passing through point P. The curvature $\kappa(\theta)$ of this curve at P is called the *normal curvature* for orientation θ. Let $\kappa_1 = \kappa(\theta_1)$ and $\kappa_2 = \kappa(\theta_2)$ be the maximum and minimum values of the normal curvature $\kappa(\theta)$ over $0 \leqq \theta < 2\pi$. These values are called the *principal curvatures* at P, and the corresponding orientations θ_1, θ_2 are called the *principal directions* at P. The *Gaussian curvature* K and the *mean curvature* H at P are defined by

$$K = \kappa_1\kappa_2 , \qquad H = \frac{1}{2}(\kappa_1 + \kappa_2) . \tag{2.7.33}$$

The surface is called *elliptic* if $K > 0$, *hyperbolic* if $K < 0$, and *parabolic* if $K = 0$ (Fig. 2.14). If $H = 0$, the surface is called a *minimal surface*—a surface of minimum area for a fixed boundary.

The coefficient matrix of the surface equation (2.7.31) is diagonalized by appropriately rotating the XYZ coordinate system around the Z-axis. It is easy to prove that the principal curvatures κ_1, κ_2 are the eigenvalues of the coefficient matrix. Thus, the surface equation (2.7.31) reduces to

$$Z = \frac{1}{2}(X' \quad Y')\begin{pmatrix} \kappa_1 & \\ & \kappa_2 \end{pmatrix}\begin{pmatrix} X' \\ Y' \end{pmatrix} + \dots . \tag{2.7.34}$$

Since the determinant and the trace are invariants under coordinate rotations, we have $ac - b^2 = \kappa_1\kappa_2$ and $c + a = \kappa_1 + \kappa_2$. Hence, the Gaussian

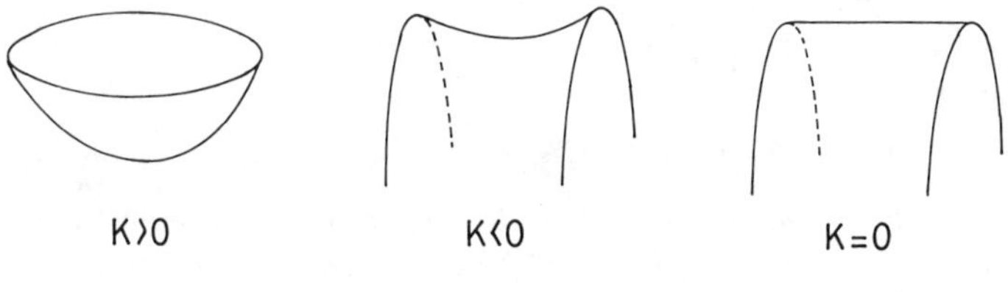

Fig. 2.14. A surface is elliptic, hyperbolic, or parabolic depending on whether the Gaussian curvature K is positive, negative, or zero

curvature K and the mean curvature H are computed by

$$K = ac - b^2 , \qquad H = \frac{a + c}{2} . \tag{2.7.35}$$

All these relations are simple because we chose a very special Cartesian XYZ coordinate system. We can, however, compute the principal curvature, the Gaussian curvature, and the mean curvature with reference to a general (not necessarily Cartesian) coordinate system or for a more general parametrization of the surface. All we need is to compute the *first fundamental form* (Chaps. 5, 9) and the *second fundamental form* (cf. Exercise 2.19).

The important fact to note is that the principal curvatures, the Gaussian curvature, and the mean curvatures are *intrinsic quantities* of the surface, and their values do not depend on the choice of the coordinate system or surface parametrization for which the computation is done.

Exercises

2.1 From (2.3.2, 3), conclude that $T_i(\theta) = e^{in\theta}$ for some integer n. *Hint*: What is $\lim_{\alpha \to 0} [T_i(\theta + \alpha) - T_i(\theta)]/\alpha$?

2.2 Show that the matrix $R(\theta)$ of (2.4.3) satisfies (2.4.4).

2.3 Compute each matrix element of (2.5.1) and prove (2.5.2).

2.4 Show that $(AB)^{\mathrm{T}} = B^{\mathrm{T}}A^{\mathrm{T}}$ holds for arbitrary matrices A, B if the product AB can be defined.

2.5 Show that $\mathrm{Tr}\{AB\} = \mathrm{Tr}\{BA\}$ holds for arbitrary square matrices A, B whose dimensions are the same.

2.6 Prove (2.5.16).

2.7 Prove (2.5.30).

2.8 Prove (2.5.31, 32).

2.9 (a) Derive simultaneous linear equations—called *normal equations*—to determine a, b, A, B, C, D that minimize M of (2.6.6).
(b) Show that if $\hat{a}$, $\hat{b}$, $\hat{A}$, $\hat{B}$, $\hat{C}$, $\hat{D}$ are the flow parameters that minimize M of (2.6.6), the value M, caled the *residual*, is given by

$$M = \sum_i [(\hat{a} + \hat{A}x_i + \hat{B}y_i - u_i)u_i + (\hat{v}_0 + \hat{C}x_i + \hat{D}y_i - v_i)v_i] ,$$

where $u_i = u(x_i, y_i)$ and $v_i = v(x_i, y_i)$.

2.10* (a) Show that if one of the two solutions of (2.6.36) is the true solution ω_3, the other is

$$\tilde{\omega}_3 = \omega_3 - \omega_1 + \omega_2 .$$

(b) Let $\tilde{P}$, $\tilde{W}$ be the spurious solutions obtained by replacing the true ω_3 with the spurious $\tilde{\omega}_3$. Using (2.6.34b), show that the spurious solutions satisfy

$$\tilde{P}\tilde{W}^* = -2\omega_3 + (R - \mathrm{i}T) , \qquad \tilde{P}\tilde{W} = S .$$

(c) Show that $\tilde{W}/\tilde{W}^* = P/P^*$ and hence

$$\tilde{\omega} = \arg(P) + \frac{1}{2}\pi \qquad (\text{mod } 2\pi) \ .$$

(d) Similarly, show that

$$\tilde{P} = \arg(W) + \frac{1}{2}\pi \qquad (\text{mod } 2\pi) \ .$$

2.11 Prove (2.7.2).
2.12 Prove (2.7.5, 6).
2.13 Derive (2.7.19).
2.14 Prove (2.7.20c).
2.15 Prove (2.7.21).
2.16* Suppose we observe an optical flow in the form

$$u = a + Ax + By + Ex^2 + 2Fxy + Gy^2 \ ,$$

$$v = b + Cx + Dy + Kx^2 + 2Lxy + My^2 \ .$$

Show that the following quantities are invariants:

$$\begin{aligned}
&V = a + b && (\text{weight } 1) \ , \\
&T = A + D && (\text{weight } 0) \ , \\
&R = B - C && (\text{weight } 0) \ , \\
&S = (A - D) + \mathrm{i}(B + C) && (\text{weight } 2) \ , \\
&H = (E + 2L - G) + \mathrm{i}(M + 2F - K) && (\text{weight } 1) \ , \\
&I = (E - 2L + 3G) + \mathrm{i}(M - 2F + 3K) && (\text{weight } 1) \ , \\
&J = (E - 2L - G) - \mathrm{i}(M - 2F - K) && (\text{weight } 3) \ .
\end{aligned}$$

What kind of flows do invariants H, I, J represent?

2.17 (a) Show that, by appropriately rotating and translating the xy coordinate system, a quadratic form

$$\frac{1}{2}(ax^2 + 2bxy + cy^2) + px + qy + r \qquad (1)$$

reduces to the form

$$\frac{1}{2}(a'x'^2 + c'y'^2) + r'$$

if $\Delta \neq 0$, where Δ is the Hessian $\Delta = ac - b^2$.

(b) If $\Delta = 0$ but not $a = b = c = 0$, show that, by appropriately rotating and translating the xy coordinate system, the quadratic form (1) reduces to the form

$$\frac{1}{2}c'y'^2 + p'x' + r' \ .$$

2.18 (a) Consider a surface

$$Z = \frac{1}{2}(aX^2 + 2bXY + cY^2) + \ldots,$$

where ... designates higher-order terms in X and Y. If this surface is cut by plane $Y = X \tan\theta$, show that the intersection curve is given by

$$Z = \frac{1}{2}r^2(a\cos^2\theta + 2b\cos\theta\sin\theta + c\sin^2\theta) + \ldots,$$

where r designates the distance from the Z-axis.

(b) Show that the normal curvature $\kappa(\theta)$ for this cut is given by

$$\kappa(\theta) = a\cos^2\theta + 2b\cos\theta\sin\theta + c\sin^2\theta\ .$$

(c) By differentiating the above expression, show that the two principal directions θ_1, θ_2 are given by the two solutions of

$$\tan 2\theta = \frac{2b}{a - b}\ .$$

(d) Show that if κ_1, κ_2 are the principal curvatures, then

$$\kappa_1\kappa_2 = ac - b^2\ , \qquad \kappa_1 + \kappa_2 = a + c\ .$$

2.19* (a) Consider a surface parametrized by (u, v), and let

$$Edu^2 + 2Fdudv + Gdv^2\ , \qquad Ldu^2 + 2Mdudv + Ndv^2$$

be the first fundamental form and the second fundamental form, respectively. Show that the normal curvature κ along orientation du/dv = const is given by

$$\kappa = \frac{Ldu^2 + 2Mdudv + Ndv^2}{Edu^2 + 2Fdudv + Gdv^2}\ .$$

(b) Let κ_1, κ_2 be the principal curvatures. Prove the following relations:

$$\kappa_1\kappa_2 = \frac{LN - M^2}{EG - F^2}\ , \qquad \kappa_1 + \kappa_2 = \frac{EN + GL - 2FM}{EG - F^2}\ .$$

3. 3D Rotation and Its Irreducible Representations

In Chap. 2, we studied invariance properties of image characteristics under image coordinate rotation. In this chapter, we study invariance properties under the camera rotation we introduced in Chap. 1. Just as invariants for the image coordinate rotation are obtained by irreducibly reducing representations of $SO(2)$, invariants for the camera rotation are obtained by irreducibly reducing representations of $SO(3)$. However, direct analysis is very difficult due to the fact that $SO(3)$ is not Abelian. Fortunately, a powerful tool is available: we only need to analyze "infinitesimal transformations". This is because the structure of a compact Lie group is completely determined by its Lie algebra. To demonstrate our technique, we analyze the transformation of optical flow under camera rotation.

3.1 Invariance for the Camera Rotation Transformation

In Chap. 2, we studied invariance properties of image characteristics under image coordinate rotations. We were motivated by the fact that no image coordinate system is inherent to the image; we can arbitrarily rotate the image coordinate system around the center of the image. This observation suggested that a set of image characteristics $c_1, \ldots, c_m$ may be rearranged into new parameters $C_1, \ldots, C_m$ in such a way that each C_i, $i = 1, \ldots, m$, has an "invariant meaning"—a meaning that is invariant under image coordinate rotations. We showed that if the characteristics undergo a linear transformation under coordinate rotations, this process of rearrangement is, in mathematical terms, the *irreducible reduction* of a representation of $SO(2)$. We also interpreted this process according to a general statement, which we called *Weyl's thesis.*

In Chap. 1, we defined the *camera rotation transformation*, and observed that it, too, does not alter the information contained in the image.[1] Hence, it seems, by analogy, that a set of image characteristics $c_1, \ldots, c_m$ may be rearranged into new parameters $C_1, \ldots, C_m$ in such a way that each has an invariant meaning. Also, it is expected that this rearrangement is attained by irreducible reduction of representations of $SO(3)$. However, some difficulties arise.

[1] We ignore the existence of the image boundary, assuming that the camera rotation is not very large and that the part we are interested in is always seen in the field of view. A more detailed discussion will be given in Chap. 5.

Given a set of image characteristics $c_1, \ldots, c_m$, we must first check how they are transformed into new values $c'_1, \ldots, c'_m$ under the camera rotation. In particular, we must check whether or not they are transformed linearly so that they define a representation of $SO(3)$. For image coordinate rotations, this is easily checked by direct calculation. By contrast, the camera rotation transformation (1.3.2) is nonlinear, and direct calculation is almost impossible. How can we test whether given image characteristics define a representation of $SO(3)$?

Suppose we have already checked that $c_1, \ldots, c_m$ indeed define a representation of $SO(3)$. But then how can we reduce it into irreducible representations? Or can it be reduced at all? As we discussed in Chap. 2, any representation of $SO(2)$ is reducible to one-dimensional irreducible representations. In contrast, irreducible representations of $SO(3)$ are not necessarily one dimensional. This difference arises from the fact that $SO(3)$ *is not Abelian*.

Fortunately, a powerful tool is available. We need not consider the *entire* camera rotation; we need only consider *infinitesimal* camera rotations. Surprisingly, consideration of infinitesimal camera rotations automatically leads to *all* aspects of finite camera rotations. Illustrating this fact is the main purpose of this chapter.

In Sect. 3.2 we study representations of *Lie groups* and their *Lie algebras*. A Lie group is a group specified by a finite number of parameters that take real values continuously from some fixed domain of values. (The precise definition is given in the Appendix.) The number of these parameters is called the *dimension*. For example, the 2D rotation group $SO(2)$ is a one-dimensional Lie group, and the 3D rotation group $SO(3)$ is a three-dimensional Lie group.

For a given representation of a Lie group, the set of "infinitesimal transformations"—transformations corresponding to group elements in a small neighborhood of the unit element—forms an algebraic structure called a *Lie algebra*. It can be shown that *almost all properties of a global representation can be deduced from its Lie algebra*. This remarkable fact was first established by the Norwegian mathematician Sophus Lie (1842–1899). We will demonstrate this for $SO(3)$, and apply our technique to the transformation rule of optical flow under the camera rotation.

3.1.1 $SO(3)$ Is Three Dimensional and Not Abelian

Any rotation can be realized as a rotation around an axis through some angle. This fact is known as *Euler's theorem* (Chap. 6). If we accept this, a 3D rotation is specified by the angle of rotation and the orientation of the axis. The axis orientation is specified by two parameters, say its spherical coordinates $\{\theta, \varphi\}$. Hence, a 3D rotation is determined by three parameters. (There exist many other ways of specifying a 3D rotation. They will be discussed in Chap. 6.)

Two 3D rotations do not necessarily commute, i.e., $SO(3)$ is a non-Abelian group. Indeed, let $R_X(\theta)$ and $R_Y(\theta)$ be rotations by angle θ around the X- and

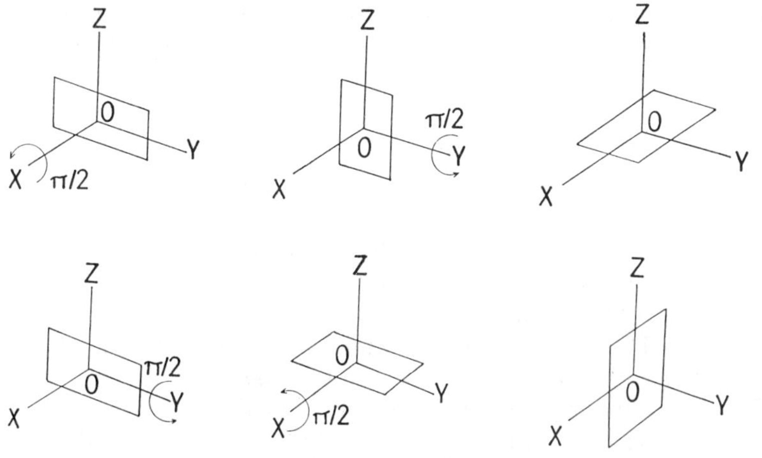

Fig. 3.1. Two 3D rotations do not commute: $\boldsymbol{R}_Y(\pi/2)\boldsymbol{R}_X(\pi/2) \neq \boldsymbol{R}_X(\pi/2)\boldsymbol{R}_Y(\pi/2)$. Here, $\boldsymbol{R}_X(\theta)$ and $\boldsymbol{R}_Y(\theta)$ are, respectively, rotations around the X- and Y-axes by angle θ screw-wise

Y-axes, respectively. Figure 3.1 shows that $R_Y(\pi/2)R_X(\pi/2) \neq R_X(\pi/2)R_Y(\pi/2)$. If $SO(3)$ were Abelian, a 3D rotation would conveniently be specified by the angles of rotation around the X-, Y-, and Z-axes, but, $SO(3)$ being non-Abelian, different orders of application result in different rotations. Thus, the non-commutativity of $SO(3)$ makes the rule of group composition very complicated.

3.2 Infinitesimal Generators and Lie Algebra[2]

Suppose we are observing some phenomenon for which a Lie group G of transformations—such as the group of rotations—is defined. Let $c_1, \ldots, c_m$ be the parameters that specify the phenomenon. Let $c'_1, \ldots, c'_m$ be their new values after a transformation $g \in G$ is applied. Suppose the induced transformation $\boldsymbol{T}(g)$ of $c_1, \ldots, c_m$ is linear):

$$\begin{pmatrix} c'_1 \\ \vdots \\ c'_m \end{pmatrix} = \boldsymbol{T}(g) \begin{pmatrix} c_1 \\ \vdots \\ c_m \end{pmatrix} . \tag{3.2.1}$$

[2] The argument in this section may be too abstract for readers without a mathematical background. However, mathematical rigor is not necessary in the subsequent applications. Interested readers are recommended to read books on Lie group theory for details. Also, see the Appendix, Sect. A.9.

As discussed in Sect. 2.2, this linear transformation defines a *representation* of group G (a *homomorphism* from group G into the group of matrices):

$$\boldsymbol{T}(g')\boldsymbol{T}(g) = \boldsymbol{T}(g' \circ g) , \qquad g', g \in G . \tag{3.2.2}$$

$$\boldsymbol{T}(e) = \boldsymbol{I}, \qquad \boldsymbol{T}(g)^{-1} = \boldsymbol{T}(g^{-1}) , \tag{3.2.3}$$

where e is the unit element of group G, and $\boldsymbol{I}$ is the unit matrix.

Suppose we can define a parameter t that measures a kind of "distance" of an element $g \in G$ from the unit element e. We require that if an element $g \in G$ whose "distance" is t is applied twice, the "distance" of $g^2 \in G$ should be $2t$. (To be precise, we ought to give the exact definition of *one-parameter subgroups* $g(t)$ of a Lie group G. However, we avoid the complications and ask readers to resort to their intuition.) An element $g \in G$ is said to be *small* if the "distance" of $g \in G$ is small. This "distance" is always measured from the unit element e, i.e., the "distance" between two arbitrary elements g, $g' \in G$ is not defined. For $SO(3)$, a 3D rotation is specified by the angle Ω of rotation around some axis, so the angle Ω is a natural choice for this "distance". For $SO(2)$, the angle θ of rotation is a natural choice.

If an element $g \in G$ is small, the corresponding matrix $\boldsymbol{T}(g)$ is also close to the unit matrix $\boldsymbol{I}$. Hence, we may expand the matrix $\boldsymbol{T}(g)$ in the form

$$\boldsymbol{T}(g) = \boldsymbol{I} + t\boldsymbol{A} + \mathrm{O}(t^2) , \tag{3.2.4}$$

where $\mathrm{O}(t^2)$ denotes terms of order equal to or higher than 2 in t. (Whether the terms $\mathrm{O}(\ldots)$ are expressions in scalars, vectors or matrices can be determined from the context.) The set

$$L = \{\boldsymbol{A} \,|\, \boldsymbol{T}(g) = \boldsymbol{I} + t\boldsymbol{A} + \mathrm{O}(t^2) \quad \text{for small } g \in G\} \tag{3.2.5}$$

is called the set of *infinitesimal transformations*.[3]

Let us study the properties of the set L. First, we can easily see that if $\boldsymbol{A}$, $\boldsymbol{B} \in L$, then $\boldsymbol{A} + \boldsymbol{B} \in L$. In fact, if

$$T(g) = \boldsymbol{I} + t\boldsymbol{A} + \mathrm{O}(t^2) , \qquad T(g') = \boldsymbol{I} + t\boldsymbol{B} + \mathrm{O}(t^2) , \tag{3.2.6}$$

then

$$\begin{aligned} \boldsymbol{T}(g \circ g') = \boldsymbol{T}(g)\boldsymbol{T}(g') &= [\boldsymbol{I} + t\boldsymbol{A} + \mathrm{O}(t^2)][\boldsymbol{I} + t\boldsymbol{B} + \mathrm{O}(t^2)] \\ &= \boldsymbol{I} + t(\boldsymbol{A} + \boldsymbol{B}) + \mathrm{O}(t^2) . \end{aligned} \tag{3.2.7}$$

We can aso easily see that if $\boldsymbol{A} \in L$, then $c\boldsymbol{A} \in L$ for any real number c. Evidently, this is true when c is a nonnegative integer, since we can apply the above

[3] To be precise, these matrices are defined as *derivatives* $d\boldsymbol{T}(g(t))/dt$ at $t = 0$. An alternative definition which is free from the parameter t is that they are the *left invariant tangent spaces*. Such an elegant definition may be preferable for mathematicians, but for practical applications, our treatment is most convenient.

relationship c times to the same $\boldsymbol{A}$. Due to the "smoothness" of the Lie group G, this relationship is extended to an arbitrary real number (we omit the rigorous proof). If we admit this fact, we find:

Lemma 3.1. *The set L of infinitesimal transformations forms a (real) linear space:*

$$\boldsymbol{A}, \boldsymbol{B} \in L \ , \qquad c \in \boldsymbol{R} \rightarrow \boldsymbol{A} + \boldsymbol{B} \in L \ , \qquad c\boldsymbol{A} \in L \ , \qquad (c \in \mathbb{R}). \tag{3.2.8}$$

The *commutator* of two square matrices $\boldsymbol{A}$, $\boldsymbol{B}$ is defined by

$$[\boldsymbol{A}, \boldsymbol{B}] \equiv \boldsymbol{AB} - \boldsymbol{BA} \ . \tag{3.2.9}$$

We can immediately see that the commutator is *anticommutative*:

$$[\boldsymbol{A}, \boldsymbol{B}] = [\boldsymbol{B}, \boldsymbol{A}] \ . \tag{3.2.10}$$

It is also a *bilinear operation*

$$[\boldsymbol{A} + \boldsymbol{B}, \boldsymbol{C}] = [\boldsymbol{A}, \boldsymbol{C}] + [\boldsymbol{B}, \boldsymbol{C}] \ , \qquad [c\boldsymbol{A}, \boldsymbol{B}] = c[\boldsymbol{A}, \boldsymbol{B}] \ , \qquad (c \in \mathbb{R}) \tag{3.2.11}$$

and satisfies the *Jacobi identity*

$$[\boldsymbol{A}, [\boldsymbol{B}, \boldsymbol{C}]] + [\boldsymbol{B}, [\boldsymbol{C}, \boldsymbol{A}]] + [\boldsymbol{C}, [\boldsymbol{A}, \boldsymbol{B}]] = \boldsymbol{O} \ , \tag{3.2.12}$$

where $\boldsymbol{O}$ denotes a matrix whose elements are all 0. See Exercise 3.1.

The crucial fact about Lie groups is the following (we omit the proof):

Lemma 3.2. *The set of infinitesimal transformations is closed under commutation:*

$$\boldsymbol{A}, \boldsymbol{B} \in L \rightarrow [\boldsymbol{A}, \boldsymbol{B}] \in L \ . \tag{3.2.13}$$

An operation $[\cdot, \cdot]$ that is anticommutative, bilinear, and satisfies the Jacobi identity is called a *Lie bracket*. (The term *Poisson bracket* is also used, but the definition may be different depending on the authors.) The commutator (3.2.9) is a Lie bracket. A linear space equipped with a Lie bracket is called a *Lie algebra*. Thus, from Lemmas 3.1 and 3.2, we conclude:

Theorem 3.1. *The set of infinitesimal transformations L associated with a Lie group G is a Lie algebra.*

Since the set L is a linear space, we can find a basis $\{\boldsymbol{A}_1, \ldots, \boldsymbol{A}_r\}$ such that any infinitesimal transformation A is expressed uniquely as a linear combination of these basis matrices:

$$\boldsymbol{A} = \alpha_1 \boldsymbol{A}_1 + \ldots + \alpha_r \boldsymbol{A}_r \ . \tag{3.2.14}$$

The dimension r of L as a linear space is also called the *dimension* of the Lie algebra L. It can be proved that *the dimension of a Lie algebra L is equal to the* *dimension of the associated Lie group G*. The basis matrices $\{A_1, \ldots, A_r\}$ are called the *infinitesimal generators* of the representation $T(g)$.

Since all elements of L can be expressed as linear combinations of the infinitesimal generators $\{A_1, \ldots, A_r\}$, the commutator $[A, B]$ of any two elements $A, B \in L$ is uniquely determined once the commutators of the infinitesimal generators $\{A_1, \ldots, A_r\}$ are determined. Since L is closed under commutation, we can write

$$[A_i, A_j] = \sum_{k=1}^{r} c_{ij}^k A_k , \qquad i, j = 1, \ldots, r . \tag{3.2.15}$$

These relations are called the *commutation relations*. The constants c_{ij}^k, $i, j, k = 1, \ldots, r$, are called the *structure constants* of the Lie algebra L. If all the structure constants are zero ($[A_i, A_j] = O$, $i, j = 1, \ldots, r$), the Lie algebra L is said to be *commutative*. Then, it can be proved that the corresponding Lie group G is an Abelian group.

What makes the theory of Lie agebra indispensable is the fact that the Lie algebra L of any (finite dimensional) representation $T(g)$ *uniquely* determines the original representation $T(g)$ if the underlying Lie group G is connected and simply connected. A Lie group G is *connected* if we can traverse between any two elements $g, g' \in G$ by continuously changing parameters. It is *simply connected* if any loop traversing elements within the Lie group G can be continuously shrunk into one element. (See the Appendix, Sect. A.8, for precise definitions.)

If the original Lie group G is connected and simply connected, the infinitesimal transformations of L can be extended continuously and uniquely onto the entire set of finite transformations $T(g)$. This is done by the following process (not necessarily rigorously stated).

Equation (3.2.4) implies $dT(g(t))/dt|_{t=0} = A$. This suggests that $T(g(t))$ is obtained by "integrating" the differential equation

$$\frac{d}{dt} T(g(t)) = A T(g(t)) \tag{3.2.16}$$

under the initial condition $T(g(0)) = I$.[4] The solution is given by

$$T(g(t)) = e^{tA} , \tag{3.2.17}$$

where the exponential of a matrix is defined by

$$e^{tA} = \sum_{k=0}^{\infty} t^k A^k / k! , \tag{3.2.18}$$

which always converges absolutely (Exercise 3.3).

---

[4] Strictly speaking, (3.2.16) is not the only way to extend the relation $A = dT(g(t))/dt|_{t=0}$ globally, but by (3.2.16) we can obtain a desired *one-parameter subgroup* of $T(g)$. In fact, $T(g)$. for *any* $g \in G$ is defined by integrating (3.2.16) for *some* $A \in L$. We omit the proof.

Condition (3.2.13), or (3.2.15), is the *integrability condition* that guarantees that this process defines $\boldsymbol{T}(g)$ consistently, i.e., in such a way that elements thus defined are closed under the group operation and have the desired group properties (at least locally). The connectedness and simply connectedness of G assure that this local definition can be uniquely extended over the entire group G. If a simply connected Lie group G and a non-simply-connected Lie group G' have the same Lie algebra L, then Lie group G is said to *cover* Lie group G'. The simply connected Lie group G is called the *universal covering group* of Lie group G'. (See Sect. 6.9 and the Appendix, Sect. A.8).

Suppose we are given a set of parameters $c_1, \ldots, c_m$, but we do not know how they are transformed under the group G of transformations—in particular, we do not know whether or not they are transformed linearly. Let us assume that we can compute their infinitesimal transformations. In other words, suppose we can compute their transformation up to the first order of t by neglecting $O(t^2)$. If the transformation is infinitesimal, the parameters $c_1, \ldots, c_m$ are also infinitesimally transformed into new values $c_1 + \delta c_1, \ldots, c_m + \delta c_m$. Hence, we can check whether the variations $\delta c_1, \ldots, \delta c_m$ are expressed linearly in $c_1, \ldots, c_m$:

$$\delta\begin{pmatrix} c_1 \\ \vdots \\ c_m \end{pmatrix} = t\,\boldsymbol{A}(g)\begin{pmatrix} c_1 \\ \vdots \\ c_m \end{pmatrix} + O(t^2)\ . \tag{3.2.19}$$

If the variations $\delta c_1, \ldots, \delta c_m$ are not expressed linearly in $c_1, \ldots, c_m$, the globally integrated transformation cannot be linear, and hence cannot define a representation of G.

If all infinitesimal transformations have the form of (3.2.19), we can obtain the set L of infinitesimal transformations by collecting the matrix $\boldsymbol{A}(g)$ for all small elements $g \in G$. Then, we can check whether the set L forms a Lie algebra. If not, (3.2.19) cannot be integrated globally, and hence the parameters $c_1, \ldots, c_m$ cannot define a representation of G.

We do not give the proof, but a small element g of an r-dimensional Lie group G can be specified by r "coordinates" $(t_1, \ldots, t_r)$ in such a way that the unit element e has "coordinates" $(0, \ldots, 0)$.[5] In terms of these "coordinates", (3.2.19) is rewritten as

$$\delta\begin{pmatrix} c_1 \\ \vdots \\ c_m \end{pmatrix} = (t_1\boldsymbol{A}_1 + \ \ldots\ + t_r\boldsymbol{A}_r)\begin{pmatrix} c_1 \\ \vdots \\ c_m \end{pmatrix} + O(t_1, \ldots, t_1)^2\ , \tag{3.2.20}$$

[5] They may be thought of as "distances in r different directions" like the Cartesian coordinates near the origin. An element g is reached from the unit element e by going "distance" t_1 in the first direction, then "distance" t_2 in the second direction, and so on. The order of these r steps does not affect the destination g *to a first approximation* due to the commutation relations.

where the last term denotes terms of degree 2 or higher in $(t_1, \ldots, t_r)$. The matrices $\{A_1, \ldots, A_r\}$ are the infinitesimal generators with respect to this "coordinate system"; they define a natural basis of L. Let their commutation relations be

$$[A_i, A_j] = \sum_{k=1}^{r} c_{ij}^k A_k , \qquad i, j = 1, \ldots, r . \tag{3.2.21}$$

The structure constants c_{ij}^k, $i, j, k = 1, \ldots, r$, are determined solely by the underlying Lie group G and its "coordinatization" $(t_1, \ldots, t_r)$. Suppose we have another Lie algebra $\tilde{L}$ of the same dimension r. If the structure constants, $\tilde{c}_{ij}^k$, $i, j, k = 1, \ldots, r$, of $\tilde{L}$ are equal to the structure constants of L, i.e.,

$$\tilde{c}_{ij}^k = c_{ij}^k , \qquad i, j, k = 1, \ldots, r , \tag{3.2.22}$$

the two Lie algebras are *isomorphic* to each other.[6] Then, both define representations of the same (connected) Lie group G. Furthermore, if the infinitesimal generators themselves are identical, the resulting representations are also identical.

Thus, we can check transformation of given parameters by analyzing infinitesimal transformations. Once we have successfully confirmed that a given set of parameters defines a representation of a Lie group G, the next step is to see whether the representation is reducible or irreducible. Again, this can be checked by analyzing infinitesimal transformations: the reducibility or irreducibility of a representation is completely determined from its Lie algebra L. One simple test is to compute the *Casimir operator* defined by

$$\boldsymbol{H} \equiv 2 \sum_{i,j=1}^{r} \alpha^{ij} A_i A_j , \tag{3.2.23}$$

where (α^{ij}) is the inverse matrix of (α_{ij}) defined by

$$\alpha_{ij} = \sum_{k,l=1}^{r} c_{ik}^l c_{jl}^k , \qquad i, j = 1, \ldots, r , \tag{3.2.24}$$

and c_{ij}^k are the structure constants of L.

It can be proved that the Casimir operator $\boldsymbol{H}$ commutes with all the infinitesimal generators $\{A_1, \ldots, A_r\}$:

$$[\boldsymbol{H}, A_i] = 0 , \qquad i = 1, \ldots, r . \tag{3.2.25}$$

Since $\{A_1, \ldots, A_r\}$ are the basis of the Lie algebra L, this means that the Casimir operator $\boldsymbol{H}$ commutes with all the elements of L. Consequently, if

---

6 Two Lie algebras can be isomorphic even if the structure constants are not identical. The values of the structure constants are defined with reference to a particular "coordinatization" of infinitesimal elements and the associated infinitesimal generators. Two Lie algebras are said to be *isomorphic* to each other if there *exist* bases for which the structure constants become equal. Mathematically speaking, this "abstract structure" common to all isomorphic Lie algebras is *the* Lie algebra of the underlying Lie group G, and a particular Lie algebra is a *realization* of it.

the representation is irreducible, the Casimir operator $\boldsymbol{H}$ must be a multiple of the unit matrix due to *Schur's lemma* (Appendix, Sect. A.7). The multiple characterizes the irreducible representation. This fact provides a very powerful tool for studying transformation properties under $SO(3)$: for $SO(3)$, irreducibility is determined by the Casimir operator alone, as shown in the next section.

3.3 Lie Algebra and Casimir Operator of the 3D Rotation Group

Take the viewer-centered coordinate system (Sect. 1.2), and consider the camera rotation transformation (Sect. 1.3). If we define a set of image characteristics $c_1, \ldots, c_m$ on the image xy-plane, they change their values when the camera is rotated. If they are transformed linearly in the form

$$\begin{pmatrix} c'_1 \\ \vdots \\ c'_m \end{pmatrix} = \boldsymbol{T}(\boldsymbol{R}) \begin{pmatrix} c_1 \\ \vdots \\ c_m \end{pmatrix} \tag{3.3.1}$$

the transformation matrix defines a representation of $SO(3)$:

$$\boldsymbol{T}(\boldsymbol{R}')\,\boldsymbol{T}(\boldsymbol{R}) = \boldsymbol{T}(\boldsymbol{R}\boldsymbol{R}') \ , \qquad \boldsymbol{R}', \boldsymbol{R} \in SO(3) \ , \tag{3.3.2}$$

$$\boldsymbol{T}(\boldsymbol{I}) = \boldsymbol{I} \ , \qquad \boldsymbol{T}(\boldsymbol{R}^{-1}) = \boldsymbol{T}(\boldsymbol{R})^{-1} \ . \tag{3.3.3}$$

(Note that in (3.3.2), the order of $\boldsymbol{R}$ and $\boldsymbol{R}'$ is reversed between the left-hand and right-hand sides. See Sect. 3.3.2.)

Suppose a set of image characteristics $c_1, \ldots, c_m$ is given, and we want to know how they are transformed—in particular, we want to know if they are transformed linearly.

A 3D rotation is specified by an axis $\boldsymbol{n} = (n_1, n_2, n_3)$, which is taken to be a unit vector, and an angle Ω (rad) of rotation screw-wise around it (*Euler's theorem*, Chap. 6). The angle Ω is a natural choice for the "distance" from the unit element (i.e. the identity). (This choice defines a *one-parameter subgroup of* $SO(3)$ and hence satisfies all mathematical requirements, but the details are omitted here.) We say that a rotation is *infinitesimal* if the rotation angle Ω is infinitesimal.

Consider a camera rotation by a small angle Ω around a fixed orientation $\boldsymbol{n}$. Suppose the given image characteristics $c_1, \ldots, c_m$ change their values into $c_1 + \delta c_1, \ldots, c_m + \delta c_m$, and suppose the variations are written, *to a first approximation*, in the linear form

$$\delta \begin{pmatrix} c_1 \\ \vdots \\ c_m \end{pmatrix} = \Omega\, \boldsymbol{A}(\boldsymbol{n}) \begin{pmatrix} c_1 \\ \vdots \\ c_m \end{pmatrix} + \mathrm{O}(\Omega^2) \ . \tag{3.3.4}$$

The matrix $A(\boldsymbol{n})$ designates the infinitesimal transformation for axis $\boldsymbol{n}$. If we consider all orientations of $\boldsymbol{n}$, we obtain a set L of infinitesimal transformations: $L = \{A(\boldsymbol{n}) | \|\boldsymbol{n}\| = 1\}$.

Since $SO(3)$ is a three-dimensional Lie group, the set L must be a three-dimensional linear space. We can choose any three linearly independent infinitesimal transformations as a basis, but a natural choice may be $\{\boldsymbol{R}_X(\Omega), \boldsymbol{R}_Y(\Omega), \boldsymbol{R}_Z(\Omega)\}$—the rotations by angle Ω around the X-, Y-, and Z-axes, respectively. The associated infinitesimal transformations $\{A_1, A_2, A_3\}$ are the *infinitesimal generators* for this basis.

Although $SO(3)$ is not Abelian, *two infinitesimal* 3D *rotations do commute to a first approximation* (see Sect. 3.3.1). Hence, the infinitesimal rotation around axis $\boldsymbol{n} = (n_1, n_2, n_3)$ (unit vector) by angle Ω (infinitesimal) is equivalent to the composite of three rotations $\boldsymbol{R}_X(\Omega n_1)$, $\boldsymbol{R}_Y(\Omega n_2)$, $\boldsymbol{R}_Z(\Omega n_2)$. Thus, (3.3.4) is written as

$$\delta\begin{pmatrix} c_1 \\ \vdots \\ c_m \end{pmatrix} = \Omega\,(n_1 A_1 + n_2 A_2 + n_3 A_3)\begin{pmatrix} c_1 \\ \vdots \\ c_m \end{pmatrix} + \mathrm{O}(\Omega^2)\ . \tag{3.3.5}$$

In other words, $\Omega n_1, \Omega n_2, \Omega n_3$ play the role of the "coordinate" for infinitesimal rotations.

The fundamental theorem about the necessary and sufficient condition for infinitesimal transformations to be extended globally is as follows:[7]

Theorem 3.2. *A set L of infinitesimal transformations defines a (global) representation of $SO(3)$ if and only if the infinitesimal generators $\{A_1, A_2, A_3\}$ associated with rotations around the X-, Y-, and Z-axes satisfy the following commutation relations:*

$$[A_1, A_2] = -A_3\ ,\quad [A_2, A_3] = -A_1\ ,\quad [A_3, A_1] = -A_2\ . \tag{3.3.6}$$

If (3.3.6) is satisfied, the image characteristics $c_1, \ldots, c_m$ define a representation of $SO(3)$. The next question is whether this representation is reducible or irreducible. Again, this is determined by the Lie algebra L. From the commutation relations (3.3.6), the Casimir operator (3.2.8) becomes

$$\boldsymbol{H} = -(A_1^2 + A_2^2 + A_3^2)\ . \tag{3.3.7}$$

Here, again, we state an important theorem without proof.

---

[7] We omit the proof. These are the *integrability conditions* for the Lie algebra L viewed as a *vector field* over the Lie group $SO(3)$. (See also the Appendix, Sect. A.9). Note that the minus signs appear on the right-hand sides, whereas in most of the literature they do not. This is because the camera (or the coordinate system) is rotated relative to a stationary scene, while in most of the literature the scene is rotated relative to a fixed coordinate system.

Theorem 3.3. *A representation of $SO(3)$ is irreducible if and only if, for some integer or half-integer l, its Casimir operator $\boldsymbol{H}$ is $l(l+1)$ times the $(2l+1)$-dimensional unit matrix:*

$$\boldsymbol{H} = l(l+1)\boldsymbol{I} \ . \tag{3.3.8}$$

The integer or half-integer l is called the *weight* of the resulting irreducible representation, which is denoted by $\mathscr{D}_l$. If the representation is not irreducible, we can reduce it into the direct sum of irreducible representations by constructing a new basis of the Lie algebra L in such a way that the Casimir operator $\boldsymbol{H}$ is diagonalized. Namely:

Corollary 3.1. *If there exists a basis of the Lie algebra L of a representation $\boldsymbol{T}(\boldsymbol{R})$ of $SO(3)$ such that its Casimir operator $\boldsymbol{H}$ has the diagonal form*

$$\boldsymbol{H} = \begin{pmatrix} * & & & \\ & * & & \\ & & \ddots & \\ & & & * \end{pmatrix}, \tag{3.3.9}$$

where each submatrix $(*)$ is $l_i(l_i+1)$ times the $(2l_i+1)$-dimensional unit matrix for some integer or half-integer l_i, $i = 1, 2, \ldots, s$, and if $l_1, \ldots, l_s$ are all distinct, then the representation $\boldsymbol{T}(\boldsymbol{R})$ is equivalent to the direct sum of the irreducible representations $\mathscr{D}_{l_i}$, of weights l_i, $i = 1, \ldots, s$:

$$\boldsymbol{T}(\boldsymbol{R})) \cong \mathscr{D}_{l_1} \oplus \ldots \oplus \mathscr{D}_{l_s} \ . \tag{3.3.10}$$

Again, we omit the proof, but this corollary is fundamental in finding bases for irreducible representations. (Note that if some of the weights $l_1, \ldots, l_s$ are the same, the corresponding subspaces are intermingled and cannot be separated by this method alone.)

3.3.1 Infinitesimal Rotations Commute

The matrix $\boldsymbol{R}$ of rotation around axis $\boldsymbol{n} = (n_1, n_2, n_3)$ (unit vector) by angle Ω is given by

$$\boldsymbol{R} = \begin{pmatrix} \cos\Omega + n_1^2(1-\cos\Omega) & n_1n_2(1-\cos\Omega) - n_3\sin\Omega & n_1n_3(1-\cos\Omega) + n_2\sin\Omega \\ n_2n_1(1-\cos\Omega) + n_3\sin\Omega & \cos\Omega + n_2^2(1-\cos\Omega) & n_2n_3(1-\cos\Omega) - n_1\sin\Omega \\ n_3n_1(1-\cos\Omega) - n_2\sin\Omega & n_3n_2(1-\cos\Omega) + n_1\sin\Omega & \cos\Omega + n_3^2(1-\cos\Omega) \end{pmatrix}. \tag{3.3.11}$$

(The proof is given in Chap. 6.) In particular, the screw-wise rotations by angle θ around the X-, Y-, and Z-axes are respectively given by

$$R_X(\theta) = \begin{pmatrix} 1 & 0 & 0 \\ 0 & \cos\theta & -\sin\theta \\ 0 & \sin\theta & \cos\theta \end{pmatrix}, \quad R_Y(\theta) = \begin{pmatrix} \cos\theta & 0 & \sin\theta \\ 0 & 1 & 0 \\ -\sin\theta & 0 & \cos\theta \end{pmatrix},$$

$$R_Z(\theta) = \begin{pmatrix} \cos\theta & -\sin\theta & 0 \\ \sin\theta & \cos\theta & 0 \\ 0 & 0 & 1 \end{pmatrix}, \tag{3.3.12}$$

If the angle Ω is small, Taylor expansion of (3.3.11) yields

$$R = I + \Omega A(n) + O(\Omega^2), \tag{3.3.13}$$

where I is the unit matrix. The matrix $A(n)$ is given by

$$A(n) = \begin{pmatrix} 0 & -n_3 & n_2 \\ n_3 & 0 & -n_1 \\ -n_2 & n_1 & 0 \end{pmatrix}. \tag{3.3.14}$$

Thus, a point (X, Y, Z) is mapped by (3.3.13) onto point $(X + \delta X,\ Y + \delta Y,\ Z + \delta Z)$, where δX, δY, δZ are given by

$$\delta \begin{pmatrix} X \\ Y \\ Z \end{pmatrix} = \Omega \begin{pmatrix} n_2 Z - n_3 Y \\ n_3 X - n_1 Z \\ n_1 Y - n_2 X \end{pmatrix} + O(\Omega^2) = \Omega \begin{pmatrix} n_1 \\ n_2 \\ n_3 \end{pmatrix} \times \begin{pmatrix} X \\ Y \\ Z \end{pmatrix} + O(\Omega^2), \tag{3.3.15}$$

where $\times$ denotes the vector product.

If R and R' are two matrices of infinitesimal rotations around axes n and n' by angles Ω and Ω' respectively, we observe that

$$\begin{aligned} RR' &= [I + \Omega A(n) + O(\Omega^2)][I + \Omega' A(n') + O(\Omega'^2)] \\ &= I + (\Omega A(n) + \Omega' A(n')) + O(\Omega, \Omega')^2 \\ &= I + A(\Omega n + \Omega' n') + O(\Omega, \Omega')^2 = R'R, \end{aligned} \tag{3.3.16}$$

where $O(\Omega, \Omega')^2$ denotes terms of degree 2 or higher in Ω, Ω'. Thus, two infinitesimal rotations do commute *to a first approximation* in the rotation angles. Furthermore, we see that if we define the "rotation vector" by $\Omega n = (\Omega n_1, \Omega n_2, \Omega n_3)$, we can add the two corresponding "rotation vectors" as if they were ordinary vectors *as long as the rotations are infinitesimal.* In this sense, we can say that *infinitesimal rotations form a three-dimensional linear space.*

If we divide the "rotation vector" by the time Δt it took to realize that rotation, and take the limit $\Delta t \to 0$, we obtain the *rotation velocity* $(\omega_1, \omega_2, \omega_3)$. Hence, *the rotation velocity can be treated as an ordinary vector*; we can add two rotation vectors when superposing two rotations. By definition, the orientation of the rotation velocity indicates the instantaneous axis of rotation. Its magnitude $\sqrt{\omega_1^2+\omega_2^2+\omega_3^2}$ indicates the angular velocity screw-wise around it. Equation (3.3.15) now becomes

$$\begin{pmatrix} \dot{X} \\ \dot{Y} \\ \dot{Z} \end{pmatrix} = \begin{pmatrix} \omega_1 \\ \omega_2 \\ \omega_3 \end{pmatrix} \times \begin{pmatrix} X \\ Y \\ Z \end{pmatrix} . \tag{3.3.17}$$

3.3.2 Reciprocal Representations

If we compare (3.2.2) and (3.3.2), we find that the order of $\boldsymbol{R}$ and $\boldsymbol{R}'$ is reversed between the left- and right-hand sides. This seeming irregularity arises from our convention that the camera, not the scene, is rotated by $\boldsymbol{R}$.

Strictly speaking, the group element $g \in G$ in Sect. 3.2 does not correspond to the rotation $\boldsymbol{R}$ itself but rather the image transformation M_R defined in Sect. 1.3. Hence, the direct counterpart of (3.2.2) for the camera rotation is given by

$$T(M_{R'})T(M_R) = T(M_{R'} \circ M_R) , \qquad \boldsymbol{R}', \boldsymbol{R} \in SO(3) . \tag{3.3.18}$$

From (1.3.7), the right-hand side becomes $\boldsymbol{T}(M_{RR'})$. If we abbreviate $\boldsymbol{T}(M_R)$ by $\boldsymbol{T}(\boldsymbol{R})$, we obtain (3.3.2).

The relationship (3.3.2) is an *anti-homomorphism* from $SO(3)$ into the group of matrices under multiplication, and $\boldsymbol{T}(g)$ is called a *reciprocal representation* of $SO(3)$. However, the (usual) representation and the reciprocal representation need not be treated separately. If $\boldsymbol{T}(\boldsymbol{R})$ is a reciprocal representation, put

$$\boldsymbol{S}(\boldsymbol{R}) = \boldsymbol{T}(\boldsymbol{R}^{\mathrm{T}}) . \tag{3.3.19}$$

Then,

$$\boldsymbol{S}(\boldsymbol{R}')\boldsymbol{S}(\boldsymbol{R}) = \boldsymbol{S}(\boldsymbol{R}'\boldsymbol{R}) , \qquad \boldsymbol{R}', \boldsymbol{R} \in SO(3) , \tag{3.3.20}$$

$$\boldsymbol{S}(\boldsymbol{I}) = \boldsymbol{I} , \qquad \boldsymbol{S}(\boldsymbol{R}^{-1}) = \boldsymbol{S}(\boldsymbol{R})^{-1} \tag{3.3.21}$$

(Exercise 3.7). Thus, $\boldsymbol{S}(\boldsymbol{R})$ is a (usual) representation of $SO(3)$. The representation $\boldsymbol{S}(\boldsymbol{R})$ is called the *transposed* (or *dual*) *representation* of $\boldsymbol{T}(\boldsymbol{R})$, and a reciprocal representation is converted into a (usual) representation by interchanging the role of $\boldsymbol{R}$ and $\boldsymbol{R}^{\mathrm{T}}$. In the following, however, we save on notation by treating a reciprocal representation as if it is a (usual) representation without constructing its transposed representation.

If the scene, instead of the camera, is rotated by $\boldsymbol{R}$ relative to a fixed coordinate system, we must interchange $\boldsymbol{R}$ and $\boldsymbol{R}^{\mathrm{T}}$ and at the same time reverse the signs of infinitesimal transformations; since $\boldsymbol{R}$ and $\boldsymbol{R}^{\mathrm{T}}$ ($= \boldsymbol{R}^{-1}$) are rotations of opposite senses, infinitesimal transformations proceed in the opposite direction. Hence, the commutation relations (3.3.6) become

$$[A_1, A_2] = A_3 , \qquad [A_2, A_3] = A_1 , \qquad [A_3, A_1] = A_2 . \tag{3.3.22}$$

3.3.3 Spinors

It can be shown that all irreducible representations of $SO(3)$ are indexed by the *weight* l, which is an integer or a half-integer. The irreducible representation of $SO(3)$ of weight l is denoted by $\mathscr{D}_l$.

In image understanding applications, we need not consider representations of half-integer weights. If the weight is a half-integer, the signs of the image characteristics are reversed by the transformation associated with a rotation of angle 2π (we omit the proof).[8] Since a camera rotation by angle 2π around an arbitrary axis is the same as no rotation, the values of any image characteristics must be the same as before. Thus, representations of half-integers do not occur for image characteristics.

In general, a set of parameters is called a *tensor* with respect to $SO(3)$ if it undergoes an irreducible representation of an integer weight, while it is called a *spinor* if it undergoes an irreducible representation of a half-integer weight. In image understanding applications, we need not consider spinors.

The term "spinor" comes from quantum mechanics. The *state* of a particle which has spherical symmetry (or, to be precise, has a *Hamiltonian* that *commutes* with rotations) is characterized by an irreducible representation of $SO(3)$, and its weight indexes the discrete values of the *spin* angular momentum, A particle of an integer spin is called a *boson*, whereas a particle of a half integer spin is called a *fermion*.

The term "tensor" comes from elasticity theory, meaning something that causes "tension" in materials, and was originally used to describe the state of tension inside a material under external loading. Today, this tensor is known as the *stress tensor*.

3.4 Representation of a Scalar and a Vector

If an image characteristic c does not change its value under camera rotation, i.e.,

$$c' = c \,, \tag{3.4.1}$$

it is called a *scalar*. A scalar defines a trivial representation of $SO(3)$, called the *identity representation*. Evidently, the infinitesimal generators $\{A_1, A_2, A_3\}$ are all zero, and the commutation relations (3.3.6) are trivially satisfied. Obviously, it is an irreducible representation.

Theorem 3.4. *The identity representation is equivalent to the irreducible representation $\mathscr{D}_0$ of weight* 0.

[8] This is because $SO(3)$ is *not* simply connected. If the Lie algebra of infinitesimal transformations is integrated globally, $SO(3)$ is covered twice, defining a *covering space* of $SO(3)$ homeomorphic to $SU(2)$ (Chap. 6).

Proof. The Casimir operator H defined by (3.3.7) is 0. This is $0(0+1)$ times the $(2\times 0+1)$-dimensional unit matrix 1. The theorem follows from Theorem 3.3.

A set of three image characteristics $\{a_1, a_2, a_3\}$ is called a *vector* if it is transformed as

$$\begin{pmatrix} a_1' \\ a_2' \\ a_3' \end{pmatrix} = \boldsymbol{R}^{\mathrm{T}} \begin{pmatrix} a_1 \\ a_2 \\ a_3 \end{pmatrix} . \tag{3.4.2}$$

This is a linear mapping, and hence it defines a representation of $SO(3)$, called the *vector representation*. (If the scene is rotated by $\boldsymbol{R}$ relative to the camera, the transformation matrix is simply $\boldsymbol{R}$ instead of $\boldsymbol{R}^{\mathrm{T}}$.)

Let $\boldsymbol{n}$ be the axis and Ω the angle of rotation $\boldsymbol{R}$. As was shown in Sect. 3.3.1, if Ω is small, the matrix $\boldsymbol{R}$ takes the form

$$\boldsymbol{R} = \boldsymbol{I} + \Omega \boldsymbol{A}(\boldsymbol{n}) + \mathrm{O}(\Omega^2) . \tag{3.4.3}$$

The values of $\{a_1, a_2, a_3\}$ are also infinitesimally changed into new values $\{a_1 + \delta a_1, a_2 + \delta a_2, a_3 + \delta a_3\}$. Substituting (3.4.3) together with (3.3.14) into (3.4.2), we obtain

$$\delta \begin{pmatrix} a_1 \\ a_2 \\ a_3 \end{pmatrix} = \Omega \begin{pmatrix} 0 & n_3 & -n_2 \\ -n_3 & 0 & n_1 \\ n_2 & -n_1 & 0 \end{pmatrix} \begin{pmatrix} a_1 \\ a_2 \\ a_3 \end{pmatrix} + \mathrm{O}(\Omega^2)$$

$$= \Omega(n_1 \boldsymbol{A}_1 + n_2 \boldsymbol{A}_2 + n_3 \boldsymbol{A}_3) \begin{pmatrix} a_1 \\ a_2 \\ a_3 \end{pmatrix} + \mathrm{O}(\Omega^2) , \tag{3.4.4}$$

where the infinitesimal generators $\{\boldsymbol{A}_1, \boldsymbol{A}_2, \boldsymbol{A}_3\}$ are given by

$$\boldsymbol{A}_1 = \begin{pmatrix} & & \\ & & 1 \\ & -1 & \end{pmatrix}, \quad \boldsymbol{A}_2 = \begin{pmatrix} & & -1 \\ & & \\ 1 & & \end{pmatrix}, \quad \boldsymbol{A}_3 = \begin{pmatrix} & 1 & \\ -1 & & \\ & & \end{pmatrix} . \tag{3.4.5}$$

These infinitesimal generators satisfy the commutation relations

$$[\boldsymbol{A}_1, \boldsymbol{A}_2] = -\boldsymbol{A}_3 , \quad [\boldsymbol{A}_2, \boldsymbol{A}_3] = -\boldsymbol{A}_1 , \quad [\boldsymbol{A}_3, \boldsymbol{A}_1] = -\boldsymbol{A}_2 . \tag{3.4.6}$$

Hence, the vector representation is indeed a representation of $SO(3)$. Moreover:

Theorem 3.5. *The vector representation is equivalent to the irreducible representation $\mathscr{D}_1$ of weight* 1.

Proof. The Casimir operator $\boldsymbol{H}$ of (3.3.7) becomes

$$\boldsymbol{H} = \begin{pmatrix} 2 & & \\ & 2 & \\ & & 2 \end{pmatrix} . \tag{3.4.7}$$

This is $1(1+1)$ times the $(2 \times 1 + 1)$-dimensional unit matrix. The theorem follows from Theorem 3.3.

3.5 Irreducible Reduction of a Tensor

A set of image characteristics a_{ij}, $i, j = 1, 2, 3$, is called a *tensor* if it is transformed as

$$\begin{pmatrix} a'_{11} & a'_{12} & a'_{13} \\ a'_{21} & a'_{22} & a'_{23} \\ a'_{31} & a'_{32} & a'_{33} \end{pmatrix} = \boldsymbol{R}^{\mathrm{T}} \begin{pmatrix} a_{11} & a_{12} & a_{13} \\ a_{21} & a_{22} & a_{23} \\ a_{31} & a_{32} & a_{33} \end{pmatrix} \boldsymbol{R} . \tag{3.5.1}$$

This is a linear mapping from a_{ij}, $i, j = 1, 2, 3$, onto a'_{ij}, $i, j = 1, 2, 3$, and hence it defines a representation of $SO(3)$—called the *tensor representation* (of degree 2). (The two matrices $\boldsymbol{R}^{\mathrm{T}}$ and $\boldsymbol{R}$ are interchanged if the scene is rotated by $\boldsymbol{R}$ relative to the camera.)

The tensor representation is not irreducible. The proof is similar to that of the two-dimensional case (Sect. 2.5), because most of the proofs given there do not depend on the dimensionality. First, a tensor is decomposed into its *symmetric part* and *antisymmetric part*:

$$\begin{pmatrix} a_{11} & a_{12} & a_{13} \\ a_{21} & a_{22} & a_{23} \\ a_{31} & a_{32} & a_{33} \end{pmatrix} = \begin{pmatrix} a_{11} & (a_{12} + a_{21})/2 & (a_{13} + a_{31})/2 \\ (a_{21} + a_{13})/2 & a_{22} & (a_{23} + a_{32})/2 \\ (a_{31} + a_{13})/2 & (a_{32} + a_{23})/2 & a_{33} \end{pmatrix}$$

$$+ \begin{pmatrix} 0 & (a_{12} - a_{21})/2 & -(a_{31} - a_{13})/2 \\ -(a_{12} - a_{21})/2 & 0 & (a_{23} - a_{32})/2 \\ (a_{31} - a_{13})/2 & -(a_{23} - a_{32})/2 & 0 \end{pmatrix} . \tag{3.5.2}$$

This decomposition is unique. The proof is the same as that of Lemma 2.1. Similarly, we can prove that this decomposition is *invariant*: the symmetric and antisymmetric parts are transformed separately as tensors. The proof is the same as that of Proposition 2.2. Thus, the tensor representation is reduced to the direct sum of two separate representations.

The antisymmetric part has three independent elements $(a_{23} - a_{32})/2$, $(a_{31} - a_{13})/2$, $(a_{12} - a_{21})/2$. They define a three-dimensional representation, which should be $\mathscr{D}_1$.

Lemma 3.3. *The antisymmetric part defines the vector representation* $\mathscr{D}_1$:

$$\begin{pmatrix} (a'_{23} - a'_{32})/2 \\ (a'_{31} - a'_{13})/2 \\ (a'_{12} - a'_{21})/2 \end{pmatrix} = \boldsymbol{R}^{\mathrm{T}} \begin{pmatrix} (a_{23} - a_{32})/2 \\ (a_{31} - a_{13})/2 \\ (a_{12} - a_{21})/2 \end{pmatrix} . \tag{3.5.3}$$

Proof. Let $B = (b_{ij})$, $i, j = 1, 2, 3$, be an antisymmetric tensor. The transformation rule (3.5.1) is written as

$$\boldsymbol{B}' = \boldsymbol{R}^{\mathrm{T}} \boldsymbol{B} \boldsymbol{R} . \tag{3.5.4}$$

Suppose the rotation $\boldsymbol{R}$ is infinitesimal:

$$\boldsymbol{R} = \boldsymbol{I} + \Omega A(\boldsymbol{n}) + O(\Omega^2) . \tag{3.5.5}$$

Substitution of this together with (3.3.14) into (3.5.4) yields

$$\boldsymbol{B}' = \boldsymbol{B} - \Omega(A(\boldsymbol{n})\boldsymbol{B} - \boldsymbol{B}A(\boldsymbol{n})) + O(\Omega^2) . \tag{3.5.6}$$

Hence, the increment $\delta \boldsymbol{B} \equiv \boldsymbol{B}' - \boldsymbol{B}$ is given by

$$\delta \boldsymbol{B} = \Omega[-A(\boldsymbol{n})\boldsymbol{B} - \boldsymbol{B}A(\boldsymbol{n})] + O(\Omega^2) . \tag{3.5.7}$$

If we pick out the three nonzero independent elements $\{b_{23}, b_{31}, b_{12}\}$, equation (3.5.7) is written as

$$\delta \begin{pmatrix} b_{23} \\ b_{31} \\ b_{12} \end{pmatrix} = \Omega \begin{pmatrix} 0 & n_3 & -n_2 \\ -n_3 & 0 & n_1 \\ n_2 & -n_1 & 0 \end{pmatrix} \begin{pmatrix} b_{23} \\ b_{31} \\ b_{12} \end{pmatrix} + O(\Omega^2) .$$

$$= \Omega(\boldsymbol{n}_1 \boldsymbol{A}_1 + \boldsymbol{n}_2 \boldsymbol{A}_2 + \boldsymbol{n}_3 \boldsymbol{A}_3) \begin{pmatrix} b_{23} \\ b_{31} \\ b_{12} \end{pmatrix} + O(\Omega^2) , \tag{3.5.8}$$

where the infinitesimal generators $\{\boldsymbol{A}_1, \boldsymbol{A}_2, \boldsymbol{A}_3\}$ are given by

$$\boldsymbol{A}_1 = \begin{pmatrix} & & \\ & & 1 \\ & -1 & \end{pmatrix}, \quad \boldsymbol{A}_2 = \begin{pmatrix} & & -1 \\ & & \\ 1 & & \end{pmatrix}, \quad \boldsymbol{A}_3 = \begin{pmatrix} & 1 & \\ -1 & & \\ & & \end{pmatrix} . \tag{3.5.9}$$

Thus, the infinitesimal generators are exactly those of a vector. Hence, the global transformation must also be the same, defining the irreducible representation $\mathscr{D}_1$ of weight 1.

On the other hand, the representation defined by the symmetric part is not irreducible. As in the two-dimensional case, the symmetric part is further decomposed into its *scalar part* (multiple of the unit matrix) and *deviator part* (matrix of trace 0):

$$\begin{pmatrix} a_{11} & (a_{12}+a_{21})/2 & (a_{13}+a_{31})/2 \\ (a_{21}+a_{12})/2 & a_{22} & (a_{23}+a_{32})/2 \\ (a_{31}+a_{13})/2 & (a_{32}+a_{23})/2 & a_{33} \end{pmatrix} = \frac{a_{11}+a_{22}+a_{33}}{3}\begin{pmatrix} 1 & & \\ & 1 & \\ & & 1 \end{pmatrix}$$

$$+\begin{pmatrix} (2a_{11}-a_{22}-a_{33})/3 & (a_{12}+a_{21})/2 & (a_{13}+a_{31})/2 \\ (a_{21}+a_{12})/2 & (2a_{22}-a_{11}-a_{33})/3 & (a_{23}+a_{32})/2 \\ (a_{31}+a_{13})/2 & (a_{32}+a_{23})/2 & (2a_{33}-a_{11}-a_{22})/3 \end{pmatrix} . \tag{3.5.10}$$

Again, this decomposition is unique and *invariant*: the scalar and deviator parts are transformed separately as tensors. The proof is the same as those of Lemma 2.2 and Proposition 2.3.

The scalar part has only one independent element $a_{11}+a_{22}+a_{33}$. It defines a one-dimensional representation, which should be $\mathscr{D}_0$. In fact, it is easy to prove that $a_{11}+a_{22}+a_{33}$ is a scalar (Exercise 3.13). Hence, we need to consider only the remaining deviator part. Put the deviator part to be $D=(d_{ij})$, $i, j = 1, 2, 3$. Since it is a symmetric tensor of trace 0, it contains only five independent elements, say $\{d_{11}, d_{22}, d_{12}, d_{31}, d_{32}\}$, and they define a five-dimensional representation. Is this representation irreducible? If so, which irreducible representation is it equivalent to? These questions can be answered by analyzing infinitesimal transformations.

The transformation rule (3.5.1) is written as

$$\boldsymbol{D}' = \boldsymbol{R}^{\mathrm{T}}\boldsymbol{D}\boldsymbol{R} . \tag{3.5.11}$$

Suppose the rotation $\boldsymbol{R}$ is infinitesimal:

$$\boldsymbol{R} = \boldsymbol{I} + \Omega \boldsymbol{A}(\boldsymbol{n}) + \mathrm{O}(\Omega^2) . \tag{3.5.12}$$

Substitution of this together with (3.3.14) into (3.5.11) yields

$$\boldsymbol{D}' = \boldsymbol{D} - \Omega[\boldsymbol{A}(\boldsymbol{n})\boldsymbol{D} - \boldsymbol{D}\boldsymbol{A}(\boldsymbol{n})] + \mathrm{O}(\Omega^2) . \tag{3.5.13}$$

Hence, the increment $\delta\boldsymbol{D} = \boldsymbol{D}' - \boldsymbol{D}$ is given by

$$\delta\boldsymbol{D} = \Omega[-\boldsymbol{A}(\boldsymbol{n})\boldsymbol{D} - \boldsymbol{D}\boldsymbol{A}(\boldsymbol{n})] + \mathrm{O}(\Omega^2) . \tag{3.5.14}$$

If we pick out the five independent elements $\{d_{11}, d_{22}, d_{12}, d_{31}, d_{32}\}$, equation (3.5.14) is written as

$$\delta\begin{pmatrix} d_{11} \\ d_{22} \\ d_{12} \\ d_{31} \\ d_{32} \end{pmatrix} = \Omega \begin{pmatrix} & & 2n_3 & -2n_2 & \\ & & -2n_3 & & 2n_1 \\ -n_3 & n_3 & & n_1 & -n_2 \\ 2n_2 & n_2 & -n_1 & & n_3 \\ -n_1 & -2n_1 & n_2 & -n_3 & \end{pmatrix} \begin{pmatrix} d_{11} \\ d_{22} \\ d_{12} \\ d_{31} \\ d_{32} \end{pmatrix} + \mathrm{O}(\Omega^2)$$

$$= \Omega(\boldsymbol{n}_1 A_1 + \boldsymbol{n}_2 A_2 + \boldsymbol{n}_3 A_3) \begin{pmatrix} d_{11} \\ d_{22} \\ d_{12} \\ d_{31} \\ d_{32} \end{pmatrix} + \mathrm{O}(\Omega^2) \ . \tag{3.5.15}$$

where the infinitesimal generators $\{A_1, A_2, A_3\}$ are given by

$$A_1 = \begin{pmatrix} & & & & \\ & & & & 2 \\ & & & 1 & \\ & & -1 & & \\ -1 & -2 & & & \end{pmatrix}, \quad A_2 = \begin{pmatrix} & & & -2 & \\ & & & & \\ & & & & -1 \\ 2 & 1 & & & \\ & & 1 & & \end{pmatrix},$$

$$A_3 = \begin{pmatrix} & & 2 & & \\ & & -2 & & \\ -1 & 1 & & & \\ & & & & 1 \\ & & & -1 & \end{pmatrix}. \tag{3.5.16}$$

These infinitesimal generators satisfy the commutation relations

$$[A_1, A_2] = -A_3 \ , \ [A_2, A_3] = -A_1 \ , \ [A_3, A_1] = -A_2 \ . \tag{3.5.17}$$

Hence, the deviator part indeed defines a representation of $SO(3)$. Moreover:

Lemma 3.4. *The representation defined by a deviator tensor is equivalent to the irreducible representation $\mathscr{D}_2$ of weight 2.*

Proof. The Casimir operator of (3.3.7) becomes

$$H = \begin{pmatrix} 6 & & & & \\ & 6 & & & \\ & & 6 & & \\ & & & 6 & \\ & & & & 6 \end{pmatrix}. \tag{3.5.18}$$

This is $2(2+1)$ times the $(2 \times 2 + 1)$-dimensional unit matrix. The lemma follows from Theorem 3.3.

Thus, we conclude the following:

Theorem 3.6. *The tensor representation is equivalent to the direct sum of the irreducible representations:* $\mathscr{D}_0 \oplus \mathscr{D}_1 \oplus \mathscr{D}_2$.

This process of decomposing a tensor into its symmetric and antisymmetric parts and then decomposing the symmetric part into its scalar and deviator parts is the same as in the two-dimensional case (Sect. 2.5). Indeed, *Weyl's theorem* (Sect. 2.5) asserts that *all irreducible representations of a tensor representation of any dimension and any degree can be attained by operating on the symmetry and trace of indices* (Sect. 2.5.2, Chap. 5).

3.5.1 Canonical Form of Infinitesimal Generators

In Sect. 3.4 we identified a vector $\{a_1, a_2, a_3\}$ as a column vector of $\mathbb{R}^3$ in the straightforward way, and regarded $\mathbb{R}^3$ as the representation space. However, we can arbitrarily change the basis of the representation space. Changing the basis means considering a representation defined by some linear combinations of the original three quantities—say $\{a_1 + a_2, a_2 + a_3, a_3 + a_1\}$, for example.

We also represented a deviator tensor $\boldsymbol{D} = (d_{ij})$, $i, j = 1, 2, 3$, by a five-dimensional column vector of $\mathbb{R}^5$ whose components are $\{d_{11}, d_{22}, d_{12}, d_{31}, d_{32}\}$. However, we may as well consider five arbitrary (but independent) linear combinations of them. This means changing the basis of the representation space $\mathbb{R}^5$.

In general, the basis of the representation space $\mathbb{R}^n$ can be changed arbitrarily, and the matrix elements of the infinitesimal generators $\{\boldsymbol{A}_1, \boldsymbol{A}_2, \boldsymbol{A}_3\}$ change accordingly. However, for the purpose of comparing different representations, it is desirable to fix a special basis. Such a basis is called the *canonical basis*, and resulting expressions are called *canonical forms*. In most of the literature, an orthonormal system constructed from eigenvectors of $\boldsymbol{A}_3$ is used as the canonical basis. (This means that the representation space is regarded as $\mathbb{C}^n$, since eigenvectors do not necessarily exist within $\mathbb{R}^n$.) Then, the infinitesimal generators $\{\boldsymbol{A}_1, \boldsymbol{A}_2, \boldsymbol{A}_3\}$ of $\mathscr{D}_l$ are expressed as

$$
A_1 = \frac{\mathrm{i}}{2}\begin{pmatrix}
0 & f(-l+1) & 0 & \cdots & 0 & 0 \\
f(-l+1) & 0 & f(-l+2) & \cdots & 0 & 0 \\
\cdot\cdot\cdot\cdot & \cdot\cdot\cdot\cdot & \cdot\cdot\cdot\cdot & \cdot\cdot\cdot\cdot & \cdot\cdot\cdot\cdot & \cdot\cdot\cdot\cdot \\
\cdot\cdot\cdot\cdot & \cdot\cdot\cdot\cdot & \cdot\cdot\cdot\cdot & \cdot\cdot\cdot\cdot & \cdot\cdot\cdot\cdot & \cdot\cdot\cdot\cdot \\
0 & 0 & 0 & \cdots & 0 & f(l) \\
0 & 0 & 0 & \cdots & f(l) & 0
\end{pmatrix},
$$

$$
A_2 = \frac{1}{2}\begin{pmatrix}
0 & -f(-l+1) & 0 & \cdots & 0 & 0 \\
f(-l+1) & 0 & -f(-l+2) & \cdots & 0 & 0 \\
\cdot\cdot\cdot\cdot & \cdot\cdot\cdot\cdot & \cdot\cdot\cdot\cdot & \cdot\cdot\cdot\cdot & \cdot\cdot\cdot\cdot & \cdot\cdot\cdot\cdot \\
\cdot\cdot\cdot\cdot & \cdot\cdot\cdot\cdot & \cdot\cdot\cdot\cdot & \cdot\cdot\cdot\cdot & \cdot\cdot\cdot\cdot & \cdot\cdot\cdot\cdot \\
0 & 0 & 0 & \cdots & 0 & -f(l) \\
0 & 0 & 0 & \cdots & f(l) & 0
\end{pmatrix},
$$

$$
A_3 = \mathrm{i}\begin{pmatrix}
-l & 0 & 0 & \cdots & 0 & 0 \\
0 & -(l-1) & 0 & \cdots & 0 & 0 \\
0 & 0 & -(l-2) & \cdots & 0 & 0 \\
\cdot\cdot\cdot\cdot & \cdot\cdot\cdot\cdot & \cdot\cdot\cdot\cdot & \cdot\cdot\cdot\cdot & \cdot\cdot\cdot\cdot & \cdot\cdot\cdot\cdot \\
\cdot\cdot\cdot\cdot & \cdot\cdot\cdot\cdot & \cdot\cdot\cdot\cdot & \cdot\cdot\cdot\cdot & \cdot\cdot\cdot\cdot & \cdot\cdot\cdot\cdot \\
0 & 0 & 0 & \cdots & l-1 & 0 \\
0 & 0 & 0 & \cdots & 0 & l
\end{pmatrix}, \tag{3.5.19}
$$

where $f(m) = \sqrt{(l+m)(l-m+1)}$. For $\mathscr{D}_1$, we have

$$
A_1 = \frac{\mathrm{i}}{\sqrt{2}}\begin{pmatrix} & 1 & \\ 1 & & 1 \\ & 1 & \end{pmatrix}, \qquad A_2 = \frac{1}{\sqrt{2}}\begin{pmatrix} & -1 & \\ 1 & & -1 \\ & 1 & \end{pmatrix},
$$

$$
A_3 = \mathrm{i}\begin{pmatrix} -1 & & \\ & & \\ & & 1 \end{pmatrix}. \tag{3.5.20}
$$

For $\mathscr{D}_2$, we have

$$
A_1 = \frac{\mathrm{i}}{\sqrt{2}}\begin{pmatrix}
 & \sqrt{2} & & & \\
\sqrt{2} & & \sqrt{3} & & \\
 & \sqrt{3} & & \sqrt{3} & \\
 & & \sqrt{3} & & \sqrt{2} \\
 & & & \sqrt{2} &
\end{pmatrix},
$$

$$A_2 = \frac{1}{\sqrt{2}}\begin{pmatrix} & -\sqrt{2} & & & \\ \sqrt{2} & & -\sqrt{3} & & \\ & \sqrt{3} & & -\sqrt{3} & \\ & & \sqrt{3} & & -\sqrt{2} \\ & & & \sqrt{2} & \end{pmatrix},$$

$$A_3 = \mathrm{i}\begin{pmatrix} -2 & & & & \\ & -1 & & & \\ & & 0 & & \\ & & & 1 & \\ & & & & 2 \end{pmatrix}. \tag{3.5.21}$$

This canonical basis originates in quantum mechanics. In quantum mechanics, a vector of a representation space of $SO(3)$ represents a quantum state of angular momentum. In the above canonical form, the generator A_3 is diagonalized. This means that the Z-axis undergoes special treatment. This is due to the established convention that the orientation along which the angular momentum is measured is taken as the Z-axis (along which, say, a magnetic field is applied). Then, the resulting discrete quantum states are indexed by the eigenvalue m of $\mathrm{i}A_3$, which is called the *magnetic quantum number*. The weight l is called the *azimuthal quantum number* and indexes the total (or spin) angular momentum.

For this reason, many books present analysis of equations expressed with respect to this canonical basis, and many formulae have been developed to convert given expressions into canonical forms. In image analysis, the above canonical basis has no special significance.

3.5.2 Alibi vs Alias

We have occasionally pointed out that some differences arise in equations depending on whether the camera or the scene is rotated. The difference originates in the interpretation of a transformation written in terms of coordinates. A transformation of a space equipped with a coordinate system (x_i), $i = 1, \ldots, n$, is written in the form

$$x'_i = f_i(x_1, \ldots, x_n), \qquad i = 1, \ldots, n. \tag{3.5.22}$$

Here, two interpretations are possible. One is that a point which was located at the position (x_i) "moves" into the new position (x'_i). In other words, the coordinate system is fixed, but all the points "change their locations". This interpretation is called an *alibi*.

The other interpretation is that a point which was "referred to" by coordinates (x_i) is now "referred to" by the new coordinates (x'_i). In other words, all the points are fixed in the space, but the coordinate system is altered and all the points are "renamed". This interpretation is called an *alias*.

In this book, we are dealing with image transformations in the alias sense, seeking properties that are invariant under renaming. However, most books on mathematics treat representations of $SO(2)$ and $SO(3)$ (or $SO(n)$ in general) in the alibi sense. As a result, appropriate modifications are necessary when we use the results in those books.

In Chap. 1, we rotated the image coordinate system by angle θ clockwise. This means that the points on the image plane rotate relative to the coordinate system by angle $-\theta$ clockwise. Hence, we used $e^{-in\theta}$ to define the weight n rather than $e^{in\theta}$. If the image, not the coordinate system, is rotated by angle θ clockwise, all $\boldsymbol{R}(\theta)$ must be replaced by its transpose $\boldsymbol{R}(\theta)^{\mathrm{T}}$.

Similarly, rotation of the camera by $\boldsymbol{R}$ is equivalent to rotation of the scene by $\boldsymbol{R}^{\mathrm{T}}$. Hence, if the scene, not the camera, is rotated by $\boldsymbol{R}$, we must interchange $\boldsymbol{R}$ and $\boldsymbol{R}^{\mathrm{T}}$ in all equations. The signs of infinitesimal generators $\{\boldsymbol{A}_1, \boldsymbol{A}_2, \boldsymbol{A}_3\}$ are also reversed, and the commutation relations become

$$[\boldsymbol{A}_1, \boldsymbol{A}_2] = \boldsymbol{A}_3 \ , \qquad [\boldsymbol{A}_2, \boldsymbol{A}_3] = \boldsymbol{A}_1 \ , \qquad [\boldsymbol{A}_3, \boldsymbol{A}_1] = \boldsymbol{A}_2 \ . \tag{3.5.23}$$

3.6 Restriction of $SO(3)$ to $SO(2)$

The camera rotation transformation is a generalization of the image coordinate rotation, since coordinate rotation is realized by camera rotation around the Z-axis. From (3.3.12), the matrix of rotation around the Z-axis by angle θ screw-wise is given by

$$\boldsymbol{R} = \left(\begin{array}{cc|c} \cos\theta & -\sin\theta & \\ \sin\theta & \cos\theta & \\ \hline & & 1 \end{array}\right) , \tag{3.6.1}$$

which we call the *restriction of* $SO(3)$ to $SO(2)$. In order to make a distinction between invariance for $SO(3)$ and invariance for $SO(2)$, let us use such terms as *3D scalars, 3D vectors*, and *3D tensors* as opposed to *2D scalars, 2D vectors*, and *2D tensors* for $SO(2)$.

Evidently, a 3D scalar c is also a 2D scalar:

$$c' = c \ . \tag{3.6.2}$$

Hence, we obtain:

Theorem 3.7. *By restricting $SO(3)$ to $SO(2)$, a 3D scalar becomes a 2D scalar.*

Consider a 3D vector $\{a_1, a_2, a_3\}$:

$$\begin{pmatrix} a_1' \\ a_2' \\ a_3' \end{pmatrix} = \boldsymbol{R}^{\mathrm{T}} \begin{pmatrix} a_1 \\ a_2 \\ a_3 \end{pmatrix} . \tag{3.6.3}$$

If (3.6.1) is substituted into this, the equation is split into two parts:

$$\begin{pmatrix} a'_1 \\ a'_2 \end{pmatrix} = \begin{pmatrix} \cos\theta & \sin\theta \\ -\sin\theta & \cos\theta \end{pmatrix} \begin{pmatrix} a_1 \\ a_2 \end{pmatrix}, \qquad a'_3 = a_3 . \tag{3.6.4}$$

Thus, we obtain:

Theorem 3.8. *By restricting $SO(3)$ to $SO(2)$, a 3D vector is split into a 2D vector and a 2D scalar.*

Hence, we can construct the following (2D) invariants:

$$V = a_1 + \mathrm{i}a_2 , \qquad c = a_3 . \tag{3.6.5}$$

(V is a relative invariant of weight 1, and c is an absolute invariant.)

Consider a 3D tensor a_{ij}, $i, j = 1, 2, 3$:

$$\begin{pmatrix} a'_{11} & a'_{12} & a'_{13} \\ a'_{21} & a'_{22} & a'_{23} \\ a'_{31} & a'_{32} & a'_{33} \end{pmatrix} = \boldsymbol{R}^{\mathrm{T}} \begin{pmatrix} a_{11} & a_{12} & a_{13} \\ a_{21} & a_{22} & a_{23} \\ a_{31} & a_{32} & a_{33} \end{pmatrix} \boldsymbol{R} . \tag{3.6.6}$$

If (3.6.1) is substituted into this, the equation is split into four parts:

$$\begin{pmatrix} a'_{13} \\ a'_{23} \end{pmatrix} = \begin{pmatrix} \cos\theta & \sin\theta \\ -\sin\theta & \cos\theta \end{pmatrix} \begin{pmatrix} a_{13} \\ a_{23} \end{pmatrix}, \quad \begin{pmatrix} a'_{31} \\ a'_{32} \end{pmatrix} = \begin{pmatrix} \cos\theta & \sin\theta \\ -\sin\theta & \cos\theta \end{pmatrix} \begin{pmatrix} a_{31} \\ a_{32} \end{pmatrix},$$

$$a'_{33} = a_{33} . \tag{3.6.7}$$

Hence, a_{ij}, $i, j = 1, 2$, are transformed as a 2D tensor; $\{a_{13}, a_{23}\}$ and $\{a_{31}, a_{32}\}$ as 2D vectors; a_{33} as a 2D scalar. Thus, we conclude:

Theorem 3.9. *By restricting $SO(3)$ to $SO(2)$, a 3D tensor is split into a 2D tensor, two 2D vectors, and a 2D scalar.*

Hence, we can construct the following (2D) invariants:

$$T = a_{11} + a_{22} , \quad R = a_{12} - a_{21} , \quad S = (a_{11} - a_{22}) + \mathrm{i}(a_{12} + a_{21}) ,$$

$$U = a_{13} + \mathrm{i}a_{23} , \qquad V = a_{31} + \mathrm{i}a_{32} , \qquad c = a_{33} . \tag{3.6.8}$$

(T, R and c are absolute invariants, U and V are relative invariants of weight 1, and S is a relative invariant of weight 2.)

In general, if the transformations are restricted to its subgroup, we observe a "finer" structure—a set of quantities that was invariant as a whole is now split into separately invariant quantities. In the extreme, if the group of transformations is restricted to the identity transformation, all quantities become invariants. In the opposite extreme, if we consider all one-to-one and onto continuous mappings, the only remaining invariant quantities are *topological invariants* (Appendix, Sect. A.8) such as the number of connected components, the number of connected boundaries, and the number of holes.

3.6.1 Broken Symmetry

The hierarchy of structures derived from the hierarchy of underlying groups of transformations plays an essential role in theoretical physics. The emergence of fine structures in materials induced by changes of external parameters—temperature, pressure, electric field, magnetic field, etc.—is often explained as a result of restricting some group of transformations into its subgroup.

To be more precise, many physical laws are formulated as some kind of *variational principle*: a physically possible state is the state which minimizes (or maximizes) some quantity—the Lagrangian, Hamiltonian, energy, free energy, entropy, etc. Let us tentatively call it the Lagrangian. Consider one physically possible state (at which the Lagrangian is minimized). If the application of a group of transformations—translations, rotations, etc.—to this state does not change the value of the Lagrangian, the group of transformations is called the *symmetry* of this state. A state is said to be *homogeneous* if it has translations as its symmetry, and *isotropic* if it has rotations as its symmetry.

If the values of the external parameters change, the group of transformations which did not affect the value of the Lagrangian may now affect it. As a result, the Lagrangian may remain unaffected only by a subgroup of the original symmetry. Then, we say that the symmetry is *broken*. This phenomenon is associated with structural change of the material—loss of homogeneity, isotropy, etc. A simple example is a circular elastic plate placed in the XY-plane centered at the origin O undergoing uniform compression from its circumference. If the compressing force is small, the plate has the symmetry of rotations around the Z-axis and reflections with respect to the XY-plane. However, if the compressing force is above a critical value, the plate buckles to one side. It still retains the symmetry of rotations around the Z-axis, but has lost the symmetry of reflections.

A similar formulation is applied in elementary particle theory and grand unified field theory, but the concept of symmetry is extended to include more abstract transformations such as the *gauge group*. In the study of image understanding, there is no natural law about which types of transformations we should consider; that is completely determined by our interest.

3.7 Transformation of Optical Flow

So far, we have implicitly assumed that the scene is stationary and the camera is rotated. However, the camera rotation transformation can be applied to motion as well; if we are observing the motion of an object, we can predict how the motion would look if the camera were oriented differently. In this section, we show how our theory works for motion.

Let us take the viewer-centered coordinate system (Sect. 1.2), and consider a planar surface

$$Z = pX + qY + r \ . \tag{3.7.1}$$

Under perspective projection, a point (X, Y, Z) in the scene is projected onto the point (x, y) on the image plane given by the projection equations

$$x = \frac{fX}{Z} \ , \qquad y = \frac{fY}{Z} \ . \tag{3.7.2}$$

Combining the surface equation (3.7.1) and the projection equations (3.7.2), the scene coordinates (X, Y, Z) of a point on the surface are expressed in terms of the image coordinates (x, y) as follows (Exercise 3.19):

$$X = \frac{rx}{f - px - qy} \ , \qquad Y = \frac{ry}{f - px - qy} \ , \qquad Z = \frac{rf}{f - px - qy} \ . \tag{3.7.3}$$

Suppose the surface is moving rigidly in the scene. As discussed in Sect. 2.6, an instantaneous rigid motion is specified by the *translation velocity* (a, b, c) at an arbitrarily fixed reference point and the *rotation velocity* $(\omega_1, \omega_2, \omega_3)$ around it. We choose, as our reference point, the intersection $(0, 0, r)$ of the Z-axis with the surface (Fig. 3.2). Then, the optical flow induced on the image plane is given as follows:

Proposition 3.1 (*Flow equations*). *The optical flow induced by perspective projection of planar surface motion has the form*

$$u(x, y) = u_0 + Ax + By + (Ex + Fy)x \ ,$$

$$v(x, y) = v_0 + Cx + Dy + (Ex + Fy)y \ , \tag{3.7.4}$$

where

$$u_0 = \frac{fa}{r} \ , \qquad v_0 = \frac{fb}{r} \ ,$$

$$A = p\omega_2 - \frac{pa + c}{r} \ , \qquad B = q\omega_2 - \omega_3 - \frac{qa}{r} \ ,$$

$$C = -p\omega_1 + \omega_3 - \frac{pb}{r} \ , \qquad D = -q\omega_1 - \frac{qb + c}{r} \ ,$$

$$E = \frac{1}{f}\left(\omega_2 + \frac{pc}{r}\right) \ , \qquad F = \frac{1}{f}\left(-\omega_1 + \frac{qc}{r}\right) \ . \tag{3.7.5}$$

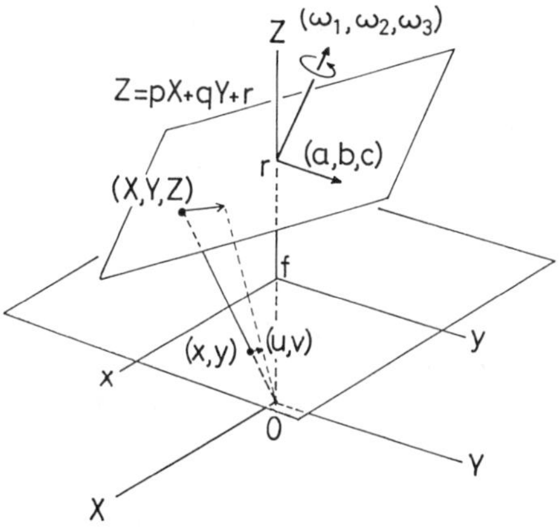

Fig. 3.2. A planar surface whose equation is $Z = pX + qY + r$ is moving with translation velocity (a, b, c) at $(0, 0, r)$ and rotation velocity $(\omega_1, \omega_2, \omega_3)$ around it. An optical flow is induced on the image xy plane by perspective projection from the viewpoint O

Proof. The velocity $(X, \dot{Y}, \dot{Z})$ of point (X, Y, Z) is given by

$$\begin{pmatrix} \dot{X} \\ \dot{Y} \\ \dot{Z} \end{pmatrix} = \begin{pmatrix} a \\ b \\ c \end{pmatrix} + \begin{pmatrix} \omega_1 \\ \omega_2 \\ \omega_3 \end{pmatrix} \times \begin{pmatrix} X \\ Y \\ Z - r \end{pmatrix} . \tag{3.7.6}$$

Substituting the surface equation (3.7.1) into this, we obtain

$$\dot{X} = a + p\omega_2 X + (q\omega_2 - \omega_3)Y , \quad \dot{Y} = b + (\omega_3 - p\omega_1)X - q\omega_1 Y ,$$
$$\dot{Z} = c - \omega_2 X + \omega_1 Y . \tag{3.7.7}$$

Differentiating both sides of (3.7.2), we obtain

$$\dot{x} = \frac{f\dot{X}}{Z} - \frac{fX\dot{Z}}{Z^2} = \frac{f\dot{X}}{Z} - \frac{x\dot{Z}}{Z} , \qquad \dot{y} = \frac{f\dot{Y}}{Z} - \frac{fY\dot{Z}}{Z^2} = \frac{f\dot{Y}}{Z} - \frac{x\dot{Y}}{Z} . \tag{3.7.8}$$

From (3.7.7, 2), we obtain

$$\frac{f\dot{Y}}{Z} = \frac{fb}{Z} + (\omega_3 - p\omega_1)x - q\omega_1 y ,$$

$$\frac{f\dot{X}}{Z} = \frac{fa}{Z} + p\omega_2 x + (q\omega_2 - \omega_3)y ,$$

$$\frac{f\dot{Z}}{Z} = \frac{fc}{Z} - \omega_2 x + \omega_1 y . \tag{3.7.9}$$

Equations (3.7.4, 5) are obtained if we substitute (3.7.9) into (3.7.8) and eliminate Z by (3.7.3c).

Thus, the optical flow is is completely specified by eight parameters u_0, v_0, A, B, C, D, E, F, which we call the *flow parameters*. These parameters are estimated by fitting the flow equations (3.7.4) to the observed optical flow, say by the least squares method (Sect. 2.6). Hence, these eight flow parameters are the "image characteristics". The "object parameters" are p, q, r, a, b, c, ω_1, ω_2, ω_3. Equations (3.7.5a–h) are the "3D recovery equations" of this problem.

Now, what flow would be observed if the camera were oriented differently? One thing is certain: the flow equations still have the same form

$$u' = u'_0 + A'x' + B'y' + (E'x' + F'y')x' \;,$$
$$v' = v'_0 + C'x' + D'y' + (E'x' + F'y')y' \;. \tag{3.7.10}$$

This is because *a planar surface is always a planar surface wherever it is viewed from.* Hence, the form of the optical flow must always be the same.

Suppose the camera is oriented differently by rotation $\boldsymbol{R}$ from the original position. The following questions naturally arise.

1. How are the new flow parameters u'_0, v'_0, A', B', C', D', E', F' expressed in terms of the original flow parameters u_0, v_0, A, B, C, D, E, F and the rotation $\boldsymbol{R}$?
2. In particular, is the transformation from the old flow parameters to the new ones a linear mapping?
3. If the transformation is linear and defines a representation, is it reducible or irreducible?
4. If the representation is irreducible, which representation $\mathscr{D}_l$ is it equivalent to?

All these questions can be answered by analyzing infinitesimal transformations. We show this by a series of Lemmas, Propositions, and Theorems.

Lemma 3.5. *The infinitesimal camera rotation transformation* $x' = x + \delta x$, $y' = x + \delta y$ *for axis* $\boldsymbol{n}$ *and angle* Ω *is given by*

$$\delta x = \Omega[-fn_2 + n_3 y + \frac{1}{f}(-n_2 x + n_1 y)x] + O(\Omega^2) \;,$$
$$\delta y = \Omega[-fn_1 - n_3 y + \frac{1}{f}(-n_2 x + n_1 y)y] + O(\Omega^2) \;. \tag{3.7.11}$$

Proof. From (1.3.5), the camera rotation transformation is given by

$$x = f\frac{r_{11}x' + r_{12}y' + r_{13}f}{r_{31}x' + r_{32}y' + r_{33}f} \;, \qquad y = f\frac{r_{21}x' + r_{22}y' + r_{23}f}{r_{31}x' + r_{32}y' + r_{33}f} \;. \tag{3.7.12}$$

If we substitute

$$R = I + \Omega A(n) + O(\Omega^2) \ , \quad A(n) = \begin{pmatrix} 0 & -n_3 & n_2 \\ n_3 & 0 & -n_1 \\ -n_2 & n_1 & 0 \end{pmatrix} , \tag{3.7.13}$$

and apply a Taylor expansion to the right-hand sides of (3.7.12), we obtain $x' = x + \delta x$, $y' = y + \delta y$, where δx, δy are given by (3.7.11).

Lemma 3.6. *The infinitesimal transformation of the flow parameters u_0, v_0, A, B, C, D, E, F is given by $u_0' = u_0 + \delta u_0, \ldots, F' = F + \delta F$ with*

$$\delta \begin{pmatrix} u_0 \\ v_0 \\ A \\ B \\ C \\ D \\ E \\ F \end{pmatrix} = \Omega$$

$$\times \begin{pmatrix} & n_3 & n_2 & -fn_1 & & & & \\ -n_3 & & & & fn_2 & -fn_1 & & \\ -2n_2/f & n_1/f & & n_3 & n_3 & & 2fn_2 & -fn_1 \\ n_1 & & -n_3 & & & n_3 & & fn_2 \\ & -n_2/f & -n_3 & & & n_3 & -fn_1 & \\ -n_2/f & 2n_1/f & & -n_3 & -n_3 & & fn_2 & -2fn_1 \\ & & -n_2/f & & n_1/f & & & n_3 \\ & & & -n_2/f & & n_1/f & -n_3 & \end{pmatrix}$$

$$\times \begin{pmatrix} u_0 \\ v_0 \\ A \\ B \\ C \\ D \\ E \\ F \end{pmatrix} + O(\Omega^2) \ . \tag{3.7.14}$$

Proof. Under transformation $x \to x'$, $y \to y'$ the velocity components (u, v) are transformed as

$$\begin{pmatrix} u' \\ v' \end{pmatrix} = \begin{pmatrix} \partial x'/\partial x & \partial x'/\partial y \\ \partial y'/\partial x & \partial y'/dy \end{pmatrix} \begin{pmatrix} u \\ v \end{pmatrix} . \tag{3.7.15}$$

(The transformation matrix is called the *Jacobi matrix*.) If we substitute $x' = x + \delta x$, $y' = y + \delta y$ and use (3.7.11) and the flow equations (3.7.4), we obtain $u' = u + \delta u$, $v' = v + \delta v$ with

$$\begin{aligned}
\delta u = \Omega\Bigg[& n_3 v_0 + \left(-\frac{1}{f}(2n_2 u_0 - n_1 v_0) + n_3 C \right) x + \left(\frac{1}{f} n_1 u_0 + u_3 D \right) y \\
& - \frac{1}{f}(2n_2 A - n_1 C)x^2 - \left(\frac{1}{f}(n_1 A - 2n_2 B + n_1 D) + n_3 E \right) xy \\
& + \left(\frac{1}{f} n_1 B + n_3 F \right) y^2 - \frac{2}{f} n_2 E x^3 + \frac{2}{f}(n_1 E - n_2 F) x^2 y \\
& + \frac{2}{f} n_1 F x y^2 \Bigg] + \mathrm{O}(\Omega^2) , \\
\delta v = \Omega\Bigg[& - n_3 u_0 - \left(\frac{1}{f} n_2 v_0 + n_3 A \right) x \\
& + \left(-\frac{1}{f}(n_2 u_0 - 2n_1 v_0) - n_3 B \right) y - \left(\frac{1}{f} n_2 C + n_3 E \right) x^2 \\
& + \left(-\frac{1}{f}(n_2 A - 2n_1 C + n_2 D) - n_3 E \right) xy \\
& - \frac{1}{f}(n_2 B - 2n_1 D) y^2 - \frac{2}{f} n_2 E x^2 y + \frac{2}{f}(n_1 E - n_2 F) x y^2 \\
& + \frac{2}{f} n_1 F y^3 \Bigg] + \mathrm{O}(\Omega^2) .
\end{aligned} \tag{3.7.16}$$

We know that the new optical flow must have the form of (3.7.10). If we substitute $u' = u + \delta u$, $v' = v + \delta v$ into the left-hand sides, and $x' = x + \delta x$, $y' = y + \delta y$, $u'_0 = u_0 + \delta u_0, \ldots, F' = F + \delta F$ into the right-hand sides, we obtain

$$\begin{aligned}
\delta u &= \delta u_0 + \delta A x + A \delta x + \delta B y + B \delta y + \delta E x^2 + 2 E x \delta x + \delta F x y \\
&\quad + F y \delta x + F x \delta y + \mathrm{O}(\Omega^2) , \\
\delta v &= \delta v_0 + \delta C_x + C \delta x + \delta D y + D \delta y + \delta E x y + E y \delta x + E x \delta y \\
&\quad + \delta F y^2 + 2 F y \delta y + \mathrm{O}(\Omega^2) .
\end{aligned} \tag{3.7.17}$$

Substituting (3.7.11), we obtain

$$\begin{aligned}
\delta u &= \delta u_0 - f\Omega(n_2 A - n_1 B) + (\delta A - \Omega n_3 B - f(2n_2 E - n_1 F))x \\
&\quad + (\delta B + \Omega n_3 A - f\Omega n_2 F)y + \left(\delta E - \frac{1}{f}\Omega n_2 A - \Omega n_3 F \right) x^2
\end{aligned}$$

$$+\left(\delta F+\frac{1}{f}\Omega(n_1A-n_2B)+2\Omega n_3E\right)xy+\left(\frac{1}{f}\Omega n_1B+\Omega n_3F\right)y^2$$

$$-\frac{2}{f}\Omega n_2Ex^3+\frac{2}{f}\Omega(n_1E-n_2F)x^2y+\frac{2}{f}\Omega n_1Fxy^2+O(\Omega^2)\ ,$$

$$\delta v=\delta v_0-f\Omega(n_2C-n_1D)+(\delta C-\Omega n_3D+f\Omega n_1E)x$$

$$+(\delta D+\Omega n_3C-f(n_2E-2n_1F))y-\left(\frac{1}{f}\Omega n_2C+\Omega n_3E\right)x^2$$

$$-\left(\delta E+\frac{1}{f}\Omega(n_1C-n_2D)-2\Omega n_3F\right)xy$$

$$+\left(\delta F+\frac{1}{f}\Omega n_1D+\Omega n_3E\right)y^2-\frac{2}{f}\Omega n_2Ex^2y$$

$$+\frac{2}{f}\Omega(n_1E-n_2F)xy^2+\frac{2}{f}\Omega n_1Fy^3+O(\Omega^2)\ . \qquad (3.7.18)$$

Comparing (3.7.16) and (3.7.18), we obtain

$$\delta u_0=\Omega[n_3v_0+fn_2A-fn_1B]+O(\Omega^2)\ .$$

$$\delta v_0=\Omega[-n_3u_0+fn_2C-fn_1D]+O(\Omega^2)\ .$$

$$\delta A=\Omega\left[-\frac{2}{f}n_2u_0+\frac{1}{f}n_1v_0+n_3B+n_3C+2fn_2E-fn_1F\right]$$

$$+O(\Omega^2)\ ,$$

$$\delta B=\Omega\left[\frac{1}{f}n_1u_0-n_3A+n_3D+fn_2F\right]+O(\Omega^2)\ ,$$

$$\delta C=\Omega\left[-\frac{1}{f}n_2v_0-n_3A+n_3D-fn_1E\right]+O(\Omega^2)\ ,$$

$$\delta D=\Omega\left[-\frac{1}{f}n_2u_0+\frac{2}{f}n_1v_0-n_3B-n_3C+fn_2E-2fn_1F\right]+O(\Omega^2)\ ,$$

$$\delta E=\Omega\left[-\frac{1}{f}n_2A+\frac{1}{f}n_1C+n_3F\right]+O(\Omega^2)\ ,$$

$$\delta F=\Omega\left[-\frac{1}{f}n_2B+\frac{1}{f}n_1D-n_3E\right]+O(\Omega^2)\ , \qquad (3.7.19)$$

from which follows (3.7.12).

Proposition 3.2. *The flow parameters u_0, v_0, A, B, C, D, E, F are transformed linearly by the camera rotation transformation, and define a representation of $SO(3)$:*

$$\begin{pmatrix} u_0' \\ \vdots \\ F' \end{pmatrix} = \boldsymbol{T}(\boldsymbol{R}) \begin{pmatrix} u_0 \\ \vdots \\ F \end{pmatrix} . \tag{3.7.20}$$

Proof. Equations (3.7.15) is linear in the flow parameters. Hence, it defines a global linear transformation by integration. Equation (3.7.15) is rewritten as

$$\delta \begin{pmatrix} u_0 \\ \vdots \\ F \end{pmatrix} = \Omega\,(n_1 \boldsymbol{A}_1 + n_2 \boldsymbol{A}_2 + n_3 \boldsymbol{A}_3) \begin{pmatrix} u_0 \\ \vdots \\ F \end{pmatrix} + \mathrm{O}(\Omega^2) , \tag{3.7.21}$$

where the infinitesimal generators $\{\boldsymbol{A}_1, \boldsymbol{A}_2, \boldsymbol{A}_3\}$ are given by

$$\boldsymbol{A}_1 = \begin{pmatrix} & & & -f & & & & \\ & & & & & -f & & \\ & 1/f & & & & & & -f \\ 1/f & & & & & & & \\ & & & & & & -f & \\ & 2f & & & & & & -2f \\ & & & & 1/f & & & \\ & & & & & 1/f & & \end{pmatrix},$$

$$\boldsymbol{A}_2 = \begin{pmatrix} & & f & & & & & \\ & & & & & f & & \\ -2/f & & & & & & 2f & \\ & & & & & & & f \\ & -1/f & & & & & & \\ -1/f & & & & & & f & \\ & & -1/f & & & & & \\ & & & -1/f & & & & \end{pmatrix},$$

$$\boldsymbol{A}_3 = \begin{pmatrix} & 1 & & & & & & \\ -1 & & & & & & & \\ & & & 1 & 1 & & & \\ & & -1 & & & -1 & & \\ & & -1 & & & -1 & & \\ & & & -1 & -1 & & & \\ & & & & & & & 1 \\ & & & & & & -1 & \end{pmatrix}, \tag{3.7.22}$$

It is easy to see that they satisfy the commutation relations

$$[A_1, A_2] = -A_3 \ , \quad [A_2, A_3] = -A_1 \ , \ [A_3, A_1] = -A_2 \ . \tag{3.7.23}$$

Proposition 3.3. *The representation $\boldsymbol{T}(\boldsymbol{R})$ defined by the flow parameters u_0, v_0, A, B, C, D, E, F is equivalent to the direct sum of the irreducible representations $\mathscr{D}_1$ and $\mathscr{D}_2$:*

$$\boldsymbol{T}(\boldsymbol{R}) \cong \mathscr{D}_1 \oplus \mathscr{D}_2 \ . \tag{3.7.24}$$

Proof. If we use the infinitesimal generators $\{A_1, A_2, A_3\}$ of (3.7.22), the Casimir operator $\boldsymbol{H}$ of (3.3.7) becomes

$$\boldsymbol{H} = \begin{pmatrix} 4 & & & & & & -2f^2 & \\ & 4 & & & & & & -2f \\ & & 6 & & & & & \\ & & & 4 & 2 & & & \\ & & & 2 & 4 & & & \\ & & & & & 6 & & \\ -2/f^2 & & & & & & 4 & \\ & -2/f^2 & & & & & & 4 \end{pmatrix} . \tag{3.7.25}$$

The characteristic polynomial of matrix $\boldsymbol{H}$ is

$$\det(\lambda \boldsymbol{I} - \boldsymbol{H}) = (\lambda - 2)^3 (\lambda - 6)^5 \ , \tag{3.7.26}$$

where $\boldsymbol{I}$ is the unit matrix of dimension 8. Thus, the eigenvalues of matrix $\boldsymbol{H}$ are 2 (multiplicity 3) and 6 (multiplicity 5). This means that the eight-dimensional representation space $\mathbb{R}^8$ is resolved into two eigenspaces of dimensions 3 and 5, in which the restriction of $\boldsymbol{H}$ acts as multiplication by 2 and 6, respectively. Since $1(1+1) = 2$, $2 \times 1 + 1 = 3$, and $2(2+1) = 6$, $2 \times 2 + 1 = 5$, the representation $\boldsymbol{T}(\boldsymbol{R})$ of (3.7.20) is equivalent to the direct sum $\mathscr{D}_1 \oplus \mathscr{D}_2$ (Theorem 3.3 and Corollary 3.1).

Theorem 3.10. *If parameters a_1, a_2, a_3 are defined by*

$$a_1 \equiv -\frac{1}{2}(v_0/f + fF) \ , \ a_2 \equiv \frac{1}{2}(u_0/f + fE) \ , \ a_3 \equiv \frac{1}{2}(C - B) \ , \tag{3.7.27}$$

they are transformed as a vector, and if we define

$$b_{11} \equiv \frac{1}{3}(2A - D) \ , \qquad b_{22} \equiv \frac{1}{3}(2D - A) \ , \qquad b_{33} \equiv -\frac{1}{3}(A + D) \ ,$$

$$b_{12} = b_{21} \equiv \frac{1}{2}(B + C) \ , \quad b_{23} = b_{32} \equiv \frac{1}{2}(v_0/f - fF) \ ,$$

$$b_{31} = b_{13} \equiv \frac{1}{2}(u_0/f - fE) \ , \tag{3.7.28}$$

they are transformed as a deviator tensor. Namely,

$$\begin{pmatrix} a'_1 \\ a'_2 \\ a'_3 \end{pmatrix} = \boldsymbol{R}^{\mathrm{T}} \begin{pmatrix} a_1 \\ a_2 \\ a_3 \end{pmatrix} \ , \tag{3.7.29}$$

$$\begin{pmatrix} b'_{11} & b'_{12} & b'_{13} \\ b'_{11} & b'_{12} & b'_{13} \\ b'_{11} & b'_{12} & b'_{13} \end{pmatrix} = \boldsymbol{R}^{\mathrm{T}} \begin{pmatrix} b_{11} & b_{12} & b_{13} \\ b_{11} & b_{12} & b_{13} \\ b_{11} & b_{12} & b_{13} \end{pmatrix} \begin{pmatrix} \boldsymbol{R} \end{pmatrix} \ . \tag{3.7.30}$$

Proof. Take the following set of vectors as a basis of the eight-dimensional representation space $\mathbb{R}^8$:

$$\begin{pmatrix} \\ -f \\ \\ \\ \\ \\ \\ -1/f \end{pmatrix} \begin{pmatrix} f \\ \\ \\ \\ \\ \\ 1/f \\ \\ \end{pmatrix} \begin{pmatrix} \\ \\ \\ -1 \\ 1 \\ \\ \\ \\ \end{pmatrix}$$

$$\begin{pmatrix} \\ \\ 2 \\ \\ \\ 1 \\ \\ \\ \end{pmatrix} \begin{pmatrix} \\ \\ 1 \\ \\ \\ 2 \\ \\ \\ \end{pmatrix} \begin{pmatrix} \\ \\ \\ 1 \\ 1 \\ \\ \\ \\ \end{pmatrix} \begin{pmatrix} f \\ \\ \\ \\ \\ \\ -1/f \\ \\ \end{pmatrix} \begin{pmatrix} \\ f \\ \\ \\ \\ \\ \\ -1/f \end{pmatrix} . \tag{3.7.31}$$

With respect to this basis, the Casimir operator is diagonalized as

$$
\boldsymbol{H}' = \begin{pmatrix} 2 &&&&&&& \\ & 2 &&&&&& \\ && 2 &&&&& \\ &&& 6 &&&& \\ &&&& 6 &&& \\ &&&&& 6 && \\ &&&&&& 6 & \\ &&&&&&& 6 \end{pmatrix}. \tag{3.7.32}
$$

With respect to this basis, the vector constructed from the flow parameters $u_0, \ldots, F$ is expressed as

$$
\begin{pmatrix} u_0 \\ v_0 \\ A \\ B \\ C \\ D \\ E \\ F \end{pmatrix} = \frac{v_0/f + fF}{2} \begin{pmatrix} \\ -f \\ \\ \\ \\ \\ \\ -1/f \end{pmatrix} + \frac{u_0/f + fE}{2} \begin{pmatrix} f \\ \\ \\ \\ \\ \\ 1/f \\ \\ \end{pmatrix}
$$

$$
+ \frac{C - B}{2} \begin{pmatrix} \\ \\ \\ -1 \\ 1 \\ \\ \\ \\ \end{pmatrix} + \frac{2A - D}{3} \begin{pmatrix} \\ \\ 2 \\ \\ \\ 1 \\ \\ \\ \end{pmatrix} + \frac{2D - A}{3} \begin{pmatrix} \\ \\ 1 \\ \\ \\ 2 \\ \\ \\ \end{pmatrix}
$$

$$
+ \frac{B + C}{2} \begin{pmatrix} f \\ \\ \\ \\ \\ \\ -1/f \\ \\ \end{pmatrix} + \frac{u_0/f - fE}{2} \begin{pmatrix} f \\ \\ \\ \\ \\ \\ -1/f \\ \\ \end{pmatrix} + \frac{v_0/f - fF}{2} \begin{pmatrix} \\ f \\ \\ \\ \\ \\ \\ -1/f \end{pmatrix}. \tag{3.7.33}
$$

Hence, the new components are given by a_1, a_2, a_3 of (3.7.27), and $b_{11}, b_{22}, b_{12}, b_{31}, b_{32}$, of (3.7.28). In terms of these, (3.7.14) is rewritten as

$$\begin{pmatrix} a_1 \\ a_2 \\ a_3 \\ b_{11} \\ b_{22} \\ b_{13} \\ b_{31} \\ b_{32} \end{pmatrix} = \Omega$$

$$\times \left(\begin{array}{ccc|ccccc} & n_3 & -n_2 & & & & & \\ -n_3 & & n_1 & & & & & \\ n_2 & -n_1 & & & & & & \\ \hline & & & & & 2n_3 & -2n_2 & \\ & & & & & -2n_3 & & 2n_1 \\ & & & -n_3 & n_3 & & n_1 & -n_2 \\ & & & 2n_2 & n_2 & -n_1 & & n_3 \\ & & & -n_1 & -2n_1 & n_2 & -n_3 & \end{array}\right)$$

$$\times \begin{pmatrix} a_1 \\ a_2 \\ a_3 \\ b_{11} \\ b_{22} \\ b_{13} \\ b_{31} \\ b_{32} \end{pmatrix} + O(\Omega^2). \tag{3.7.34}$$

The coefficient matrix is written as $n_1 A'_1 + n_2 A'_2 + n_3 A'_3$, where the infinitesimal generators $\{A'_1, A'_2, A'_3\}$ are given by

$$A'_1 = \left(\begin{array}{ccc|ccccc} & & & & & & & \\ & & 1 & & & & & \\ & -1 & & & & & & \\ \hline & & & & & & & \\ & & & & & & & 2 \\ & & & & & & 1 & \\ & & & & & -1 & & \\ & & & -1 & -2 & & & \end{array}\right),$$

$$
A'_2 = \left(\begin{array}{ccc|ccccc}
 & & -1 & & & & & \\
 & & & & & & & \\
1 & & & & & & & \\
\hline
 & & & & & & -2 & \\
 & & & & & & & \\
 & & & & & & & -1 \\
 & & & 2 & 1 & & & \\
 & & & & & 1 & &
\end{array}\right) ,
$$

$$
A'_3 = \left(\begin{array}{ccc|ccccc}
 & 1 & & & & & & \\
-1 & & & & & & & \\
 & & & & & & & \\
\hline
 & & & & & 2 & & \\
 & & & & & -2 & & \\
 & & & -1 & 1 & & & \\
 & & & & & & & 1 \\
 & & & & & & -1 &
\end{array}\right) . \tag{3.7.35}
$$

Since these infinitesimal generators are the direct sums of those for a vector (3.4.4) and those for a tensor (3.5.16), parameters $\{a_1, a_2, a_3\}$ are transformed as a vector and b_{ij}, $i, j = 1, 2, 3$, are transformed as a deviator tensor (the rest of the elements are defined so that (b_{ij}) becomes a traceless symmetric tensor).

Hence, we can conclude:

Corollary 3.2 (*Transformation rule of flow parameters*). *The flow parameters* u_0, v_0, A, B, C, D, E, F *are transformed by the camera rotation as follows. First, compute* a_i, $i = 1, 2, 3$, *by* (3.7.27), *and* b_{ij}, $i, j = 1, 2, 3$, *by* (3.7.28). *Apply the vector transformation rule* (3.7.29) *to* a_i, $i = 1, 2, 3$, *and tensor transformation rule* (3.7.30) *to* b_{ij}, $i, j = 1, 2, 3$. *The transformed values are given by*

$$
\begin{aligned}
u'_0 &= f(b'_{13} + a'_2) , & v'_0 &= f(b'_{23} - a'_1) , \\
A' &= b'_{11} - b'_{33} , & B' &= b'_{12} - a'_2 , \\
C' &= b'_{21} + a'_3 , & D' &= b'_{22} - b'_{33} ,
\end{aligned}
$$

$$
E' = -\frac{1}{f}(b'_{31} - a'_2) , \quad F' = -\frac{1}{f}(b'_{32} + a'_1) . \tag{3.7.36}
$$

Thus, we have established the transformation rule of the flow parameters by analyzing infinitesimal transformations. Group theoretical considerations played an essential role in this analysis. However, the merit of group theoretical analysis is not limited to constructing the transformation rule; we can also see various invariant properties. We will discuss this in the next chapter in more detail. Here, we point out one immediate consequence in the following section.

3.7.1 Invariant Decomposition of Optical Flow

In terms of the parameters a_i, $i = 1, 2, 3$ and b_{ij}, $i, j = 1, 2, 3$, the flow equation (3.7.4) are rewritten as

$$u = f(b_{13} + a_2) + (b_{11} - b_{33})x + (b_{12} - a_3)y - \frac{1}{f}[(b_{31} - a_2)x + (b_{32} + a_1)y]x \ ,$$

$$v = f(b_{23} - a_1) + (b_{21} + a_3)x + (b_{22} - b_{33})y - \frac{1}{f}[(b_{31} - a_2)x + (b_{32} + a_1)y]y \ . \tag{3.7.37}$$

This means that the optical flow is decomposed into two parts $u = u_a + u_b$, $v = v_a + v_b$, where

$$u_a = fa_2 - a_3y + \frac{1}{f}(a_2x - a_1y)x \ ,$$

$$v_a = -fa_1 + a_3x + \frac{1}{f}(a_2x - a_1y)y \ , \tag{3.7.38}$$

$$u_b = fb_{31} + (b_{11} - b_{33})x + b_{12}y + \frac{1}{f}(-b_{31}x - b_{23}y)x \ ,$$

$$v_b = fb_{32} + b_{12}x + (b_{22} - b_{33})y + \frac{1}{f}(-b_{31}x - b_{23}y)y \ . \tag{3.7.39}$$

This decomposition is unique because a_i, $i = 1, 2, 3$, and b_{ij}, $i, j = 1, 2, 3$, are computed from the original flow parameters u_0, v_0, A, B, C, D, E, F by (3.7.27, 28). Moreover, this decomposition is *invariant*: each flow is transformed independently of the other (Theorem 3.10). At the same time, this decomposition is *irreducible*: no further decomposition is possible, since the transformation rules (3.7.29, 30) define irreducible representations $\mathscr{D}_1$ and $\mathscr{D}_2$. Hence, according to Weyl's thesis, each of these flows should have some invariant meaning. Let us call the flow (u_a, v_a) the *vector part* and the flow (u_b, v_b) the *tensor part.*

Suppose the entire scene, including all the objects in it, is stationary, while the camera is rotating around the center of the lens with rotation velocity $(\omega_1, \omega_2, \omega_3)$. Then, what optical flow do we observe on the image plane? As discussed in Sect. 3.3, the rotations velocity is obtained by dividing the rotation vector $(\Omega n_1, \Omega n_2, \Omega n_3)$ by the time Δt the rotation took and taking the limit $\Delta t \to 0$. Dividing both sides of (3.7.11) by Δt, and taking the limit $\Delta t \to 0$, we obtain

$$u = -f\omega_2 + \omega_3y + \frac{1}{f}(-\omega_2x + \omega_1y)x \ ,$$

$$v = -f\omega_1 - \omega_3x + \frac{1}{f}(-\omega_2x + \omega_1y)y \ . \tag{3.7.40}$$

Comparing these equations with (3.7.38), we see that the vector part (u_a, v_a) is the flow we will observe if the camera is rotated with rotation velocity $(-a_1, -a_2, -a_3)$ relative to a stationary scene, or equivalently if the planar surface (or any other object or scene) is *orbiting* with rotation velocity (a_1, a_2, a_3) around the center of the lens with the configuration relative to the camera kept fixed. This means that *the vector part does not contain any information about the 3D structure and motion of the surface*; the information is contained in the tensor part (u_b, v_b). The decomposition of the flow into two parts as (3.7.38, 39) is simply the separation of the component due to the orbiting motion from the remaining components.

This observation implies that *3D recovery from optical flow is performed from the tensor part alone*. We will show the analytical solution of this problem in Chap. 7. As we will see there, the solution is essentialy determined from the tensor part alone. Namely, suppose we compute the gradient (p, q), the translation velocity $(a/r, b/r, c/r)$ scaled by the depth r, and the rotation velocity $(\omega_1, \omega_2, \omega_3)$ by assuming that the vector part is zero. Then, addition of a nonzero vector part only modifies the solution in the form

$$a/r \rightarrow a/r + a_2 \ , \qquad b/r \rightarrow b/r - a_1 \ , \tag{3.7.41}$$

$$\omega_1 \rightarrow \omega_1 + a_1 \ , \qquad \omega_2 \rightarrow \omega_2 + a_2 \ , \qquad \omega_3 \rightarrow \omega_3 + a_3 \ . \tag{3.7.42}$$

In other words, the vector part simply adds the effect of the orbiting motion around the center of the lens; the surface gradient (p, q) and the scaled translation velocity component c/r are not affected.

Finally, note that if the camera is rotating with some angular velocity, the flow of (3.7.40) is superimposed onto the original optical flow. However, if we decompose the flow into its vector and tensor parts, the tensor part is not altered at all; the superimposition only adds to the vector part.

Exercises

3.1 Show that the commutator defined by (3.2.9) satisfies anticommutativity (3.2.10), linearity (3.2.11), and the Jacobi identity (3.2.12).

3.2 Derive equations that the structure constants c_{ij}^k (defined by (3.2.15)) must satisfy so that anticommutativity (3.2.10) and the Jacobi identity (3.2.12) are always satisfied.

3.3* A matrix series $\sum_{k=0}^{\infty} A_k$ is said to *absolutely converge* if $\sum_{k=0}^{\infty} |A_k|$ converges, where

$$|A| \equiv \max_i \sum_{j=1}^{n} |a_{ij}|$$

for a matrix $A = (a_{ij})$, $i, j = 1, \ldots, n$. Prove that

$$e^A \equiv \sum_{k=0}^{\infty} A^k/k!$$

absolutely converges for any square matrix A. *Hint*: First, prove $|AB| \leqq |A| \cdot |B|$ for any square matrices A, B. Then, using the fact that the exponential function $e^x \equiv \sum_{k=0}^{\infty} x^k/k!$ converges, show that the sequence $S_n = \sum_{k=0}^{n} |A^k|/k!$, $n = 0, 1, 2, \ldots$ is a *Cauchy* (or *fundamental*) *sequence*.

3.4* Consider simultaneous linear ordinary differential equations $d\boldsymbol{x}/dt = A\boldsymbol{x}$, where $\boldsymbol{x}$ is an n-dimensional column vector and A is an n-dimensional constant square matrix. Show that the solution that satisfies the initial condition $\boldsymbol{x}(0) = \boldsymbol{x}$ is given by $\boldsymbol{x}(t) = e^{At}\boldsymbol{x}_0$.

3.5* Prove (3.2.25), i.e., prove that the Casimir operator $\boldsymbol{H}$ defined by (3.2.23) commutes with the infinitesimal operators $\{A_1, A_2, A_3\}$.

3.6 Derive (3.3.12–14) from (3.3.11).

3.7 Prove (3.3.20, 21).

3.8 Derive (3.4.4), and show that the infinitesimal generators $\{A_1, A_2, A_3\}$ of (3.4.5) satisfy the commutation relations of (3.4.6).

3.9 Show that if we use the vector product notation, (3.4.4) is written as

$$\delta \begin{pmatrix} a_1 \\ a_2 \\ a_3 \end{pmatrix} = -\Omega \begin{pmatrix} n_1 \\ n_2 \\ n_3 \end{pmatrix} \times \begin{pmatrix} a_1 \\ a_2 \\ a_3 \end{pmatrix} + O(\Omega^2) .$$

3.10 Show that for the infinitesimal generators $\{A_1, A_2, A_3\}$ of (3.4.5), the Casimir operator $\boldsymbol{H}$ takes the form of (3.4.7).

3.11 Check directly from the tensor transformation rule (3.5.1) that the anti-symmetric components $(a_{23} - a_{32})/2$, $(a_{31} - a_{13})/2$, $(a_{12} - a_{21})/2$ of a tensor a_{ij}, $i, j = 1, 2, 3$, are transformed as a vector as shown in (3.5.3).

3.12 Prove (3.5.6–9).

3.13 Check that the trace $a_{11} + a_{22} + a_{33}$ of a symmetric tensor a_{ij}, $i, j = 1, 2, 3$, is a scalar.

3.14 Show that the infinitesimal generators $\{A_1, A_2, A_3\}$ of (3.5.16) satisfy the commutation relations of (3.5.17).

3.15 Prove (3.5.15, 16).

3.16 Show that for the infinitesimal generators $\{A_1, A_2, A_3\}$ of (3.5.16), the Casimir operator $\boldsymbol{H}$ takes the form of (3.5.17).

3.17 Prove (3.6.4).

3.18 Prove (3.6.7).

3.19 Derive (3.7.3).

3.20 Derive (3.7.4, 5) from (3.7.8, 9, 3c).

3.21 Derive (3.7.11).

3.22* Prove (3.7.15).

3.23 Derive (3.7.16).

3.24 Derive (3.7.17, 18).

3.25 Confirm that (3.7.19), and hence (3.7.14), holds.

3.26 Show that the infinitesimal generators $\{A_1, A_2, A_3\}$ of (3.7.22) satisfy the commutation relations of (3.7.23).

3.27 Show that for the infinitesimal generators $\{\boldsymbol{A}_1, \boldsymbol{A}_2, \boldsymbol{A}_3\}$ of (3.7.22), the Casimir operator $\boldsymbol{H}$ takes the form of (3.7.25).

3.28 Derive the characteristics polynomial of (3.7.26).

3.29 Check that vectors (3.7.31) are eigenvectors of the Casimir operator $\boldsymbol{H}$ of (3.7.25) and that they diagonalize it in the form of (3.7.32).

3.30 Prove (3.7.33, 34).

3.31 Derive (3.7.36, 37).

3.32 (a) Show that the matrix

$$\begin{pmatrix} & -a_3 & a_2 \\ a_3 & & -a_1 \\ -a_2 & a_1 & \end{pmatrix}$$

constructed from a vector $\{a_1, a_2, a_3\}$ is a tensor.

(b) Let (a_{ij}), $i, j = 1, 2, 3$, be the matrix constructed from a_1, a_2, a_3 of (3.7.27) as shown in (a). Let $c_{ij} = a_{ij} + b_{ij}$, $i, j = 1, 2, 3$, where b_{ij}, $i, j = 1, 2, 3$, are defined by (3.7.28). Show that

$$c_{11} = \frac{1}{3}(2A - D)\ , \qquad c_{12} = B\ , \qquad c_{13} = \frac{1}{f}u_0\ ,$$

$$c_{21} = C\ , \qquad c_{22} = \frac{1}{3}(2D\text{-}A)\ , \qquad c_{23} = -fF\ ,$$

$$c_{31} = -fE\ , \qquad c_{32} = \frac{1}{f}v_0\ , \qquad c_{33} = -\frac{1}{3}(A + D)\ ,$$

and prove that $C = (c_{ij})$, $i, j = 1, 2, 3$, is also a tensor.

(c) Show that

$$u_0 = fc_{13}\ , \qquad v_0 = fc_{32}\ , \qquad A = c_{11} - c_{33}\ , \qquad B = c_{12}\ ,$$

$$C = c_{21}\ , \quad D = c_{22} - c_{33}\ , \quad E = -\frac{1}{f}c_{31}\ , \quad F = -\frac{1}{f}c_{23}\ ,$$

and hence the optical flow of (3.7.4) is expressed in terms of c_{ij}, $i, j = 1, 2, 3$, as

$$u = fc_{13} + (c_{11} - c_{33})x + c_{12}y - \frac{1}{f}(c_{31}x + c_{32}y)x\ ,$$

$$v = fc_{23} + c_{21}x + (c_{22} - c_{33})y - \frac{1}{f}(c_{31}x + c_{32}y)y\ ,$$

(e) Argue that the optical flow of (3.7.4) is represented by a single tensor C of trace 0 and that the invariant decomposition into its vector part (3.7.38) and tensor part (3.7.39) corresponds to the decomposition of tensor C into its antisymmetric and symmetric parts.

3.33 Derive (3.7.40) from (3.7.11) as explained in the text.

3.34 (a) If the camera is rotated around the center of the lens with rotation velocity $(\omega_1, \omega_2, \omega_3)$ relative to a stationary scene, all the points in the scene look as if they are rotating around the center of the lens with a rotation velocity $(-\omega_1, -\omega_2, -\omega_3)$ relative to the camera. Show that a point (X, Y, Z) has the instantaneous velocity

$$\begin{pmatrix} \dot{X} \\ \dot{Y} \\ \dot{Z} \end{pmatrix} = -\begin{pmatrix} \omega_1 \\ \omega_2 \\ \omega_3 \end{pmatrix} \times \begin{pmatrix} X \\ Y \\ Z \end{pmatrix}$$

relative to the camera. See Exercise 3.9.

(b) Derive (3.7.40) directly by substituting the above equation into (3.7.8).

4. Algebraic Invariance of Image Characteristics

In the preceding chapter, we considered image characteristics that were transformed linearly under camera rotation. Such linear transformations defined representations of $SO(3)$. By taking linear combinations, we rearranged such image characteristics into groups such that each had independent transformation properties. In mathematical terms, this process is the *reduction* of the representation. In this chapter, we remove the restriction of linearity. We consider image characteristics whose new values are algebraic expressions in the original values. By taking algebraic combinations, we rearrange them into groups such that each has independent transformation properties. Then, we construct algebraic expressions that do not change their values under camera rotation. Such expressions are called *scalar invariants*. We will also show that if two images depict one and the same scene viewed from two different camera angles, the camera rotation that transforms one image into the other can be reconstructed from a small number of image characteristics.

4.1 Algebraic Invariants and Irreducibility

In the preceding chapter, we considered image characteristics $c_1, \ldots, c_m$ that were *linearly* transformed into new values $c'_1, \ldots, c'_m$ under camera rotation; each c'_i was a *linear* combination of $c_1, \ldots, c_m$. Then, we *linearly* rearranged $c_1, \ldots, c_m$ into new parameters $C_1, \ldots, C_m$ so that they could be split into groups and each had independent transformation properties.

If image characteristics $c_1, \ldots, c_m$ are transformed *linearly* under the camera rotation, the linear transformation defines a *representation* of $SO(3)$. In mathematical terms, the above rearrangement process is the *reduction* of the representation. This observation provides us with a powerful tool: we need only analyze *infinitesimal transformations*. In Chap. 3, we introduced the notion of *Lie algebra*, whose mathematical structure is determined by the *structure constants* (or equivalently the *commutation relations*). We also showed that we can attain *irreducible representations* by computing the *Casimir operator* from *infinitesimal generators*. In particular, we closely examined image characteristics that are transformed as a *scalar*, a *vector*, and a *tensor*. We demonstrated our technique by examining the transformation properties of the optical flow parameters.

However, the geometrical meaning of the irreducible decomposition was not so very clear. For example, we do not know what kind of image properties are transformed as vectors or tensors. Such questions cannot be fully answered as long as we deal only with *linear* transformations. Recall our original aim. Our motivation was the observation that if given image characteristics can be rearranged into groups such that each of them has independent transformation properties, the resulting groups of parameters can be regarded as indicating separate properties of the image. We termed this philosophy *Weyl's thesis.* However, the rearrangement need not necessarily be restricted to linear operations; we may as well take arbitrary algebraic expressions.

In this chapter, we remove the restriction of "linearity". We consider image characteristics $c_1, \ldots, c_m$ that are transformed under camera rotation into new values $c'_1, \ldots, c'_m$ such that c'_i is an *algebraic* expression in $c_1, \ldots, c_m$. By taking *algebraic* combinations, we rearrange such image characteristics $c_1, \ldots, c_m$ into new parameters $C_1, \ldots, C_m$ so that they split into groups and each has independent transformation properties.[1] This process enables us to understand the geometrical meanings of the image characteristics in more detail.

However, we will lose a powerful mathematical tool if the operation is extended to arbitrary algebraic manipulations: we can no longer rely on group representation theory. In particular, we cannot easily construct global transformations from their infinitesimal transformations. However, we can use the results of the preceding chapter as an intermediate step, as we will show.

Consider a set of image characteristics $\{c_1, \ldots, c_m\}$, and let $\{c'_1, \ldots, c'_m\}$ be their new values after the camera rotation transformation. Let R be the matrix of the camera rotation. We say that a set of image characteristics $\{c_1, \ldots, c_m\}$ is *invariant* if the new values $\{c'_1, \ldots, c'_m\}$ are completely determined by the original values $\{c_1, \ldots, c_m\}$ and the camera rotation R. For instance, the image coordinates (x, y) of a particular feature point (the brightest spot, for example) is an invariant set, but the x-coordinate (or the y-coordinate) alone is not invariant; the new value x' (or y') depends on both x and y. Since, an invariant set of image characteristics is "closed" under the camera rotation, we can regard it as indicating some aspects of the image inherent to the scene and independent of the camera orientation.

Let $\{c_1, \ldots, c_m\}$ be an invariant set of image characteristics. We say that this set is *reducible* if it splits, after an appropriate (generally nonlinear) rearrangement, into two or more sets such that each is itself an invariant set. We say that the set of image characteristics is *irreducible* if no further reduction is possible. As we argued in the preceding chapters, we may regard an irreducible set of image characteristics as indicating a "single" property of the scene, while a reducible set can be thought of as indicating two or more different properties

[1] In order to do this, we must give a precise definition of an *algebraic expression*, specifying what kind of operations are allowed. This, however, gives rise to some complications. For the time being, let us simply understand it to be a "nonlinear" expression. We will discuss this in Sect. 4.7.

at the same time (*Weyl's thesis*). If we consider image characteristics that are transformed linearly, this irreducible decomposition is unique except for degeneracy. However, if we consider nonlinear transformations, the decomposition is no longer unique. In the following, we always treat transformations of image characteristics only in the *real* domain; we do not introduce complex variables.

One important class of irreducible invariant sets is those consisting of a single element whose value is not changed by the camera rotation. We call such an element a *scalar invariant*. Let $C_1, \ldots, C_l$ be scalar invariants constructed from image characteristics $c_1, \ldots, c_m$. Then, each C_i is an algebraic expression in $c_1, \ldots, c_m$, and does not change its value under camera rotation. We say that scalar invariants $\{C_1, \ldots, C_l\}$ are an *invariant basis* if *any* scalar invariant of $c_1, \ldots, c_m$ is expressed as an algebraic expression in $\{C_1, \ldots, C_l\}$.

We say that two images are *equivalent* if one can be mapped onto the other by some camera rotation. In other words, equivalent images are views of the same scene observed from different camera angles. By definition, all scalar invariants must take the same values for equivalent images. We will show that, if two images are equivalent, the camera rotation that maps one image onto the other can be easily reconstructed from a small number of image characteristics. This means that we *need not know the correspondences of points between the two images*. This problem will also be discussed from a different viewpoint in Chap. 5.

4.2 Scalars, Points, and Lines

As shown in Sect. 1.3, the camera rotation transformation defined by camera rotation $\boldsymbol{R} = (r_{ij})$, $i, j = 1, 2, 3$, is given by

$$x' = f\frac{r_{11}x + r_{21}y + r_{31}f}{r_{13}x + r_{23}y + r_{33}f}, \qquad y' = f\frac{r_{12}x + r_{22}y + r_{32}f}{r_{13}x + r_{23}y + r_{33}f}. \tag{4.2.1}$$

An image characteristic c is called a *scalar* if its value is not changed under camera rotation (Sect. 3.4):

$$c' = c\ . \tag{4.2.2}$$

Obviously, a scalar is itself an invariant and irreducible set. Hence, it should have a meaning inherent to the scene.

If a pair $\{a, b\}$ of numbers is transformed as $\{x, y\}$ of (4.2.1), i.e.,

$$a' = f\frac{r_{11}a + r_{21}b + r_{31}f}{r_{13}a + r_{23}b + r_{33}f}, \qquad b' = f\frac{r_{12}a + r_{22}b + r_{32}f}{r_{13}a + r_{23}b + r_{33}f}, \tag{4.2.3}$$

we call it a *point*. As the transformation rule (2.3) suggests, a point is an invariant set. We also see:

Proposition 4.1. *A point is irreducible.*

Proof. Let $\{a, b\}$ be a point. We can identify a, b as the coordinates on the image plane. Suppose the pair $\{a, b\}$ is not irreducible. Then, there is a one-to-one correspondence between $\{a, b\}$ and another pair $\{u, v\}$ (except for a finite number of singularities) such that u and v are separately invariant sets. The transformation of u and v can be written as $u' = U(u)$ and $v' = V(v)$, where $U(u)$ and $V(v)$ are algebraic expressions in u and v. Now, curves $u = \text{const.}$ and $v = \text{const.}$ define a curvilinear coordinate system (possibly with a finite number of singularities). Since u and v are separately invariant sets, the u-curves and v-curves are always mapped onto themselves by all camera rotations. This is impossible, since the orientation of a line passing through a point on the image plane can be rotated arbitrarily by an appropriate camera rotation.

Hence, a point should have a meaning inherent to the scene—while any pair of numbers can be interpreted as a position on the image plane, it can indicate a position in the scene *if and only if it is transformed as a point.*

A line on the image plane is expressed in the form

$$Ax + By + C = 0 \, . \tag{4.2.4}$$

However, only the ratio $A{:}B{:}C$ has a geometrical meaning; A, B, C and cA, cB, cC for a nonzero scalar c define one and the same line. Hence, one of these three coefficients can be normalized to 1, and the line can be specified by the remaining two parameters. However, we cannot fix the parameter to be normalized (e.g., if B is normalized to 1, lines parallel to the y-axis cannot be expressed). In the following, we always write $A{:}B{:}C$ to express a line. The transformation of a line is given as follows:

Lemma 4.1. *A line $A{:}B{:}C$ is transformed into a line*

$$A'{:}B'{:}C' = \left(r_{11}A + r_{21}B + \frac{1}{f}r_{31}C\right){:}$$
$$\left(r_{12}A + r_{22}B + \frac{1}{f}r_{32}C\right){:}\left(f(r_{13}A + r_{23}B) + r_{33}C\right) \, . \tag{4.2.5}$$

Proof. If the projection equations

$$x = \frac{fX}{Z} \, , \qquad y = \frac{fY}{Z} \, , \tag{4.2.6}$$

are substituted into (4.2.4), we obtain

$$Af\frac{X}{Z} + Bf\frac{Y}{Z} + C = 0 \, . \tag{4.2.7}$$

(This is the equation of the plane in the scene that passes through the viewpoint O and the line (4.2.4) on the image plane.) This equation is rewritten as

$$(A\ B\ C/f)\begin{pmatrix} X \\ Y \\ Z \end{pmatrix} = 0 \ . \tag{4.2.8}$$

Similarly, the line

$$A'x' + B'y' + C' = 0 \tag{4.2.9}$$

is rewritten as

$$(A'\ B'\ C'/f)\begin{pmatrix} X' \\ Y' \\ Z' \end{pmatrix} = 0 \ . \tag{4.2.10}$$

The rotation of the camera by $\boldsymbol{R}$ is equivalent to the rotation of the scene around the camera by $\boldsymbol{R}^{-1}$ $(= \boldsymbol{R}^{\mathrm{T}})$. If point (X, Y, Z) in the scene moves to point (X', Y', Z'), the relationship between these two points is given by

$$\begin{pmatrix} X \\ Y \\ Z \end{pmatrix} = \begin{pmatrix} r_{11} & r_{12} & r_{13} \\ r_{21} & r_{22} & r_{23} \\ r_{31} & r_{32} & r_{33} \end{pmatrix}\begin{pmatrix} X' \\ Y' \\ Z' \end{pmatrix} . \tag{4.2.11}$$

Substituting this into (4.2.8), we have

$$(A\,B\,C/f)\begin{pmatrix} r_{11} & r_{12} & r_{13} \\ r_{21} & r_{22} & r_{23} \\ r_{31} & r_{32} & r_{33} \end{pmatrix}\begin{pmatrix} X' \\ Y' \\ Z' \end{pmatrix} = 0 \ , \tag{4.2.12}$$

or

$$\left[\begin{pmatrix} r_{11} & r_{21} & r_{31} \\ r_{12} & r_{22} & r_{32} \\ r_{13} & r_{23} & r_{33} \end{pmatrix}\begin{pmatrix} A \\ B \\ C/f \end{pmatrix}\right]^{\mathrm{T}}\begin{pmatrix} X' \\ Y' \\ Z' \end{pmatrix} = 0 \ . \tag{4.2.13}$$

Comparing this with (4.2.10), we conclude that

$$\begin{pmatrix} A' \\ B' \\ C'/f \end{pmatrix} = \begin{pmatrix} r_{11} & r_{21} & r_{31} \\ r_{12} & r_{22} & r_{32} \\ r_{13} & r_{23} & r_{33} \end{pmatrix}\begin{pmatrix} A \\ B \\ C/f \end{pmatrix} . \tag{4.2.14}$$

Equation (4.2.5) is obtained by taking the ratio $A'{:}B'{:}C'$.

If the ratio of three quantities A, B, C is transformed by (4.2.5), we call them a *line* and write it as $A{:}B{:}C$. Since we consider only the ratio, this is essentially a set of two elements. We can alternatively write A/C, B/C if $C \neq 0$. However, in order to avoid a separate treatment, we simply write $A:B:C$ and regard it as essentially expressing two quantities.

A line is an invariant set because it is transformed by itself. As for irreducibility, note the following two Lemmas, which establish the *duality* between points and lines.

Lemma 4.2. *If $A:B:C$ is a line, the pair $\{f^2 A/C, f^2 B/C\}$ is a point.*

Proof. From (2.5), we see that

$$f^2 \frac{A'}{C'} = f^2 \frac{r_{11}A + r_{21}B + r_{31}C/f}{f(r_{13}A + r_{23}B) + r_{33}C}$$
$$= f\frac{r_{11}(f^2A/C) + r_{21}(f^2B/C) + r_{31}f}{r_{13}(f^2A/C) + r_{23}(f^2B/C) + r_{33}f} ,$$
$$f^2 \frac{B'}{C'} = f^2 \frac{r_{12}A + r_{22}B + r_{32}C/f}{f(r_{13}A + r_{23}B) + r_{33}C}$$
$$= f\frac{r_{12}(f^2A/C) + r_{22}(f^2B/C) + r_{32}f}{r_{13}(f^2A/C) + r_{23}(f^2B/C) + r_{33}f} . \tag{4.2.15}$$

Lemma 4.3. *If the pair $\{a, b\}$ is a point, the ratio $a:b:f^2$ is a line.*

Proof. From (4.2.3), we see that

$$a':b':f^2 = f\frac{r_{11}a + r_{21}b + r_{31}f}{r_{13}a + r_{23}b + r_{33}f} : f\frac{r_{12}a + r_{22}b + r_{32}f}{r_{13}a + r_{23}b + r_{33}f} : f^2$$
$$= \left(r_{11}a + r_{21}b + \frac{1}{f}r_{31}f^2\right) : \left(r_{12}a + r_{22}b\right.$$
$$\left. + \frac{1}{f}r_{32}f^2\right) : \left((r_{13}a + r_{23}b) + r_{33}f^2\right) . \tag{4.2.16}$$

We say that point $\{f^2A/C, f^2B/C\}$ *is dual* to line $A:B:C$, and line $a:b:f^2$ is *dual* to point $\{a, b\}$. A line can be expressed in the *Hessian normal form*

$$x\cos\theta + y\sin\theta = h , \qquad h \geqq 0 , \tag{4.2.17}$$

where h is the distance of this line from the image origin and θ is the angle made by the normal to this line with the x-axis. The point dual to this line is

$$\left(-\frac{f^2}{h}\cos\theta, \ -\frac{f^2}{h}\sin\theta\right) . \tag{4.2.18}$$

Hence, the duality is geometrically interpreted as follows. Given a line l, draw a line passing through the image origin O and perpendicular to line l. If h is the

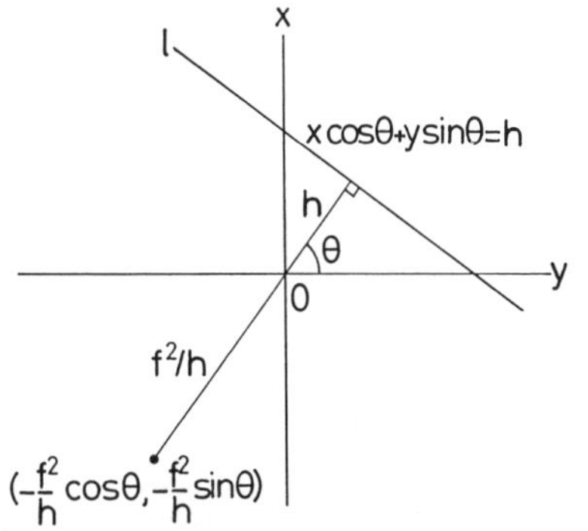

Fig. 4.1. Let l be a line whose distance from the image origin O is h. A point dual to line l is located on the other side of the line perpendicular to l and at distance f^2/h from the image origin O

distance between O and line l, the dual point P is located on the other side of the perpendicular line at a distance f^2/h from O (Fig. 4.1). If $h = 0$ (the line l passes through the image origin O), the dual point P is interpreted as being located at infinity. Similarly, the line at infinity is regarded as the dual of the image origin O.

Since a point and a line have essentially the same transformation property, and a point is irreducible (Proposition 4.1), we conclude:

Proposition 4.2. *A line is irreducible.*

Hence, a line should have a meaning inherent to the scene—any triplet of numbers can be interpreted as a line on the image plane, but it is interpreted as a line inherent to the scene *if and only if it is transformed as a line.*

4.3 Irreducible Decomposition of a Vector

In Chap. 3, we called three quantities $\{a_1, a_2, a_3\}$ a *(3D) vector* if they are transformed as

$$\begin{pmatrix} a_1' \\ a_2' \\ a_3' \end{pmatrix} = \begin{pmatrix} r_{11} & r_{21} & r_{31} \\ r_{12} & r_{22} & r_{32} \\ r_{13} & r_{23} & r_{33} \end{pmatrix} \begin{pmatrix} a_1 \\ a_2 \\ a_3 \end{pmatrix} . \tag{4.3.1}$$

Since a vector is transformed by itself, it is an invariant set. However, it is not irreducible.[2] For example, the "length" $\sqrt{a_1^2 + a_2^2 + a_3^2}$ is a scalar (Corollary 4.1). The remaining invariant set is given by the following lemma.

Lemma 4.4. *If the triplet* $\{a_1, a_2, a_3\}$ *is a vector, the pair* $\{fa_1/a_3, fa_2/a_3\}$ *is a point.*

Proof. From (4.3.1), we see that

$$f\frac{a_1'}{a_3'} = f\frac{r_{11}a_1 + r_{21}a_2 + r_{31}a_3}{r_{13}a_1 + r_{23}a_2 + r_{33}a_3} = f\frac{r_{11}(fa_1/a_3) + r_{21}(fa_2/a_3) + r_{31}f}{r_{13}(fa_1/a_3) + r_{23}(fa_2/a_3) + r_{33}f},$$

$$f\frac{a_2'}{a_3'} = f\frac{r_{12}a_1 + r_{22}a_2 + r_{32}a_3}{r_{13}a_1 + r_{23}a_2 + r_{33}a_3} = f\frac{r_{12}(fa_1/a_3) + r_{22}(fa_2/a_3) + r_{32}f}{r_{13}(fa_1/a_3) + r_{23}(fa_2/a_3) + r_{33}f}. \tag{4.3.2}$$

Since a point is irreducible, a vector is irreducibly decomposed into a scalar and a point.[3] According to the duality between points and lines (Lemmas 4.2 and 4.3), a point is equivalently represented as a line. Hence:

Lemma 4.5. *If the triplet* $\{a_1, a_2, a_3\}$ *is a vector, the ratio* $a_1{:}a_2{:}fa_3$ *is a line.*

Thus, we conclude:

Theorem 4.1. *A vector* $\{a_1, a_2, a_3\}$ *is irreducibly decomposed into a scalar* $\sqrt{a_1^2 + a_2^2 + a_3^2}$ *and either a point* $\{fa_1/a_3, fa_2/a_3\}$ *or a line* $a_1{:}a_2{:}fa_3$ *dual to it.*

The geometrical meaning of the point $P(fa_1/a_3, fa_2/a_3)$ and the line l: $a_1x + a_2y + fa_3 = 0$ represented by vector $\boldsymbol{a} = (a_1, a_2, a_3)$ is simple (Fig. 4.2): if the vector $\boldsymbol{a}$ is placed at the viewpoint O, the "ray" it defines intersects the image plane at the point $P(fa_1/a_3, fa_2/a_3)$. The line l: $a_1x + a_2y + fa_3 = 0$ is the intersection of the image plane with the plane passing through the viewpoint O and perpendicular to the vector $\boldsymbol{a}$. If $a_3 = 0$, the point $(fa_1/a_3, fa_2/a_3)$ is regarded as located at infinity. Similarly, if $a_1 = a_2 = 0$, the line $a_1x + a_2y + fa_3 = 0$ is be regarded as located at infinity.

[2] A vector was irreducible in Chap. 3, but it is no longer irreducible because we have extended the definition of irreducibility by allowing arbitrary algebraic manipulation of components.

[3] Strictly speaking, if we construct a scalar $\sqrt{a_1^2 + a_2^2 + a_3^2}$ and a point $\{fa_1/a_3, fa_2/a_3\}$ from a vector $\{a_1, a_2, a_3\}$, information about the "orientation" is lost: two vectors $\{a_1, a_2, a_3\}$, $\{-a_1, -a_2, -a_3\}$ have the same scalar and point. This ambiguity can be avoided, for example, by associating a scalar $\sqrt{a_1^2 + a_2^2 + a_3^2}$ if $a_3 > 0$ and a scalar $-\sqrt{a_1^2 + a_2^2 + a_3^2}$ if $a_3 < 0$. (The case of $a_3 = 0$ needs special treatment.) In the following, however, we ignore anomalies like this for the sake of simplicity, since we can always find a remedy whenever it is necessary.

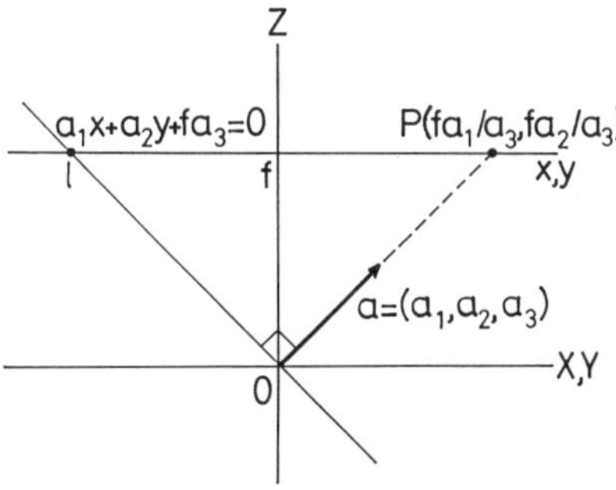

Fig. 4.2. A vector $\boldsymbol{a} = (a_1, a_2, a_3)$ starting from the viewpoint O intersects the image plane $Z = f$ at the point $P(fa_1/a_3, fa_2/a_3)$. Line l: $a_1x + a_2y + fa_3 = 0$ is the intersection of the image plane with the plane passing through the viewpoint O and perpendicular to vector $\boldsymbol{a}$.

Example 4.1. We showed in Sect. 3.7 that the vector part of the optical flow has the form

$$u = fa_2 - a_3y + \frac{1}{f}(a_2x - a_1y)x \ ,$$

$$v = fa_1 + a_3x + \frac{1}{f}(a_2x - a_1y)y \ . \tag{4.3.3}$$

We also showed that the coefficients $\{a_1, a_2, a_3\}$ are transformed as a vector. Hence, the pair $\{fa_1/a_3, fa_2/a_3\}$ is a point (Lemma 4.4), and the ratio $a_1{:}a_2{:}fa_3$ is a line dual to it (Lemma 4.5). Since both are irreducible sets of image characteristics, Weyl's thesis suggests that they should have meanings inherent to the scene.

Consider the point $(fa_1/a_3, fa_2/a_3)$. From (4.3.3), we find that the flow is zero at $x = fa_1/a_3$, $y = fa_2/a_3$ (Exercise 4.2). Namely, the point $(fa_1/a_3, fa_2/a_3)$ is a *fixed point* (or *singularity*) of the flow (Fig. 4.3).

Consider the line $a_1x + a_2y + fa_3 = 0$. If we substitute $y = -a_1x/a_2 - fa_3/a_2$ into (4.3.3), we find that along this line the ratio of the velocity components is constant: $u/v = -a_2/a_1$ (Exercise 4.3). Namely, the flow takes place "along" this line, and this line is a *fixed line* of the flow (Fig. 4.3).

Then, what is the geometrical meaning of the scalar $\sqrt{a_1^2 + a_2^2 + a_3^2}$? Since this is also irreducible, Weyl's thesis suggests that it too should have a meaning inherent to the scene. We saw in Sect. 3.7 that a flow in the form of (4.3.3) results if an object (not necessarily a planar surface) is "orbiting" around the viewer. The scalar $\sqrt{a_1^2 + a_2^2 + a_3^2}$ is exactly the *angular velocity* of this orbiting motion.

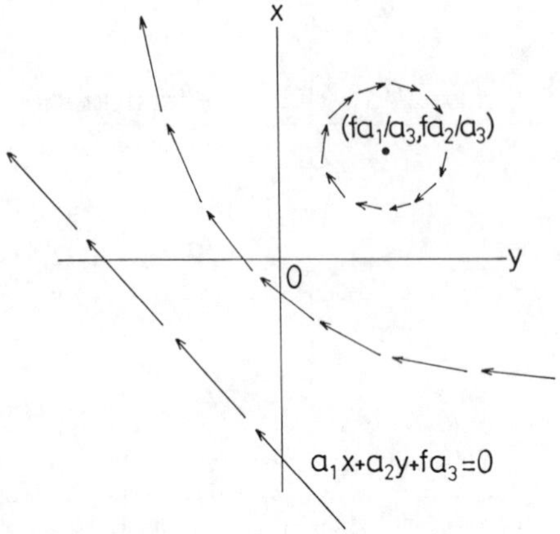

Fig. 4.3. Point $(fa_1/a_3, fa_2/a_3)$ is a fixed point and line $a_1x + a_2y + fa_3 = 0$ is a fixed line of the flow

4.4 Irreducible Decomposition of a Tensor

In Chap. 3, a *tensor* was defined as a set of nine quantities a_{ij}, $i, j = 1, 2, 3$, that is transformed as

$$\begin{pmatrix} a'_{11} & a'_{12} & a'_{13} \\ a'_{21} & a'_{22} & a'_{23} \\ a'_{31} & a'_{32} & a'_{33} \end{pmatrix} = \boldsymbol{R}^{\mathrm{T}} \begin{pmatrix} a_{11} & a_{12} & a_{13} \\ a_{21} & a_{22} & a_{23} \\ a_{31} & a_{32} & a_{33} \end{pmatrix} \boldsymbol{R} . \tag{4.4.1}$$

Since a tensor is transformed by itself, it is an invariant set. However, it is not irreducible. As was shown in Sect. 3.5, a tensor is uniquely decomposed into its *symmetric* and *antisymmetric* parts

$$\begin{pmatrix} a_{11} & a_{12} & a_{13} \\ a_{21} & a_{22} & a_{23} \\ a_{31} & a_{32} & a_{33} \end{pmatrix} = \begin{pmatrix} a_{11} & (a_{12}+a_{21})/2 & (a_{13}+a_{31})/2 \\ (a_{21}+a_{12})/2 & a_{22} & (a_{23}+a_{32})/2 \\ (a_{31}+a_{13})/2 & (a_{32}+a_{23})/2 & a_{33} \end{pmatrix}$$
$$+ \begin{pmatrix} 0 & (a_{12}-a_{21})/2 & -(a_{13}+a_{31})/2 \\ -(a_{21}-a_{12})/2 & 0 & (a_{23}-a_{32})/2 \\ (a_{31}-a_{13})/2 & -(a_{32}-a_{23})/2 & 0 \end{pmatrix} , \tag{4.4.2}$$

and the three independent components $\{(a_{23} - a_{32})/2,\ (a_{31} - a_{13})/2,\ (a_{12} - a_{21})/2\}$ of the antisymmetric part are a vector. This vector is decomposed into a scalar and either a point or a line (Theorem 4.1).

Thus, we need to consider only the symmetric part. Let $A = (a_{ij})$, $i, j = 1, 2, 3$, be a symmetric tensor. The following fact is well known in linear algebra:

Lemma 4.6 *A real symmetric matrix is specified by the three mutually perpendicular real unit vectors* $\boldsymbol{e}_1, \boldsymbol{e}_2, \boldsymbol{e}_3$, *that indicate its three principal axes, and the corresponding real principal values* $\sigma_1, \sigma_2, \sigma_3$, *in the form*

$$A = \sigma_1 \boldsymbol{e}_1 \boldsymbol{e}_1^{\mathrm{T}} + \sigma_2 \boldsymbol{e}_2 \boldsymbol{e}_2^{\mathrm{T}} + \sigma_3 \boldsymbol{e}_3 \boldsymbol{e}_3^{\mathrm{T}} \ . \tag{4.4.3}$$

Proof. The principal values of a real symmetric matrix are all real, and the corresponding principal axes can always be chosen to be mutually orthogonal real unit vectors (Exercise 4.4). Conversely, suppose a matrix A is expressed as in (4.4.3), and

$$(\boldsymbol{e}_i, \boldsymbol{e}_j) = \delta_{ij} \ , \qquad i, j = 1, 2, 3 \ , \tag{4.4.4}$$

where δ_{ij} is the *Kronecker delta* taking the value 1 if $i = j$ and 0 otherwise, and $(.\,,.)$ denotes the (Euclidean) inner product. Equation (4.4.3) defines a real symmetric matrix, since each $\boldsymbol{e}_i \boldsymbol{e}_i^{\mathrm{T}}$, $i = 1, 2, 3$, is a real symmetric matrix. It is easy to see that $\boldsymbol{e}_1, \boldsymbol{e}_2, \boldsymbol{e}_3$ are its eigenvectors and $\sigma_1, \sigma_2, \sigma_3$ are the corresponding eigenvalues:

$$A\boldsymbol{e}_i = \left(\sum_{k=1}^{3} \sigma_k \boldsymbol{e}_k \boldsymbol{e}_k^{\mathrm{T}} \right) \boldsymbol{e}_i = \sum_{k=1}^{3} \sigma_k \boldsymbol{e}_k (\boldsymbol{e}_k, \boldsymbol{e}_i) = \sum_{k=1}^{3} \sigma_k \boldsymbol{e}_k \delta_{ki} = \sigma_i \boldsymbol{e}_i \ . \tag{4.4.5}$$

The matrix A in (4.4.3) does not change if any $\boldsymbol{e}_i$, $i = 1, 2, 3$, is replaced by $-\boldsymbol{e}_i$. Hence, a symmetric real matrix is decomposed into three scalars (its principal values) and three mutually orthogonal *undirected* axes (its principal axes).

Let us call an undirected line passing through the viewpoint simply an *axis*. Let $\boldsymbol{e}$ be a unit vector along an axis L. Since an axis is not directed, $-\boldsymbol{e}$ also defines the same axis. Since $\boldsymbol{e}$ is a vector, it represents a point P on the image plane (Theorem 4.1). However, vector $-\boldsymbol{e}$ also represents the same point P. This means that *an axis is uniquely represented by a point.* (The point can be at infinity. In terms of projective geometry, representing a point by an axis means converting *inhomogeneous coordinates* into *homogeneous coordinates.*)

In order to prove the next lemma, we now introduce the following two definitions. Given a line l and a point P on it on the image plane, let H be the foot of the perpendicular line drawn from the image origin O to line l. Let h be the distance of line l from image origin O, and let d be the distance between points P and H. Take a point Q on the other side of line l at distance $(f^2 + h^2)/d$ from point H (Fig. 4.4). We say that point Q is *conjugate* to point P, and conversely point P is conjugate to point Q. If line l passess through the image origin O, point H is chosen to be O. If $d = 0$, point Q is regarded as located at

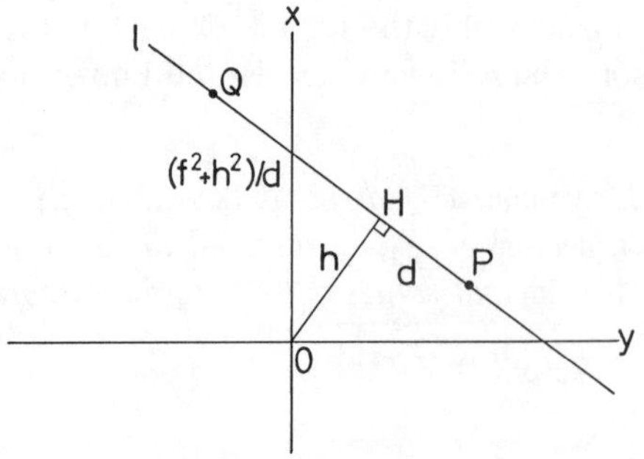

Fig. 4.4. Let H be the foot of the perpendicular line drawn from the image origin O to line l, and let h be the distance of l from O. If point P is on line l at distance d from point H, its conjugate point Q is located on the other side of l at distance $(f^2 + h^2)/d$ from H

infinity. It is easy to prove that the line dual to point P passes through point Q, and the line dual to point Q passes through point P (see also Exercise 4.7).

Lemma 4.7. *Three mutually orthogonal axes are represented by a point and a line passing through it.*

Proof. Let L_1, L_2, L_3 be three mutually orthogonal axes, and let P_1, P_2, P_3 be the respective points representing them. Let l_1 be the line on the image plane passing through points P_2 and P_3. Similarly, let l_2 and l_3 be the lines passing, respectively, through points P_3 and P_1 and through points P_1 and P_2.

It is easy to see that the orthogonality of axes L_1 and L_2 implies that points P_1 and P_2 are conjugate to each other (Exercise 4.7). Similarly, points P_2 and P_3, and points P_3 and P_1 are conjugate to each other. Hence, point P_1 and line l_1 are dual to each other. Similarly, point P_2 and line l_2 are dual to each other, and so are point P_3 and line l_3.

From these observations, we conclude that the configuration of these points and lines is completely specified by a point and line passing through it. For example, suppose we are given point P_3 and line l_1. Line l_3 is obtained as the dual of point P_3, and point P_1 is obtained as the dual of line l_1. Line l_2 is obtained as the line passing through points P_3 and P_1, and point P_2 is obtained either as the intersection of lines l_3 and l_1 or as the dual of line l_2 (Fig. 4.5). At the same time, it is also clear that neither a single point nor a single line can generate the rest. (The choice is not limited to a point and a line passing through it. We can alternatively choose two points conjugate to each other, or two lines. However, our choice is the simplest to state.)

From these lemmas, we finally conclude:

Theorem 4.2. *A tensor is invariantly decomposed into its symmetric and antisymmetric parts. The antisymmetric part is irreducibly decomposed into*

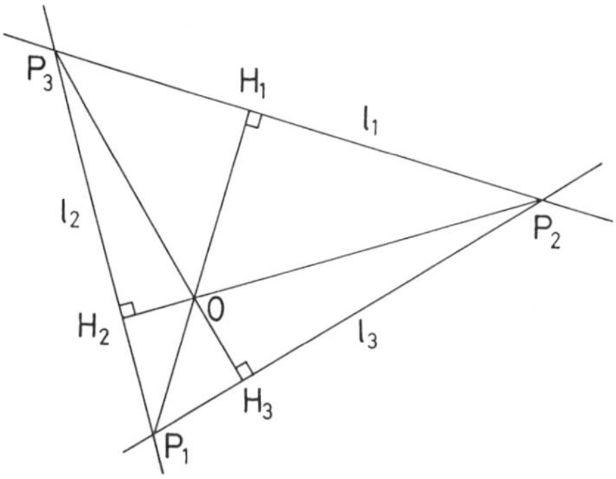

Fig. 4.5. If points P_1, P_2, P_3 and lines l_1, l_2, l_3 represent three mutually orthogonal axes, then point P_1 and line l_1, point P_2 and line l_2, and point P_3 and line l_3 are dual to each other. At the same time, points P_2, P_3, points P_3, P_1, and points P_1, P_2 are conjugate to each other

a scalar and either a point or a line dual to it. The symmetric part is irreducibly decomposed into three scalars, a point, and a line passing through it.

4.5 Invariants of Vectors

Now we consider *scalar invariants*—algebraic expressions that do not change their values under camera rotation. We first begin with a single vector $\boldsymbol{a} = (a_1, a_2, a_3)$. It has the following scalar invariant (Corollary 4.1):

$$\|\boldsymbol{a}\|^2 = a_1^2 + a_2^2 + a_3^2 \ . \tag{4.5.1}$$

[In the remainder of this chapter, $(.,\ .)$ and $\|\ \|$ respectively denote the (Euclidean) inner product and norm.] According to Theorem 4.1, a vector is irreducibly decomposed into a scalar and a point. Since a point cannot have a scalar invariant (because it is irreducible), $\|\boldsymbol{a}\|^2$ is essentially the *only* scalar invariant; any other scalar invariant is expressed as an algebraic expression in it. Hence, $\|\boldsymbol{a}\|^2$ is an *invariant basis* of vector $\boldsymbol{a}$. Let us put this as a proposition.

Proposition 4.3. *An invariant basis of a vector* $\boldsymbol{a}$ *is given by* $\|\boldsymbol{a}\|^2$.

Next, consider two vectors $\boldsymbol{a} = (a_1, a_2, a_3)$, $\boldsymbol{b} = (b_1, b_2, b_3)$. Although $\|\boldsymbol{a}\|^2$ is an invariant basis of $\boldsymbol{a}$ and $\|\boldsymbol{b}\|^2$ is an invariant basis of $\boldsymbol{b}$, the pair $\{\|\boldsymbol{a}\|^2, \|\boldsymbol{b}\|^2\}$ is *not* an invariant basis of $\boldsymbol{a}$, $\boldsymbol{b}$. The following lemma is easy to prove.

Lemma 4.8. *The inner product*

$$(\boldsymbol{a}, \boldsymbol{b}) = a_1 b_1 + a_2 b_2 + a_3 b_3 \tag{4.5.2}$$

of two vectors $\boldsymbol{a} = (a_1, a_2, a_3)$, $\boldsymbol{b} = (b_1, b_2, b_3)$ *is a scalar invariant.*

Proof. From the transformation rule (4.3.1), we have

$$(\boldsymbol{a}', \boldsymbol{b}') = \left[\boldsymbol{R}^{\mathrm{T}} \begin{pmatrix} a_1 \\ a_2 \\ a_3 \end{pmatrix} \right]^{\mathrm{T}} \boldsymbol{R}^{\mathrm{T}} \begin{pmatrix} b_1 \\ b_2 \\ b_3 \end{pmatrix}$$

$$= (a_1\ a_2\ a_3)\, \boldsymbol{R}\boldsymbol{R}^{\mathrm{T}} \begin{pmatrix} b_1 \\ b_2 \\ b_3 \end{pmatrix} = (\boldsymbol{a}, \boldsymbol{b}) \ , \tag{4.5.3}$$

because $\boldsymbol{R}$ is an orthogonal matrix: $\boldsymbol{R}\boldsymbol{R}^{\mathrm{T}} = \boldsymbol{I}$.

In particular, if we put $\boldsymbol{a} = \boldsymbol{b}$, we have

Corollary 4.1. *The squared norm*

$$\|\boldsymbol{a}\|^2 = a_1^2 + a_2^2 + a_3^2 \tag{4.5.4}$$

for a vector $\boldsymbol{a} = (a_1, a_2, a_3)$ *is a scalar invariant.*

We now prove that, for two vectors $\boldsymbol{a}$, $\boldsymbol{b}$, the invariants $\{\|\boldsymbol{a}\|^2, (\boldsymbol{a}, \boldsymbol{b}), \|\boldsymbol{b}\|^2\}$ are sufficient to generate all invariants. Namely:

Proposition 4.4. *For two vectors* $\boldsymbol{a}$, $\boldsymbol{b}$, *the three scalar invariants*

$$\|\boldsymbol{a}\|^2 \ , \qquad (\boldsymbol{a}, \boldsymbol{b}) \ , \qquad \|\boldsymbol{b}\|^2 \tag{4.5.5}$$

are an invariant basis.

Proof. Let

$$J = g(a_1, a_2, a_3, b_1, b_2, b_3) \tag{4.5.6}$$

be an arbitrary scalar invariant of nonzero vectors $\boldsymbol{a}$, $\boldsymbol{b}$, where the right-hand side is an algebraic expression in $a_1, a_2, a_3, b_1, b_2, b_3$. If vectors $\boldsymbol{a}$, $\boldsymbol{b}$ are placed at the coordinate origin O in 3D space, we can rotate the coordinate system so that vector $\boldsymbol{a}$ lies along the Z-axis and vector $\boldsymbol{b}$ lies on the YZ-plane. Let $\boldsymbol{i}, \boldsymbol{j}, \boldsymbol{k}$ be the unit basis vectors along the X-, Y-, and Z-axes, respectively. Then, we can put

$$\boldsymbol{a} = \alpha\boldsymbol{k} \ , \qquad \boldsymbol{b} = \beta\boldsymbol{j} + \gamma\boldsymbol{k} \ , \tag{4.5.7}$$

where α, β, γ are real numbers. We can assume $\alpha > 0$, $\beta \geqq 0$ by appropriately rotating the coordinate system (Fig. 4.6). With reference to this coordinate system, the components of vector $\boldsymbol{a}$ are $(0, 0, \alpha)$, and the components of vector $\boldsymbol{b}$ are $(0, \beta, \gamma)$. Since J is a scalar invariant, we have

$$J = g(0, 0, \alpha, 0, \beta, \gamma) \ . \tag{4.5.8}$$

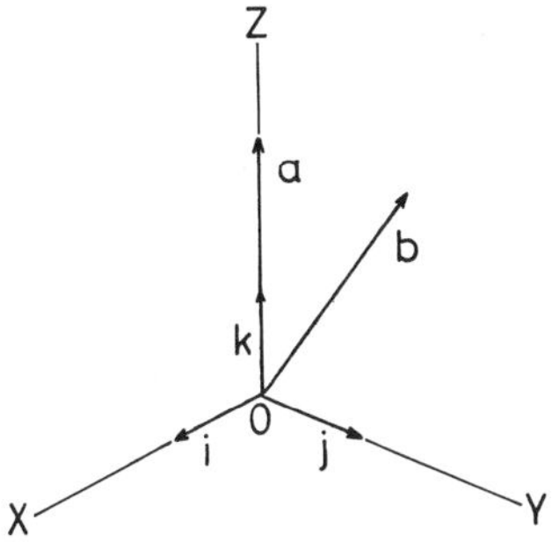

Fig. 4.6. The coordinate system can be rotated so that vector $\boldsymbol{a}$ is along the Z-axis and vector $\boldsymbol{b}$ is on the YZ-plane

From (4.5.7), we see that

$$\|\boldsymbol{a}\|^2 = \alpha^2 \ , \qquad (\boldsymbol{a}, \boldsymbol{b}) = \alpha\gamma \ , \qquad \|\boldsymbol{b}\|^2 = \beta^2 + \gamma^2 \ . \tag{4.5.9}$$

Since $\alpha > 0$, $\beta \geqq 0$, we obtain

$$\alpha = \sqrt{\|\boldsymbol{a}\|^2} \ , \qquad \beta = \frac{\sqrt{\|\boldsymbol{a}\|^2 \|\boldsymbol{b}\|^2 - (\boldsymbol{a}, \boldsymbol{b})^2}}{\sqrt{\|\boldsymbol{a}\|^2}} \ , \qquad \gamma = \frac{(\boldsymbol{a}, \boldsymbol{b})}{\sqrt{\|\boldsymbol{a}\|^2}} \ . \tag{4.5.10}$$

Thus, α, β, γ are algebraic expressions in $\{\|\boldsymbol{a}\|^2, (\boldsymbol{a}, \boldsymbol{b}), \|\boldsymbol{b}\|^2\}$. By (4.5.8), J is an algebraic expression in α, β, γ. Hence, J is also algebraically expressed in $\{\|\boldsymbol{a}\|^2, (\boldsymbol{a}, \boldsymbol{b}), \|\boldsymbol{b}\|^2\}$.

Note that no "equations" constrain the three scalar invariants, i.e., the value of one is not numerically determined from the values of the others. However, this does not ean that their values can be assigned *arbitrarily*. For example, their values are constrained by *Schwarz's inequality*: $|(\boldsymbol{a}, \boldsymbol{b})| \leqq \|\boldsymbol{a}\| \cdot \|\boldsymbol{b}\|$ (Exercise 4.8).

Now, consider three vectors $\boldsymbol{a}$, $\boldsymbol{b}$, $\boldsymbol{c}$. From Lemma 4.8 and Corollary 4.1, it is clear that $\|\boldsymbol{a}\|^2$, $\|\boldsymbol{b}\|^2$, $\|\boldsymbol{c}\|^2$, $(\boldsymbol{a}, \boldsymbol{b})$, $(\boldsymbol{b}, \boldsymbol{c})$, $(\boldsymbol{c}, \boldsymbol{a})$ are scalar invariants. There is another scalar invariant—the *scalar triple product*—defined by

$$|\boldsymbol{abc}| \equiv (\boldsymbol{a} \times \boldsymbol{b}, \boldsymbol{c}) = (\boldsymbol{b} \times \boldsymbol{c}, \boldsymbol{a}) = (\boldsymbol{c} \times \boldsymbol{a}, \boldsymbol{b}) \ . \tag{4.5.11}$$

As is well known in vector calculus, this quantity is numerically equal to the determinant of the matrix consisting of vectors $\boldsymbol{a}$, $\boldsymbol{b}$, $\boldsymbol{c}$ as its three columns (or rows) in this order (Exercise 4.10). Its value is also equal to the volume of a parallelepiped defined by $\boldsymbol{a}$, $\boldsymbol{b}$, $\boldsymbol{c}$ as its three edges (Fig. 4.7). Its sign is positive or negative depending on whether $\{\boldsymbol{a}, \boldsymbol{b}, \boldsymbol{c}\}$ are a right-hand or left-hand system. We can prove that these seven invariants are sufficient to generate all invariants. Namely:

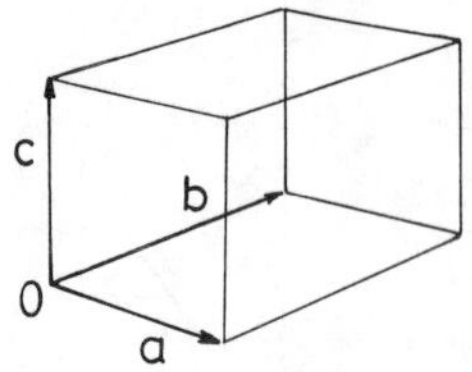

Fig. 4.7. The scalar triple product $|\boldsymbol{abc}|$ is numerically equal to the volume of the parallelepiped with $\boldsymbol{a}$, $\boldsymbol{b}$, $\boldsymbol{c}$ as its three edges. Its sign is positive or negative depending on whether $\{\boldsymbol{a}, \boldsymbol{b}, \boldsymbol{c}\}$ are a right-hand or left-hand system

Proposition 4.5. *For three vectors* $\boldsymbol{a}$, $\boldsymbol{b}$, $\boldsymbol{c}$, *the seven invariants*

$$\|\boldsymbol{a}\|^2, \quad \|\boldsymbol{b}\|^2, \quad \|\boldsymbol{c}\|^2 ,$$
$$(\boldsymbol{a}, \boldsymbol{b}) , \quad (\boldsymbol{b}, \boldsymbol{c}) , \quad (\boldsymbol{c}, \boldsymbol{a}) , \quad |\boldsymbol{abc}| \tag{4.5.12}$$

are an invariant basis.

We omit the rigorous proof of Proposition 4.5. We will give an intuitive explanation in the next subsection, but we must point out that if $O(3)$, which is obtained from $SO(3)$ by adding reflections, is the group of admissible transformations (Sect. 4.5.1), the scalar triple product (4.5.11) can be removed. The scalar triple product is necessary only when the admissible transformations are restricted to $SO(3)$. However, we can prove that $|\boldsymbol{abc}|^2$ is expressed in terms of the rest of the invariants. This type of relation is called a *syzygy* (Sect. 4.7).

Similar arguments can be made for four or more vectors. From Proposition 4.5, we obtain:

Proposition 4.6. *For N vectors* $\boldsymbol{a}_1, \ldots, \boldsymbol{a}_N$, $N \geqq 3$, invariants

$$\|\boldsymbol{a}_i\|^2, \quad (\boldsymbol{a}_i, \boldsymbol{a}_j) , \quad |\boldsymbol{a}_i\boldsymbol{a}_j\boldsymbol{a}_k| , \quad i \neq j , \quad j \neq k , \quad k \neq i , \tag{4.5.13}$$

for $i, j, k = 1, \ldots, N$, *are an invariant basis.*

If $O(3)$ is the group of admissible transformations, the scalar triple products can be removed. Syzygies also exist.

Example 4.2. Consider two vectors $\boldsymbol{a} = (a_1, a_2, a_3)$, $\boldsymbol{b} = (b_1, b_2, b_3)$. Since the vector product

$$\boldsymbol{a} \times \boldsymbol{b} = (a_2b_3 - a_3b_2, a_3b_1 - a_1b_3, a_1b_2 - a_2b_1) \tag{4.5.14}$$

is also a vector (Exercise 4.11), we can compute its scalar invariant

$$\|\boldsymbol{a} \times \boldsymbol{b}\|^2 = (a_2b_3 - a_3b_2)^2 + (a_3b_1 - a_1b_3)^2 + (a_1b_2 - a_2b_1)^2 \ . \tag{4.5.15}$$

Since $\{\|\boldsymbol{a}\|^2, (\boldsymbol{a}, \boldsymbol{b}), \|\boldsymbol{b}\|^2\}$ is an invariant basis for vectors $\boldsymbol{a}$ and $\boldsymbol{b}$, equation (4.5.15) must be expressed in terms of these. In fact, if we faithfully follow the proof of Proposition 4.4, we obtain the following expression (Exercise 4.11):

$$\|\boldsymbol{a} \times \boldsymbol{b}\|^2 = \|\boldsymbol{a}\|^2 \ \|\boldsymbol{b}\|^2 - (\boldsymbol{a}, \boldsymbol{b})^2 \ . \tag{4.5.16}$$

4.5.1 Interpretation of Invariants

Here, we consider in intuitive terms how to interpret invariants and how to construct them. We start with general definitions. Consider a geometrical object S in a space X. Suppose a group G of transformations acts on space X. Let us call G the group of *admissible transformations* of X. A *scalar invariant* of object S is a quantity that does not change its value when object S undergoes any of the admissible transformations. Its invariance implies that it characterizes some intrinsic property of the object (*Weyl's thesis*).

Suppose object S has scalar invariants $\alpha_1, \ldots, \alpha_N$. Consider another object S' for which the same scalar invariants can be defined. If object S' is obtained by appropriately transforming object S by an admissible transformation, the values of these scalar invariants must be identical. However, the converse is not true in general; the values of these scalar invariants can be the same for different objects. Hence, we can define an *invariant basis* by requiring the converse. We say that a set of scalar invariants is an invariant basis if *coincidence of their values for two objects implies existence of an admissible transformation that maps one to the other.*

It is intuitively clear that this definition is equivalent to the one given earlier, because if object S cannot be mapped to object S' by any admissible transformation and hence "discrepancies" always arise between them whichever admissible transformation we choose, we should be able to define some scalar invariants which account for those discrepancies. These extra scalar invariants cannot be expressed in terms of the other invariants, which have the same values for S and S'. Thus, a set of scalar invariants is an invariant basis if and only if *any* scalar invariant can be expressed in terms of them.

Now, consider 3D Euclidean space, and let $SO(3)$ be the group of admissible transformations. A *scalar invariant* of object S for $SO(3)$ is a quantity whose value does not change if object S is rotated around the coordinate origin O—for example, the "size" (if appropriately defined) of object S.

Consider a vector. A vector can be represented by an arrow starting from the coordinate origin O. Consider two vectors $\boldsymbol{a}$, $\boldsymbol{a}'$ starting from the origin O. Evidently, we can make one vector overlap the other by an appropriate rotation around the origin O if and only if they have the same length. Hence, $\|\boldsymbol{a}\|$ (or its square) alone constitutes an invariant basis.

Next, consider two vectors $\boldsymbol{a}$, $\boldsymbol{b}$. When they are rotated around the origin O, they must be moved "rigidly" without changing the relative configuration. If there is another set of two vectors $\boldsymbol{a}'$, $\boldsymbol{b}'$, we can make the two sets overlap by a rotation if and only if the triangles defined by the origin O and the endpoints of the two vectors are congruent—for example, if the lengths of the two vectors and the angle between them are equal. Since the cosine of the angle θ between $\boldsymbol{a}$ and $\boldsymbol{b}$ is given by

$$\cos\theta = \frac{(\boldsymbol{a}, \boldsymbol{b})}{\|\boldsymbol{a}\| \cdot \|\boldsymbol{b}\|}, \tag{4.5.17}$$

we conclude that $\{\|\boldsymbol{a}\|^2, (\boldsymbol{a}, \boldsymbol{b}), \|\boldsymbol{b}\|^2\}$ are an invariant basis.

Consider three vectors $\boldsymbol{a}$, $\boldsymbol{b}$, $\boldsymbol{c}$. These three vectors define a parallelepiped, $\boldsymbol{a}$, $\boldsymbol{b}$, $\boldsymbol{c}$ constituting three edges starting from the origin O. Evidently, we can make two sets of three vectors overlap by a rotation if and only if the parallelepipeds they define are congruent. If the values of $\|\boldsymbol{a}\|^2$, $\|\boldsymbol{b}\|^2$, $\|\boldsymbol{c}\|^2$, $(\boldsymbol{a}, \boldsymbol{b})$, $(\boldsymbol{b}, \boldsymbol{c})$, $(\boldsymbol{c}, \boldsymbol{a})$ are specified, all the lengths of the edges and all the angles between two edges can be fixed. However, this is not sufficient: if two parallelepipeds are the "mirror images" of each other, we cannot make them overlap by a *proper* rotation, i.e., a rotation without any reflection. If we add the scalar triple product $|\boldsymbol{abc}|$, no indeterminacy remains. Thus, $\{\|\boldsymbol{a}\|^2, \|\boldsymbol{b}\|^2, \|\boldsymbol{c}\|^2, (\boldsymbol{a}, \boldsymbol{b}), (\boldsymbol{b}, \boldsymbol{c}), (\boldsymbol{c}, \boldsymbol{a}), |\boldsymbol{abc}|\}$ are an invariant basis for $\boldsymbol{a}$, $\boldsymbol{b}$, $\boldsymbol{c}$.

From the above argument, it is clear that if $O(3)$, which consists of both *proper* and *improper* 3D rotations, is the group of admissible transformations, the six invariants $\{\|\boldsymbol{a}\|^2, \|\boldsymbol{b}\|^2, \|\boldsymbol{c}\|^2, (\boldsymbol{a}, \boldsymbol{b}), (\boldsymbol{b}, \boldsymbol{c}), (\boldsymbol{c}, \boldsymbol{a})\}$ are an invariant basis.

Similar arguments can be made for four or more vectors $\boldsymbol{a}_1, \ldots, \boldsymbol{a}_N$. If $O(3)$ is the group of admissible transformations, the inner products of all pairs $(\boldsymbol{a}_i, \boldsymbol{a}_j)$, $i, j = 1, \ldots, N$, are an invariant basis. If the group is restricted to $SO(3)$, the scalar triple product of all three vectors must also be added. However, the N^2 invariants $(\boldsymbol{a}_i, \boldsymbol{a}_j)$, $i, j = 1, \ldots, N$, are not algebraically independent. First, we have $(\boldsymbol{a}_i, \boldsymbol{a}_j) = (\boldsymbol{a}_j, \boldsymbol{a}_i)$ for $i \neq j$. Moreover, we have relations

$$\begin{vmatrix} (\boldsymbol{a}_i, \boldsymbol{a}_i) & (\boldsymbol{a}_i, \boldsymbol{a}_j) & (\boldsymbol{a}_i, \boldsymbol{a}_k) & (\boldsymbol{a}_i, \boldsymbol{a}_l) \\ (\boldsymbol{a}_j, \boldsymbol{a}_i) & (\boldsymbol{a}_j, \boldsymbol{a}_j) & (\boldsymbol{a}_j, \boldsymbol{a}_k) & (\boldsymbol{a}_j, \boldsymbol{a}_l) \\ (\boldsymbol{a}_k, \boldsymbol{a}_i) & (\boldsymbol{a}_k, \boldsymbol{a}_j) & (\boldsymbol{a}_k, \boldsymbol{a}_k) & (\boldsymbol{a}_k, \boldsymbol{a}_l) \\ (\boldsymbol{a}_l, \boldsymbol{a}_i) & (\boldsymbol{a}_l, \boldsymbol{a}_j) & (\boldsymbol{a}_l, \boldsymbol{a}_k) & (\boldsymbol{a}_l, \boldsymbol{a}_l) \end{vmatrix} = 0 \tag{4.5.18}$$

for all four distinct indices i, j, k, l. This is a direct consequence of the fact that any four vectors in 3D space are linearly dependent. Equations which constrain scalar invariants, such as (4.5.18), are called *syzygies*. It can be proved that (4.5.18) and $(\boldsymbol{a}_i, \boldsymbol{a}_j) = (\boldsymbol{a}_j, \boldsymbol{a}_i)$ are essentially the *only* relationships that constrain the inner products: all syzygies are generated by combining them.

Equation (4.5.18) is proved as follows. Consider vectors $\boldsymbol{a}_1$, $\boldsymbol{a}_2$, $\boldsymbol{a}_3$, $\boldsymbol{a}_4$, for example. Since they are linearly dependent, we can find four real numbers x_1, x_2, x_3, x_4, which are not all zero, such that $\sum_{j=1}^{4} x_j \boldsymbol{a}_j = 0$. Taking the inner product with vector $\boldsymbol{a}_i$, $i = 1, 2, 3, 4$, we obtain

$$\sum_{j=1}^{4} (\boldsymbol{a}_i, \boldsymbol{a}_j) x_j = 0 , \qquad i = 1, 2, 3, 4 . \tag{4.5.19}$$

Since this set of simultaneous linear equations has a nontrivial solution x_j, $j = 1, \ldots, 4$, the determinant of the coefficient matrix must vanish.

4.6 Invariants of Points and Lines

In this section, we consider scalar invariants of points and lines. Some complications will arise due to the fact that only objects in front of the camera (i.e., $Z > 0$) can be visible. In order to avoid such complications, let us confine ourselves to small camera rotations. The question of how small the rotation should be depends on the object image. We say that a rotation $\boldsymbol{R}$ is *small* for the given object image if there exists a smooth parametrization $\boldsymbol{R}(t)$, $0 \leqq t \leqq 1$, of rotations such that $\boldsymbol{R}(0) = \boldsymbol{I}$, $\boldsymbol{R}(1) = \boldsymbol{R}$ and *if the object image does not pass through points at infinity for any rotations* $\boldsymbol{R}(t)$, $0 \leqq t \leqq 1$. In other words, a rotation is small if the object image can be moved *within the image plane.*

First, consider a point $P(a, b)$. Since it is irreducible, it has no scalar invariants. This is interpreted as follows. As long as we consider small rotations, the point P on the image plane can be represented by the unit vector

$$\boldsymbol{m} = \left(\frac{a}{\sqrt{a^2 + b^2 + f^2}}, \frac{b}{\sqrt{a^2 + b^2 + f^2}}, \frac{f}{\sqrt{a^2 + b^2 + f^2}} \right) \tag{4.6.1}$$

starting from the viewpoint O and pointing toward point P on the image plane (Fig. 4.8). From the argument of Sect. 4.5.1, it is clear that a unit vector does not have scalar invariants. Hence:

Proposition 4.7. *A point has no scalar invariants.*

Next, consider two points $P_1(a_1, b_1)$, $P_2(a_2, b_2)$. The unit vectors representing them are

$$\boldsymbol{m}_1 = \left(\frac{a_1}{\sqrt{a_1^2 + b_1^2 + f^2}}, \frac{b_1}{\sqrt{a_1^2 + b_1^2 + f^2}}, \frac{f}{\sqrt{a_1^2 + b_1^2 + f^2}} \right),$$

$$\boldsymbol{m}_2 = \left(\frac{a_2}{\sqrt{a_2^2 + b_2^2 + f^2}}, \frac{b_2}{\sqrt{a_2^2 + b_2^2 + f^2}}, \frac{f}{\sqrt{a_2^2 + b_2^2 + f^2}} \right). \tag{4.6.2}$$

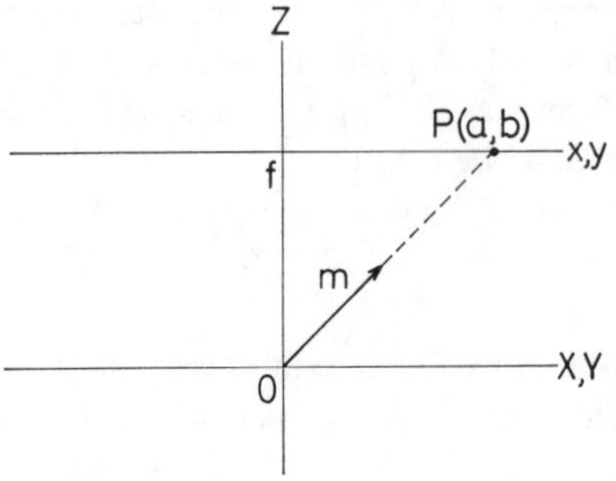

Fig. 4.8. Point P on the image plane is represented by the unit vector $\boldsymbol{m}$ starting from the viewpoint O and pointing toward P

The only scalar invariant is the inner product of these two vectors:

$$(\boldsymbol{m}_1, \boldsymbol{m}_2) = \frac{a_1 a_2 + b_1 b_2 + f^2}{\sqrt{a_1^2 + b_1^2 + f^2}\sqrt{a_2^2 + b_2^2 + f^2}} . \tag{4.6.3}$$

Let us define the *invariant distance* $\rho(P_1, P_2)$ as

$$\rho(P_1, P_2) = f \cos^{-1} \frac{a_1 a_2 + b_1 b_2 + f^2}{\sqrt{a_1^2 + b_1^2 + f^2}\sqrt{a_2^2 + b_2^2 + f^2}} . \tag{4.6.4}$$

Since this is constructed from the scalar invariant (4.6.3), this is also a scalar invariant. It can be proved that this distance approaches the usual (Euclidean) distance if both points come close to the image origin. Since this distance is algebraically equivalent to the scalar invariant of (4.6.3), we conclude:

Proposition 4.8. *For two points P_1, P_2, an invariant basis is given by the invariant distance $\rho(P_1, P_2)$ between them.*

Now, consider three points P_1, P_2, P_3. From the arguments in the preceding section, we see:

Proposition 4.9. *For three points P_1, P_2, P_3, an invariant basis is given by*

$$\rho(P_1, P_2) , \quad \rho(P_2, P_3) , \quad \rho(P_3, P_1) , \quad \text{sign}\,(P_1, P_2, P_3) . \tag{4.6.5}$$

Here, sign (P_1, P_2, P_3) is the *signature* of the three points, defined to be 1 if P_1, P_2, P_3 are positioned clockwise in this order, 0 if any two of P_1, P_2, P_3 coincide, and -1 if P_1, P_2, P_3 are positioned counterclockwise. This result can be extended to four or more points as follows.

Proposition 4.10. *For N points $P_1, \ldots, P_N$, $N \geqq 3$, the invariants*

$$\rho(P_i, P_j)\,, \qquad \mathrm{sign}\,(P_i, P_j, P_k)\,, \qquad i \neq j\,, \qquad j \neq k\,, \qquad k \neq i\,, \tag{4.6.6}$$

for i, j, k = 1, . . . , N, are an invariant basis.

Now let us consider lines. The argument runs almost parallel to the case of points. However, some complications arise even if rotations are restricted to be small. Consider a line $A{:}B{:}C$ on the image plane. Consider in the XYZ-space a plane passing through the viewpoint O and intersecting the image plane $Z = f$ along the line $l{:}\, Ax + By + C = 0$. Then, line l is represented by a unit vector $\boldsymbol{n}$ normal to that plane. From Fig. 4.9, the vector n is given by

$$\boldsymbol{n} = \left(\frac{A}{\sqrt{A^2 + B^2 + C^2/f^2}}\,, \frac{B}{\sqrt{A^2 + B^2 + C^2/f^2}}\,, \frac{C/f}{\sqrt{A^2 + B^2 + C^2/f^2}}\right)\,. \tag{4.6.7}$$

However, we must note that if $\boldsymbol{n}$ is a unit vector normal to the plane, so is $-\boldsymbol{n}$. In other words, vector $\boldsymbol{n}$ must be treated as an *axis* (Sect. 4.4). If we keep this in mind, the argument runs similar to the case of points. Since a unit vector $\boldsymbol{n}$ (or $-\boldsymbol{n}$) does not have any scalar invariants, a line does not have any scalar invariants.

Proposition 4.11. *A line has no scalar invariants.*

Consider two lines $A_1{:}B_1{:}C_1$ and $A_2{:}B_2{:}C_2$. Let $\boldsymbol{n}_1$ and $\boldsymbol{n}_2$ be the unit vectors (regarded as axes) representing them. They are given by

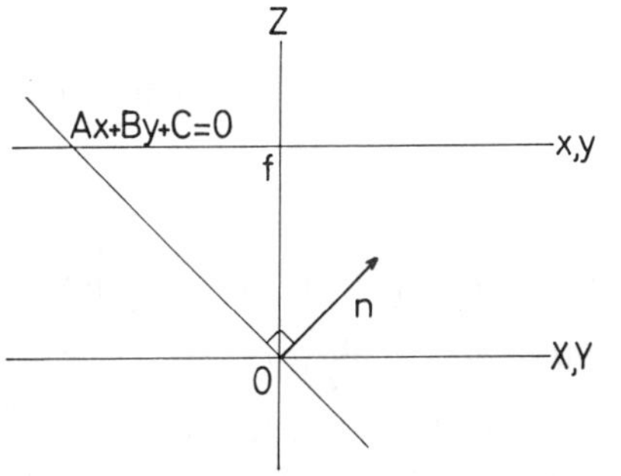

Fig. 4.9. A line l on the image plane is represented by a unit vector $\boldsymbol{n}$ normal to the plane passing through the viewpoint O and intersecting the image plane along the line l

$$\boldsymbol{n}_1 = \left(\frac{A_1}{\sqrt{A_1^2 + B_1^2 + C_1^2/f^2}}, \frac{B_1}{\sqrt{A_1^2 + B_1^2 + C_1^2/f^2}}, \frac{C_1/f}{\sqrt{A_1^2 + B_1^2 + C_1^2/f^2}} \right),$$

$$\boldsymbol{n}_2 = \left(\frac{A_2}{\sqrt{A_2^2 + B_2^2 + C_2^2/f^2}}, \frac{B_2}{\sqrt{A_2^2 + B_2^2 + C_2^2/f^2}}, \frac{C_2/f}{\sqrt{A_2^2 + B_2^2 + C_2^2/f^2}} \right). \tag{4.6.8}$$

If they were ordinary vectors, the only scalar invariant would be the inner product $(\boldsymbol{n}_1, \boldsymbol{n}_2)$. Since they are axes, their signs are irrelevant. Hence, we must take its absolute value $|(\boldsymbol{n}_1, \boldsymbol{n}_2)|$. Namely, the only scalar invariant for two lines is

$$|(\boldsymbol{n}_1, \boldsymbol{n}_2)| = \frac{|A_1 A_2 + B_1 B_2 + C_1 C_2/f^2|}{\sqrt{A_1^2 + B_1^2 + C_1^2/f^2}\sqrt{A_2^2 + B_2^2 + C_2^2/f^2}}. \tag{4.6.9}$$

Let us define the *invariant angle* $\theta(l_1, l_2)$ between two lines l_1: $A_1x + B_1y + C_1 = 0$ and l_2: $A_2x + B_2y + C_2 = 0$ by

$$\theta(l_1, l_2) = \cos^{-1} \frac{|A_1 A_2 + B_1 B_2 + C_1 C_2/f^2|}{\sqrt{A_1^2 + B_1^2 + C_1^2/f^2}\sqrt{A_2^2 + B_2^2 + C_2^2/f^2}}. \tag{4.6.10}$$

Since this angle is constructed from the scalar invariant (4.6.9), it is also a scalar invariant. It can be proved that this angle $\theta(l_1, l_2)$ coincides with the usual Euclidean angle (the smaller of the two) between the two lines l_1, l_2 when their intersection coincides with the image origin. Since this angle is algebraically equivalent to the scalar invariant of (4.6.9), we conclude:

Proposition 4.12. *For two lines l_1, l_2, the invariant angle $\theta(l_1, l_2)$ is an invariant basis.*

Consider three lines l_1, l_2, l_3. By an argument similar to the case of three points, we conclude:

Proposition 4.13. *For three lines l_1, l_2, l_3, an invariant basis is given by*

$$\theta(l_1, l_2), \quad \theta(l_2, l_3), \quad \theta(l_3, l_1), \quad \mathrm{sign}(l_1, l_2, l_3). \tag{4.6.11}$$

Here, $\mathrm{sign}(l_1, l_2, l_3)$ is the *signature* of the three lines, defined to be 1 if l_1, l_2, l_3 intersect in pairs clockwise, 0 if any two of l_1, l_2, l_3 coincide, and -1 if l_1, l_2, l_3 intersect in pairs counterclockwise. This result can be extended to four or more lines:

Proposition 4.14. *For N lines $l_1, \ldots, l_N$, $N \geqq 3$, the invariants*

$$\rho(l_i, l_j) , \qquad \text{sign}\,(l_i, l_j, l_k) , \qquad i \neq j , j \neq k, k \neq i , \tag{4.6.12}$$

for i, j, k = 1, . . . N, are an invariant basis.

Finally, consider a point $P(a, b)$ and a line l: $Ax + By + C = 0$. Point P is represented by vector $\boldsymbol{m}$ of (4.6.1), and line l is represented by *axis* $\boldsymbol{n}$ of (4.6.7). Hence, their only scalar invariant is the absolute value of the inner product

$$|(\boldsymbol{a}, \boldsymbol{n})| = \frac{|aA + bB + C|}{\sqrt{a^2 + b^2 + f^2}\sqrt{A^2 + B^2 + C^2/f^2}} . \tag{4.6.13}$$

Define angle $\theta(P, l)$ by

$$\theta(P, l) = \cos^{-1}\frac{|aA + bB + C|}{\sqrt{a^2 + b^2 + f^2}\sqrt{A^2 + B^2 + C^2/f^2}} . \tag{4.6.14}$$

Proposition 4.15. *For a point P and a line l, an invariant basis is given by the angle $\theta(P, l)$.*

Example 4.3. Consider two lines l_1, l_2. If both are infinitely long, we can always rotate the camera so that l_1 passes through the image origin and l_2 is parallel to l_2. (Line l_2 may be at infinity, but we ignore that case.) Let $h(l_1, l_2)$ be the usual (Euclidean) distance between them in this configuration. Let us call this the *invariant distance* of line l_2 from line l_1. By definition, it is a scalar invariant. From Proposition 4.12, the invariant angle $\theta(l_1, l_2)$ of (4.6.10) is an invariant basis. Hence, the invariant distance $h(l_1, l_2)$ must be expressed in terms of the invariant angle $\theta(l_1, l_2)$. From Fig. 4.10, it is easy to see that

$$h(l_1, l_2) = f \tan \theta(l_1, l_2) . \tag{4.6.15}$$

Furthermore, we also see that $h(l_1, l_2) = h(l_2, l_1)$.

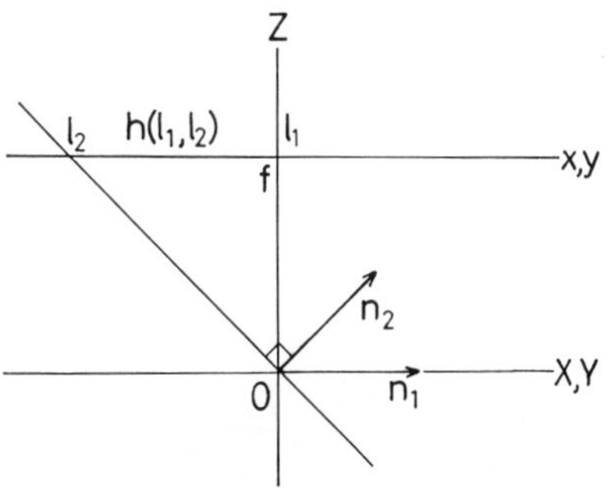

Fig. 4.10. The invariant distance $h(l_1, l_2)$ between two lines l_1, l_2 is defined as the usual distance between them when the camera is rotated so that one passes through the image origin and the other is parallel to it

Example 4.4. Consider a point P and a line l. Let $h(P, l)$ be the usual (Euclidean) distance from point P to line l when point P is brought to the image origin by an appropriate camera rotation. Line l may be at infinity, but we ignore that case. Let us call $h(P, l)$ the *invariant distance* between point P and line l. By definition, it is a scalar invariant. From Proposition 4.15, the angle $\theta(P, l)$ of (4.6.14) is an invariant basis. Hence, the invariant distance $h(P, l)$ must be expressed in terms of the angle $\theta(P, l)$. From Fig. 4.11, it is easy to see that

$$h(P, l) = f \cot \theta(P, l) \ . \tag{4.6.16}$$

4.6.1 Spherical Geometry

By now, readers may have realized that what we are doing is in fact *spherical geometry*—geometry on a sphere. Recall that our aim is to derive invariant quantities whose values do not change when the camera is rotated. If the scene is projected onto the image plane by perspective projection, the projected object shape inevitably undergoes distortion when the camera is rotated. Consider, on the other hand, an *image sphere* centered at the viewpoint O with radius f. A point (X, Y, Z) in the scene is projected onto the intersection P of the image sphere with the ray connecting the point (X, Y, Z) and the viewpoint O (Fig. 4.12). If we deal with images projected onto the image sphere, the study of camera rotation invariance is trivial, because images simply move rigidly as the camera is rotated. Hence, if we define angles and lengths on the image sphere, all properties described in terms of them are invariant.

Then, why did we not start with the image sphere? The reason comes from practical considerations. In computer vision applications, it is most convenient to treat images on a plane equipped with an xy-coordinate system; images are usually printed out on flat paper or projected onto (almost) flat display screens. When digital images are stored in a computer, it is most convenient to use rectangular arrays. (At times, triangular or hexagonal arrays are also used.)

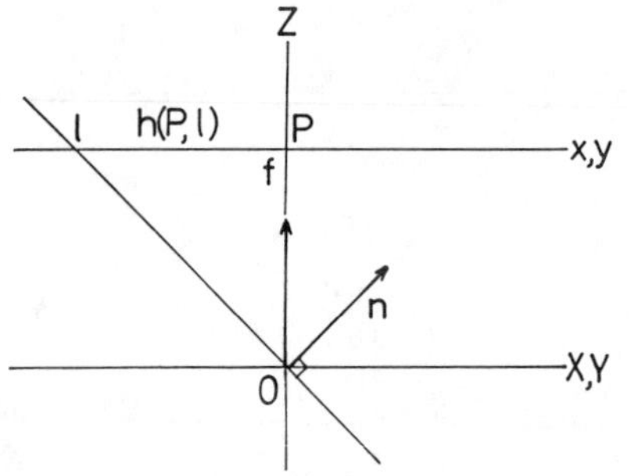

Fig. 4.11. The invariant distance $h(P, l)$ between a point P and a line l is defined as the usual distance between them when the camera is rotated so that point P coincides with the image origin

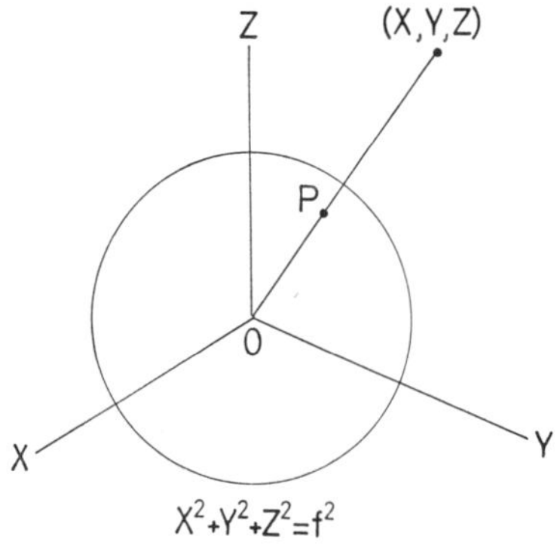

Fig. 4.12. A point (X, Y, Z) in the scene is projected onto the intersection P of the image sphere with the ray connecting the point (X, Y, Z) and the viewpoint O

There have been some attempts to simulate the image sphere in computer memories. The use of the image sphere representation has a big advantage when we deal with image data ranging over a very wide angle of view—say, of solid angle almost 2π; if we use the image plane representation, the size of the array becomes infinite. However, there is no obvious way to tessellate the sphere's surface in a homogeneous and periodic manner, and devising easy memory access is a challenging problem. Hence, except for special purposes, the image plane representation is the most feasible way.

The xy image plane is not only easy to handle technically: *We can do on the xy image plane all that can be done on the image sphere.* The demonstration of this is one of the aims of this section. For example, it is easy to see that the invariant distance $\rho(P_1, P_2)$ of (4.6.4) is equal to the *arc length* between the corresponding points on the image sphere along the *great circle* passing through them. Hence, this distance approaches the usual (Euclidean) distance if both points come close to the image origin. The crucial fact is that *we can compute it in terms of image coordinates*; there is no need to generate or compute the image sphere.

Consider three distinct points P_0, P_1, P_2 on the image plane. Let us define the *invariant angle* θ between the line segments P_0P_1 and P_0P_2 as follows (Fig. 4.13):

$$\theta = \cos^{-1}\frac{\cos(\rho(P_1, P_2)/f) - \cos(\rho(P_0, P_1)/f)\cos(\rho(P_0, P_2)/f)}{\sin(\rho(P_0, P_1)/f)\,\sin(\rho(P_0, P_2)/f)}\,. \tag{4.6.17}$$

Since θ is defined in terms of scalar invariants, it is also a scalar invariant. In fact, this is equal to the angle defined on the image sphere (Exercise 4.16). Hence, this angle coincides with the usual (Euclidean) angle if point P_0 is at the image origin.

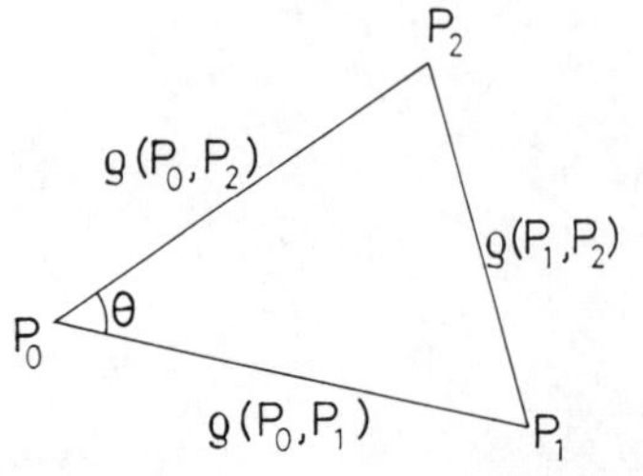

Fig. 4.13. The invariant angle at vertex P_0 of triangle $\Delta P_0 P_1 P_2$ is defined in terms of the invariant distances $\rho(P_0, P_1)$, $\rho(P_0, P_2)$, $\rho(P_1, P_2)$ of its three sides

Now, define the *invariant area* of the triangle defined by three points P_1, P_2, P_3 on the image plane as follows (Fig. 4.14):

$$S = f^2(\theta_1 + \theta_2 + \theta_3 + \pi) \ . \tag{4.6.18}$$

Here, θ_1, θ_2, θ_3 are the invariant angles between two line segments at vertices P_1, P_2, P_3, respectively. Since S is defined in terms of scalar invariants, this is also a scalar invariant. In fact, according to *spherical trigonometry*, this is equal to the area of the *spherical triangle* defined by the three great circles on the image sphere corresponding to the three lines on the image plane. The angle $\theta_1 + \theta_2 + \theta_3 - \pi$ is called the *spherical excess*. The invariant area S approaches the usual (Euclidean) area as the three points come closer to the image origin.

In summary, all quantities defined with reference to the image sphere can be computed — without constructing the image sphere — in terms of xy image coordinates. In mathematical terms, we can *realize* the spherical geometry on the xy-plane by introducing "distorted" distances and angles, or a *non-Euclidean metric*.

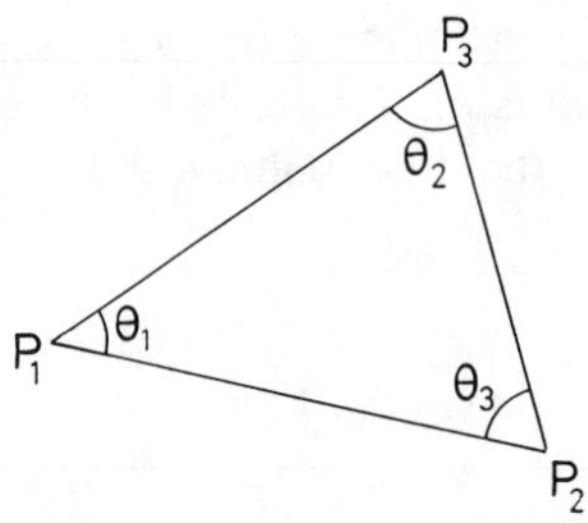

Fig. 4.14. The invariant area $S(P_1, P_2, P_3)$ of triangle $\Delta P_1 P_2 P_3$ is defined in terms of the invariant angles θ_1, θ_2, θ_3 at its vertices P_1, P_2, P_3

4.7 Invariants of Tensors

Consider a tensor. As discussed in Sect. 4.4, a tensor is invariantly decomposed into its symmetric and antisymmetric parts. Let $A = (a_{ij})$, $i, j = 1, 2, 3$ be a symmetric tensor. As we saw in Sect. 4.4, a real symmetric tensor has three scalar invariants: the three principal values σ_1, σ_2, σ_3, all of which are real. However, we need an ordering of the three principal values. Otherwise, we do not know which of these three values σ_1 indicates. A natural ordering may be $\sigma_1 \leqq \sigma_2 \leqq \sigma_3$. However, we can completely do away with such an ordering if we define *symmetric polynomials*. The basic expressions are the *fundamental symmetric polynomials*:

$$S_1 = \sigma_1 + \sigma_2 + \sigma_3 \ , \quad S_2 = \sigma_1\sigma_2 + \sigma_2\sigma_3 + \sigma_3\sigma_1 \ , \quad S_3 = \sigma_1\sigma_2\sigma_3 \ . \tag{4.7.1}$$

Lemma 4.9. *The set* $\{S_1, S_2, S_3\}$ *is algebraically equivalent to the set* $\{\sigma_1, \sigma_2, \sigma_3\}$.

Proof. We only need to show that the three principal values σ_1, σ_2, σ_3 are computed from the values of S_1, S_2, S_3. Indeed, $\sigma_1, \sigma_2, \sigma_3$ are given as the three roots of the cubic equation

$$(\lambda - \sigma_1)(\lambda - \sigma_2)(\lambda - \sigma_3) = 0 \ , \tag{4.7.2}$$

which is rewritten as

$$\lambda^3 - S_1\lambda^2 + S_2\lambda - S_3 = 0 \ . \tag{4.7.3}$$

Thus, σ_1, σ_2, σ_3 are computed if the values of S_1, S_2, S_3 are given[4].

The significance of these symmetric polynomials lies in the fact that we can compute them *without computing the three principal values*; they are given directly in terms of the original tensor components a_{ij}, $i, j = 1, 2, 3$, as follows.

Proposition 4.16.

$$S_1 = a_{11} + a_{22} + a_{33} \ , \tag{4.7.4}$$

$$S_2 = \begin{vmatrix} a_{11} & a_{12} \\ a_{21} & a_{22} \end{vmatrix} + \begin{vmatrix} a_{22} & a_{23} \\ a_{32} & a_{33} \end{vmatrix} + \begin{vmatrix} a_{33} & a_{31} \\ a_{13} & a_{11} \end{vmatrix} \ , \tag{4.7.5}$$

$$S_3 = \begin{vmatrix} a_{11} & a_{12} & a_{13} \\ a_{21} & a_{22} & a_{23} \\ a_{31} & a_{32} & a_{33} \end{vmatrix} \ . \tag{4.7.6}$$

[4] This does not mean that we can assign *arbitrary* values to S_1, S_2, S_3. The values of S_1, S_2, S_3 must be assigned so that the cubic equation (4.7.3) has three real roots. This condition is given by $(2S_1^3 - 9S_1S_2 + 27S_3^2) - 4(S_1^2 - 3S_2)^3 \leqq 0$ (Exercise 4.27).

Proof. Let $\bar{S}_1$, $\bar{S}_2$, $\bar{S}_3$ be the right-hand sides of (4.7.4–6), respectively. The three principal values $\sigma_1, \sigma_2, \sigma_3$ are the three roots of the *characteristic equation*

$$(\det(\lambda I - A) =)\, \lambda^3 - \bar{S}_1\lambda^2 + \bar{S}_2\lambda - \bar{S}_3 = 0 \ . \tag{4.7.7}$$

Comparing this equation with (4.7.3), we obtain $\bar{S}_1 = S_1$, $\bar{S}_2 = S_2$, $\bar{S}_3 = S_3$.

Computation of S_1, S_2, S_3 involves the computation of determinants. There exists another choice that is computationally more convenient. Consider the following three symmetric polynomials:

$$T_1 = \sigma_1 + \sigma_2 + \sigma_3 \ , \quad T_2 = \sigma_1^2 + \sigma_2^2 + \sigma_3^2 \ , \quad T_3 = \sigma_1^3 + \sigma_2^3 + \sigma_3^3 \ . \tag{4.7.8}$$

Lemma 4.10. *The set* $\{T_1, T_2, T_3\}$ *is algebraically equivalent to the set* $\{\sigma_1, \sigma_2, \sigma_3\}$.

Proof. We only need to show that the three principal values σ_1, σ_2, σ_3 are computed from the values of T_1, T_2, T_3. Note that the fundamental symmetric polynomials S_1, S_2, S_3 are expressed in terms of T_1, T_2, T_3 as follows (Exercise 4.19):

$$S_1 = T_1 \ , \quad S_2 = \frac{1}{2}(T_1^2 - T_2) \ , \quad S_3 = \frac{1}{6}(T_1^3 - 3T_1T_2 + 2T_3) \ . \tag{4.7.9}$$

Since the principal values σ_1, σ_2, σ_3 are computed from the fundamental symmetric polynomials S_1, S_2, S_3, we can compute σ_1, σ_2, σ_3 from T_1, T_2, T_3.

The three symmetric polynomials T_1, T_2, T_3 are computed from the original tensor components a_{ij}, i, $j = 1, 2, 3$, as follows:

Proposition 4.17.

$$T_1 = \mathrm{Tr}\{A\} \ , \qquad T_2 = \mathrm{Tr}\{A^2\} \ , \qquad T_3 = \mathrm{Tr}\{A^3\} \ . \tag{4.7.10}$$

In order to prove this, we need the following two lemmas.

Lemma 4.11. *If matrix* A *is a tensor, so are its powers* A^k, $k = 1, 2, \ldots$.

Proof. A three-dimensional matrix $\boldsymbol{A}$ is a tensor if it is transformed as

$$\boldsymbol{A}' = \boldsymbol{R}^{\mathrm{T}}\boldsymbol{A}\boldsymbol{R} \tag{4.7.11}$$

when the XYZ coordinate system is rotated by $\boldsymbol{R}$. Since $\boldsymbol{R}$ is an orthogonal matrix ($\boldsymbol{R}\boldsymbol{R}^{\mathrm{T}} = \boldsymbol{I}$), we have

$$\boldsymbol{A}'^k = (\boldsymbol{R}^{\mathrm{T}}\boldsymbol{A}\boldsymbol{R})(\boldsymbol{R}^{\mathrm{T}}\boldsymbol{A}\boldsymbol{R}) \ldots (\boldsymbol{R}^{\mathrm{T}}\boldsymbol{A}\boldsymbol{R}) = \boldsymbol{R}^{\mathrm{T}}\boldsymbol{A}^k\boldsymbol{R} \tag{4.7.12}$$

for $k = 1, 2, \ldots$. Thus, $\boldsymbol{A}^k$, $k = 1, 2, \ldots$, are also transformed as tensors.

Lemma 4.12. *The trace of a tensor is a scalar invariant.*

Proof. Since $\mathrm{Tr}\{\boldsymbol{AB}\} = \mathrm{Tr}\{\boldsymbol{BA}\}$ for arbitrary square matrices $\boldsymbol{A}$, $\boldsymbol{B}$ of the same dimension, we see from (4.7.11) that

$$\mathrm{Tr}\{\boldsymbol{A}'\} = \mathrm{Tr}\{\boldsymbol{R}^{\mathrm{T}}\boldsymbol{A}\boldsymbol{R}\} = \mathrm{Tr}\{\boldsymbol{A}\boldsymbol{R}\boldsymbol{R}^{\mathrm{T}}\} = \mathrm{Tr}\{\boldsymbol{A}\} . \tag{4.7.13}$$

Proof of Proposition 4.17. A symmetric tensor $\boldsymbol{A}$ is transformed, by an appropriate rotation $\boldsymbol{R}$ of the XYZ coordinate system, into a diagonal matrix (its *canonical form*) whose diagonal elements are the principal values $\sigma_1, \sigma_2, \sigma_3$. (The resulting coordinate system is called the *canonical frame* of tensor $\boldsymbol{A}$.) Equations (4.7.8, 10) are identical with reference to this canonical frame. However, they must hold for *any* coordinate system, because the right-hand sides of (4.7.10a-c) are scalar invariants (Lemmas 4.11, 12), and hence their values do not change if the coordinate system is rotated.

In summary, we observe:

Proposition 4.18. *An invariant basis for a symmetric tensor* $\boldsymbol{A}$ *is given by*

$$\mathrm{Tr}\{\boldsymbol{A}\} , \qquad \mathrm{Tr}\{\boldsymbol{A}^2\} , \qquad \mathrm{Tr}\{\boldsymbol{A}^3\} . \tag{4.7.14}$$

Now, we consider a general non-symmetric tensor. A tensor is invariantly decomposed into its symmetric and antisymmetric parts, and the antisymmetric part is represented by a vector (Sect. 4.4). Hence, a tensor is equivalently represented by a symmetric tensor $\boldsymbol{A} = (a_{ij})$, $i, j = 1, 2, 3$ (representing the symmetric part) and a vector $\boldsymbol{a} = (a_i)$, $i = 1, 2, 3$ (representing the antisymmtric part).

Lemma 4.13. *If* $\boldsymbol{A}$ *is a tensor and* $\boldsymbol{a}$ *is a vector, then* $\boldsymbol{Aa}$ *is a vector.*

Proof. If $\boldsymbol{A}$ is a tensor and $\boldsymbol{a}$ is a vector, they are transformed as

$$\boldsymbol{A}' = \boldsymbol{R}^{\mathrm{T}}\boldsymbol{A}\boldsymbol{R} , \qquad \boldsymbol{a}' = \boldsymbol{R}^{\mathrm{T}}\boldsymbol{a} . \tag{4.7.15}$$

Since $\boldsymbol{R}$ is an orthogonal matrix ($\boldsymbol{R}\boldsymbol{R}^{\mathrm{T}} = \boldsymbol{I}$), the product $\boldsymbol{Aa}$ is transformed as a vector:

$$\boldsymbol{A}'\boldsymbol{a}' = (\boldsymbol{R}^{\mathrm{T}}\boldsymbol{A}\boldsymbol{R})(\boldsymbol{R}^{\mathrm{T}}\boldsymbol{a}) = \boldsymbol{R}^{\mathrm{T}}(\boldsymbol{Aa}) . \tag{4.7.16}$$

Proposition 4.19. *An invariant basis for a tensor whose symmetric part is* $\boldsymbol{A}$ *and whose antisymmetric part is represented by vector* $\boldsymbol{a}$ *is given by*[5]

$$\|\boldsymbol{a}\|^2 , \qquad \mathrm{Tr}\{\boldsymbol{A}\} , \qquad \mathrm{Tr}\{\boldsymbol{A}^2\} , \qquad \mathrm{Tr}\{\boldsymbol{A}^3\} .$$

$$(\boldsymbol{a}, \boldsymbol{Aa}) , \qquad \|\boldsymbol{Aa}\|^2 , \qquad |\boldsymbol{a}\ \boldsymbol{Aa}\ \boldsymbol{A}^2\boldsymbol{a}| . \tag{4.7.17}$$

---

[5] If $O(3)$ is the group of admissible transformations, the scalar triple product can be removed from (4.7.17), and the remaining scalar invariants are an invariant basis.

Proof. The vector $\boldsymbol{a}$ has an invariant basis $\|\boldsymbol{a}\|^2$ (Proposition 4.3), and the symmetric tensor $\boldsymbol{A}$ has an invariant basis $\{\mathrm{Tr}\{\boldsymbol{A}\}, \mathrm{Tr}\{\boldsymbol{A}^2\}, \mathrm{Tr}\{\boldsymbol{A}^3\}\}$ (Proposition 4.18). Hence, we need only consider scalar invariants that specify the "relative configuration" of $\boldsymbol{a}$ and $\boldsymbol{A}$.

The XYZ coordinate system can be rotated into the canonical frame so that the three principal axes of $\boldsymbol{A}$ coincide with the X-, Y-, and Z-axes. Tensor $\boldsymbol{A}$ now takes its canonical form: a diagonal matrix whose diagonal elements are the principal values $\sigma_1, \sigma_2, \sigma_3$. Let $\bar{a}_1, \bar{a}_2, \bar{a}_3$ be the components of vector $\boldsymbol{a}$ with reference to this canonical frame. Since its "length" is already counted, we must find scalar invariants which specify the "orientation" of vector $\boldsymbol{a}$ relative to the principal axes of $\boldsymbol{A}$. An orientation in 3D space is specified by two parameters (e.g., its spherical coordinates θ, φ). Hence, we need two scalar invariants to specify the orientation of vector $\boldsymbol{a}$ relative to the XYZ coordinate system (Fig. 4.15).

Consider $(\boldsymbol{a}, \boldsymbol{A}\boldsymbol{a})$ and $(\boldsymbol{a}, \boldsymbol{A}^2\boldsymbol{a})$. The square $\boldsymbol{A}^2$ is a tensor (Lemma 4.11), and $\boldsymbol{A}\boldsymbol{a}$ and $\boldsymbol{A}^2\boldsymbol{a}$ are both vectors (Lemma 4.13). Since the inner product of two vectors is a scalar invariant (Lemma 4.8), both $(\boldsymbol{a}, \boldsymbol{A}\boldsymbol{a})$ and $(\boldsymbol{a}, \boldsymbol{A}^2\boldsymbol{a})$ are scalar invariants. With reference to the canonical frame, they are expressed as

$$(\boldsymbol{a}, \boldsymbol{A}\boldsymbol{a}) = \sigma_1\bar{a}_1^2 + \sigma_2\bar{a}_2^2 + \sigma_3\bar{a}_3^2 , \tag{4.7.18}$$

$$(\boldsymbol{a}, \boldsymbol{A}^2\boldsymbol{a}) = \sigma_1^2\bar{a}_1^2 + \sigma_2^2\bar{a}_2^2 + \sigma_3^2\bar{a}_3^2 . \tag{4.7.19}$$

Evidently, they are algebraically independent. Now, we must take into account the fact that *the principal axes are not directed*; they are *axes*. Hence, if the three principal axes are rotated rigidly by angle π around any of them, the resulting configuration must be regarded as identical to the original one. Alternatively, we

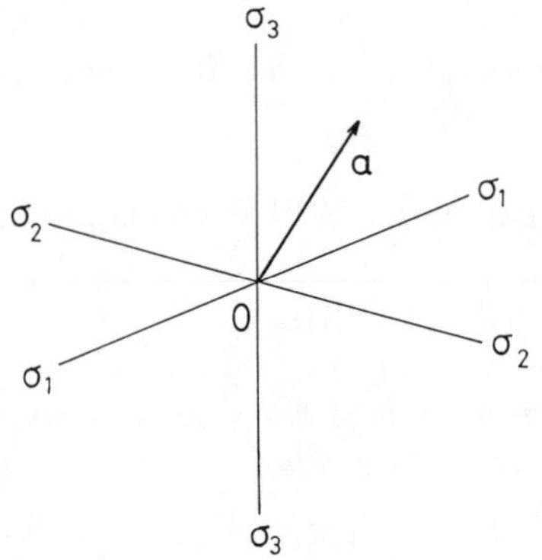

Fig. 4.15. A vector $\boldsymbol{a}$ is represented by a (directed) arrow. A symmetric tensor $\boldsymbol{A}$ is represented by three mutually perpendicular principal axes (not directed) and their corresponding principal values $\sigma_1, \sigma_2, \sigma_3$. If a vector $\boldsymbol{a}$ and a symmetric tensor $\boldsymbol{A}$ are given, there exist scalar invariants that specify the orientation of vector $\boldsymbol{a}$ relative to the principal axes of tensor $\boldsymbol{A}$

may fix the principal axes, and identify all the configurations obtained by rotating vector $\boldsymbol{a}$ by angle π around any of the principal axes. In this canonical frame, a rotation of $\boldsymbol{a}$ by angle π around a coordinate axis means changing the signs of two of its components. Equations (4.7.18, 19) have exactly this property: they are quadratic in a_1, a_2, a_3, and hence their values do not change if the signs of two of the components are reversed.

However, transformations of $\boldsymbol{a}$ that do not change the values of $(\boldsymbol{a}, \boldsymbol{A}\boldsymbol{a})$ and $(\boldsymbol{a}, \boldsymbol{A}^2\boldsymbol{a})$ are not restricted to rotations of $\boldsymbol{a}$ by angle π around each principal axis. Since (4.7.18, 19) are quadratic in a_1, a_2, a_3, *any* change of their signs will not alter the values of (4.7.8, 19). In other words, the values of $(\boldsymbol{a}, \boldsymbol{A}\boldsymbol{a})$ and $(\boldsymbol{a}, \boldsymbol{A}^2\boldsymbol{a})$ do not change if the *mirror image* of $\boldsymbol{a}$ is taken with respect to the planes defined by two principal axes. In order to forbid odd numbers of reflections, we need to incorporate another scalar invariant. One way is to add $|\boldsymbol{a}\,\boldsymbol{A}\boldsymbol{a}\,\boldsymbol{A}^2\boldsymbol{a}|$, which has in this canonical frame the following form:

$$|\boldsymbol{a}\,\boldsymbol{A}\boldsymbol{a}\,\boldsymbol{A}^2\boldsymbol{a}| = \begin{vmatrix} \bar{a}_1 & \sigma_1\bar{a}_1 & \sigma_1^2\bar{a}_1 \\ \bar{a}_2 & \sigma_2\bar{a}_2 & \sigma_2^2\bar{a}_2 \\ \bar{a}_3 & \sigma_3\bar{a}_3 & \sigma_3^2\bar{a}_3 \end{vmatrix} . \tag{4.7.20}$$

To put it in mathematical terms, if we fix the values of σ_1, σ_2, σ_3, the group of transformations of vector $\boldsymbol{a}$ that do not change the values of $(\boldsymbol{a}, \boldsymbol{A}\boldsymbol{a})$ and $(\boldsymbol{a}, \boldsymbol{A}^2\boldsymbol{a})$—let us call it the *transformation group associated with these invariants*—is the group known as the *octahedral group* O_h in crystrallography. If the scalar invariant of (4.7.20) is added, the associated transformation group shrinks into the *tetrahedral group* T.

Example 4.5. Let A be a symmetric tensor. From Lemma 4.11, powers A^4, $A^5, \ldots$ are also tensors. From Lemma 4.12, their traces $T_4 = \mathrm{Tr}\{A^4\}$, $T_5 = \mathrm{Tr}\{A^5\}, \ldots$ are scalar invariants. From Proposition 4.18, $T_1 = \mathrm{Tr}\{A\}$, $T_2 = \mathrm{Tr}\{A^2\}$, $T_3 = \mathrm{Tr}\{A^3\}$ are an invariant basis of tensor A. Hence, invariants $T_4, T_5, \ldots$ must be expressed in terms of $\{T_1, T_2, T_3\}$. This is easily done by applying the *Cayley–Hamilton theorem*

$$A^3 - S_1A^2 + S_2A - S_3\boldsymbol{I} = 0 , \tag{4.7.21}$$

where S_1, S_2, S_3 are the fundamental symmetric polynomials (4.7.1) in the principal values of A. If we multiply (4.7.21) by A, we have

$$A^4 - S_1A^3 + S_2A^2 - S_3A = \boldsymbol{0} . \tag{4.7.22}$$

Taking the trace of both sides, we have

$$T_4 - S_1T_3 + S_2T_2 - S_3T_1 = 0 . \tag{4.7.23}$$

Combining this with (4.7.9), we obtain

$$T_4 = \frac{1}{6}(T_1^4 + 3T_2^2 - 6T_1^2T_2 + 8T_1T_3) . \tag{4.7.24}$$

Invariants $T_5, T_6, \ldots$ are also expressed in terms of $\{T_1, T_2, T_3\}$ in a similar way.

Example 4.6. Consider the optical flow resulting from planar motion (Sect. 3.7):

$$u = u_0 + Ax + By + (Ex + Fy)x \,,$$
$$v = v_0 + Cx + Dy + (Ex + Fy)y \,. \tag{4.7.25}$$

In Chap. 3, we proved that if we put

$$\boldsymbol{a} = \begin{pmatrix} -(v_0/f + fF)/2 \\ (u_0/f + fE)/2 \\ (C - B)/2 \end{pmatrix},$$

$$\boldsymbol{B} = \begin{pmatrix} (2A - D)/3 & (B + C)/2 & (u_0/f - fE)/2 \\ (B + C)/2 & (2D - A)/3 & (v_0/f - fF)/2 \\ (u_0/f - fE)/2 & (v_0/f - fF)/2 & -(A + D)/3 \end{pmatrix}, \tag{4.7.26}$$

then $\boldsymbol{a}$ is a vector and $\boldsymbol{B}$ is a tensor (Theorem 3.10). Since tensor $\boldsymbol{B}$ is a deviator (its trace is zero), an invariant basis for this optical flow is given by

$$\|\boldsymbol{a}\|^2 \,, \quad \mathrm{Tr}\{\boldsymbol{B}^2\} \,, \quad \mathrm{Tr}\{\boldsymbol{B}^3\} \,, \quad (\boldsymbol{a}, \boldsymbol{Ba}) \,, \quad \|\boldsymbol{Ba}\|^2 \,, \quad |\boldsymbol{a}\ \boldsymbol{Ba}\ \boldsymbol{B}^2\boldsymbol{a}| \,. \tag{4.7.27}$$

In other words, *any* scalar invariant of optical flow (4.7.25) is expressed in terms of these quantities.

4.7.1 Symmetric Polynomials

Symmetric polynomials play a fundamental role in the theory of invariants. The following polynomials are called the *fundamental symmetric polynomials* in n variables $x_1, \ldots, x_n$:

$$S_1 = x_1 + x_2 + \ldots + x_n,$$
$$S_2 = x_1x_2 + x_1x_3 + \ldots + x_{n-1}x_n,$$
$$\ldots\ldots\ldots\ldots\ldots\ldots$$
$$S_{n-1} = x_2 \ldots x_n + x_1x_3 \ldots x_n + \ldots + x_1 \ldots x_{n-1},$$
$$S_n = x_1x_2 \ldots x_n. \tag{4.7.28}$$

A function

$$J = g(x_1, \ldots, x_n) \tag{4.7.29}$$

of $x_1, \ldots, x_n$ is said to be *symmetric* if the value J does not change under arbitrary permutations of $x_1, \ldots, x_n$. The *fundamental theorem on symmetric functions* states that *any symmetric function is expressed in terms of the fundamental symmetric polynomials*:

$$J = G(S_1, \ldots, S_n) . \tag{4.7.30}$$

The proof is simple. The value of J is computed from $S_1, \ldots, S_n$ if we recall that equation

$$(\lambda - x_1)(\lambda - x_2) \ldots (\lambda - x_n) = 0 \tag{4.7.31}$$

is rewritten as

$$\lambda^n - S_1\lambda^{n-1} + S_2\lambda^{n-2} - \ \ldots\ + (-1)^{n-2}S_{n-1}\lambda + (-1)^{n-1}S_n = 0 . \tag{4.7.32}$$

Namely, given the values of $S_1, \ldots, S_n$, we only need to compute the n roots $x_1, \ldots, x_n$ of (4.7.32), and substitute them into (4.7.29). The ordering of the n roots is irrelevant because (4.7.29) is symmetric in them.

This proof is rigorous. However, we must be careful about one thing that has not been explicitly stated so far. We have used the term *algebraic expression* without specifically mentioning what the *basic operations* are. If we closely examine our previous reasoning, we find that we have tacitly assumed that the basic operations are addition, subtraction, multiplication, division, and *computing roots of polynomials.*

On the other hand, most of our results did not involve roots. For example, we can directly compute invariants $\|\boldsymbol{a}\|^2$, $(\boldsymbol{a}, \boldsymbol{b})$, $|\boldsymbol{abc}|$, $\mathrm{Tr}\{\boldsymbol{A}\}$, $\mathrm{Tr}\{\boldsymbol{A}^2\}$, $\mathrm{Tr}\{\boldsymbol{A}^3\}$, etc. without computing roots. The proofs of these results can also be rewritten without using roots. For instance, we can modify the proof of Proposition 4.4 so that it does not involve square roots (Exercise 4.9). In general, however, this process requires many intricate techniques.

As an example of such intricacy, let us state the *strong form* of the fundamental theorem on symmetric functions. Let the right-hand side of (4.7.29) be a symmetric polynomial in $x_1, \ldots, x_n$. Let us prove that there exists a polynomial $g(S_1, \ldots S_n)$ for (4.7.30). The proof is given by induction based on the degree of the polynomial. Let the *degree* of term $x_1^{k_1}x_2^{k_2} \ldots x_n^{k_n}$ be defined as an n-tuple $(k_1, k_2, \ldots, k_n)$. For example, the degree of x_1x_2 is $(1, 1, 0, \ldots, 0)$. Degrees are ordered *lexicographically*: the one with the larger k_1 is higher; if k_1 is the same, the one with the larger k_2 is higher; if both k_1 and k_2 are the same, we look at k_3, and so on. The degree of a polynomials is defined by the highest degrees of its constituent terms.

The lowest-degree symmetric polynomial is a constant, whose degree is $(0, \ldots, 0)$. The next lowest is const. $\times\ S_1$ + const., whose degree is $(1, 0, \ldots, 0)$. They are obviously polynomials in $S_1, \ldots, S_n$. Assume that the theorem holds for all symmetric polynomials with degrees lower than $(k_1, \ldots, k_n)$. Let $g(x_1, \ldots, x_n)$ be a symmetric polynomial of degree $(k_1, \ldots, k_n)$. Then,

$$\tilde{g}(x_1, \ldots, x_n) = g(x_1, \ldots, x_n) - cS_1^{k_1-k_2}S_2^{k_2-k_3} \ldots S_{n-1}^{k_{n-1}-k_n}S_n^{k_n} \tag{4.7.33}$$

is also a symmetric polynomial but with a lower degree, where c is the coefficient of the term of the highest degree in $g(x_1, \ldots, x_n)$. *Hint*: The degree of $S_1^{k_1} \ldots S_n^{k_n}$ is $\left(\sum_{i=1}^n k_i, \sum_{i=2}^n k_i, \ldots, \sum_{i=n}^n k_i\right)$ (Exercise 4.25). Hence, from the induction hypothesis, it is expressed as a polynomial $\tilde{G}(S_1, \ldots, S_n)$ in the fundamental symmetric polynomials $S_1, \ldots, S_n$. Consequently, $g(x_1, \ldots, x_n)$ is expressed as a polynomial in $S_1, \ldots, S_n$ as follows:

$$g(x_1, \ldots, x_n) = \tilde{G}(S_1, \ldots, S_n) + S_1^{k_1-k_2} S_2^{k_2-k_3} \ldots S_{n-1}^{k_{n-1}-k_n} S_n^{k_n} . \tag{4.7.34}$$

Consider, for example,

$$J = x^3 + y^3 + z^3 + (x + y + z)^3 - (y + z)^3 - (z + x)^3 - (x + y)^3 . \tag{4.7.35}$$

According to the theorem we have just proved, it must be expressed in the form

$$J = \alpha S_1^3 + \beta S_1 S_2 + \gamma S_3 , \tag{4.7.36}$$

where α, β, γ are constants. These constants are determined by comparing (4.7.35) and (4.7.36). Alternatively, we can substitute particular values for x, y, z. For example, if we put $x = 1$, $y = z = 0$, we obtain $J = 0$ from (4.7.35) and $J = \alpha$ from (4.7.36). Hence, $\alpha = 0$. Similarly, from $x = y = 1$, $x = 0$, we obtain $\beta = 0$, and from $x = y = z = 1$, we obtain $\gamma = 6$. Thus, we obtain $J = 6S_3$.

Another technique for constructing expressions in the fundamental symmetric polynomials is the use of *Euler's formulae* (Exercise 4.26).

4.7.2 Classical Theory of Invariants

Let S be a set of geometrical quantities such as vectors and tensors, and let the group G of admissible transformations be the *general linear group* $GL(n)$, or its subgroup such as the *orthogonal group* $O(n)$ and the *special orthogonal group* $SO(n)$. Suppose each element of S changes its value in a specified way when a transformation $g \in G$ is applied.

An *algebraic expression* over S is a root of an *algebraic equation* $X^n + A_1 X^{n-1} + \ldots + A_n = 0$ for some n, where the coefficients $A_1, \ldots, A_n$ are all rational expressions in the elements of S. An *algebraic invariant* of S is an algebraic expression over S that does not change its value under the group G of admissible transformations.

The classical theory of invariants focuses on *polynomial invariants*—polynomials in the elements of S that do not change their values under the group G of admissible transformations. The justification is that an algebraic invariant is represented by *rational invariants*, since an algebraic invariant is a root of an algebraic equation with rational coefficients. It is easy to prove that these rational coefficients themselves are also invariants if the algebraic equation is *irreducible* (i.e., if no factorization is possible), which can be always assumed without losing generality. It is also easy to prove that *a rational invariant is a quotient of two polynomial invariants.*

Let A be a set of polynomial invariants of S. An element of A is said to be *reducible* if it is expressed as a polynomial in the rest of the elements of A. If no element is reducible, the set A is said to be *irreducible*. Irreducibility does *not* mean that the members of A are algebraically independent. Even if some elements of A are mutually related by a polynomial, it is not always possible to solve for one as a polynomial in the others. Such a polynomial relationship is called a *syzygy*.

If *any* polynomial invariant of S is expressed as a polynomial in the elements of A, the set A is called an *integrity* (or *polynomial*) *basis* of S. An integrity basis may contain redundant elements, or some of its elements are related by syzygies. *Hilbert's theorem*, states that *for any finite system of vectors and tensors, there exists an integrity basis that consists of a finite number of invariants.*

If we remove the restriction of an algebraic expression and allow any functional relationships, we are talking about *functional invariants*. A set of functional invariants of S is called a *functional basis* if any functional invariant of S is expressed as a function of its members.

At first sight, there two types of invariants are quite different. In reality, they are not so very different. In fact, it can be proved that *an integrity basis is also a functional basis*. The intuitive explanation in Sect. 4.5 is a discussion from the viewpoint of functional invariance, but the term *invariant basis* has been used sometimes in the sense of integrity basis and sometimes in the sense of functional basis.

Another issue is the *minimality* of an invariant basis, i.e., to test whether or not any of its elements can be removed in such a way that the remaining invariants are also an invariant basis. This is in general a very difficult problem, since even if some elements are related by an equation (a polynomial or function equation), the relationship may be a syzygy: one element may not be expressed in terms of the others in a required form (a polynomial or function form).

4.8 Reconstruction of Camera Rotation

As mentioned earlier, we say that two images are *equivalent* if one image can be transformed into the other by some camera rotation transformation. If two images are equivalent, all scalar invariants must have identical values. Suppose two images are known to be equivalent. We now show that we can reconstruct the camera rotation that transforms one image into the other. Let $c_1, \ldots, c_m$ be image characteristics of the first image, and let $c'_1, \ldots, c'_m$ be their values for the second image. We want to determine the camera rotation $\boldsymbol{R}$ that transforms the second image to the first image. (In the following, whenever vectors and tensors are mentioned, they are assumed to be nonzero.)

First, consider the case of a vector. We can easily observe the following fact:

Theorem 4.3. *A vector* $\boldsymbol{a}$ *is mapped onto another vector* $\boldsymbol{a}'$ *by a rotation around unit vector* $\boldsymbol{n}$ *by angle* Ω *screw-wise if*

$$\boldsymbol{n} = \frac{\boldsymbol{a} \times \boldsymbol{a}'}{\|\boldsymbol{a} \times \boldsymbol{a}'\|}, \qquad \Omega = \cos^{-1}\frac{(\boldsymbol{a}, \boldsymbol{a}')}{\|\boldsymbol{a}\| \cdot \|\boldsymbol{a}'\|}. \tag{4.8.1}$$

Proof. Vector $\boldsymbol{n}$ is perpendicular to both $\boldsymbol{a}$ and $\boldsymbol{a}'$ such that $\{\boldsymbol{a}, \boldsymbol{a}', \boldsymbol{n}\}$ are a right-hand system. Since Ω is the angle between vectors $\boldsymbol{a}$ and $\boldsymbol{a}'$, if vector $\boldsymbol{a}$ is rotated around axis $\boldsymbol{n}$ screw-wise by angle Ω, it coincides with vector $\boldsymbol{a}'$ (Fig. 4.16).

The matrix $\boldsymbol{R}$ which realizes this rotation is given by (3.3.11). From this theorem, we see that if the camera is rotated around axis $\boldsymbol{n}$ screw-wise by angle Ω, the second image (vector $\boldsymbol{a}'$) coincides with the first image (vector $\boldsymbol{a}$). This means that the camera rotation can be computed once we observe, for two equivalent images, some image characteristics that are transformed as vectors. The rotation is not unique; after this rotation, the camera can be rotated arbitrarily around vector $\boldsymbol{a}'$.

Consider two or more vectors. Let us call a set of three mutually orthogonal unit vectors an *orthonormal system.*

Lemma 4.14. *An orthonormal system* $\{\boldsymbol{e}_1, \boldsymbol{e}_2, \boldsymbol{e}_3\}$ *is mapped onto another orthonormal system* $\{\boldsymbol{e}'_1, \boldsymbol{e}'_2, \boldsymbol{e}'_3\}$ *by rotation*[6]

$$\boldsymbol{R} = \boldsymbol{e}'_2\boldsymbol{e}_1^{\mathrm{T}} + \boldsymbol{e}'_2\boldsymbol{e}_2^{\mathrm{T}} + \boldsymbol{e}'_3\boldsymbol{e}_3^{\mathrm{T}} . \tag{4.8.2}$$

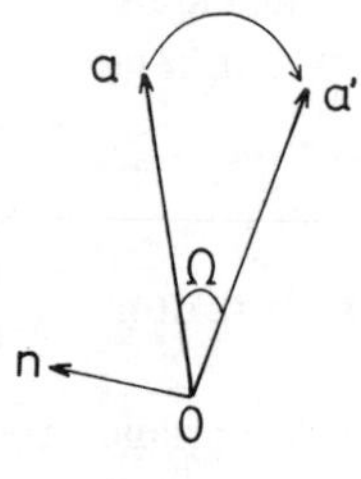

Fig. 4.16. Vector $\boldsymbol{a}$ is mapped onto vector $\boldsymbol{a}'$ by a rotation of angle Ω screw-wise around unit vector $\boldsymbol{n} = \boldsymbol{a} \times \boldsymbol{a}'/\|\boldsymbol{a} \times \boldsymbol{a}'\|$, where $\cos\Omega = (\boldsymbol{a}, \boldsymbol{b})/\|\boldsymbol{a}\| \cdot \|\boldsymbol{a}'\|$

[6] If the two systems have different senses, e.g., one is a right-hand system and the other is a left-hand system, the resulting $\boldsymbol{R}$ is an improper rotation of determinant -1.

Proof. From the orthonormality

$$(e_i, e_j) = \delta_{ij} , \qquad i, j = 1, 2, 3 , \tag{4.8.3}$$

we can easily confirm that

$$Re_j = \sum_{i=1}^{3} e'_i e_i^{\mathrm{T}} e_j = \sum_{i=1}^{3} e'_i (e_i, e_j) = \sum_{i=1}^{3} e'_i \delta_{ij} = e'_j , \qquad j = 1, 2, 3 . \tag{4.8.4}$$

Hence, if the camera is rotated by R, the second image (vectors $\{e'_1, e'_2, e'_3\}$) coincides with the first image (vectors $\{e_1, e_2, e_3\}$).

This result is extended to non-orthonormal vectors. Let $\{a, b, c\}$ be three linearly independent vectors. The three vectors $\{\tilde{a}, \tilde{b}, \tilde{c}\}$ are called their *reciprocal system* if they satisfy

$$\begin{aligned} &(a, \tilde{a}) = 1 , \quad &(a, \tilde{b}) = 0 , \quad &(a, \tilde{c}) = 0 , \\ &(b, \tilde{a}) = 0 , \quad &(b, \tilde{b}) = 1 , \quad &(b, \tilde{c}) = 0 , \\ &(c, \tilde{a}) = 0 , \quad &(c, \tilde{b}) = 0 , \quad &(c, \tilde{c}) = 1 , \end{aligned} \tag{4.8.5}$$

and if the sense is also the same (i.e., if $\{a, b, c\}$ are a right-hand system, so are $\{\tilde{a}, \tilde{b}, \tilde{c}\}$). The reciprocal system of $\{a, b, c\}$ is easily constructed by the following vectors:

$$\tilde{a} = \frac{b \times c}{|abc|} , \qquad \tilde{b} = \frac{c \times a}{|abc|} , \qquad \tilde{c} = \frac{a \times b}{|abc|} . \tag{4.8.6}$$

The reciprocal system of the reciprocal system is the original three vectors (Exercise 4.28), and the reciprocal of an orthonormal system is itself an orthonormal system: it is *self-reciprocal.*

If we use the reciprocal system, Lemma 4.14 is extended to non-orthonormal systems as follows:

Lemma 4.15. *Three linearly independent vectors* $\{a, b, c\}$ *are mapped onto vectors* $\{a', b', c'\}$ *by*

$$R = a'\tilde{a}^{\mathrm{T}} + b'\tilde{b}^{\mathrm{T}} + c'\tilde{c}^{\mathrm{T}} , \tag{4.8.7}$$

where $\{\tilde{a}, \tilde{b}, \tilde{c}\}$ *are the reciprocal system of* $\{a, b, c\}$.

The proof is exactly the same as that of Lemma 4.14. Hence, if the camera is rotated by R, the second image (vectors $\{a', b', c'\}$) coincides with the first image (vectors $\{a, b, c\}$).

This result can be applied to two pairs of vectors as follows:[7]

[7] In (4.8.7, 8), the two systems of vectors need not be congruent to each other. The resulting R still defines a transformation between these two systems, but it is no longer a rotation.

Theorem 4.4. *Two non-collinear vectors* $\{\boldsymbol{a}, \boldsymbol{b}\}$ *are mapped into vectors* $\{\boldsymbol{a}', \boldsymbol{b}'\}$ *by*

$$\boldsymbol{R} = \boldsymbol{a}'\tilde{\boldsymbol{a}}^{\mathrm{T}} + \boldsymbol{b}'\tilde{\boldsymbol{b}}^{\mathrm{T}} + \boldsymbol{c}'\tilde{\boldsymbol{c}}^{\mathrm{T}} , \tag{4.8.8}$$

where

$$\boldsymbol{c} = \boldsymbol{a} \times \boldsymbol{b} , \qquad \boldsymbol{c}' = \boldsymbol{a}' \times \boldsymbol{b}' , \tag{4.8.9}$$

and $\{\tilde{\boldsymbol{a}}, \tilde{\boldsymbol{b}}, \tilde{\boldsymbol{c}}\}$ *are the reciprocal system of* $\{\boldsymbol{a}, \boldsymbol{b}, \boldsymbol{c}\}$.

Hence, if the camera is rotated by $\boldsymbol{R}$, the second image (vectors $\{\boldsymbol{a}', \boldsymbol{b}'\}$) coincides with the first image (vectors $\{\boldsymbol{a}, \boldsymbol{b}\}$). If we observe three or more vectors, we can choose from among them two arbitrary non-collinear vectors. Any choice yields the same camera rotation as long as the two imges are equivalent.

Now, consider points. If the camera rotation is small (Sect. 4.6), a point is represented by a unit vector, (4.6.1). As a result, if we observe two points on the image plane, we can compute the camera rotation $\boldsymbol{R}$ that maps one to the other by Theorem 4.3. If two or more pairs of points should coincide, we apply Theorem 4.4.

Next, consider lines. They are also represented by unit vectors, (4.6.7). However, their signs must be disregarded—the vector $\boldsymbol{n}$ must be treated as an *axis*. This fact causes some complications. Suppose we observe a line on the two images. Let $\boldsymbol{n}$ be the unit vector representing the line for the first image. The sign of $\boldsymbol{n}$ can be chosen arbitrarily. Let $\boldsymbol{n}'$ be the unit vector for the second image. Since vector $\boldsymbol{n}'$ must be treated as an axis, we must consider $\boldsymbol{n}'$ and $-\boldsymbol{n}'$ separately. Hence, we obtain two solutions for $\boldsymbol{R}$. (The second solution is obtained from the first by adding a rotation by angle π around an axis starting from the viewpoint and passing through the line on the image plane.)

Suppose we observe a pair of lines on the first image and another pair on the second image. Let $\{\boldsymbol{n}_1, \boldsymbol{n}_2\}$ be the unit vectors representing the lines on the first image. Their signs can be chosen arbitrarily. Let $\{\boldsymbol{n}'_1, \boldsymbol{n}'_2\}$ be the corresponding unit vectors for the second image. Since their signs are indeterminate, we must consider the four cases $\{\boldsymbol{n}'_1, \boldsymbol{n}'_2\}$, $\{\boldsymbol{n}'_1, -\boldsymbol{n}'_2\}$, $\{-\boldsymbol{n}'_1, \boldsymbol{n}'_2\}$, $\{-\boldsymbol{n}'_1, -\boldsymbol{n}'_2\}$. However, if $\boldsymbol{n}_1$ and $\boldsymbol{n}_2$ are not perpendicular to each other, only two out of these four pairs are congruent to $\{\boldsymbol{n}_1, \boldsymbol{n}_2\}$. For example, if $\boldsymbol{n}_1$ and $\boldsymbol{n}_2$ make an acute angle, we can choose the two pairs that define acute angles. Applying Theorem 4.4, we have two solutions for $\boldsymbol{R}$, but only one is the true solution: the other is not a "small" rotation. (The false solution is the rotation that brings the scene behind the camera to the front.) If $\boldsymbol{n}_1$ and $\boldsymbol{n}_2$ are perpendicular to each other, we have two solutions. The condition $(\boldsymbol{n}_1, \boldsymbol{n}_2) = 0$ means that the two lines are perpendicular to each other if the intersection is brought to the image origin by a camera rotation. The second solution is obtained from the first by adding a rotation by angle π around an axis starting from the viewpoint and passing through the intersection of the two lines. The case of three or more lines can be treated similarly.

Suppose a point and a line not passing through it are observed for two images. Let $\boldsymbol{a}$ be the unit vector representing the point and $\boldsymbol{n}$ the unit vector representing the line for the first image. Let $\boldsymbol{a}'$ and $\boldsymbol{n}'$ be those for the second image. Since the point is not on the line, the two vectors $\boldsymbol{a}$, $\boldsymbol{n}$ are not mutually perpendicular. Hence, we can decide which of $\{\boldsymbol{a}', \boldsymbol{n}'\}$ and $\{\boldsymbol{a}', -\boldsymbol{n}'\}$ is congruent to $\{\boldsymbol{a}, \boldsymbol{n}\}$. Then, we apply Theorem 4.4.

Now, consider tensors. Suppose we observe a tensor on two images, and suppose the two images are equivalent. Then, the camera rotation $\boldsymbol{R}$ that brings the second image to the first is reconstructed as follows. As discussed earlier, a tensor is represented by its symmetric and antisymmtric parts. The latter is represented by a vector. Let $\boldsymbol{A}$ be the symmetric part for the first image, and $\boldsymbol{A}'$ that for the second.

First, suppose tensor $\boldsymbol{A}$ (hence $\boldsymbol{A}'$ as well) has three distinct principal values. Let $\boldsymbol{e}_1, \boldsymbol{e}_2, \boldsymbol{e}_3$ be the unit vectors indicating its principal axes, and let $\boldsymbol{e}'_1, \boldsymbol{e}'_2, \boldsymbol{e}'_3$ be those for tensor $\boldsymbol{A}'$. Since they are orthonormal systems, we can apply Lemma 4.14. However, we must treat the principal axes as *axes*. For $\boldsymbol{A}$, we can always take its eigenvectors $\{\boldsymbol{e}_1, \boldsymbol{e}_2, \boldsymbol{e}_3\}$ so that they form a right-hand orthonormal system. For $\boldsymbol{A}'$, we obtain, from $\pm \boldsymbol{e}'_1$, $\pm \boldsymbol{e}'_2$, $\pm \boldsymbol{e}'_3$, eight different combinations, and four of them are left-hand systems. Applying Theorem 4.4, we obtain four solutions for $\boldsymbol{R}$. (Three of them are obtained from the remaining one by adding rotations around each of the three principal axes by angle π.) From among the four, we can choose the correct one if we incorporate the vector representing the antisymmetric part of the original tensor. Let $\boldsymbol{a}$ be the vector representing the antisymmetric part for the first image, and $\boldsymbol{a}'$ that for the second. From the four rotations computed above, we choose the one that maps $\boldsymbol{a}$ to $\boldsymbol{a}'$. If $\boldsymbol{a}$ is parallel to one of the principal axes of $\boldsymbol{A}$, it is mapped into only two different vectors, and we have two solutions. If the vector $\boldsymbol{a}$ is zero, we obtain four solutions.

Suppose the symmetric part $\boldsymbol{A}$ (and hence $\boldsymbol{A}'$) has only two distinct principal values (a single root and a pair of multiple roots). Let $\boldsymbol{e}$ be the unit vector indicating the principal axes of $\boldsymbol{A}$ for the single root, and let $\boldsymbol{e}'$ be that of $\boldsymbol{A}'$. Application of Theorem 4.3 gives a solution, but arbitrary rotations around $\boldsymbol{e}'$ can be added. The solution becomes unique if we incorporate the vectors $\boldsymbol{a}$ and $\boldsymbol{a}'$ representing the antisymmetric parts. If $\boldsymbol{a}$ is not perpendicular to $\boldsymbol{e}$, we can take $\boldsymbol{e}$ such that $\boldsymbol{a}$ and $\boldsymbol{e}$ make an acute angle. The sign of $\boldsymbol{e}'$ is chosen so that $\boldsymbol{a}'$ and $\boldsymbol{e}'$ also make an acute angle. If $\boldsymbol{a}$ is perpendicualr to $\boldsymbol{e}$, we must consider two cases $\{\boldsymbol{a}', \boldsymbol{e}'\}$, $\{\boldsymbol{a}', -\boldsymbol{e}'\}$, and we obtain two solutions.

If the symmetric part $\boldsymbol{A}$ has only one principal value, $\boldsymbol{A}$ is a multiple of the unit vector $\boldsymbol{I}$, and the rotation is completely indeterminate. If we incorporate the vectors $\boldsymbol{a}$ and $\boldsymbol{a}'$ representing the antisymmetric parts, Theorem 4.3 gives one solution, but arbitrary rotations can be added around $\boldsymbol{a}'$.

Example 4.7. Consider the optical flow given in Example 4.6. Since the invariants (4.7.27) constitute an invariant basis, *two flows are equivalent if and* *only if these scalar invariants have the same values.* If two flows are equivalent, we can reconstruct the camera rotation $\boldsymbol{R}$ by computing vector $\boldsymbol{a}$ and tensor $\boldsymbol{B}$ of (4.7.26), and applying the procedure described above.

Exercises

4.1 Show that the point and line representing the same vector are mutually dual.

4.2 Show that $(fa_1/a_3, fa_2/a_3)$ is a fixed point of the optical flow of (4.3.3).

4.3 Show that $a_1x + a_2y + fa_3 = 0$ is a fixed line of the optical flow of (4.3.3).

4.4 (a) Prove that all eigenvalues of a real symmetric matrix are real. (They are called the *principal values*.)
(b) Prove that the eigenvectors of a real symmetric matrix can be chosen to be real vectors. (They define the *principal axes*.)
(c) Prove that eigenvectors of a real symmetric matrix for distinct eigenvectors are mutually orthogonal.

4.5 (a) Show that if $\{\boldsymbol{e}_1, \boldsymbol{e}_2, \boldsymbol{e}_3\}$ are an orthonormal system, the matrix $\boldsymbol{R} = (\boldsymbol{e}_1\boldsymbol{e}_2\boldsymbol{e}_3)$ consisting of vectors $\boldsymbol{e}_1, \boldsymbol{e}_2, \boldsymbol{e}_3$ as its three columns in this order is an orthogonal matrix: $\boldsymbol{R}\boldsymbol{R}^{\mathrm{T}} = \boldsymbol{R}^{\mathrm{T}}\boldsymbol{R} = \boldsymbol{I}$.
(b) Show the converse: If $\boldsymbol{R} = (\boldsymbol{e}_1\boldsymbol{e}_2\boldsymbol{e}_3)$ is an orthogonal matrix, its columns $\{\boldsymbol{e}_1, \boldsymbol{e}_2, \boldsymbol{e}_3\}$ are an orthonormal system.
(c) Show that the determinant of an orthogonal matrix $\boldsymbol{R} = (\boldsymbol{e}_1\boldsymbol{e}_2\boldsymbol{e}_3)$ is 1 if and only if its columns $\{\boldsymbol{e}_1, \boldsymbol{e}_2, \boldsymbol{e}_3\}$ are a right-hand system.

4.6 Confirm (4.4.5).

4.7 (a) Show that two vectors representing two points on the image plane are mutually orthogonal if and only if the two points are conjugate to each other.
(b) Show that a vector representing a point and an axis representing a line are mutually orthogonal if and only if the point and the line are dual to each other.

4.8 Prove *Schwarz's inequality*:

$$|(\boldsymbol{a}, \boldsymbol{b})| \leqq \|\boldsymbol{a}\| \cdot \|\boldsymbol{b}\|.$$

Hint: The quadratic polynomial $f(t) = \|\boldsymbol{a} - t\boldsymbol{b}\|^2$ in t is nonnegative for all values of t. Consider the discriminant.

4.9 Prove the following strong form of Proposition 4.4: Any *polynomial invariant* of two vectors $\{\boldsymbol{a}, \boldsymbol{b}\}$ is expressed as a *rational* expression in $\{\|\boldsymbol{a}\|^2, (\boldsymbol{a}, \boldsymbol{b}), \|\boldsymbol{b}\|^2\}$. In other words, if (4.5.6) is a scalar invariant and the right-hand side is a polynomial in $a_1, a_2, a_3, b_1, b_2, b_3$, it can be expressed in terms of additions, subtractions, multiplications, and divisions of $\{\|\boldsymbol{a}\|^2, (\boldsymbol{a}, \boldsymbol{b}), \|\boldsymbol{b}\|^2\}$ without involving square roots. Give the proof by showing the following facts.
(i) If the XYZ coordinate system is further rotated from the configuration of Fig. 4.6 by angle π around the Z-axis, and if $\boldsymbol{i}', \boldsymbol{j}', \boldsymbol{k}'$ are the unit

basis vectors along the new coordinate axes, vectors $\boldsymbol{a}$, $\boldsymbol{b}$ are expressed as $\boldsymbol{a} = \alpha \boldsymbol{k}'$, $\boldsymbol{b} = -\beta \boldsymbol{j}' + \gamma \boldsymbol{k}'$. Hence, $J = g(0, 0, \alpha, 0, -\beta, \gamma)$.
(ii) If the XYZ coordinate system is further rotated from the configuration of Fig. 4.6 by angle π around the X'-axis, and if $\boldsymbol{i}''$, $\boldsymbol{j}''$, $\boldsymbol{k}''$ are the unit basis vectors along the new coordinate axes, vectors $\boldsymbol{a}$, $\boldsymbol{b}$ are expressed as $\boldsymbol{a} = -\alpha \boldsymbol{k}''$, $\boldsymbol{b} = \gamma \boldsymbol{j}'' - \gamma \boldsymbol{k}''$. Hence, $J = g(0, 0, -\alpha, 0, \beta, -\gamma)$.
(iii) From (i) and (ii), it is concluded that if J consists of terms of the form of $\alpha^p \beta^q \gamma^r$, the exponent q must be even, and the exponents p and r must have the same parity, i.e., both even or both odd.
(iv) Hence, the invariant J is a polynomial in $\alpha^2 = \|\boldsymbol{a}\|^2$, $\beta^2 = \|\boldsymbol{b}\|^2 - /\|\boldsymbol{a}\|^2$, $\gamma^2 = (\boldsymbol{a}, \boldsymbol{b})^2/\|\boldsymbol{a}\|^2$, and $\alpha\gamma = (\boldsymbol{a}, \boldsymbol{b})$.

4.10 (a) Let $(\boldsymbol{abc})$ be the matrix consisting of three vectors $\boldsymbol{a}$, $\boldsymbol{b}$, $\boldsymbol{c}$ as its columns, and let $|\boldsymbol{abc}|$ be its determinant. Show that $|\boldsymbol{abc}|$ is computed by inner and vector products as (4.5.11).
(b) Show that $|\boldsymbol{abc}|$ is the volume of the parallelepiped constructed from $\boldsymbol{a}$, $\boldsymbol{b}$, $\boldsymbol{c}$ as the three edges if $\{\boldsymbol{a}, \boldsymbol{b}, \boldsymbol{c}\}$ are a right-hand system in this order. Show also that $|\boldsymbol{abc}|$ is the negative of the parallelepiped of the volume if $\{\boldsymbol{a}, \boldsymbol{b}, \boldsymbol{c}\}$ are a left-hand system.

4.11 (a) Show that if $\boldsymbol{a}$ and $\boldsymbol{b}$ are vectors, the vector product $\boldsymbol{a} \times \boldsymbol{b}$ is also transformed as a vector.
(b) Prove (4.5.16) by following the proof of Proposition 4.4.
(c) If $\boldsymbol{a}$, $\boldsymbol{b}$, $\boldsymbol{c}$, $\boldsymbol{d}$ are vectors, the inner product $(\boldsymbol{a} \times \boldsymbol{b}, \boldsymbol{c} \times \boldsymbol{d})$ is a scalar invariant. Hence, it must be expressed in terms of the invariant basis of (4.5.13). Show that

$$(\boldsymbol{a} \times \boldsymbol{b}, \boldsymbol{c} \times \boldsymbol{d}) = (\boldsymbol{a}, \boldsymbol{c})(\boldsymbol{b}, \boldsymbol{d}) - (\boldsymbol{a}, \boldsymbol{d})(\boldsymbol{b}, \boldsymbol{c}) \ .$$

4.12 Show that a point (a, b) is represented by the unit vector (4.6.1).

4.13 Show that a line $Ax + By + C = 0$ is represented by the unit vector (4.6.7).

4.14 Prove (4.6.15) of Example 4.3.

4.15 Prove (4.6.16) of Example 4.4.

4.16* *Spherical trigonometry.* Consider a sphere of unit radius. A *spherical triangle* is a triangle defined by three great circles. Let α, β, γ be the angles at its three vertices, and let A, B, C be the respective arc lengths of the sides opposite to those vertices. Let S be the area of the triangle (the area of the portion of the sphere's surface inside the triangle).
(a) Prove the *cosine law*:

$$\cos A = \cos B \cos C + \sin B \sin C \cos \alpha \ ,$$

and explain the geometrical meaning of the angle θ defined by (4.6.17).
(b) Prove the *sine law*:

$$\frac{\sin \alpha}{\sin A} = \frac{\sin \beta}{\sin B} = \frac{\sin \gamma}{\sin C} \ .$$

(c) Prove that the area S is given by the *spherical excess*:

$$S = \alpha + \beta + \gamma - \pi \ .$$

4.17 Show that (4.7.2) can be rewritten as (4.7.3).
4.18 Show that the characteristic equation of a symmetric matrix is given by (4.7.7), where $\bar{S}_1$, $\bar{S}_2$, $\bar{S}_3$ are defined by the right-hand sides (4.7.4–6).
4.19 Prove the expressions (4.7.9)
4.20 Determine the possible orientations of vector $\boldsymbol{a}$ when the values of the invariants (7.18–20) are specified, and show that the subgroup associated with these invariants is the *trihedral group* (the transformations that map a trihedron onto itself).
4.21* Prove the *Cayley–Hamilton theorem* (4.7.21).
4.22 Derive (4.7.24) from (4.7.9, 23).
4.23 (a) Prove that the determinant $|\boldsymbol{A}|$ of a symmetric tensor $\boldsymbol{A}$ is a scalar invariant.
(b) Hence, the determinant $|\boldsymbol{A}|$ may be expressed in terms of the invariant basis $\{\mathrm{Tr}\{\boldsymbol{A}\}, \mathrm{Tr}\{\boldsymbol{A}^2\}, \mathrm{Tr}\{\boldsymbol{A}^3\}\}$. Show that

$$|\boldsymbol{A}| = \frac{1}{6}(\mathrm{Tr}\{\boldsymbol{A}\}^3 - 3\mathrm{Tr}\{\boldsymbol{A}\}\mathrm{Tr}\{\boldsymbol{A}^2\} + 2\mathrm{Tr}\{\boldsymbol{A}^3\}).$$

Hint: Consider the expression in the canonical frame, see (4.7.9c).
4.24 Show that (4.7.31) can be rewritten as (4.7.32).
4.25 (a) Show that the degree of $S_1^{k_1} \dots S_n^{k_n}$ is $\left(\sum_{i=1}^n k_i, \sum_{i=2}^n k_i, \dots, \sum_{i=n}^n k_i\right)$.
(b) Show that the degree of (4.7.33) is lower than that of $g(x_1, \dots, x_n)$.
4.26 (a) *Euler's formulae*. Let T_k be the following symmetric polynomial in n variables $x_1, \dots, x_n$:

$$T_k = x_1^k + \dots + x_n^k .$$

Since this is a symmetric polynomial, it must be expressed in terms of the fundamental symmetric polynomials $\{S_1, \dots, S_n\}$. This is done by applying the following recurrence relationships to $T_1, T_2, \dots$ successively:

$$T_k = S_1 T_{k-1} - S_2 T_{k-2} + \dots + (-1)^{k-2} S_{k-1} T_1 + (-1)^{k-1} k S_k \ , \quad k = 1, 2, \dots, n \ ,$$

$$T_k = S_1 T_{k-1} - S_2 T_{k-2} + \dots + (-1)^{n-1} S_n T_{k-n} \ , \qquad k > n \ .$$

Prove these by induction based on the number n of variables.
(b) The fundamental symmetric polynomials $\{S_1, \dots, S_n\}$ can be ex-

pressed in terms of $\{T_1, \ldots, T_n\}$ by applying the following recurrence relationship successively:

$$S_k = \frac{1}{k}(T_1 S_{k-1} - T_2 S_{k-2} + \ldots + (-1)^{k-2} T_{k-1} S_1$$

$$+ (-1)^{k-1} T_k), \qquad k = 1, 2, \ldots, n .$$

Derive this relationship from the relationships given in (a).

4.27* (a) Consider a cubic equation $x^3 + ax^2 + bx + c = 0$, and let α, β, γ be its three roots. Its *discriminant* D is defined by

$$D = [(\alpha - \beta)(\beta - \gamma)(\gamma - \alpha)]^2 .$$

Show that the three roots are all real if $D \geqq 0$, while they are one real root and a complex conjugate pair if $D < 0$. *Hint*: Show that if α is a real, $\beta = p + iq$, and $\gamma = p - iq$ $(q \neq 0)$, then $D = -4q^2[(\alpha - p)^2 + q^2]$.

(b) Since the discriminant D is a symmetric polynomial in α, β, γ, it must be expressed in terms of their fundamental symmetric polynomials. From the relation between roots and coefficients of algebraic equations, we find

$$a = -(\alpha + \beta + \gamma), \qquad b = \alpha\beta + \beta\gamma + \gamma\alpha, \qquad c = \alpha\beta\gamma .$$

Hence, the discriminant D must be expressed as a polynomial in a, b, c. First, consider the case of $a = 0$: $x^3 + bx + c = 0$. Show that

$$D = -4b^3 - 27c^3 .$$

Hint: Infer that the final form must be $D = Ab^3 + Bc^2$, where A, B are constants. Then, determine the constants A and B.

(c) Show that by substituting $x = x' - a/3$, the cubic equation $x^3 + ax^2 + bx + c = 0$ is converted into

$$x'^3 + b'x' + c' = 0 ,$$

$$a' = -\frac{1}{3}(a^2 - 3b), \quad b' = \frac{1}{27}(2a^2 - 9ab + 27c) .$$

(d) Show that the discriminant D of cubic equation $x^3 + ax^2 + bx + c = 0$ is given by

$$D = a^2b^2 + 18abc - 4b^3 - 4a^3c - 27c^2 .$$

4.28 (a) Show that the reciprocal system of $\{\boldsymbol{a}, \boldsymbol{b}, \boldsymbol{c}\}$ is given by (4.8.6).

(b) Show that the reciprocal system of the reciprocal system is the original system.

(c) Show that an orthonormal system is self-reciprocal, i.e., the reciprocal of an orthonormal system is itself.

(d) Let $(\boldsymbol{abc})$ be the matrix consisting of three vectors $\boldsymbol{a}$, $\boldsymbol{b}$, $\boldsymbol{c}$ as its three columns. Show that

$$(\boldsymbol{abc})(\tilde{\boldsymbol{a}}\,\tilde{\boldsymbol{b}}\,\tilde{\boldsymbol{c}})^{\mathrm{T}} = \boldsymbol{I} ,$$

and hence

$$(\boldsymbol{abc})^{-1} = (\tilde{\boldsymbol{a}}\,\tilde{\boldsymbol{b}}\,\tilde{\boldsymbol{c}})^{\mathrm{T}} ,$$

where $\{\tilde{\boldsymbol{a}}, \tilde{\boldsymbol{b}}, \tilde{\boldsymbol{c}}\}$ are the reciprocal system of $\{\boldsymbol{a}, \boldsymbol{b}, \boldsymbol{c}\}$.

5. Characterization of Scenes and Images

In this chapter, we characterize continuous images by discrete parameters. Noting that the input to the camera is a collection of rays of light to which intensity values are attached, we define "scenes" as functions of orientations. Images are perspective projections of these functions onto the image plane. The camera rotation transformation of scenes and images induces representations of $SO(3)$ in the space of functions known as *spherical harmonics*. We study the geometrical structures of the "scene space" and the "image space" by means of tensor calculus. Then, we analyze invariance properties of "linear functionals" of images, which we call *image features*, and show how they are used for identification and classification of shapes.

5.1 Parametrization of Scenes and Images

In the previous chapters, we have studied transformation properties of a finite number of image characteristics, assuming that they are given through appropriate measurements. In this chapter, we enquire into the question of how to characterize continuous images by discrete parameters. One approach is to assume a parameterized model of the object. Then, the model parameters are determined by fitting a parametrized function to the image. If no parametrized model is available, as in the case of a natural scene, the image must be regarded as a two-dimensional function, taking the intensity value $F(x, y)$ at each point (x, y). Then, the following questions naturally arise: Is there a general procedure to characterize a continuous image $F(x, y)$ by discrete parameters? How do we know whether or not the parameters have meanings inherent to the scene? How do we know their transformation rule under camera rotation?

In order to answer such questions, we must recall the imaging geometry of perspective projection (Sect. 1.2). The input to the camera is *rays of light* to which intensity values are attached. Let $s(\boldsymbol{m})$ be the intensity of the ray in the orientation specified by a unit vector $\boldsymbol{m}$. To the viewer (or the camera), the function $s(\boldsymbol{m})$ is all that is observed. We call such a function a *scene*, and call the space of such scenes the *scene space*. This is a linear space, over which the camera rotation induces representations of $SO(3)$. Employing tensor calculus, we will actually construct basis functions of all irreducible representations. These functions are called *spherical harmonics*.

An image $F(x, y)$ is a perspective projection of a scene $s(\boldsymbol{m})$. We define the *projection operator* Π, and study how it preserves the properties of the scene space. We will see that the space consisting of all images—we call it the *image space*—also defines representations of $SO(3)$. We will show that the irreducible representations in the image space are in one-to-one correspondence to the irreducible representations in the scene space through the projection operator Π.

Then, we go on to characterization of images by *linear functionals*, i.e., linear mapping from the image space to real numbers. Here, we consider functionals $J_i[F]$, $i = 1, \ldots, N$, defined by integration of images with weight functions $w_i(x, y)$, $i = 1, \ldots, N$:

$$J_i[F] = \int w_i(x, y) F(x, y) dx\, dy\ , \qquad i = 1, \ldots, N\ . \tag{5.1.1}$$

The integration is carried out over the entire image plane. We call such quantities *image features.*

The following question naturally arises: How can we choose the weight functions $w_i(x, y)$, $i = 1, \ldots, N$, so that the resulting image features $J_i[F]$, $i = 1, \ldots, N$, have desirable properties? For example, how can we make them a vector or a tensor? We will show that the study of the scene space $\mathscr{S}$ and the image space $\mathscr{I}$ from the group theoretical viewpoint automatically leads to the answer to this question. We will also examine the geometrical meanings of typical image features, and show some applications to the classification and identification of shapes.

5.2 Scenes, Images, and the Projection Operator

As pointed out in the previous section, the outside world is recognized by the camera (or the viewer) as *rays of light* to which intensity values are attached; depth information is completely lost. A ray is specified by the unit vector $\boldsymbol{m}$ starting from the viewpoint. Let $s(\boldsymbol{m})$ be its intensity value—it may be the brightness or the darkness, the R–G–B values (the intensities of the *R*ed, *G*reen and *B*lue components) when colors are involved, or anything else. Since the function $s(\boldsymbol{m})$ represents a distribution of the intensity over the rays coming into the camera, we call it a *scene*. In this chapter, we use the viewer-centered XYZ coordinate system (Sect. 1.2), and regard $Z = f$ as the image plane (Fig. 5.1). We also assume the existence of an ideal camera that records the exact intensity value of each ray at its intersection with the image plane. We call a function $F(x, y)$ thus obtained an *image*.

Proposition 5.1. *A scene $s(\boldsymbol{m})$ produces an image*

$$F(x, y) = s\left(\frac{x}{\sqrt{x^2 + y^2 + f^2}}, \frac{y}{\sqrt{x^2 + y^2 + f^2}}, \frac{f}{\sqrt{x^2 + y^2 + f^2}}\right)\ . \tag{5.2.1}$$

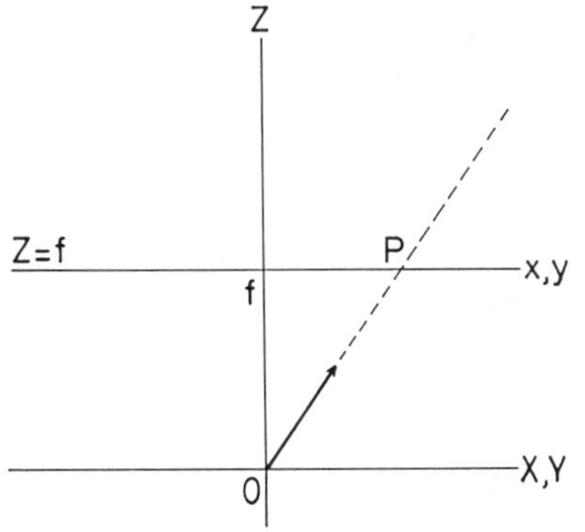

Fig. 5.1. The viewer-centered coordinate system. A point P on the image plane $Z = f$ is specified by the unit vector $\boldsymbol{m}$ along the ray starting from the viewpoint O and passing through the point P on the image plane

Proof. As shown in Fig. 5.1, a point (x, y) on the image plane defines a ray or orientation $\boldsymbol{m} = (m_1, m_2, m_3)$ given by

$$m_1 = \frac{x}{\sqrt{x^2 + y^2 + f^2}}, \; m_2 = \frac{y}{\sqrt{x^2 + y^2 + f^2}}, \; m_3 = \frac{f}{\sqrt{x^2 + y^2 + f^2}}. \tag{5.2.2}$$

The image (5.2.1) results from our definition of the ideal camera.

Let $\mathscr{S}$ be the set of all scenes; we call it the *scene space*. Let $\mathscr{I}$ be the set of all images; we call it the *image space*. Equation (5.2.1) defines an operator that maps a scene in $\mathscr{S}$ to an image in $\mathscr{I}$. We put

$$\Pi[s](x, y) \equiv s\left(\frac{x}{\sqrt{x^2 + y^2 + f^2}}, \frac{y}{\sqrt{x^2 + y^2 + f^2}}, \frac{f}{\sqrt{x^2 + y^2 + f^2}}\right), \tag{5.2.3}$$

and call the operator $\Pi: \mathscr{S} \to \mathscr{I}$ the *perspective projection operator* or simply *projection*. The projection Π is an onto mapping from the scene space $\mathscr{S}$ to the image space $\mathscr{I}$, but not a one-to-one mapping, since $s(\boldsymbol{m})$ can take arbitrary values for $m_3 \leqq 0$.

A ray of orientation $\boldsymbol{m} = (m_1, m_2, m_3)$ intersects with the image plane $Z = f$ if $m_3 > 0$. The intersection (x, y) is given by

$$x = f\frac{m_1}{m_3}, \qquad y = f\frac{m_2}{m_3}. \tag{5.2.4}$$

In view of this, we define the *inverse perspective projection operator* or simply *inverse projection* by

$$\Pi^{-1}[F](\boldsymbol{m}) \equiv \begin{cases} F\left(f\dfrac{m_1}{m_3}, f\dfrac{m_2}{m_3}\right), & m_3 > 0 , \\ \text{not defined} , & m_3 \leqq 0 . \end{cases} \tag{5.2.5}$$

In the following, whenever we talk about inversely projected scenes $s(\boldsymbol{m}) = \Pi^{-1}[F](\boldsymbol{m})$, we consider only the part for $m_3 > 0$.

We define the *camera rotation transformation* $T_{\boldsymbol{R}}: \mathscr{S} \to \mathscr{S}$ of scenes such that if we observe a scene $s(\boldsymbol{m})$ and if the camera is rotated around the center of the lens by $\boldsymbol{R}$, we observe a new scene $T_{\boldsymbol{R}}s(\boldsymbol{m})$. We immediately find:

Proposition 5.2.

$$T_{\boldsymbol{R}}s(\boldsymbol{m}) = s(\boldsymbol{R}\boldsymbol{m}) . \tag{5.2.6}$$

Proof. Let $\tilde{s}(\boldsymbol{m})$ be the new scene after the camera rotation by $\boldsymbol{R}$. Then, the value of $\tilde{s}(\boldsymbol{m})$ is equal to $s(\boldsymbol{R}\boldsymbol{m})$, since orientation $\boldsymbol{m}$ after the camera rotation indicates orientation $\boldsymbol{R}\boldsymbol{m}$ before the camera rotation (Fig. 5.2).

Corollary 5.1. *The set of all the camera rotation transformations $T_{\boldsymbol{R}}$ is a group of transformations, and the correspondence from $\boldsymbol{R}$ to $T_{\boldsymbol{R}}$ is an (anti-)homomorphism from $SO(3)$:*[1]

$$T_{\boldsymbol{R}'} \circ T_{\boldsymbol{R}} = T_{\boldsymbol{R}\boldsymbol{R}'} , \qquad \boldsymbol{R}', \boldsymbol{R} \in SO(3) , \tag{5.2.7}$$

$$T_{\boldsymbol{I}} = I , \qquad (T_{\boldsymbol{R}})^{-1} = T_{\boldsymbol{R}^{-1}} . \tag{5.2.8}$$

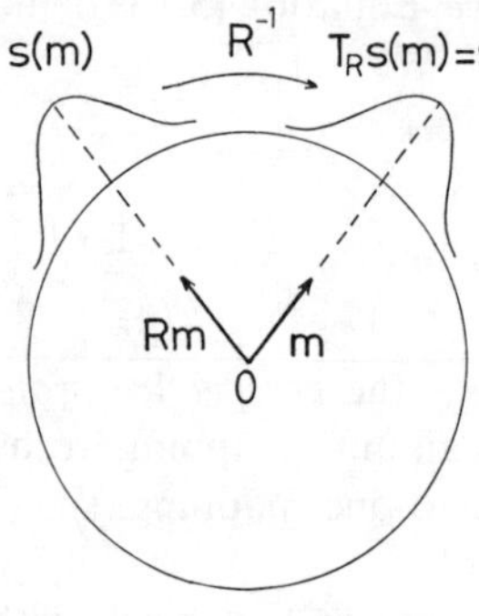

Fig. 5.2. A scene can be viewed as an intensity map on a sphere surrounding the viewpoint. Rotation of the camera by $\boldsymbol{R}$ is equivalent to rotation of the scene by $\boldsymbol{R}^{-1}$. If a new scene $T_{\boldsymbol{R}}f(\boldsymbol{m})$ is obtained by rotating a scene $f(\boldsymbol{m})$ by $\boldsymbol{R}^{-1}$, the value of the new scene in orientation $\boldsymbol{m}$ is equal to the value of the original scene in orientation $\boldsymbol{R}\boldsymbol{m}$. Hence $T_{\boldsymbol{R}}f(\boldsymbol{m}) = f(\boldsymbol{R}\boldsymbol{m})$

[1] If the scene (instead of the camera) is rotated by $\boldsymbol{R}$, the order of $\boldsymbol{R}$ and $\boldsymbol{R}'$ is reversed between the left- and the right-hand sides of (5.2.7) (Sect. 3.3).

Proof. If we put $\tilde{s}(\boldsymbol{m}) \equiv s(\boldsymbol{Rm})$, we see that

$$(T_{R'} \circ T_R)s(\boldsymbol{m}) = T_{R'}\tilde{s}(\boldsymbol{m}) = \tilde{s}(\boldsymbol{R'm}) = s(\boldsymbol{RR'm}) = T_{RR'}s(\boldsymbol{m}) \ . \tag{5.2.9}$$

As scene $s(\boldsymbol{m})$ is transformed into $T_R s(\boldsymbol{m})$ by a camera rotation $\boldsymbol{R}$, the projection image $\Pi[s](x, y)$ is transformed into $\Pi[T_R s](x, y)$. Using the same symbol T_R, we define the *camera rotation transformation* T_R: $\mathscr{I} \to \mathscr{I}$ of images as follows (Fig. 5.3):

$$T_R F(x, y) \equiv \Pi[T_R \Pi^{-1}[F]](x, y) \ . \tag{5.2.10}$$

There is, however, one problem. As shown in (5.2.5), the inverse projection Π^{-1} is not well defined; if the camera is rotated, a new part of the scene comes into view and some part goes out of view, even if the image plane is infinitely large. Hence, the transformation T_R cannot be defined solely in terms of images. In the following, we adopt the following convention: Whenever the camera rotation is mentioned, we assume that the image is assumed to have a *finite support*, i.e., has nonzero intensity values over a finite region around the image origin, and the part coming into or going out of view has zero intensity. In fact, we have implicitly assumed this in the previous chapters.

Let $\boldsymbol{R} = (r_{ij})$, $i, j = 1, 2, 3$, be the camera rotation. If (5.2.3, 6) are substituted into (5.2.10), we obtain the following relation (Exercise 5.1):

Proposition 5.3.

$$T_R F(x, y) = F\left(f\frac{r_{11}x + r_{12} + r_{13}}{r_{31}x + r_{32} + r_{33}},\ f\frac{r_{21}x + r_{22} + r_{23}}{r_{31}x + r_{32} + r_{33}}\right) . \tag{5.2.11}$$

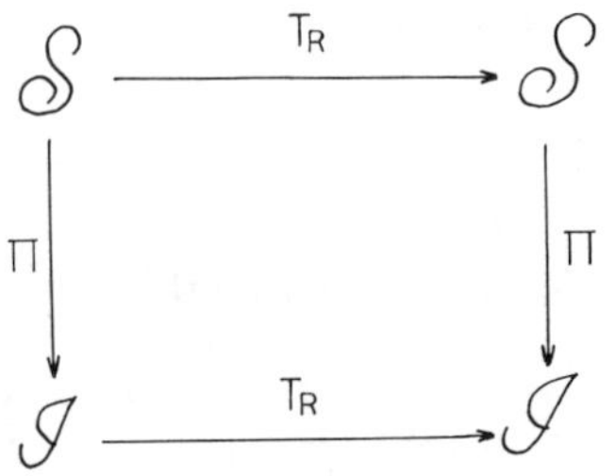

Fig. 5.3. The camera rotation transformation T_R in the image space $\mathscr{I}$ is defined so that the projection Π followed by T_R in the image space $\mathscr{I}$ is equal to T_R in the scene space $\mathscr{S}$ followed by the projction Π

5.3 Invariant Subspaces of the Scene Space

The scene space $\mathscr{S}$ is the set of all $s(\boldsymbol{m})$. For convenience, we allow negative intensity values. Then, the scene space $\mathscr{S}$ is regarded as a linear space. From Proposition 5.2, we immediately observe:

Lemma 5.1. *The camera rotation transformation* $T_R: \mathscr{S} \to \mathscr{S}$ *is a linear mapping:*

$$T_R[c_1 s_1(\boldsymbol{m}) + c_2 s_2(\boldsymbol{m})] = c_1 T_R s_1(\boldsymbol{m}) + c_2 T_R s_2(\boldsymbol{m}) \tag{5.3.1}$$

for all $s_1(\boldsymbol{m}), s_2(\boldsymbol{m}) \in \mathscr{S}$ *and all real numbers* c_1, c_2.

We say that a subspace $L \subset \mathscr{S}$ is invariant if $s(\boldsymbol{m}) \in L$ implies $T_R s(\boldsymbol{m}) \in L$ for all $\boldsymbol{R} \in SO(3)$. An invariant subspace is said to be *reducible* if it is expressed as the direct sum of two invariant subspaces. If no such reduction is possible, the invariant subspace is called *irreducible.* Since $SO(3)$ is a compact group, all invariant subspaces are *fully reducible* (Appendix, Sect. A.6). Hence, we can equivalently say that an invariant subspace is irreducible *if and only if it does not include any proper invariant subspace.*

An invariant subspace of a finite dimension defines a representation of $SO(3)$. To see this, let $\{s_1(\boldsymbol{m}), \ldots, s_n(\boldsymbol{m})\}$ be a basis of an invariant subspace L. Since subspace L is invariant, each $T_R s_i(\boldsymbol{m})$, $i = 1, \ldots, n$, is also a member of L and hence expressed as a linear combination of $\{s_1(\boldsymbol{m}), \ldots, s_n(\boldsymbol{m})\}$:

$$T_R s_i(\boldsymbol{m}) = \sum_{j=1}^{n} t_{ij}(\boldsymbol{R}) s_j(\boldsymbol{m}) \;, \qquad i = 1, \ldots, n \;. \tag{5.3.2}$$

The matrix $\boldsymbol{T}(\boldsymbol{R}) = (t_{ij})$, $i, j = 1, \ldots, n$, has the following property:

Proposition 5.4. *The matrix* $\boldsymbol{T}(\boldsymbol{R})$ *defines a representation of* $SO(3)$:

$$\boldsymbol{T}(\boldsymbol{R}\boldsymbol{R}') = \boldsymbol{T}(\boldsymbol{R})\boldsymbol{T}(\boldsymbol{R}') \;, \qquad \boldsymbol{R}, \boldsymbol{R}' \in SO(3) \;, \tag{5.3.3}$$

$$\boldsymbol{T}(\boldsymbol{I}) = \boldsymbol{I} \;, \qquad \boldsymbol{T}(\boldsymbol{R})^{-1} = \boldsymbol{T}(\boldsymbol{R}^{-1}) \;. \tag{5.3.4}$$

Proof. By definition,

$$T_{RR'} s_i(\boldsymbol{m}) = \sum_{j=1}^{n} t_{ij}(\boldsymbol{R}\boldsymbol{R}') s_j(\boldsymbol{m}) \;, \qquad i = 1, \ldots, n \;. \tag{5.3.5}$$

From (5.2.7) and (5.3.1), the left-hand side becomes

$$\begin{aligned}(T_{R'} \circ T_R) s_i(\boldsymbol{m}) &= T_{R'}\left(\sum_{k=1}^{n} t_{ik}(\boldsymbol{R}) s_k(\boldsymbol{m})\right) = \sum_{k=1}^{n} t_{ik}(\boldsymbol{R}) T_{R'} s_k(\boldsymbol{m}) \\ &= \sum_{k=1}^{n} t_{ik}(\boldsymbol{R})\left(\sum_{j=1}^{n} t_{kj}(\boldsymbol{R}')\, s_j(\boldsymbol{m})\right) = \sum_{j=1}^{n}\left(\sum_{k=1}^{n} t_{ik}(\boldsymbol{R}) t_{kj}(\boldsymbol{R}')\right) s_j(\boldsymbol{m}) \;.\end{aligned} \tag{5.3.6}$$

Comparing (5.3.5) and (5.3.6), we find that

$$t_{ij}(\boldsymbol{R}) = \sum_{k=1}^{n} t_{ik}(\boldsymbol{R}) t_{kj}(\boldsymbol{R}') \ , \qquad i, j = 1, \ldots, n \ . \tag{5.3.7}$$

From this follow (5.3.3, 4).

The dimension n of matrix $\boldsymbol{T}(\boldsymbol{R})$ is equal to the dimension of the invariant subspace L, which is equal to the number of functions that generate L. The representation thus defined in L is reducible or irreducible depending on whether the invariant subspace L is reducible or irreducible.

One trivial invariant subspace of $\mathscr{S}$ is the one-dimensional subspace $L^{(0)}$ consisting of scenes whose intensity is constant over all orientations, and the scene that takes 1 for all orientations, denoted by 1, can be chosen as its basis: all the scenes of $L^{(0)}$ are multiples of scene 1. Since the camera rotation transformation T_R does not change any of the scenes of $L^{(0)}$, we conclude:

Lemma 5.2. *Subspace $L^{(0)}$ is a one-dimensional invariant subspace of $\mathscr{S}$.*

The one-dimensional representation of $SO(3)$ defined in $L^{(0)}$ is called the *identity representation*: the transformation matrix $\boldsymbol{T}(\boldsymbol{R})$ is simply 1.

Consider a scene defined by $f(\boldsymbol{m}) = m_1$. This is the scene whose intensity at orientation $\boldsymbol{m}$ is given by m_1 (the X-component of vector $\boldsymbol{m}$). We denote such a scene by m_1. Scenes m_2 and m_3 are defined similarly. Let $L^{(1)}$ be the three-dimensional subspace of $\mathscr{S}$ generated by scenes $\{m_1, m_2, m_3\}$.

Lemma 5.3. *Subspace $L^{(1)}$ is a three-dimensional invariant subspace of $\mathscr{S}$.*

Proof. If we put $\boldsymbol{R} = (r_{ij})$, $i, j = 1,2,3$, we see from (5.2.6) that

$$T_R m_i = (\boldsymbol{R}\boldsymbol{m})_i = \sum_{j=1}^{3} r_{ij} m_j \ , \qquad i = 1, 2, 3 \ , \tag{5.3.8}$$

where $(\ldots)_i$ denotes the ith component. Thus, the camera rotation transformation T_R of each of $\{m_1, m_2, m_3\}$ is expressed as a linear combination of them.

The three-dimensional representation of $SO(3)$ defined in $L^{(1)}$ is called the *vector representation*: the transformation matrix $\boldsymbol{T}(\boldsymbol{R})$ is $\boldsymbol{R}$ itself.

Consider a scene $f(\boldsymbol{m}) = m_1 m_2$ whose intensity is $m_1 m_2$ at orientation $\boldsymbol{m}$. We denote this scene by $m_1 m_2$. Scenes $m_i m_j$, $i, j = 1, 2, 3$, are defined similarly. Let $L^{(2)}$ be the subspace generated by the nine scenes $\{m_i m_j\}$, $i, j = 1, 2, 3$.

Lemma 5.4. *The subspace $L^{(2)}$ is a six-dimensional invariant subspace of $\mathscr{S}$.*

Proof. From (5.2.6), we see that

$$T_R m_i m_j = (\boldsymbol{Rm})_i (\boldsymbol{Rm})_j = \sum_{k=1}^{3} \sum_{l=1}^{3} r_{ik} r_{jl} m_k m_l \,, \qquad i, j = 1, 2, 3 \,. \quad (5.3.9)$$

Thus, the camera rotation transformation T_R of each of the scenes $\{m_i m_j\}$, $i, j = 1, 2, 3$, is a linear combination of them. Since $m_i m_j = m_j m_i$, the subspace $L^{(2)}$ is generated by $\{m_1 m_1, m_2 m_2, m_3 m_3, m_2 m_3, m_3 m_1, m_1 m_2\}$. Hence, it is six dimensional.

The six-dimensional representation of $SO(3)$ defined in $L^{(2)}$ is called the *tensor representation of degree* 2. Equation (5.3.9) defines a nine-dimensional transformation matrix $\boldsymbol{T}(\boldsymbol{R})$—this nine-dimensional matrix is called the *tensor* (or *Kronecker*) *product* and denoted by $\boldsymbol{R} \times \boldsymbol{R}$. We need not compute this nine-dimensional matrix; it is much easier to deal with (5.3.9) directly.

Arguing similarly, we can define the subspace $L^{(l)}$ of $\mathscr{S}$ as the space of scenes generated by scenes $\{m_{i_1} \ldots m_{i_l}\}$, $i_1, \ldots, i_l = 1, 2, 3$. We observe:

Proposition 5.5. *The supspace $L^{(l)}$ is an $(l+1)(l+2)/2$-dimensional invariant subspace of $\mathscr{S}$.*

Proof. From (5.2.6), we see that

$$T_R m_{i_1} \ldots m_{i_l} = (\boldsymbol{Rm})_{i_1} \ldots (\boldsymbol{Rm})_{i_l} = \sum_{i_1=1}^{3} \ldots \sum_{i_l=1}^{3} r_{i_1 j_1} \ldots r_{i_l j_l} m_{j_1} \ldots m_{j_l} \quad (5.3.10)$$

for $i_1, \ldots, i_l = 1, 2, 3$. Thus, the camera rotation transformation T_R of each of the scenes $\{m_{i_1} \ldots m_{i_l}\}$, $i_1, \ldots, i_l = 1, 2, 3$, is a linear combination of them. Since there are $(l+1)(l+2)/2$ different scenes in $m_{i_1} \ldots m_{i_l}$, $i_1, \ldots, i_l = 1, 2, 3$ (Exercise 5.2), the invariant subspace $L^{(l)}$ is $(l+1)(l+2)/2$ dimensional.

The $(l+1)(l+2)/2$-dimensional representation of $SO(3)$ defined in $L^{(l)}$ is called the *tensor representation of degree l*. The $(l+1)(l+2)/2$-dimensional matrix $\boldsymbol{T}(\boldsymbol{R})$ defined by (5.3.10) is the *tensor* (or *Kronecker*) *product* denoted by $\boldsymbol{R} \times \ldots \times \boldsymbol{R}$. (The matrix form need not be computed; it is easier to deal with (5.3.10) directly.)

5.3.1 Tensor Calculus

The study of invariant linear subspaces and representations of $SO(3)$ becomes easy if we adopt *tensor calculus*. Tensor calculus originates from the study by Georg F.B. Riemann (1826–1866) of what was later known as *Riemannian geometry*, and has been established by such mathematicians as Eugenio

Beltrami (1835–1900), Elwin B. Christoffel (1829–1900), Rudolf O. S. Lipschitz (1832–1903), Curbastro G. Ricci (1853–1925), Luigi Bianchi (1856–1928), and Tullio Levi–Civita (1873–1941). Honoring the contribution of Ricci, tensor calculus is also known as *Ricci calculus.* Tensor calculus has also becomes popular among physicists since Albert Einstein (1879–1955) used it in the description of his general theory of relativity.

Tensors can be defined in a very general framework, but we restrict ourselves to *Cartesian tensors,* taking $SO(n)$ as the admissible transformations. An n-dimensional *tensor* of degree r is a quantity $A_{i_1,\ldots i_r}$ with n indices on which group element $\boldsymbol{R} \in SO(n)$ acts as a transformation T_R defined by

$$T_R A_{i_1 \ldots i_r} = \sum_{j_1}^{n} \ldots \sum_{j_r}^{n} r_{i_1 j_1} \ldots r_{i_r j_r} A_{j_1 \ldots j_r}\,, \quad i_1, \ldots, i_r = 1, \ldots, n\,, \tag{5.3.11}$$

where $\boldsymbol{R} = (r_{ij})$, $i, j = 1, \ldots, n$. In particular, a tensor of degree 0 (with no indices) is called a *scalar*. A scalar, by definition, does not change its value by the action of T_R. A tensor of degree 1 (with one index) is called a *vector*. We find from (5.3.10) that scenes $m_{i_1} \ldots m_{i_l}$, $i_1, \ldots, i_l = 1, 2, 3$, are tensors of degree l.

In order to simplify equations, tensor calculus adopts the convention that the summation symbol $\sum$ is omitted for indices that appear repeatedly in one expression. This is called *Einstein's summation convention.* We also agree to understand that primed indices indicate transformed values. Namely, we denote the tensor resulting from $A_{i_1 \ldots i_r}$ by transformation T_R simply by $A_{i'_1 \ldots i'_r}$. Thus, the tensor indices play a symbolic role. With these conventions, (5.3.11) is simply written as

$$A_{i'_1 \ldots i'_r} = r_{i'_1 i_1} \ldots r_{i'_r i_r} A_{i_1 \ldots i_r}\,. \tag{5.3.12}$$

The following is a list of fundamental properties of tensors. All the assertions are easily proved: the quantity in question is proved to be a tensor if it is shown to satisfy (5.3.12).

(i) **Scalar multiplication of tensors.** If $A_{i \ldots j}$ is a tensor, then

$$C_{i \ldots j} \equiv c A_{i \ldots j} \tag{5.3.13}$$

is also a tensor of the same degree, where c is an arbitrary real constant.

(ii) **Sum/difference of tensors.** If $A_{i \ldots j}$ and $B_{i \ldots j}$ are tensors of the same degree, then

$$C_{i \ldots j} \equiv A_{i \ldots j} + B_{i \ldots j} \tag{5.3.14}$$

is also a tensor of the same degree. This is called the *sum* of the two tensors. The *difference* of tensors can be defined similarly. Note that *only tensors of the same degree can be added or subtracted.*

(iii) **Product of tensors.** If $A_{i\ldots j}$ is a tensor of degree r, and $B_{k\ldots l}$ is a tensor of degree s, then

$$C_{i\ldots l} \equiv A_{i\ldots j} B_{k\ldots l} \tag{5.3.15}$$

is a tensor of degree $r + s$. This is called the *product* of the two tensors.

(iv) **Contraction of tensors.** If $A_{i\ldots j}$ is a tensor of degree r, then

$$C_{i\ldots j} \equiv A_{i\ldots k\ldots k\ldots j} \tag{5.3.16}$$

is also a tensor of degree $r - 2$. (Recall Einstein's summation convention. Summation is taken over two indices k.) This process is called *contraction* (or *rejuvenation*). Applying contraction repeatedly, we can successively create tensors of lower degrees.

(v) **Quotient rule.** Consider a quantity $A_{i\ldots j}$ with r indices. Let $B_{k\ldots l}$ be an arbitrary tensor of degree s. Define

$$C_{i\ldots l} \equiv A_{i\ldots m_1\ldots m_2\ldots m_t\ldots j} B_{k\ldots m_1\ldots m_2\ldots m_t\ldots l} \, , \tag{5.3.17}$$

by taking summations over t common indices. If $C_{i\ldots l}$ is a tensor of degree $r + s - 2t$ for an *arbitrary* tensor $B_{k\ldots l}$ of degree s, then $A_{i\ldots j}$ is a tensor of degree r. This property is called the *quotient rule.*

(vi) **Symmetrization.** If $A_{i_1\ldots i_r}$ is a tensor of degree r, then

$$A_{(i_1\ldots i_r)} \equiv \frac{1}{r!} \sum_\sigma A_{i_{\sigma(1)}\ldots i_{\sigma(r)}} \tag{5.3.18}$$

is also a tensor of degree r. Here, σ denotes a *permutation*, and the summation is taken over all permutations $(1, 2, \ldots, r) \to (\sigma(1), \sigma(2), \ldots, \sigma(r))$. For example,

$$A_{(ij)} \equiv \frac{1}{2}(A_{ij} + A_{ji}) \, ,$$

$$A_{(ijk)} \equiv \frac{1}{6}(A_{ijk} + A_{ikj} + A_{jik} + A_{jki} + A_{kij} + A_{kji}) \, , \ldots . \tag{5.3.19}$$

The resulting tensor has the property that its value is not changed by any permutation of the indices. A tensor that has this property is called a *symmetric tensor. A symmetric tensor is always transformed into a symmetric tensor* (Exercise 5.8). The above process of constructing a symmetric tensor out of a given tensor is called *symmetrization.*

(vii) **Antisymmetrization.** If $A_{i_1\ldots i_r}$ is a tensor of degree r, then

$$A_{[i_1\ldots i_r]} \equiv \frac{1}{r!} \sum_\sigma \operatorname{sign} \sigma A_{i_{\sigma(1)}\ldots i_{\sigma(r)}} \tag{5.3.20}$$

is also a tensor of degree r. Here, sign σ denotes the *signature* of permutation σ, taking value 1 if the permutation $(1, 2, \ldots, r) \rightarrow (\sigma(1), \sigma(2), \ldots, \sigma(r))$ is realized by an even number of transpositions (such a permutation is called an *even permutation*), and taking value -1 if an odd number of transpositions are required (such a permutation is called an *odd permutation*). For example,

$$A_{[ij]} \equiv \frac{1}{2}(A_{ij} - A_{ji}) \ ,$$

$$A_{[ijk]} \equiv \frac{1}{6}(A_{ijk} - A_{ikj} - A_{jik} + A_{jki} + A_{kij} - A_{kji}) \ , \ldots \quad (5.3.21)$$

The resulting tensor has the property that its sign is altered by odd permutations of the indices but not altered by even permutations. A tensor that has this property is called an *antisymmetric* (or *skew-symmetric*) *tensor. An antisymmetric tensor is always transformed into an antisymmetric tensor* (Exercise 5.9). The above process of constructing an antisymmetric tensor out of a given tensor is called *antisymmetrization.*

(viii) **Kronecker delta.** The Kronecker delta δ_{ij}, which takes value 1 if $i = j$ and value 0 otherwise, is a tensor of degree 2. Tensor indicces are "renamed" by the Kronecker delta:

$$A_{i \ldots m_1 \ldots m_2 \ldots \ldots m_s \ldots j}$$
$$= \delta_{m_1 n_1} \delta_{m_2 n_2} \ldots d_{m_s n_s} A_{i \ldots n_1 \ldots n_2 \ldots \ldots n_s \ldots j} \ . \quad (5.3.22)$$

Contraction is also expressed by the Kronecker delta:

$$A_{i \ldots k \ldots k \ldots j} = \delta_{lm} A_{i \ldots l \ldots m \ldots j} \ . \quad (5.3.23)$$

5.4 Spherical Harmonics

A scene $s(\boldsymbol{m})$ is regarded as a function over a unit sphere if the unit vector $\boldsymbol{m}$ is identified with a point on the unit sphere surrounding the viewpoint. Hence, a scene $s(\boldsymbol{m})$ can be integrated over the unit sphere. We define the *invariant integration* by

$$\int s(\boldsymbol{m})\, d\Omega \equiv \int_0^{2\pi} \int_0^{\pi} s(\sin\theta\cos\varphi, \sin\theta\sin\varphi, \cos\theta) \sin\theta \, d\theta \, d\varphi \ , \quad (5.4.1)$$

where (θ, φ) are the spherical coordinates of the unit vector $\boldsymbol{m}$, and $d\Omega$ denotes the differential solid angle. This integration is invariant to the camera rotation transformation, since scenes are rotated rigidly around the viewpoint. Namely:

Lemma 5.5.

$$\int T_R s(\boldsymbol{m})\, d\Omega = \int s(\boldsymbol{m})\, d\Omega. \tag{5.4.2}$$

Given two scenes $s_1(\boldsymbol{m})$, $s_2(\boldsymbol{m})$, we define the *invariant inner product* by

$$(s_1(\boldsymbol{m}), s_2(\boldsymbol{m})) \equiv \int s_1(\boldsymbol{m}) s_2(\boldsymbol{m})\, d\Omega\ . \tag{5.4.3}$$

From Lemma 5.5, we see that this is indeed invariant, namely:

Lemma 5.6.

$$(T_R s_1(\boldsymbol{m}), T_R s_2(\boldsymbol{m})) = (s_1(\boldsymbol{m}), s_2(\boldsymbol{m}))\ . \tag{5.4.4}$$

We can also confirm easily that this inner product satisfies the following necessary requirements on an inner product (the axioms of an inner products, see Appendix, Sect. A.4):

Lemma 5.7. *For all* $s(\boldsymbol{m})$, $s_1(\boldsymbol{m})$, $s_2(\boldsymbol{m}) \in \mathscr{S}$ *and all real numbers* c_1, c_2,

$$\text{(i)}\ (s_1(\boldsymbol{m}), s_2(\boldsymbol{m})) = (s_2(\boldsymbol{m}), s_1(\boldsymbol{m}))\ , \tag{5.4.5}$$

$$\begin{aligned}\text{(ii)}\ &(c_1 s_1(\boldsymbol{m}) + c_2 s_2(\boldsymbol{m}), s_3(\boldsymbol{m}))\\ &= c_1 (s_1(\boldsymbol{m}), s_3(\boldsymbol{m})) + c_2 (s_2(\boldsymbol{m}), s_3(\boldsymbol{m}))\ ,\end{aligned} \tag{5.4.6}$$

$$\text{(iii)}\ (s(\boldsymbol{m}), s(\boldsymbol{m})) \geqq 0, \textit{ and equality holds only for } s(\boldsymbol{m}) = 0\ . \tag{5.4.7}$$

Hence, the scene space $\mathscr{S}$ is a *metric space* (Appendix, Sect. A.4). The following Lemma plays an important role.

Lemma 5.8. *If n is odd, then*

$$\int m_{i_1} \ldots m_{i_n}\, d\Omega = 0\ . \tag{5.4.8}$$

Proof. If n is odd, at least one of m_1, m_2, m_3 must appear in the integrand an odd number of times. Suppose m_1 appears an odd number of times. For any fixed values of m_2 and m_3, the variable m_1 takes two values, one being the negative of the other. These values cancel out on integration. Hence, the net result is zero.

From this lemma, we can conclude that two invariant suspaces $L^{(m)}$, $L^{(n)}$ are *orthogonal* with respect to the metric (5.4.3) unless m and n have the same *parity* (i.e., unless both are even or both are odd). Namely:

Proposition 5.6.

$$L^{(2m)} \perp L^{(2n+1)}\ , \qquad m, n = 0, 1, 2, \ldots\ . \tag{5.4.9}$$

Proof. Invariant subspace $L^{(2m)}$ is generated by $\{m_{i_1} \ldots m_{i_{2m}}\}$, $i_1, \ldots, i_{2m} = 1$, 2, 3, and invariant subspace $L^{(2n+1)}$ is generated by $\{m_{j_1} \ldots m_{j_{2n+1}}\}$, $j_1, \ldots, j_{2n+1} = 1, 2, 3$. From Lemma 5.8, we have

$$(m_{i_1} \ldots m_{i_{2m}}, m_{j_1} \ldots m_{j_{2n+1}}) = \int m_{i_1} \ldots m_{i_{2m}} m_{j_1} \ldots m_{j_{2n+1}} \, d\Omega = 0 \ . \tag{5.4.10}$$

Since the generators of $L^{(2m)}$ are orthogonal to those of $L^{(2n+1)}$, the resulting subspaces $L^{(2m)}$, $L^{(2n+1)}$ are also orthogonal.

Mutually orthogonal subspaces cannot have nonzero elements in common (Appendix, Sect. A.4). Hence:

Corollary 5.2.

$$L^{(2m)} \cap L^{(2n+1)} = \{0\} \ , \qquad m, n = 0, 1, 2, \ldots \ . \tag{5.4.11}$$

Subspace $L^{(2)}$ includes scenes m_1^2, m_2^2, m_3^2. Since $\boldsymbol{m}$ is a unit vector, the sum of these scenes is 1: $m_1^2 + m_2^2 + m_3^2 = 1$. This means $L^{(0)} \subset L^{(2)}$. Subspace $L^{(3)}$ includes scenes m_1^3, $m_1 m_2^2$, $m_1 m_3^2$, and hence it also includes their sum $m_1^3 + m_1 m_2^2 + m_1 m_3^2 = m_1$. Similarly, it includes m_2 and m_3. Hence, $L^{(1)} \subset L^{(3)}$. In general:

Proposition 5.7.

$$L^{(l-2)} \subset L^{(l)} \ , \qquad l = 2, 3, 4, \ldots \ , \tag{5.4.12}$$

and hence

$$L^{(0)} \subset L^{(2)} \subset L^{(4)} \subset \ldots \subset L^{(2m)} \subset \ldots \subset \mathscr{S} \ ,$$
$$L^{(1)} \subset L^{(3)} \subset L^{(5)} \subset \ldots \subset L^{(2n+1)} \subset \ldots \subset \mathscr{S} \ . \tag{5.4.13}$$

Proof.

$$m_{i_1} \ldots m_{i_{l-2}} = m_{i_1} \ldots m_{i_{l-2}} m_k m_k \in L^{(l)} \ . \tag{5.4.14}$$

(Einstein's summation convention is used.)

Let $D^{(l)}$ be the *orthogonal complement* $L^{(l-2)\perp}$ of subspace $L^{(l-2)}$ in subspace $L^{(l)}$. We can easily observe:

Lemma 5.9. *$D^{(l)}$ is a $(2l+1)$-dimensional invariant subspace of $\mathscr{S}$.*

Proof. Evidently $D^{(l)}$ is a subspace, since any linear combination of elements of $(L^{(l-2)})^{\perp}$ is also orthogonal to $L^{(l-2)}$. Consider an arbitrary scene $s_1(\boldsymbol{m}) \in D^{(l)} \subset L^{(l)}$. From the invariance of $L^{(l)}$ (Proposition 5.5), $T_R s_1(\boldsymbol{m})$ is also an element of $L^{(l)}$. Since the camera rotation transformations T_R form a group of transformations (Corollary 5.1), the mappings are one-to-one and onto. Since $L^{(l-2)}$ is also an invariant subspace, we have

$$\{T_R s(\boldsymbol{m}) | s(\boldsymbol{m}) \in L^{(l-2)}\} = L^{(l-2)} . \tag{5.4.15}$$

Now, $s_1(\boldsymbol{m}) \in D^{(l)}$ means that $(s_1(\boldsymbol{m}), s_2(\boldsymbol{m})) = 0$ for all $s_2(\boldsymbol{m}) \in L^{(l-2)}$. The invariance of the inner product (Lemma 5.6) implies that $(T_R s_1(\boldsymbol{m}), T_R s_2(\boldsymbol{m})) = 0$. In other words, $T_R s_1(\boldsymbol{m})$ is orthogonal to all the elements of $\{T_R s_2(\boldsymbol{m}) | s_2(\boldsymbol{m}) \in L^{(l-2)}\} = L^{(l-2)}$. This means that $T_R s_1(\boldsymbol{m}) \in D^{(l)}$. Thus, $D^{(l)}$ is an invariant subspace. Since $L^{(l)}$ is $(l+1)(l+2)/2$ dimensional and $L^{(l-2)}$ is $(l-1)l/2$ dimensional, the dimension of subspace $D^{(l)}$ is

$$\frac{1}{2}(l+1)(l+2) - \frac{1}{2}l(l-1) = 2l+1 . \tag{5.4.16}$$

Thus, $L^{(l)}$ is *invariantly* and *orthogonally* decomposed into the direct sum of $L^{(l-2)}$ and $D^{(l)}$:

$$L^{(l)} = D^{(l)} \oplus L^{(l-2)} . \tag{5.4.17}$$

Elements of $D^{(l)}$ are called *spherical harmonics* (or *spherical functions*) of degree l. Applying the same decomposition to $L^{(l-2)}$ and repeating this process, we obtain the following *invariant orthogonal decomposition* of $L^{(l)}$:

Proposition 5.8 (*Invariance orthogonal decomposition*).

$$D^{(0)} \oplus D^{(2)} \oplus D^{(4)} \oplus \ldots \oplus D^{(2m)} = L^{(2m)} ,$$
$$D^{(1)} \oplus D^{(3)} \oplus D^{(5)} \oplus \ldots \oplus D^{(2n+1)} = L^{(2n+1)} . \tag{5.4.18}$$

From Proposition 5.6, all $D^{(l)}$ for even l are orthogonal to all $D^{(l)}$ for odd l. Hence:

Corollary 5.3 (*Orthogonality of spherical harmonics*).

$$D^{(l)} \perp D^{(m)} , \qquad l \neq m . \tag{5.4.19}$$

Since each $D^{(l)}$ is an $(2l+1)$-dimensional invariant subspace, the camera rotation transformation T_R defines a $(2l+1)$-dimensional representation of $SO(3)$ in $D^{(l)}$. This representation is irreducible. In fact, it is equivalent to $\mathscr{D}_l$ of weight l. We will prove this is Sect. 5.6.

It is also known that spherical harmonics are *complete* in the scene space $\mathscr{S}$. Namely, *any* scene $s(\boldsymbol{m}) \in \mathscr{S}$ (if it is not pathological) can be uniquely expressed as an infinite series in spherical harmonics:

$$s(\boldsymbol{m}) = s_0(\boldsymbol{m}) + s_1(\boldsymbol{m}) + s_2(\boldsymbol{m}) + \ldots , \qquad s_i(\boldsymbol{m}) \in D^{(i)} ,$$
$$i = 0, 1, 2, \ldots . \tag{5.4.20}$$

This is called the *spherical harmonics expansion.* We symbolically write:

Theorem 5.1 (*Spherical harmonic expansion*).

$$D^{(0)} \oplus D^{(1)} \oplus D^{(2)} \oplus D^{(3)} \oplus \ldots \oplus D^{(l)} \oplus \ldots = \mathscr{S} \ . \tag{5.4.21}$$

If we want to give a rigorous proof, we must remove unrealistic (but possible) pathological members. We must also give a precise definition of convergence and many other mathematical preliminaries. All of these are beyond the scope of this book.

5.5 Tensor Expressions of Spherical Harmonics

In this section, we derive an explicit expression of spherical harmonics by means of tensor calculus. Let A_{ij}, $i, j = 1, 2, 3$, be a symmetric tensor ($A_{ij} = A_{ji}$). If we contract it over indices i, j, we obtain a scalarA_{kk}. Consider

$$A_{ij} + c_1^{(2)}\delta_{ij}A_{kk} \ . \tag{5.5.1}$$

where $c_1^{(2)}$ is a constant. This is also a symmetric tensor of degree 2. We want to determine the value of the constant $c_1^{(2)}$ so that the resulting tensor has the property that contraction over indices i, j yields zero. (If the tensor A_{ij} is identified as a matrix, this means that its *trace* is zero.) Contraction of tensor (5.5.1) yields $(1 + 3c_1^{(2)})A_{kk}$, so we put $c_1^{(2)} = -1/3$. We denote the resulting tensor by

$$A_{\{ij\}} \equiv A_{ij} - \frac{1}{3}\delta_{ij}A_{kk} \ . \tag{5.5.2}$$

Next, consider a symmetric tensor A_{ijk}, $i, j, k = 1, 2, 3$, of degree 3. From this tensor, we also want to construct a symmetric tensor of degree 3 by adding and subtracting tensors obtained by contracting A_{ij} so that the resulting tensor has the property that contraction over any pair of indices yield zero. To do so, we put

$$A_{ijk} + \frac{1}{3}c_1^{(3)}(\delta_{ij}A_{kll} + \delta_{jk}A_{ill} + \delta_{ki}A_{jll}) \ , \tag{5.5.3}$$

and determine the value of the constant $c_1^{(2)}$ so that contraction over pair i, j, pair j, k, and pair k, i yields zero. Since the tensor (5.5.3) is also symmetric, contraction over any pair is equivalent. If we choose, say, the pair i, j, the contraction yields $(1 + 5c_1^{(2)}/3)A_{kll}$, so we put $c_1^{(2)} = -3/5$ and define

$$A_{\{ijk\}} \equiv A_{ijk} - \frac{3}{5}\delta_{(ij}A_{k)ll} \ . \tag{5.5.4}$$

Let us call a symmetric tensor for which contraction over any pair of indices yield zero a *deviator tensor*. Let us also call the value obtained by contraction over two indices the *trace* over these indices. Then, we can say that a deviator tensor is a *symmetric tensor whose trace is zero for any pair of indices.*

In general, given a symmetric tensor $A_{i_1 \ldots i_r}$, $i_1, \ldots, i_r = 1, 2, 3$, of degree r, we can construct a deviator tensor by taking an appropriate linear combination of its traces over all pairs of indices. This is done by putting

$$A_{\{i_1 \ldots i_r\}} \equiv A_{i_1 \ldots i_r} + c_1^{(r)} \delta_{(i_1 i_2} A_{i_3 \ldots i_r) j_1 j_1} + c_2^{(r)} \delta_{(i_1 i_2} \delta_{i_3 i_4} A_{i_5 \ldots i_r) j_2 j_2 j_1 j_1} + \cdots, \tag{5.5.5}$$

and determining the values of constants $c_s^{(r)}$, $s = 1, 2, \ldots$ so that all traces are zero. The last term in (5.5.5) is $c_{r/2}^{(r)} \delta_{(i_1 i_2} \cdots \delta_{i_{r-1} i_r)} A_{j_{r/2} j_{r/2} j_{r/2-1} j_{r/2-1} \ldots j_1 j_1}$ if r is an even number, and $c_{(r-1)/2}^{(r)} \delta_{(i_1 i_2} \cdots \delta_{i_{r-2} i_{r-1}} A_{i_r) j_{(r-1)/2} j_{(r-1)/2} j_{(r-1)/2-1} j_{(r-1)/2-1} \ldots j_1 j_1}$ if r is odd. The value of $c_s^{(r)}$ is determined by deriving a double recurrence equation for $c_s^{(r)}$ over r and s. We skip the proof and give the final result.

Lemma 5.10.

$$c_s^{(r)} = (-1)^s \binom{r}{2s} \binom{r-1}{s} \Big/ \binom{2r-1}{2s}. \tag{5.5.6}$$

We call the tensor defined by (5.5.5) the *deviator part* of the symmetric tensor $A_{i_1 \ldots i_r}$. It is easy to prove that *the deviator part of a tensor is also a tensor* (Exercise 5.15). We can also prove that *a deviator tensor is always transformed into a deviator tensor* by the tensor transformation rule (5.3.11) [or (5.3.12)] (Exercise 5.15).

Example 5.1. Reinserting (5.5.2, 4) again, we can write the deviator parts of tensors of degrees 2, 3, 4, 5, 6 as follows:

$$A_{\{ij\}} \equiv A_{ij} - \frac{1}{3} \delta_{ij} A_{kk}, \tag{5.5.2}$$

$$A_{\{ijk\}} \equiv A_{ijk} - \frac{3}{5} \delta_{(ij} A_{k)ll}, \tag{5.5.4}$$

$$A_{\{ijkl\}} \equiv A_{ijkl} - \frac{6}{7} \delta_{(ij} A_{kl)mm} + \frac{3}{35} \delta_{(ij} \delta_{kl)} A_{mmnn}, \tag{5.5.7}$$

$$A_{\{ijklm\}} \equiv A_{ijklm} - \frac{10}{9} \delta_{(ij} A_{klm)nn} + \frac{5}{21} \delta_{(ij} \delta_{kl} A_{m)nnpp}, \tag{5.5.8}$$

$$A_{\{ijklmn\}} \equiv A_{ijklmn} - \frac{15}{11} \delta_{(ij} A_{klmn)pp} + \frac{5}{11} \delta_{(ij} \delta_{kl} A_{mn)ppqq}$$
$$- \frac{5}{231} \delta_{(ij} \delta_{kl} \delta_{mn)} A_{ppqqrr}. \tag{5.5.9}$$

Now, we give a fundamental theorem. The rest of this section is devoted to its proof (and can be skipped for the first reading).

Theorem 5.2. *The spherical harmonics of degree l are generated by* $m_{\{i_1}m_{i_2}\ldots m_{i_l\}}$.

Proof. An element $s(\boldsymbol{m})\in D^{(l)}\subset L^{(l)}$ has the form

$$s(\boldsymbol{m}) = c_{i_1\ldots i_l}m_{i_1}\ldots m_{i_l} \; . \tag{5.5.10}$$

We can assume that $c_{i_1\ldots i_l}$ is a symmetric tensor. This is justified as follows. Consider the product

$$c_{i_1\ldots i_r}A_{i_1\ldots i_r} \tag{5.5.11}$$

of $c_{i_1\ldots i_r}$ with a symmetric tensor $A_{i_1\ldots i_r}$. If $c_{i_1\ldots i_r}$ is not symmetric over indices m and n, we can decompose $c_{i_1\ldots i_l}$ into its symmetric and antisymmetric parts over the indices m and n as follows:

$$c_{i_1\ldots i_l} = c_{i_1\ldots(m|\ldots|n)\ldots i_l} + c_{i_1\ldots[m|\ldots|n]\ldots i_l} \; . \tag{5.5.12}$$

Here, $|\ldots|$ means that the symmetrization or antisymmetrization is applied to only the two indices m, n; those between the two vertical lines are excluded. Since $A_{i_1\ldots m\ldots n\ldots i_l}$ is symmetric over indices m, n, multiplication of it by $c_{i_1\ldots[m|\ldots|n]\ldots i_l}$ must be zero—interchanging the dummy indices m and n would reverse the sign. Hence, we do not lose generality if we assume from the beginning that $c_{i_1\ldots i_l}$ is symmetric over any pair of indices *when multiplied by a symmetric tensor.*

All elements $s(\boldsymbol{m})\in D^{(l)}$ must be orthogonal to all $L^{(m)}$, $m = 0, 1, 2, \ldots, l-1$. Suppose l is even. (The argument runs similarly if l is odd.) Since $s(\boldsymbol{m})$ must be orthogonal to $L^{(0)}$, we have $(s(\boldsymbol{m}), 1) = 0$. From (5.5.10), we have

$$c_{i_1\ldots i_l}\int m_{i_1}\ldots m_{i_l}d\Omega = 0 \; . \tag{5.5.13}$$

The left-hand side is a scalar constructed by a linear combination of tensor components $c_{i_1\ldots i_l}$. The only possible scalar constructed from $c_{i_1\ldots i_l}$ is a multiple of $c_{j_1j_1j_2j_2\ldots j_{l/2}j_{l/2}}$. (Recall that $c_{i_1\ldots i_l}$ is a symmetric tensor.) Thus, we have

$$c_{j_1j_1j_2j_2\ldots j_{l/2}j_{l/2}} = 0 \; . \tag{5.5.14}$$

Next. $s(\boldsymbol{m})$ must be orthogonal to $L^{(2)}$. From $(s(\boldsymbol{m}), m_jm_k) = 0$ and (5.5.10), we obtain

$$c_{i_1\ldots i_l}\int m_{i_1}\ldots m_{i_l}m_jm_k\,d\Omega = 0 \; , \qquad j, k = 1, 2, 3 \; . \tag{5.5.15}$$

By the same argument, the left-hand side must be a tensor of degree 2 constructed by a linear combination of components of $c_{i_1\ldots i_l}$. Hence, terms of the form of

$$(\text{a scalar constructed from } c_{i_1\ldots i_l})\delta_{jk}$$

may appear. However, a multiple of the left-hand side of (5.5.14) is the only possible scalar constructed from $c_{i_1 \dots i_l}$, which has already been shown to be zero. Hence, the left-hand side of (5.5.15) must be a multiple of $c_{i_1 i_1 i_2 i_2 \dots i_{l/2-1} i_{l/2-1} jk}$. Thus, we have

$$c_{i_1 i_1 i_2 i_2 \dots i_{l/2-1} i_{l/2-1} jk} = 0 \ . \tag{5.5.16}$$

Similarly, $s(\boldsymbol{m})$ must also be orthogonal to $L^{(4)}$. From $(s(\boldsymbol{m}), m_j m_k m_l) = 0$ and (5.5.10), we obtain

$$c_{i_1 \dots i_l} \int m_{i_1} \dots m_{i_l} m_j m_k m_l m_m \, d\Omega = 0 \ , \quad j, k, l, m = 1, 2, 3 \ . \tag{5.5.17}$$

The left-hand side must be a tensor of degree 4 obtained by a linear combination of components of $c_{i_1 \dots i_l}$. Hence, terms of the form of

(a scalar constructed from $c_{i_1 \dots i_l}$)$\delta_{jk}\delta_{lm}$

and

(a tensor of degree 2 constructed from $c_{i_1 \dots i_l}$)$_{jk}\delta_{lm}$

may appear. However, a multiple of the left-hand side of (5.5.14) is the only possible scalar constructed from $c_{i_1 \dots i_l}$, and a multiple of the left-hand side of (5.5.16) is the only possible tensor of degree 2 constructed from $c_{i_1 \dots i_l}$, both of which have already been shown to be zero. Hence, the left-hand side of (5.5.17) must be a multiple of $c_{i_1 i_1 i_2 i_2 \dots i_{l/2-2} i_{l/2-2} jklm}$. Thus, we have

$$c_{i_1 i_1 i_2 i_2 \dots i_{l/2-2} i_{l/2-2} jklm} = 0 \ . \tag{5.5.18}$$

Continuing this argument, we finally conclude that

$$c_{iij_1 j_2 \dots j_{l-2}} = 0 \ . \tag{5.5.19}$$

Due to the symmetry of $c_{i_1 \dots i_l}$, this means that contraction of $c_{i_1 \dots i_l}$ over *any* pair of indices yields zero. Hence, $c_{i_1 \dots i_l}$ is a deviator tensor.

Let $A_{i_1 \dots i_l}$ be an arbitrary symmetric tensor, and consider the scalar

$$c_{i_1 \dots i_l} A_{i_1 \dots i_l} \ . \tag{5.5.20}$$

Suppose $A_{i_1 \dots i_l}$ has two indices m and n over which the trace does not vanish. Then, $A_{i_1 \dots i_l}$ is decomposed into the deviator and the trace parts over the indices m and n as

$$A_{i_1 \dots i_l} = A_{i_1 \dots \{m| \dots |n\} \dots i_l} + \frac{1}{3}\delta_{mn} A_{i_1 \dots p \dots m_l} \ . \tag{5.5.21}$$

The second term on the right-hand side is a multiple of δ_{mn}. Since multiplication of δ_{mm} by $c_{i_1 \dots i_l}$ results in contraction of $c_{i_1 \dots i_l}$ over indices m and n (cf. (5.3.23)), and since $c_{i_1 \dots i_l}$ is a deviator tensor, the scalar (5.5.20) is equal to

$c_{i_1 \ldots i_l} A_{i_1 \ldots [m| \ldots |n] \ldots i_l}$. Since this holds for any pair of indices, we conclude that the scalar (5.5.20) is equal to

$$c_{i_1 \ldots i_l} A_{\{i_1 \ldots i_l\}} \; . \tag{5.5.22}$$

Thus, (5.5.10) is written in the form

$$s(\boldsymbol{m}) = c_{i_1 \ldots i_l} m_{\{i_1} \ldots m_{i_l\}} \; , \tag{5.5.23}$$

where $c_{i_1 \ldots i_l}$ is an arbitrary deviator tensor. However, $c_{i_1 \ldots i_l}$ can range over *all* tensors of degree l. In fact, if $c_{i_1 \ldots i_l}$ is an arbitrary tensor of degree l, then

$$c_{i_1 \ldots i_l} m_{\{i_1 \ldots i_l\}} = c_{\{i_1 \ldots i_l\}} m_{\{i_1 \ldots i_l\}} \; , \tag{5.5.24}$$

because $m_{\{i_1 \ldots i_l\}}$ is a deviator tensor. Thus, we conclude that

$$m_{\{i_1} \cdots m_{i_l\}} \in D^{(l)} \; . \tag{5.5.25}$$

It remains to be shown that elements $m_{\{i_1} \ldots m_{i_l\}}$ can generate $D^{(l)}$. To see that, we count the possible number of distinct elements written in that form. In view of the symmetry over the indices, there are $(l+1)(l+2)/2$ different ways of choosing l indices. They are constrained by the equations which require that the trace is zero over any pair of indices. Since there are $l(l-1)/2$ different ways of choosing two indices from among l indices, there are $l(l-1)/2$ such constraining equations. Hence, the number of independent elements is

$$\frac{1}{2}(l+1)(l+2) - \frac{1}{2}l(l-1) = 2l+1 \; , \tag{5.5.26}$$

which is equal to the dimension of $D^{(l)}$.

5.6 Irreducibility of Spherical Harmonics[2]

In this section, we give a proof of our claim that the invariant subspace $D^{(l)}$ is irreducible. Specifically, we show that the representation of $SO(3)$ induced in the invariant subspace $D^{(l)}$ is equivalent to the irreducible representation $\mathscr{D}_l$ of weight l. Our basic tool is the *Casimir operator H* (Sect. 3.3). The irreducibility of $D^{(l)}$ is confirmed if we can prove that all elments of $D^{(l)}$ are eigenfunction of the Casimir operator H for eigenvalue $l(l+1)$ (Theorem 3.3).

In order to construct the Casimir operator H, we must first compute the *infinitesimal generators.* To do so, we must choose appropriate independent variables. Since $\boldsymbol{m} = (m_1, m_2, m_3)$ is a unit vector, we can specify it by two parameters. One natural choice is the spherical coordinates (θ, φ), $0 \leqq \theta \leqq \pi$, $0 \leqq \varphi < 2\pi$ (Fig. 5.4), defined as

$$m_1 = \sin\theta\cos\varphi \; , \qquad m_2 = \sin\theta\sin\varphi \; , \qquad m_3 = \cos\theta \; . \tag{5.6.1}$$

[2] The results of this section are not directly connected with the subsequent part of this book. Readers who are not interested in the mathematical details can skip this section entirely.

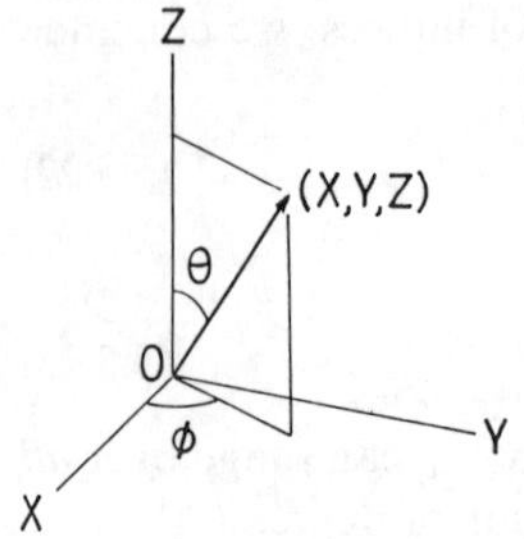

Fig. 5.4. Spherical coordinates (θ, φ)

However, there are other possible choices. For example, we may choose m_1, m_2 as independent variables and express m_3 as $m_3 = \pm\sqrt{1 - m_1^2 - m_2^2}$, though the sign must be determined by some appropriate means. Similarly, we may choose m_2, m_3 and express m_1 in terms of them, or we may choose m_3, m_1.

If the camera rotation $\boldsymbol{R}$ is infinitesimal, the matrix $\boldsymbol{R}$ is expressed in the form of (3.3.13, 14). From the transformation rule (5.2.6), a scene $s(\boldsymbol{m})$ is transformed into

$$T_R s(\boldsymbol{m}) = s\left(\left(\begin{array}{ccc} 1 & -\Omega n_3 & \Omega n_2 \\ \Omega n_3 & 1 & -\Omega n_1 \\ -\Omega n_2 & \Omega n_1 & 1 \end{array}\right)\left(\begin{array}{c} m_1 \\ m_2 \\ m_3 \end{array}\right) + \mathrm{O}(\Omega^2)\right) . \tag{5.6.2}$$

Consider an infinitesimal camera rotation around the X-axis by an infinitesimal angle Ω, and put

$$\boldsymbol{T_R} s(\boldsymbol{m}) = s(\boldsymbol{m}) + \Omega D_1 s(\boldsymbol{m}) + \mathrm{O}(\Omega^2) , \tag{5.6.3}$$

where $D_1 s(\boldsymbol{m})$ is a symbolic notation to denote the first order variation of $s(\boldsymbol{m})$ per unit Ω. Substituting $n_1 = 1$, $n_2 = n_3 = 0$ into (5.6.2), we obtain

$$T_R s(\boldsymbol{m}) = s(m_1, m_2 - \Omega m_3 + \mathrm{O}(\Omega^2), m_3 + \Omega m_2 + \mathrm{O}(\Omega^2)) . \tag{5.6.4}$$

If m_2 and m_3 are chosen as independent variables, the scene $s(\boldsymbol{m})$ is regarded as a function of m_2 and m_3:

$$\begin{aligned} T_R s(\boldsymbol{m}) = s(&\pm\sqrt{1 - m_2^2 - m_3^2}, m_2 - \Omega m_3 + \mathrm{O}(\Omega^2), \\ & m_3 + \Omega m_2 + \mathrm{O}(\Omega^2)) . \end{aligned} \tag{5.6.5}$$

This equation is rewritten as

$$\begin{aligned} T_R s(\boldsymbol{m}) = s(&\pm\sqrt{1 - (m_2 - \Omega m_3)^2 - (m_3 + \Omega m_2)^2} + \mathrm{O}(\Omega^2), \\ & m_2 - \Omega m_3 + \mathrm{O}(\Omega^2), m_3 + \Omega m_2 + \mathrm{O}(\Omega^2)) . \end{aligned} \tag{5.6.6}$$

Since $s(\boldsymbol{m}) = s(\pm\sqrt{1-m_2^2-m_3^2}, m_2, m_3)$, the right-hand side is expanded into a Taylor series in Ω by regarding m_2, m_3 as independent variables:

$$T_R s(\boldsymbol{m}) = s(\boldsymbol{m}) - \Omega\left(m_3 \frac{\partial s}{\partial m_2} - m_2 \frac{\partial s}{\partial m_3}\right) + O(\Omega^2) \ . \tag{5.6.7}$$

Comparing this with (5.6.3), we find that the symbol D_1 is identified with the differential operator

$$D_1 = -m_3 \frac{\partial}{\partial m_2} + m_2 \frac{\partial}{\partial m_3} \ . \tag{5.6.8}$$

Considering infinitesimal rotations around the Y- and Z-axes, we can similarly obtain differential operators as follows:

$$D_2 = -m_1 \frac{\partial}{\partial m_3} + m_3 \frac{\partial}{\partial m_1} \ , \qquad D_3 = -m_2 \frac{\partial}{\partial m_1} + m_1 \frac{\partial}{\partial m_2} \ . \tag{5.6.9}$$

Thus, the infinitesimal camera rotation transformation is expressed in the form

$$T_R s(\boldsymbol{m}) = s(\boldsymbol{m}) + \Omega(n_1 D_1 + n_2 D_2 + n_3 D_3) s(\boldsymbol{m}) + O(\Omega^2) \ . \tag{5.6.10}$$

The differential operators D_1, D_2, D_3 are called the *infinitesimal generators* of the camera rotation transformation T_R. However, (5.6.8, 9) cannot be substituted directly into (5.6.10). In deriving (5.6.8), we chose m_2, m_3 as independent variables; for eqns (5.6.9), we chose m_3, m_1 for D_2, and m_1, m_2 for D_3. In order to combine D_1, D_2, D_3 together, independent variables must be unified. To this end, we choose the spherical coordinates (θ, φ). Rewriting (5.6.8, 9) in terms of spherical coordinates (θ, φ), we find that the infinitesimal generators D_1, D_2, D_3 are given by

$$D_1 = -\sin\varphi \frac{\partial}{\partial\theta} - \frac{\cos\varphi}{\sin\theta}\frac{\partial}{\partial\varphi} \ , \quad D_2 = \cos\varphi \frac{\partial}{\partial\theta} - \frac{\sin\varphi}{\sin\theta} \ ,$$

$$D_3 = \frac{\partial}{\partial\varphi} \ . \tag{5.6.11}$$

We can easily confirm the *commutation relations*

$$[D_1, D_2] = -D_3 \ , \qquad [D_2, D_3] = -D_1 \ , \qquad [D_3, D_1] = -D_2 \tag{5.6.12}$$

(Excercise 5.20), where the *commutator* $[.,.]$ is defined by

$$[D_i, D_j] \equiv D_i D_j - D_j D_i \ , \qquad i, j = 1, 2, 3. \tag{5.6.13}$$

From (5.6.11), the Casimir operator $H \equiv -(D_1^2 + D_2^2 + D_3^2)$ is constructed as follows (Exercise 5.21):

$$H = -\frac{1}{\sin\theta}\frac{\partial}{\partial\theta}\sin\theta\frac{\partial}{\partial\theta} - \frac{1}{\sin^2\theta}\frac{\partial^2}{\partial\varphi^2} \ . \tag{5.6.14}$$

This operator is also known as the *Laplace-Beltrami operator*. The irreducibility of $D^{(l)}$ can be established if every element $s(\boldsymbol{m}) \in D^{(l)}$ is shown to be an eigenfunction of the Laplace-Beltrami operator H for eigenvalue $l(l+1)$:

$$Hs(\boldsymbol{m}) = l(l+1)s(\boldsymbol{m}) \tag{5.6.15}$$

(Theorem 3.3). Hence, we observe:

Lemma 5.11. *The invariant subspace $D^{(l)}$ is irreducibe if and only if all $s(\boldsymbol{m}) \in D^{(l)}$ satisfy the differential equation*

$$\frac{1}{\sin\theta}\frac{\partial}{\partial\theta}\left(\sin\theta\frac{\partial s}{\partial\theta}\right) + \frac{1}{\sin^2\theta}\frac{\partial^2 s}{\partial\varphi^2} + l(l+1)s = 0\ . \tag{5.6.16}$$

This differential equation is called the *Laplace–Beltrami equation.* In the rest of this section, we will show that all $s(\boldsymbol{m}) \in D^{(l)}$ satisfy this Laplace–Beltrami equation.

A scene $s(\boldsymbol{m}) \in D^{(l)}$ is a function of orientation $\boldsymbol{m}$. Hence, it is regarded as a function over a unit sphere centered at the coordinate origin. We now extend $s(\boldsymbol{m})$ to a function over the entire 3D space in such a way that its restriction to the unit sphere coincides with $s(\boldsymbol{m})$. Specifically, let $s(\theta, \varphi)$ be the expression in spherical coordinates (θ, φ). We put

$$\tilde{s}(r, \theta, \varphi) \equiv \frac{1}{r^{l+1}} s(\theta, \varphi)\ , \tag{5.6.17}$$

where r is the radial coordinate.

Now, we can confirm that $s(\theta, \varphi)$ satisfies the Laplace–Beltrami equation (5.6.16) if and only if the extended function $\tilde{s}(r, \theta, \varphi)$ satisfies the differential equation

$$\frac{\partial}{\partial r} r^2 \frac{\partial \tilde{s}}{\partial r} + \frac{1}{\sin\theta}\frac{\partial}{\partial\theta}\left(\sin\theta\frac{\partial \tilde{s}}{\partial\theta}\right) + \frac{1}{\sin^2\theta}\frac{\partial^2 \tilde{s}}{\partial\varphi^2} = 0 \tag{5.6.18}$$

(Exercise 5.22). The spherical coordinates (r, θ, φ) and the Cartesian coordinates (X, Y, Z) are related by

$$X = r\sin\theta\cos\varphi\ , \qquad Y = r\sin\theta\sin\varphi\ , \qquad Z = r\cos\theta\ . \tag{5.6.19}$$

In terms of X, Y, Z, equation (5.6.18) is rewritten as

$$\nabla^2 \tilde{s} = 0\ , \qquad \nabla^2 \equiv \frac{\partial^2}{\partial X^2} + \frac{\partial^2}{\partial Y^2} + \frac{\partial^2}{\partial Z^2}\ . \tag{5.6.20}$$

This equation is called the *Laplace equation.* Solutions of the Laplace equation are called *harmonics.* The operator ∇^2 is called the *Laplacian.* Thus we have obtained the following lemma.

Lemma 5.12. *A function $s(\boldsymbol{m}) \in D^l$ satisfies the Laplace–Beltrami equation* (5.6.16) *if and only if the extended function $\tilde{s}(r, \theta, \varphi)$ satisfies the Laplace equation* (5.6.20).

Hence, it remains to be shown that all $\tilde{s}(r, \theta, \varphi)$ extended from $s(\boldsymbol{m}) \in D^{(l)}$ are harmonics. Since the Laplace equation is a linear differential equation, we only need to show this for the generators $m_{\{i_1} \ldots m_{i_l\}}$ of $D^{(l)}$, which are extended over the 3D space in the form $m_{\{i_1} \ldots m_{i_l\}}/r^{l+1}$. Hence, we only need to show that $m_{\{i_1} \ldots m_{i_l\}}/r^{l+1}$ is a harmonic.

Lemma 5.13. *Function* $1/r$ *is a harmonic*:

$$\nabla^2 \frac{1}{r} = 0 \ . \tag{5.6.21}$$

This is easily confirmed by straightforward differentiation. (The origin becomes a singularity.) In physics, the function $1/r$ is known as the *Coulomb potential field* produced by a point source of charge at the origin. For notational convenience, let us put $X = X_1$, $Y = X_2$, $Z = X_3$. If we differentiate $1/r$ with respect to $X_{i_1}, \ldots, X_{i_l}$ and note that $r = \sqrt{X_1^2 + X_2^2 + X_3^2}$ and $m_i = X_i/r$, we obtain

Lemma 5.14.

$$\frac{\partial}{\partial X_{i_l}} \cdots \frac{\partial}{\partial X_{i_l}} \frac{1}{r} = \frac{(-1)^l (2l)!}{2^l l!} \frac{1}{r^{l+1}} m_{\{i_1} \ldots m_{i_l\}} \ . \tag{5.6.22}$$

Proof. Consider the left-hand side. Since differentiation is done l times, the result must be of order $-l-1$ in r. Since differentiations commute with each other, the left-hand side is a symmetric tensor of degree l. We can also see that the left-hand side is a deviator tensor, because contraction over any two indices yields the Laplacian $\nabla^2 = (\partial/\partial X_k)(\partial/\partial X_k)$, which annihilates $1/r$ (Lemma 5.13). Hence, the left-hand side of (5.6.22) is a deviator tensor of degree l whose order in r is $-l-1$. The only possible form that satisfies all these requirements is a multiple of $m_{\{i_1} \ldots m_{i_l\}}/r^{l+1}$. The multiplier is determined by computing a particular component (say, for $i_1 = \cdots = i_l = 1$). Thus, (5.6.22) is the inevitable conclusion.

Corollary 5.4.

$$\nabla^2 \left(\frac{1}{r^{l+1}} m_{\{i_1} \ldots m_{i_l\}} \right) = 0 \ . \tag{5.6.23}$$

Proof. If we take the Laplacian ∇^2 of the left-hand side of (5.6.22), the Laplacian ∇^2 commutes with all differentiations $\partial/\partial X_i$ and annihilates the harmonic $1/r$ (Lemma 5.13). Hence, the right-hand side is also a harmonic.[3]

[3] If regarded as an electrical potential field, this harmonic is called the *multipole moment field*. For $l = 0$, it is the Coulomb potential field produced by a point charge; for $l = 1$, it is the *dipole moment field* produced by a dipole at the origin; for $l = 2$, it is the *quadrupole moment field*, and so on.

This concludes our proof of the claim that the invariant subspace $D^{(l)}$ is irreducible, inducing the irreducible representation $\mathscr{D}_l$ of weight l of $SO(3)$.

5.6.1 Laplace Spherical Harmonics

We showed that spherical harmonics of different degrees are mutually orthogonal (Corollary 5.3), but the $2l + 1$ generators $m_{\{i_1} \ldots m_{i_l\}}$ of $D^{(l)}$ are not necessarily orthogonal to each other. However, we can always construct an orthogonal basis by *Schmidt orthogonalization* (Appendix, Sect. A.4). The choice of an orthogonal basis is not unique. The basis widely used in physics consists of the *Laplace spherical harmonics*.

$$P_l^m(\cos\theta)\cos m\varphi \ , \qquad P_l^m(\cos\theta)\sin m\varphi \ , \qquad m = 0, 1, \ldots, l \ , \tag{5.6.24}$$

where $P_l^m(z)$ is the *associate Legendre function* defined by

$$P_l^m(z) = (z^2 - 1)^{m/2}\frac{d^m P_l(z)}{dz^m} \ , \qquad m = 0, 1, \ldots, l \ . \tag{5.6.25}$$

The function $P_l(z)$ is a polynomial of degree l in z known as the lth *Legendre polynomial*, and defined by[4]

$$P_l(z) = \frac{1}{2^l l!}\frac{d^l(z^2 - 1)^l}{dx^l} \ . \tag{5.6.26}$$

This expression is called the *Rodrigues formula*. The Laplace spherical harmonics (5.6.24) are mutually orthogonal. Alternatively, if we construct complex-valued functions

$$Y_l^m(\theta, \varphi) = \mathrm{e}^{im\varphi} P_l^{|m|}(\cos\theta) \ , \qquad m = 0, \pm 1, \ldots, \pm l \ , \tag{5.6.27}$$

we have the following *orthogonality relation*:[5]

$$(Y_l^m, Y_{l'}^{m'}) = \frac{2\pi(l + m)!}{(l - m)!(2l + 1)}\delta_{ll'}\delta_{mm'} \ . \tag{5.6.28}$$

Namely,

$$\int_0^{2\pi}\int_0^{\pi} Y_l^m(\theta, \varphi)^* Y_{l'}^{m'}(\theta, \varphi)\sin\theta \, d\theta \, d\varphi = \frac{2\pi(l + m)!}{(l - m)!(2l + 1)}\delta_{ll'}\delta_{mm'} \ . \tag{5.6.29}$$

---

[4] The constant $1/2^l l!$ is so chosen that $P_l(0) = 1$ and consequently $P_l(-1) = (-1)^l$.

[5] If we extend the scene space $\mathscr{S}$ to complex-valued functions, the inner product must be interpreted as $(s_1(\boldsymbol{m}), s_2(\boldsymbol{m})) = \int s_1(\boldsymbol{m})^* s_2(\boldsymbol{m})\, d\Omega$, where * designates the complex conjugate.

The coefficients of the spherical harmonic expansion can be computed from this orthogonality relationship. If function $s(\theta, \varphi)$ is to be expanded in the form

$$s(\theta, \varphi) = \sum_{l=0}^{m} \sum_{m=-l}^{l} c_l^m Y_l^m(\theta, \varphi) , \tag{5.6.30}$$

the expansion coefficients c_l^* are given by

$$c_l^m = \frac{(l-m)!(2l+1)}{2\pi(l+m)!} \int_0^{2\pi} \int_0^{\pi} Y_l^m(\theta, \varphi)^* s(\varphi, \theta) \sin\theta \, d\theta \, d\varphi . \tag{5.6.31}$$

(The traditional treatment of spherical harmonics is given in the Appendix, Sect. A.10).

In physics, spherical harmonics play an important role in the description of the electron orbit state of a hydrogen-like atom equipped with spherical symmetry (of its Hamiltonian). The essential fact is that the quantum state after the Z-component of angular momentum is measured (say, by applying a magnetic field along the Z-axis) is given by one of these Laplace spherical harmonics. The reasoning, which we do not discuss here, is purely geometrical, and the actual form of the *Schrödinger equation* is irrelevant. The number l, called the *azimuthal quantum number*, indexes the discrete values of the *total angular momentum*, while the number m, called the *magnetic quantum number*, indexes the discrete values of the Z-component of the angular momentum.

The name *spherical harmonics* comes from the fact that if $s(\boldsymbol{m})$ is a spherical harmonic of degree l, then

$$r^l s(\boldsymbol{m}) , \qquad \frac{1}{r^{l+1}} s(\boldsymbol{m}) , \qquad l = 0, 1, 2, \ldots , \tag{5.6.32}$$

are harmonics. It can also be shown that functions of the form of (5.6.32) constitute a *complete* system of harmonics in the sense that all harmonics are expressed as (possibly infinite) linear combinations of them (see the Appendix, Sect. A.10). In terms of the Laplace spherical harmonics, functions

$$r^l Y_l^m(\theta, \varphi) , \quad \frac{1}{r^{l+1}} Y_l^m(\theta, \varphi) , \quad m = 0, \pm 1, \ldots, \pm l , \quad l = 0, 2, \ldots \tag{5.6.33}$$

form a complete system of harmonics.

Historically, spherical harmonics were introduced for this very purpose—for studying solutions of the Laplace equation. The Laplace equation appears in physics whenever smooth fields are involved—gravitational fields, electromagnetic fields, wave functions of quantum states, fluid flow fields, temperature and pressure fields in materials, stress and strain fields in solids, to name only a few. One reason for this is that a harmonic fields is characterized by the property that *its value at any internal point P is equal to the average of the values at the points surrounding P at an equal distance* (the *mean value theorem*). Thus, *a harmonic field cannot take its maximum or minimum value at internal points* (the *maximum principle*). Hence, if a smooth field is completely determined by the values along the boundary (the *boundary value problem*) and if there is no internal source to disturb it, it is approximated by a harmonic field *to a first approximation*. Thus, the Laplace equation appears in many linearized theories of physics.

5.7 Camera Rotation Transformation of the Image Space

Now, we turn to the image space $\mathcal{I}$. If we allow negative values for the image intensity, the image space $\mathcal{I}$ is regarded as a linear space. As for the scene space $\mathcal{S}$, we can define invariant subspaces of $\mathcal{I}$ under the camera rotation transformation T_R. Irreducible subspaces are also defined as invariant subspaces of $\mathcal{I}$ that do not include proper invariant subspaces. However, we need not develop a new theory. Recall the projection operator $\Pi: \mathcal{S} \to \mathcal{I}$, and note the following fact.

Lemma 5.15. *The projection operator* $\Pi: \mathcal{S} \to \mathcal{I}$ *is a linear operator*:

$$\Pi[c_1 s_1 + c_2 s_2](x, y) = c_1 \Pi[s_1](x, y) + c_2 \Pi[s_2](x, y) \tag{5.7.1}$$

for all $s_1(\boldsymbol{m}), s_2(\boldsymbol{m}) \in \mathcal{S}$ *and all real numbers* c_1, c_2.

Hence, all "linear structures" of the scene space $\mathcal{S}$ are homomorphically mapped onto the image plane. Application of the camera rotation transformation T_R in the scene followed by projection Π is the same as application of projection Π followed by the camera rotation operation T_R on the image plane, or in short, *projection* Π *commutes with the camera rotation transformation* T_R (Fig. 5.3). Hence, all invariant structures of the scene space $\mathcal{S}$ are also mapped onto the image plane.[6] For example, Proposition 5.7 implies

Proposition 5.9. $\Pi[L^{(l)}]$ *is an* $(l+1)(l+2)2$*-dimensional invariant subspace of the image space* $\mathcal{I}$, *and*

$$\Pi[L^{(0)}] \subset \Pi[L^{(2)}] \subset \Pi[L^{(4)}] \subset \ldots \subset \Pi[L^{(2m)}] \subset \cdots \subset \mathcal{I},$$
$$\Pi[L^{(1)}] \subset \Pi[L^{(3)}] \subset \Pi[L^{(5)}] \subset \cdots \subset \Pi[L^{(2n+1)}] \subset \cdots \subset \mathcal{I}. \tag{5.7.2}$$

---

[6] In general, the dimension of a subspace of $\mathcal{S}$ may reduce by projection Π. The *kernel* of projection Π, i.e., the set of the scenes that are mapped to 0 in $\mathcal{I}$, consists of scenes which take the value 0 for $m_3 > 0$. Since $L^{(l)}$ consists of even or odd functions in $\boldsymbol{m}$ depending on whether the degree l is even or odd, this kernel consists only of 0. Thus, as long as we consider subspaces $L^{(l)}$ or their subspaces, the dimensionality does not reduce, and the correspondence is *isomorphism*. (See the Note at the end of this section.)

If we define images

$$F_{i_1 \ldots i_l}(x, y) \equiv \Pi[m_{i_1} \ldots m_{i_l}](x, y) \in \Pi[L^{(l)}] , \tag{5.7.3}$$

they generate the subspace $\Pi[L^{(l)}]$. From Proposition 5.5, we immediately observe that for $\boldsymbol{R} = (r_{ij})$, $i, j = 1, 2, 3$:

Proposition 5.10. *$F_{i_1 \ldots i_l}(x, y)$ is transformed as a (symmetric) tensor of degree l:*

$$T_R F_{i_1 \ldots i_l}(x, y) = \sum_{i_1=1}^{3} \ldots \sum_{i_l=1}^{3} r_{i_1 j_1} \ldots r_{i_l j_l} F_{j_1 \ldots j_l}(x, y) . \tag{5.7.4}$$

Proof. From (5.2.10), (5.3.10), and (5.7.3), we see that

$$\begin{aligned} T_R F_{i_1 \ldots i_l}(x, y) &= \Pi\left[\sum_{i_1}^{3} \ldots \sum_{i_l}^{3} r_{i_1 j_1} \ldots r_{i_l j_l} m_{j_1} \ldots m_{j_l}\right](x, y) \\ &= \sum_{i_1}^{3} \ldots \sum_{i_l}^{3} r_{i_1 j_1} \ldots r_{i_l j_l} \Pi[m_{j_1} \ldots m_{j_l}](x, y) \\ &= \sum_{i_1}^{3} \ldots \sum_{i_l}^{3} r_{i_1 j_1} \ldots r_{i_l j_l} F_{j_1 \ldots j_l}(x, y) . \end{aligned} \tag{5.7.5}$$

Example 5.2. For $l = 0$, we have $F(x, y) = 1$. For $l = 1$, we have

$$F_1(x, y) = \frac{x}{\sqrt{x^2 + y^2 + f^2}} , \quad F_2(x, y) = \frac{y}{\sqrt{x^2 + y^2 + f^2}} ,$$

$$F_3(x, y) = \frac{f}{\sqrt{x^2 + y^2 + f^2}} . \tag{5.7.6}$$

$1 = 2$, we have

$$F_{11}(x, y) = \frac{x^2}{x^2 + y^2 + f^2} , \quad F_{12}(x, y) = \frac{xy}{x^2 + y^2 + f^2} ,$$

$$F_{13}(x, y) = \frac{fx}{x^2 + y^2 + f^2} , \quad F_{21}(x, y) = \frac{xy}{x^2 + y^2 + f^2} ,$$

$$F_{22}(x, y) = \frac{y^2}{x^2 + y^2 + f^2} , \quad F_{23}(x, y) = \frac{fy}{x^2 + y^2 + f^2} ,$$

$$F_{31}(x, y) = \frac{fx}{x^2 + y^2 + f^2} , \quad F_{32}(x, y) = \frac{fy}{x^2 + y^2 + f^2} ,$$

$$F_{33}(x, y) = \frac{f^2}{x^2 + y^2 + f^2} , \tag{5.7.7}$$

Proposition 5.11. $\Pi[D^{(l)}]$ *is an* $(2l+1)$*-dimensional irreducible subspace of* $\mathscr{I}$, *and*

$$\Pi[D^{(0)}] \oplus \Pi[D^{(1)}] \oplus \Pi[D^{(2)}] \oplus \Pi[D^{(3)}] \oplus \ldots \oplus \Pi[D^{(l)}] \oplus \ldots = \mathscr{I} . \tag{5.7.8}$$

Proof. This is a direct consequence of the fact that the projection $\Pi: D^{(l)} \rightarrow \Pi[D^{(l)}]$ is an isomorphism, see Sect. 5.7.1.

Since subspace $\Pi[D^{(l)}]$ is generated by

$$F_{\{i_1 \ldots i_l\}}(x, y) \equiv \Pi[m_{\{i_1} \ldots m_{i_l\}}] \in \Pi[D^{(l)}] , \tag{5.7.9}$$

we immediately observe from Proposition 5.10 that:

Proposition 5.12. $F_{\{i_1 \ldots i_l\}}$ *is transformed as a (deviator) tensor of degree* l:

$$T_R F_{\{i_1 \ldots i_l\}}(x, y) = \sum_{i_1=1}^{3} \cdots \sum_{i_l=1}^{3} r_{i_1 j_1} \ldots r_{i_l j_l} F_{\{j_1 \ldots j_l\}}(x, y) . \tag{5.7.10}$$

Example 5.3. For $l = 0$, 1, the functions $F_{\{i_1 \ldots i_l\}}(x, y)$ are the same as in Example 5.2. For $l = 2$, we have

$$F_{\{11\}}(x, y) = \frac{2x^2 - y^2 - f^2}{3(x^2 + y^2 + f^2)} , \quad F_{\{12\}}(x, y) = \frac{xy}{x^2 + y^2 + f^2} ,$$

$$F_{\{13\}}(x, y) = \frac{fx}{x^2 + y^2 + f^2} , \quad F_{\{21\}}(x, y) = \frac{xy}{x^2 + y^2 + f^2} ,$$

$$F_{\{22\}}(x, y) = \frac{-x^2 + 2y^2 - f^2}{3(x^2 + y^2 + f^2)} , \quad F_{\{23\}}(x, y) = \frac{fy}{x^2 + y^2 + f^2} ,$$

$$F_{\{31\}}(x, y) = \frac{fx}{x^2 + y^2 + f^2} , \quad F_{\{32\}}(x, y) = \frac{fy}{x^2 + y^2 + f^2} ,$$

$$F_{\{33\}}(x, y) = \frac{-x^2 - y^2 + 2f^2}{3(x^2 + y^2 + f^2)} . \tag{5.7.11}$$

Let us confirm that $F_{\{i_1 \ldots i_l\}}$ indeed defines the irreducible representation $\mathscr{D}_l$ by constructing its infinitesimal generators and Casimir operator H. We obtain:

Proposition 5.13. *For a small rotation with axis $\boldsymbol{n} = (n_1, n_2, n_3)$ and angle Ω,*

$$T_R F(x, y) = F(x, y) + \Omega(n_1 D_1 + n_2 D_2 + n_3 D_3) F(x, y) + O(\Omega^2) , \tag{5.7.12}$$

where the infinitesimal generators D_1, D_2, D_3 are given by

$$D_1 = -\frac{xy}{f}\frac{\partial}{\partial x} - \left(f + \frac{y^2}{f}\right)\frac{\partial}{\partial y} , \quad D_2 = \left(f + \frac{x^2}{f}\right)\frac{\partial}{\partial x} + \frac{xy}{f}\frac{\partial}{\partial y} ,$$

$$D_3 = -y\frac{\partial}{\partial x} + x\frac{\partial}{\partial y} . \tag{5.7.13}$$

Proof. If we substitute (3.3.13, 14) into (5.2.11) and expand its arguments into a Taylor series in Ω, we obtain

$$T_R F(x, y) = F\left(x + \Omega\left(fn_2 - n_3 y + \frac{1}{f}(n_2 x - n_1 y)x\right) + O(\Omega^2) ,\right.$$

$$\left. y + \Omega\left(-fn_1 + n_3 x + \frac{1}{f}(n_2 x - n_1 y)y\right) + O(\Omega^2)\right) . \tag{5.7.14}$$

Equations (5.7.12, 13) are obtained by further expanding the right-hand side into a Taylor series in Ω.

We can easily confirm that the infinitesimal generators D_1, D_2, D_3 satisfy the commutation relations

$$[D_1, D_2] = -D_3 , \quad [D_2, D_3] = -D_1 , \quad [D_3, D_1] = -D_2 , \tag{5.7.15}$$

and the Casimir (or Laplace–Beltrami) operator $H = -(D_1^2 + D_2^2 + D_3^2)$ takes the form

$$H = -\left(f + \frac{x^2 + y^2}{f}\right)\left[\left(f + \frac{x^2}{f}\right)\frac{\partial^2}{\partial x^2} + \frac{2xy}{f}\frac{\partial^2}{\partial x \partial y} + \left(f + \frac{y^2}{f}\right)\frac{\partial^2}{\partial y^2}\right]$$

$$- \left(f + x + \frac{2x(x^2 + y^2)}{f^2}\right)\frac{\partial}{\partial x} - \left(f + y + \frac{2y(x^2 + y^2)}{f^2}\right)\frac{\partial}{\partial y} . \tag{5.7.16}$$

As we argued earlier, the camera rotation transformation T_R defines an irreducible representation $\mathscr{D}_l$ if the subspace is generated by $2l + 1$ indepndent eigenfunctions of H for eigenvalue $l(l + 1)$ (Theorem 3.3). From (5.7.16), the *Laplace–Beltrami equation*

$$HF(x, y) = l(l + 1)F(x, y) \tag{5.7.17}$$

now becomes

$$\left(f+\frac{x^2+y^2}{f}\right)\left[\left(f+\frac{x^2}{f}\right)F_{xx}+\frac{2xy}{f}F_{xy}+\left(f+\frac{y^2}{f}\right)F_{yy}\right]$$
$$+\left(f+x+\frac{2x(x^2+y^2)}{f^2}\right)F_x+\left(f+y+\frac{2y(x^2+y^2)}{f^2}\right)F_y$$
$$+\,l(l+1)F=0\;, \tag{5.7.18}$$

where subscripts indicate partial derivatives (e.g., $F_x = \partial F/\partial x$). This is simply the Laplace–Beltrami equation (5.6.14) expressed in image coordinates (x, y) via projection Π. We can easily confirm that images $F_{\{i_1 \ldots i_l\}}(x, y)$ satisfy this for $l = 0, 1, 2$ (Exercise 5.30).

5.7.1 Parity of Scenes

Recall that we defined the camera rotation transformation T_R in the image space $\mathscr{I}$ in terms of the projection operator Π and its inverse Π^{-1} by equation (5.2.10). We noted there that the inverse projection Π^{-1} was not well-defined in general. In order to avoid this deficiency, we assumed that all images were of finite support (i.e., nonzero regions were localized near the center of the image plane). The images $F_{i_1 \ldots i_l}(x, y) \in \Pi[L^{(l)}]$ and $F_{\{i_1 \ldots i_l\}}(x, y) \in D^{(l)}$ do not satisfy this condition. Nevertheless, the camera rotation transformation T_R is well-defined for images of $\Pi[L^{(l)}]$ $(\supset \Pi[D^{(l)}])$.

Note that all scenes $s(\boldsymbol{m}) \in L^{(l)} (\supset D^{(l)})$ are even functions in $\boldsymbol{m}$, i.e.,

$$s(-\boldsymbol{m}) = s(\boldsymbol{m})\;, \tag{5.7.19}$$

if l is even, and odd functions in $\boldsymbol{m}$, i.e.,

$$s(-\boldsymbol{m}) = -\,s(\boldsymbol{m})\;, \tag{5.7.20}$$

if l is odd. Hence, if we rotate the camera, a new part comes into view and some part goes out of view, but if l is even, the part that goes out of view appears from the opposite side of the image plane, and if l is odd, the part that goes out of view appears from the opposite side of the image plane with its sign reversed.

5.8 Invariant Measure

In this section, we determine a function $\rho(x, y)$ such that the following relation holds for an arbitrary image $F(x, y)$:

$$\int T_R F(x, y)\rho(x, y)dx\,dy = \int F(x, y)\rho(x, y)dx\,dy\;. \tag{5.8.1}$$

The integration is performed over the entire image plane. This integration is well-defined if the image $F(x, y)$ has finite support and if the camera rotation $\boldsymbol{R}$ is not very large so that the nonzero part of the image is not carried into infinity. If (5.8.1) holds, we say that $\rho(x, y)dx\,dy$ is an *invariant measure.*

The easiest way to derive $\rho(x, y)$ is to project the relationship of (5.4.2),

$$\int T_R s(\boldsymbol{m})d\Omega = \int s(\boldsymbol{m})d\Omega \ , \tag{5.8.2}$$

onto the image space $\mathscr{I}$. Namely, if the differential solid angle $d\Omega$ is expressed in terms of the image coordinates (x, y) in the form of $\rho(x, y)dx\,dy$, it gives the desired invariant measure on the image plane. (We will give an alternative derivation in Sect. 5.8.2.)

As we noted earlier, scene $s(\boldsymbol{m})$ can be regarded as a function over a unit sphere centered at the coordinate origin. The vector $\boldsymbol{m} = (m_1, m_2, m_3)$ is related to the image coordinates (x, y) by

$$m_1 = \frac{x}{\sqrt{x^2 + y^2 + f^2}} \ , \quad m_2 = \frac{y}{\sqrt{x^2 + y^2 + f^2}} \ ,$$

$$m_3 = \frac{z}{\sqrt{x^2 + y^2 + f^2}} \ . \tag{5.8.3}$$

This relationship can be viewed as defining a *curvilinear surface coordinate system* (x, y) over the unit sphere.

Let $\boldsymbol{m}$ be the point on the unit sphere whose surface coordinates are (x, y). Let $\boldsymbol{m} + \boldsymbol{dm} = (m_1 + dm_1, m_2 + dm_2, m_3 + dm_3)$ be the point on the sphere whose surface coordinates are $(x + dx, y + dy)$, and let ds be the arc length between the two points $\boldsymbol{m}$ and $\boldsymbol{m} + \boldsymbol{dm}$ along the great circle passing through them. Differentiating (5.8.3), we obtain

$$dm_1 = \frac{(y^2 + f^2)dx - xydy}{\sqrt{(x^2 + y^2 + f^2)^3}} \ , \quad dm_2 = \frac{-xydx + (x^2 + f^2)dy}{\sqrt{(x^2 + y^2 + f^2)^3}} \ ,$$

$$dm_3 = -f\frac{xdx + ydy}{\sqrt{(x^2 + y^2 + f^2)^3}} \ . \tag{5.8.4}$$

Hence,

$$ds^2 = dm_1^2 + dm_2^2 + dm_3^2 = \frac{(y^2 + f^2)dx^2 + 2xydx\,dy + (x^2 + f^2)dy^2}{(x^2 + y^2 + f^2)^2} \ , \tag{5.8.5}$$

which is rewritten as

$$ds^2 = Edx^2 + 2Fdx\,dy + Gdy^2 \ , \tag{5.8.6}$$

where

$$E = \frac{y^2 + f^2}{(x^2 + y^2 + f^2)^2} , \quad F = -\frac{xy}{(x^2 + y^2 + f^2)^2} , \quad G = \frac{x^2 + f^2}{(x^2 + y^2 + f^2)^2} . \tag{5.8.7}$$

Equation (5.8.6) is called the *first fundamental form*. The differential solid angle $d\Omega$ is expressed as follows (see Sect. 5.8.1):

$$d\Omega = \sqrt{EG - F^2}\, dx\, dy = \frac{f dx\, dy}{\sqrt{(x^2 + y^2 + f^2)^3}} . \tag{5.8.8}$$

Hence, we obtain

$$\rho(x, y) = \frac{f}{\sqrt{(x^2 + y^2 + f^2)^3}} . \tag{5.8.9}$$

5.8.1 First Fundamental Form

Consider a smooth surface in 3D space parametrized by surface coordinates (u, v). Let

$$X = X(u, v) , \qquad Y = Y(u, v) , \qquad Z = Z(u, v) \tag{5.8.10}$$

be the equations of the surface. Consider an infinitesimal change of the surface coordinates from (u, v) to $(u + du,\ v + dv)$. The corresponding point on the surface moves from (X, Y, Z), to $(X + dX,\ Y + dY,\ Z + dZ)$, where the infinitesimal displacements dX, dY, dZ are given by

$$dX = \frac{\partial X}{\partial u} du + \frac{\partial X}{\partial v} dv , \quad dY = \frac{\partial Y}{\partial u} du + \frac{\partial Y}{\partial v} dv , \quad dZ = \frac{\partial Z}{\partial u} du + \frac{\partial Z}{\partial v} dv . \tag{5.8.11}$$

Let ds be the infinitesimal distance between points (X, Y, Z) and $(X + dX,$ $Y + dY,\ Z + dZ)$ along the surface. From (5.8.11), we obtain

$$ds^2 = dX^2 + dY^2 + dZ^2 = E du^2 + 2F du\, dv + F dv^2 , \tag{5.8.12}$$

where

$$E = \left(\frac{\partial X}{\partial u}\right)^2 + \left(\frac{\partial Y}{\partial u}\right)^2 + \left(\frac{\partial Z}{\partial u}\right)^2 , \quad F = \frac{\partial X}{\partial u}\frac{\partial X}{\partial v} + \frac{\partial Y}{\partial u}\frac{\partial Y}{\partial v} + \frac{\partial Z}{\partial u}\frac{\partial Z}{\partial v} ,$$

$$G = \left(\frac{\partial X}{\partial v}\right)^2 + \left(\frac{\partial Y}{\partial v}\right)^2 + \left(\frac{\partial Z}{\partial v}\right)^2 . \tag{5.8.13}$$

Equation (5.8.12) is called the *first fundamental form* of the surface.

Let $P(X, Y, Z)$ be a point on the surface whose surface coordinates are (u, v). Let P' be the point whose surface coordinates are $(u + du, v)$. The infinitesimal displacement $\overrightarrow{PP'}$ is given by

$$\overrightarrow{PP'} = \left(\frac{\partial X}{\partial u} du, \frac{\partial Y}{\partial u} du, \frac{\partial Z}{\partial u} du\right) . \tag{5.8.14}$$

Let P'' be the point whose surface coordinates are $(u, v + dv)$. The infinitesimal displacement $\overrightarrow{PP''}$ is given by

$$\overrightarrow{PP''} = \left(\frac{\partial X}{\partial v} dv, \frac{\partial Y}{\partial v} dv, \frac{\partial Z}{\partial v} dv\right) . \tag{5.8.15}$$

The area dS of the parallelogram defined by the two vectors $\overrightarrow{PP'}$, $\overrightarrow{PP''}$ is given by

$$dS = \|\overrightarrow{PP'} \times \overrightarrow{PP''}\| \tag{5.8.16}$$

(Fig. 5.5). If we substitute (5.8.14, 15) into this, and noting (4.5.14, 15), we find

$$dS = \sqrt{EG - F^2}\, du\, dv . \tag{5.8.17}$$

5.8.2 Fluid Dynamics Analogy

Here is an alternative derivation of the invariant measure. This derivation invokes physical intuition and leads to a better understanding of the geometrical meaning of the function $\rho(x, y)$. Suppose the camera is smoothly rotating with an instantaneous rotation velocity $(\omega_1, \omega_2, \omega_3)$. Then, we observe an optical flow on the image plane in the following form (cf. (3.7.40)):

$$u = -f\omega_2 + \omega_3 y + \frac{1}{f}(-\omega_2 x + \omega_1 y)x ,$$

$$\mathrm{v} = -f\omega_1 + \omega_3 x + \frac{1}{f}(-\omega_2 x + \omega_1 y)y . \tag{5.8.18}$$

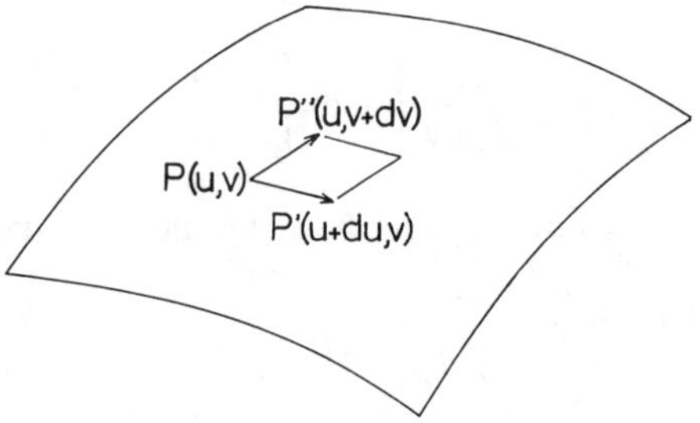

Fig. 5.5. An infinetisimal parallelogram difined by vectors $\overrightarrow{PP'}$and $\overrightarrow{PP''}$, where P, P', P'' are the points whose surface coordinates are (u, v), $(u + du, v)$, and $(u, v + dv)$, respectively

Suppose (5.8.18a, b) describe fluid flow over the image plane. Let us consider the physical meaning of (5.8.1). We do not lose generality if we assume that the function $F(x, y)$ is a characteristic function of a region S:

$$F(x, y) = \begin{cases} 1 & (x, y) \in S \ , \\ 0 & \text{otherwise} \ , \end{cases} \tag{5.8.19}$$

since all functions $F(x, y)$ (excluding extremely pathological ones) can be expressed as a (possibly infinite) linear combination of such characteristic functions. If $F(x, y)$ is the characteristic function of region S, then $T_R F(x, y)$ is the characteristic function of the region S' resulting from region S by the flow field. Then, (5.8.1) is rewritten as

$$\int_{S'} \rho(x, y) dx\, dy = \int_{S} \rho(x, y) dx\, dy \ . \tag{5.8.20}$$

Namely, the integration of the function $\rho(x, y)$ over the new region S' is the same as the integration of $\rho(x, y)$ over the original region S. This means that if we interpret the function $\rho(x, y)$ as the *density* of the fluid, (5.8.20) is interpreted as the *law of conservation of mass*: the fluid is never created or annihilated in the course of flowing.

As is well known in fluid dynamics, the integral form of (5.8.20) is converted into the differential equation

$$\frac{\partial(\rho u)}{\partial x} + \frac{\partial(\rho v)}{\partial y} = 0 \tag{5.8.21}$$

(Exercise 5.35), which is called the *equation of continuity*.

If we substitute (5.8.18) into this, we obtain

$$\begin{aligned} &\omega_1\left[\frac{3y}{f}\rho + \frac{xy}{f}\frac{\partial\rho}{\partial x} + \left(f + \frac{y^2}{f}\right)\frac{\partial\rho}{\partial y}\right] + \omega_2\left[-\frac{3x}{f}\rho \right.\\ &\left. -\left(f + \frac{x^2}{f}\right)\frac{\partial\rho}{\partial x} - \frac{xy}{f}\frac{\partial\rho}{\partial y}\right] + \omega_3\left[y\frac{\partial\rho}{\partial x} - x\frac{\partial\rho}{\partial y}\right] = 0 \ . \end{aligned} \tag{5.8.22}$$

Since this equation must hold for any flow, the density $\rho(x, y)$ must satisfy

$$\begin{aligned} &\frac{3y}{f}\rho + \frac{xy}{f}\frac{\partial\rho}{\partial x} + \left(f + \frac{y^2}{f}\right)\frac{\partial\rho}{\partial y} = 0 \ , \\ &-\frac{3x}{f}\rho - \left(f + \frac{x^2}{f}\right)\frac{\partial\rho}{\partial x} - \frac{xy}{f}\frac{\partial\rho}{\partial y} = 0 \ , \quad y\frac{\partial\rho}{\partial x} - x\frac{\partial\rho}{\partial y} = 0 \ . \end{aligned} \tag{5.8.23}$$

It is easy to confirm that the expression (5.8.9) satisfies these differential equations.

5.9 Transformation of Features

We now turn to our original objective—to characterize images by a small number of parameters. Consider the integration of image $F(x, y)$ with an appropriate weight $w(x, y)$:

$$J[F] = \int w(x, y)F(x, y)dx\,dy \ . \tag{5.9.1}$$

As before, we assume that all images are of finite support so that the integration is always defined. We call image characteristics obtained by this type of integration *image features.*

Equation (5.9.1) defines a *linear functional* over $\mathscr{I}$, i.e., a linear mapping from the image space $\mathscr{I}$ to real numbers. If the camera rotation transformation T_R is applied, the image $F(x, y)$ changes, and accordingly the image feature $J[F]$ also changes. Using the same notation T_R, we define the transformation of image features by

$$T_R J[F] \equiv \int w(x, y)T_R F(x, y)dx\,dy \quad (= J[T_R F]) \ . \tag{5.9.2}$$

Next, we define the *adjoint camera rotation transformation,* or simply *adjoint transformation, T_R^* by*

$$T_R J[F] = \int T_R^* w(x, y)F(x, y)dx\,dy \ . \tag{5.9.3}$$

In other words, we define the adjoint transformation T_R^* in such a way that the integration of the transformed image $T_R F(x, y)$ is the same as the integration of the original image $F(x, y)$ but with the new weight $T_R^* w(x, y)$.

Lemma 5.16. *The set of all the adjoint transformations T_R^* is a group of transformations, and the correspondence from $\boldsymbol{R}$ to T_R^* is a homomorphism from $SO(3)$:*

$$T_{R'}^* \circ T_R^* = T_{R'R}^* \ , \qquad \boldsymbol{R}', \boldsymbol{R} \in SO(3) \ , \tag{5.9.4}$$

$$T_I^* = I \ , \qquad (T_R^*)^{-1} = T_{R^{\mathrm{T}}}^* \ . \tag{5.9.5}$$

Proof. From (5.2.7), (5.9.2, 3), we see that

$$\int T^*_{R'R}\, w(x, y) F(x, y) dx\, dy = \int w(x, y) T_{R'R}\, F(x, y) dx\, dy$$

$$= \int w(x, y) T_R [T_{R'}\, F(x, y)] dx\, dy$$

$$= \int T^*_R\, w(x, y) T_{R'} F(x, y) dx\, dy$$

$$= \int T^*_{R'} [T^*_R w(x, y)] F(x, y) dx\, dy \ , \qquad (5.9.6)$$

which proves (5.9.4), from which (5.9.5) follows immediately,

Consider the set $\mathscr{W}$ of all weight functions $w(x, y)$. It can be regarded as a linear space. Let $\Pi[L^{(l)}]\rho(x, y)$ be the subspace of $\mathscr{W}$ generated by

$$w_{i_1 \ldots i_l}(x, y) \equiv F_{i_1 \ldots i_l}(x, y) \rho(x, y) \qquad (5.9.7)$$

for all $F_{i_1 \ldots i_l}(x, y) \in \Pi[L^{(l)}]$, where function $\rho(x, y)$ is defined by (5.8.9). We obtain the following transformation rule for camera rotation $\boldsymbol{R} = (r_{ij})$, $i, j = 1, 2, 3$.

Proposition 5.14. $\Pi[L^{(l)}]\rho(x, y)$ *is an invariant subspace of* $\mathscr{W}$ *under the adjoint transformation* T^*_R, *and* $w_{i_1 \ldots i_l}(x, y)$ *is transformed as a (symmetric) tensor of degree* l:[7]

$$T^*_R\, w_{i_1 \ldots i_l}(x, y) = \sum_{j_1=1}^{3} \cdots \sum_{j_l=1}^{3} r_{j_1 i_1} \cdots r_{j_l i_l} w_{j_1 \ldots j_l}(x, y) \ . \qquad (5.9.8)$$

Proof. From (5.8.1), (5.9.2, 3, 7), we see that

$$\int T^*_R\, w_{i_1 \ldots i_l}(x, y) F(x, y) dx\, dy = \int w_{i_1 \ldots i_l}(x, y) T_R F(x, y) dx\, dy$$

$$= \int F_{i_1 \ldots i_l}(x, y) \rho(x, y) T_R F(x, y) dx\, dy$$

$$= \int T_R [T_{R^{-1}} F_{i_1 \ldots i_l}(x, y) F(x, y)] \rho(x, y) dx\, dy$$

$$= \int T_{R^{\mathrm{T}}} F_{i_1 \ldots i_l}(x, y) F(x, y) \rho(x, y) dx\, dy \ . \qquad (5.9.9)$$

[7] If we compare (5.9.8) with (5.7.4), we see that the matrix element r_{ij} is replaced by r_{ji}. This difference comes from the difference between (5.9.2) and (5.9.3). We say that representations of $SO(3)$ defined for $F(x, y)$ and for $w(x, y)$ are *contragradient* to each other. Since the difference is whether $\boldsymbol{R}$ is transposed or not, we also call a quantity that is transformed by (5.9.8) a *tensor* of degree l. If distinction is necessary, the terms *contravariant tensor* and *covariant tensor* are sometime used.

Hence, from (5.7.4), we have

$$T_R^* w_{i_1 \dots i_l}(x, y) = T_{R^T} F_{i_1 \dots i_l}(x, y)\rho(x, y)$$
$$= \sum_{j_1=1}^{3} \dots \sum_{j_l=1}^{3} r_{j_1 i_1} \dots r_{j_l i_l} F_{j_1 \dots j_l}(x, y)\rho(x, y)$$
$$= \sum_{j_1=1}^{3} \dots \sum_{j_l=1}^{3} r_{j_1 i_1} \dots r_{j_l i_l} w_{j_1 \dots j_l}(x, y) \ . \tag{5.9.10}$$

Corollary 5.5. *The image feature defined by*

$$J_{i_1 \dots i_l}[F] = \int w_{i_1 \dots i_l}(x, y) F(x, y) dx\, dy \tag{5.9.11}$$

is transformed as a (symmetric) tensor of degree l:

$$T_R J_{i_1 \dots i_l}[F] = \sum_{j_1=1}^{3} \dots \sum_{j_l=1}^{3} r_{j_1 i_1} \dots r_{j_l i_l} J_{j_1 \dots j_l}[F] \ . \tag{5.9.12}$$

Example 5.4. For $l = 0$, we have

$$w(x, y) = \frac{f}{\sqrt{(x^2 + y^2 + f^2)^3}} \ , \tag{5.9.13}$$

which is $\rho(x, y)$ itself. For $l = 1$, we have

$$w_1(x, y) = \frac{fx}{(x^2 + y^2 + f^2)^2} \ , \quad w_2(x, y) = \frac{fy}{(x^2 + y^2 + f^2)^2} \ ,$$
$$w_3(x, y) = \frac{f^2}{(x^2 + y^2 + f^2)^2} \ . \tag{5.9.14}$$

For $l = 2$, we have

$$w_{11}(x, y) = \frac{fx^2}{\sqrt{(x^2 + y^2 + f^2)^5}} \ , \quad w_{12}(x, y) = \frac{fxy}{\sqrt{(x^2 + y^2 + f^2)^5}} \ ,$$
$$w_{13}(x, y) = \frac{f^2 x}{\sqrt{(x^2 + y^2 + f^2)^5}} \ , \quad w_{21}(x, y) = \frac{fxy}{\sqrt{(x^2 + y^2 + f^2)^5}} \ ,$$
$$w_{22}(x, y) = \frac{fy^2}{\sqrt{(x^2 + y^2 + f^2)^5}} \ , \quad w_{23}(x, y) = \frac{f^2 y}{\sqrt{(x^2 + y^2 + f^2)^5}} \ ,$$
$$w_{31}(x, y) = \frac{f^2 x}{\sqrt{(x^2 + y^2 + f^2)^5}} \ , \quad w_{32}(x, y) = \frac{f^2 y}{\sqrt{(x^2 + y^2 + f^2)^5}} \ ,$$
$$w_{33}(x, y) = \frac{f^3}{\sqrt{(x^2 + y^2 + f^2)^5}} \ . \tag{5.9.15}$$

Let $\Pi[D^{(l)}]\rho(x, y)$ be the subspace of $\mathscr{W}$ generated by $w_{\{i_1 \ldots i_l\}}(x, y)$. Then, we obtain:

Proposition 5.15. *$\Pi[D^{(l)}]\rho(x, y)$ is an irreducible subspace of $\mathscr{W}$ under the adjoint transformation T_R^*, and $w_{\{i_1 \ldots i_l\}}(x, y)$ is transformed as a (deviator) tensor of degree l:*

$$T_R^* w_{\{i_1 \ldots i_l\}}(x, y) = \sum_{j_1=1}^{3} \cdots \sum_{j_l=1}^{3} r_{j_1 i_1} \cdots r_{j_l i_l} w_{\{j_1 \ldots j_l\}}(x, y) \, . \tag{5.9.16}$$

Proof. This transformation follows from (5.9.8). Irreducibility is obvious, because the transformation of $w_{\{i_1 \ldots i_l\}}(x, y)$ is, through the relation of (5.9.7), in one-to-one correspondence with the transformation of $F_{\{i_1 \ldots i_l\}}(x, y)$, which defines the irreducible representation $\mathscr{D}_l$ (Proposition 5.11).

Corollary 5.6. *The image feature defined by*

$$J_{\{i_1 \ldots i_l\}}[F] = \int w_{\{i_1 \ldots i_l\}}(x, y) F(x, y) dx\, dy \tag{5.9.17}$$

is transformed as a (deviator) tensor of degree l:

$$T_R J_{\{i_1 \ldots i_l\}}[F] = \sum_{j_1=1}^{3} \cdots \sum_{j_l=1}^{3} r_{j_1 i_1} \cdots r_{j_l i_l} J_{\{i_1 \ldots i_l\}}[F] \, . \tag{5.9.18}$$

Example 5.5. For $l = 0, 1$, the functions $w_{\{i_1 \ldots i_l\}}$ are the same as in Example 5.4. For $l = 2$, we have

$$w_{\{11\}}(x, y) = \frac{f(2x^2 - y^2 - f^2)}{3\sqrt{(x^2 + y^2 + f^2)^5}} \, , \quad w_{\{12\}}(x, y) = \frac{fxy}{\sqrt{(x^2 + y^2 + f^2)^5}} \, ,$$

$$w_{\{13\}}(x, y) = \frac{f^2 x}{\sqrt{(x^2 + y^2 + f^2)^5}} \, , \quad w_{\{21\}}(x, y) = \frac{fxy}{\sqrt{(x^2 + y^2 + f^2)^5}} \, ,$$

$$w_{\{22\}}(x, y) = \frac{f(-x^2 + 2y^2 - f^2)}{3\sqrt{(x^2 + y^2 + f^2)^5}} \, , \quad w_{\{23\}}(x, y) = \frac{f^2 y}{\sqrt{(x^2 + y^2 + f^2)^5}} \, ,$$

$$w_{\{31\}}(x, y) = \frac{f^2 x}{\sqrt{(x^2 + y^2 + f^2)^5}} \, , \quad w_{\{32\}}(x, y) = \frac{f^2 y}{\sqrt{(x^2 + y^2 + f^2)^5}} \, ,$$

$$w_{\{33\}}(x, y) = \frac{f(-x^2 - y^2 + 2f^2)}{3\sqrt{(x^2 + y^2 + f^2)^5}} \, . \tag{5.9.19}$$

Consider the adjoint transformation T_R^* for a small camera rotation $\boldsymbol{R}$.

Proposition 5.16. *For a small rotation with axis $\boldsymbol{n} = (n_1, n_2, n_3)$ and angle Ω,*

$$T_R^* w(x, y) = w(x, y) + \Omega(n_1 D_1^* + n_2 D_2^* + n_3 D_3^*) w(x, y) + O(\Omega^2) , \tag{5.9.20}$$

where D_1^*, D_2^*, D_3^* are the adjoint infinitesimal generators given by

$$\begin{aligned} D_1^* &= \frac{3y}{f} + \frac{xy}{f}\frac{\partial}{\partial x} + \left(f + \frac{y^2}{f}\right)\frac{\partial}{\partial y} , \\ D_2^* &= -\frac{3x}{f} - \left(f + \frac{x^2}{f}\right)\frac{\partial}{\partial x} - \frac{xy}{f}\frac{\partial}{\partial y} , \\ D_3^* &= y\frac{\partial}{\partial x} - x\frac{\partial}{\partial y} . \end{aligned} \tag{5.9.21}$$

Proof. From (5.7.12), the infinitesimal change of the feature is given by

$$\begin{aligned} T_R J[F] = J[F] + \Omega \int w(x, y)(n_1 D_1 + n_2 D_2 + n_3 D_3) F(x, y) dx\, dy \\ + O(\Omega^2) , \end{aligned} \tag{5.9.22}$$

where D_1, D_2, D_3 are the infinitesimal generators defined by (5.7.13). Integration of the second term on the right-hand side by parts results in (5.9.20), where D_1^*, D_2^*, D_3^* are the differential operator given by (5.9.21).

We can easily confirm that the adjoint infinitesimal generators D_1^*, D_2^*, D_3^* satisfy the following commutation relations.

$$[D_1^*, D_2^*] = D_3^* , \qquad [D_2^*, D_3^*] = D_1^*, \qquad [D_3^*, D_1^*] = D_2^* . \tag{5.9.23}$$

(Since the transformations of $F(x, y)$ and $w(x, y)$ are contragradient to each other, the minus signs do not appear on the right-hand sides.) The Casimir (or Laplace–Beltrami) operator $H^* = -(D_1^{*2} + D_2^{*2} + D_3^{*2})$ takes the form

$$\begin{aligned} H^* = -\left(f + \frac{x^2 + y^2}{f}\right)\Bigg[\left(f + \frac{x^2}{f}\right)\frac{\partial^2}{\partial x^2} + \frac{2xy}{f}\frac{\partial^2}{\partial x \partial y} \\ + \left(f + \frac{y^2}{f}\right)\frac{\partial^2}{\partial y^2} + 8x\frac{\partial}{\partial x} + 8y\frac{\partial}{\partial y}\Bigg] - 6 - \frac{12(x^2 + y^2)}{f^2} . \end{aligned} \tag{5.9.24}$$

The adjoint transformation T_R^* defines an irreducible representation of $SO(3)$ equivalent to $\mathscr{D}_l$ if and only if the subspace is generated by $2l + 1$ independent eigenfunctions of H^* for eigenvalue $l(l + 1)$, i.e., $2l + 1$ independent solutions of

$$H^* w(x, y) = l(l + 1) w(x, y) . \tag{5.9.25}$$

From (5.9.24), this equation become

$$\left(f+\frac{x^2+y^2}{f}\right)\left[\left(f+\frac{x^2}{f}\right)w_{xx}+\frac{2xy}{f}w_{xy}+\left(f+\frac{y^2}{f}\right)w_{yy}\right.$$
$$\left.+8xw_x+8yw_y\right]+\left(l(l+1)+6+\frac{12(x^2+y^2)}{f^2}\right)w=0\ . \qquad (5.9.26)$$

We can easily confirm this for $l = 0, 1, 2$ (Exercise 5.44).[8]

5.10 Invariant Characterization of a Shape

Let S be a region on the image plane. Consider its characteristic function

$$F(x,y)=\begin{cases}1 & \text{if } (x,y)\in S\ ,\\ 0 & \text{otherwise}\ .\end{cases} \qquad (5.10.1)$$

This function $F(x, y)$ defines a binary image that satisfies the requirements we have imposed: it is of finite support, and its values (0 and 1) are not affected by the viewing orientations. In the following, we assume that the camera rotation is small so that the region S is not carried away to infinity.

The simplest characteristic of the region S may be its area

$$\bar{S}=\int dx\,dy\ . \qquad (5.10.2)$$

However, this area is *not a scalar*; this value changes as the region S moves under the camera rotation. Consequently, the area $\bar{S}$ does not have a meaning inherent to the scene.

On the other hand, we know from Corollary 5.5 that

$$C=f^3\int_S\frac{dx\,dy}{\sqrt{(x^2+y^2+f^2)^3}} \qquad (5.10.3)$$

is a scalar:

$$T_R C = C\ . \qquad (5.10.4)$$

Hence, its value has a meaning inherent to the scene. In fact, suppose region S is small and localized around the image origin. If we put $x\approx 0$, $y\approx 0$, the value of C is approximately equal to its area $\bar{S}$. Hence, we can interpret C as the area of the region when it is brought to the center of the image plane by the camera rotation. We call it the *invariant area.*

---

[8] In fact, eqs (5.8.24) are simply $D_1^*\rho(x,)y=0$, $D_2^*\rho(x,)y=0$, $D_3^*\rho(x,)y=0$. Hence, $\rho(x, y)$ is the solution of eq. (5.9.26) for $l=0$.

Another simple characteristic is the *centroid* of the region S:

$$\bar{x} = \frac{\int_S x\,dx\,dy}{\int_S dx\,dy}, \quad \bar{y} = \frac{\int_S y\,dx\,dy}{\int_S dx\,dy}, \tag{5.10.5}$$

However, the pair $\{\bar{x}, \bar{y}\}$ is *not transformed as a point*; if region S is moved to another position by a camera rotation, and if $(\bar{x}', \bar{y}')$ is the centroid there, the original centroid $(\bar{x}, \bar{y})$ is not necessarily mapped to $(\bar{x}', \bar{y}')$. Consequently, the centroid $(\bar{x}, \bar{y})$ does not have a meaning inherent to the scene.

On the other hand, we know from Corollary 5.5 that

$$a_1 = f\int_S \frac{x\,dx\,dy}{(x^2+y^2+f^2)^2}, \quad a_2 = f\int_S \frac{y\,dx\,dy}{(x^2+y^2+f^2)^2},$$
$$a_3 = f^2\int_S \frac{dx\,dy}{(x^2+y^2+f^2)^2}, \tag{5.10.6}$$

are transformed as a vector:

$$T_R\begin{pmatrix} a_1 \\ a_2 \\ a_3 \end{pmatrix} = \boldsymbol{R}^{\mathrm{T}}\begin{pmatrix} a_1 \\ a_2 \\ a_3 \end{pmatrix}. \tag{5.10.7}$$

From Lemma 4.4, the pair $\{fa_1/a_3, fa_2/a_3\}$ is transformed as a point. Hence, it has a meaning inherent to the scene. In fact, if the region S is small and localized around the image origin, we can see, by putting $x \approx 0$, $y \approx 0$, that this point is approximately the centroid of the region. Hence, it is interpreted as the centroid of the region if it is brought to the center of the image by the camera rotation. We call it the *invariant centroid.*

Another useful characteristic is the *moments* n_{ij}, $i, j = 1, 2$, defined by

$$n_{11} = \int_S (x-\bar{x})^2\,dx\,dy, \quad n_{22} = \int_S (y-\bar{y})^2\,dx\,dy,$$
$$n_{12} = n_{21} = \int_S (x-\bar{x})(y-\bar{y})\,dx\,dy. \tag{5.10.8}$$

The principal values of the matrix $\boldsymbol{N} = (n_{ij})$, $i, j = 1, 2$, indicate the extent of elongation of the region S along the corresponding principal axes (the orientations of maximum and minimum elongation). However, these moments do not have a meaning inherent to the scene; the principal values of matrix $\boldsymbol{N}$ are *not scalars*, and its principal axes are *not lines*.

On the other hand, we know from Corollary 5.5 that

$$b_{11} = f\int_S \frac{x^2\,dx\,dy}{\sqrt{(x^2+y^2+f^2)^5}}, \quad b_{12} = f\int_S \frac{xy\,dx\,dy}{\sqrt{(x^2+y^2+f^2)^5}},$$
$$b_{13} = f^2\int_S \frac{x\,dx\,dy}{\sqrt{(x^2+y^2+f^2)^5}}, \quad b_{21} = f\int_S \frac{xy\,dx\,dy}{\sqrt{(x^2+y^2+f^2)^5}},$$
$$b_{22} = f\int_S \frac{y^2\,dx\,dy}{\sqrt{(x^2+y^2+f^2)^5}}, \quad b_{23} = f^2\int_S \frac{y\,dx\,dy}{\sqrt{(x^2+y^2+f^2)^5}},$$
$$b_{31} = f^2\int_S \frac{x\,dx\,dy}{\sqrt{(x^2+y^2+f^2)^5}}, \quad b_{32} = f^2\int_S \frac{x\,dx\,dy}{\sqrt{(x^2+y^2+f^2)^5}},$$
$$b_{33} = f^3\int_S \frac{dx\,dy}{\sqrt{(x^2+y^2+f^2)^5}}, \tag{5.10.9}$$

are transformed as a (symmetric) tensor:

$$T_R\begin{pmatrix} b_{11} & b_{12} & b_{13} \\ b_{21} & b_{22} & b_{23} \\ b_{31} & b_{32} & b_{33} \end{pmatrix} = \boldsymbol{R}\begin{pmatrix} b_{11} & b_{12} & b_{13} \\ b_{21} & b_{22} & b_{23} \\ b_{31} & b_{32} & b_{33} \end{pmatrix}\boldsymbol{R}. \tag{5.10.10}$$

As long as region S is not empty, this tensor is positive definite and has three positive principal values $\sigma_1, \sigma_2, \sigma_3$. We can also see that if region S is small and localized around the image origin, the values of b_{33} dominates among the nine components. Let σ_3 be the maximum of the three principal values, and let G be the point representing the corresponding principal axis on the image plane. Let l_1, l_2 be the lines passing through point G and representing the remaining two principal axes. Thus, σ_1, σ_2 are scalars, G is a point, and l_1, l_2 are lines. Hence, these quantities have meanings inherent to the scene.

In order to understand their meanings, suppose the region S is small and localized around the center of the image. Then, the denominators of the integrands in (5.10.9) are approximately f^5. If we assume that the region S is symmetric with respect to the x- and y-axes, the matrix $\boldsymbol{B}$ becomes diagonal, and the diagonal elements are approximately equal to $\int_S x^2\,dx\,dy/f^4$, $\int_S y^2\,dx\,dy/f^4$, $\int_S dx\,dy/f^2$. Hence, $\sigma_1 \approx \int_S x^2\,dx\,dy/f^4$, $\sigma_2 \approx \int_S y^2\,dx\,dy/f^4$, $\sigma_3 \approx \int_S dx\,dy/f^2$, and lines l_1, l_2 coincide with the image coordinate axes.

From this, we observe that $f^4\sigma_1, f^4\sigma_2$ are approximately the principal values of the moment matrix $\boldsymbol{N}$, and l_1, l_2 are approximately the corresponding principal axes when the region is brought to the center of the image by the camera rotation. We call the point G the *invariant inertia center*, the values $f^4\sigma_1$, $f^4\sigma_2$ the *invariant principal values*, and the lines l_1, l_2 the corresponding *invariant principal axes*.

So far, we have assumed binary images of the form of (5.10.1), but the above procedures and interpretations also apply to gray-level images $F(x, y)$ as long as they are of finite support. We can construct the scalar C, the vector $\boldsymbol{a}$, and the tensor $\boldsymbol{B}$ in exactly the same way, and we can deduce invariant meanings from them—the invariant area, the invariant centroid, the invariant inertia center, and the invariant principal values and axes.

Since C is a scalar, $\boldsymbol{a}$ is a vector, and $\boldsymbol{B} = (b_{ij})$ is a tensor, they can be used to test for the *equivalence* of two images. Since $\mathrm{Tr}\{\boldsymbol{B}\} = C$, an invariant basis for $\{C, \boldsymbol{a}, \boldsymbol{B}\}$ is given by

$$\|\boldsymbol{a}\|^2, \quad \mathrm{Tr}\{\boldsymbol{B}\}(=C)\ , \quad \mathrm{Tr}\{\boldsymbol{B}^2\}\ , \quad \mathrm{Tr}\{\boldsymbol{B}^3\}\ ,$$
$$(\boldsymbol{a}, \boldsymbol{Ba})\ , \quad \|\boldsymbol{Ba}\|^2\ , \quad |\boldsymbol{a}\,\boldsymbol{Ba}\,\boldsymbol{B}^2\boldsymbol{a}|\ . \tag{5.10.11}$$

If two images are equivalent (i.e., one is mapped onto the other by a camera rotation), these invariant must take the same values. If the values are not identical, the two images cannot be equivalent. The usefulness of this criteria lies in the fact that we need not know the *point-to-point correspondence*—the knowledge of which point corresponds to which between the two images—which is usually very difficult to compute.

Consider the problem of shape recognition. Suppose we have a reference image. If a test image is obtained from a different camera orientation, the two images cannot be compared directly due to projective distortion. However, the invariants provide an easy test; if the invariants take different values, the equivalence is rejected. If the invariants take the same values, other characteristics must be checked, or the two images brought together by applying the predicted camera rotation transformation and then compared directly.

Example 5.6. Consider the three regions S_0, S_1, S_2 on the image plane shown in Fig. 5.6. The length is scaled by the focal length f. Computing the vector $\{a_1, a_2, a_3\}$ by (5.10.6), we find their invariant centroids as shown in Fig. 5.7. Also, computing the tensor b_{ij}, $i, j = 1, 2, 3$, by (5.10.9), we find their invariant inertia centers and invariant principal axes as shown in Fig. 5.8. By computing the scalar invariants of (5.10.11), we conclude that regions S_0 and S_1 can be equivalent, since all these scalar invariants coincide. However, S_2 cannot be equivalent to either. For example, the invariant area C is 0.144 for both S_0 and S_1 but is 0.112 for S_2.

If the three regions S_0, S_1, S_2 are known to be images of the same object, we conclude that a motion took place between S_0 and S_2 and between S_1 and S_2 independently of the camera. By the procedure described in Sect. 4.8, the camera rotation that maps region S_1 onto region S_0 is constructed as

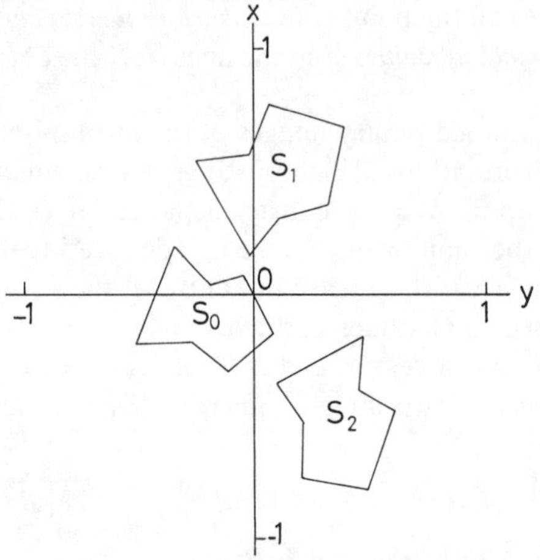

Fig. 5.6. Three regions S_0, S_1, S_2 to be tested for equivalence

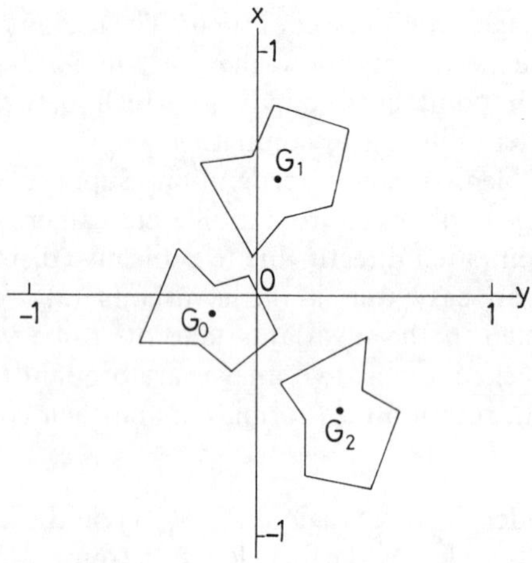

Fig. 5.7. Invariant centroids G_0, G_1, G_2 of regions S_0, S_1, S_2

$$\boldsymbol{R} = \begin{pmatrix} 0.573 & 0.567 & 0.591 \\ -0.761 & 0.631 & 0.136 \\ -0.296 & -0.530 & 0.795 \end{pmatrix}. \tag{5.10.12}$$

This is the rotation by angle 60° screw-wise around axis $\boldsymbol{n} = (-0.384, 0.512, -0.768)$, see Sect. 6.3 for the details of computation.

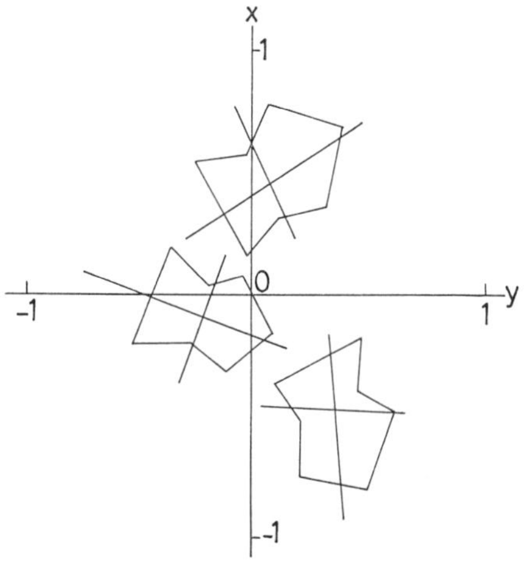

Fig. 5.8. Invariant principal axes of regions S_0, S_1, S_2

5.10.1 Invariance on the Image Sphere

It is clear from the mathematical framework of this chapter that if we use the *image sphere* (Sect. 4.6), we need not make any distinction between scenes and images, since a scene $s(\boldsymbol{m})$ is simply a function over a unit sphere centered at the viewpoint. The invariance of images simply means *congruence* on the image sphere. However, as we discussed in Sect. 4.6, what we are showing is that congruence relations on the image sphere can be characterized on the image plane in terms of the xy image coordinates.

On the other hand, it is worthwhile to consider the meanings of invariants on the image sphere. Let $X^2 + Y^2 + Z^2 = f^2$ be the image sphere centered at the coordinate origin with radius f (Fig. 5.9).

First, consider the invariant area C defined by (5.10.3). From the derivation of $\rho(x, y)$ in Sect. 5.7, we can express C as

$$C = \int_S \rho(x, y) dx\, dy = \int_{\Pi^{-1}S} d\Omega \; . \tag{5.10.13}$$

In other words, C is simply the *area of the corresponding region on the image sphere*. Evidently, this area is invariant under camera rotations. The important point is that we can compute this area on the image plane without constructing the image sphere.

Next, consider the vector $\boldsymbol{a} = (a_i)$, $i = 1, 2, 3$, defined by (5.10.6). If we recall the definition of $w_i(x, y)$, equation (10.6) becomes

$$a_i = \int_S \Pi[m_i]\rho(x, y) dx\, dy = \int_{\Pi^{-1}S} m_i d\Omega \; . \tag{5.10.14}$$

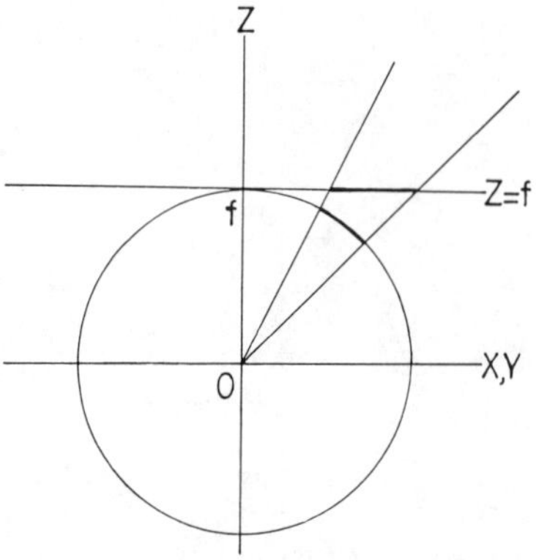

Fig. 5.9. Correspondence between the image plane $Z = f$ and the image sphere $X^2 + Y^2 + Z^2 = f^2$

A physical interpretation is given as follows. Suppose a mass is uniformly distributed within the region $\Pi^{-1}S$ on the image sphere and the density is unity per unit area. Then, the vector $\boldsymbol{a}/C$ indicates the center of mass in the 3D space. The invariant centroid is simply the intersection of the ray defined by this vector with the image plane $Z = f$.

Similarly, consider the tensor $\boldsymbol{B} = (b_{ij})$, $ij = 1, 2, 3$, defined by (5.10.9). If we recall the definition of function $w_{ij}(x, y)$, we find that

$$b_{ij} = \int_S \Pi[m_i m_j]\rho(x, y)dx\,dy = \int_{\Pi^{-1}S} m_i m_j \, d\Omega \ . \tag{5.10.15}$$

According to this physical interpretation, this tensor indicates the *moment of inertia* of the region $\Pi^{-1}S$ on the image sphere. We picked out the principal axis around which the moment of inertia is minimum (i.e., the principal value is maximum), and defined its intersection with the image plane as the invariant inertia center. The invariant principal axes are the intersections of the image plane with the two planes passing through this principal axis and the remaining principal axes respectively.

5.10.2 Further Applications

Let us say that a region on the image plane is in the *standard position* if the invariant inertia center G coincides with the image origin and the invariant principal axes coincide with the image coordinate axes. Any region can be moved into the standard position by an appropriate camera rotation. The necessary camera rotation $\boldsymbol{R}$ is computed from the requirement that the tensor $\boldsymbol{B}$ is diagonalized in the form

$$\boldsymbol{B}' = \boldsymbol{R}^{\mathrm{T}}\boldsymbol{B}\boldsymbol{R} = \begin{pmatrix} \sigma_1 & & \\ & \sigma_2 & \\ & & \sigma_3 \end{pmatrix} \qquad (5.10.16)$$

in such a way that σ_3 is the largest principal value. At the same time, the vector $\boldsymbol{a}$ must be transformed in such a way that the new vector

$$\boldsymbol{a}' = \boldsymbol{R}^{\mathrm{T}}\boldsymbol{a} = \begin{pmatrix} a_1' \\ a_2' \\ a_3' \end{pmatrix} \qquad (5.10.17)$$

has a positive value for the third component: $a_3 > 0$. Without this condition, the camera may possibly be oriented opposite to the object image.

Evidently, shape recognition becomes easier if the test shapes are always moved to the standard position (either by actually rotating the camera or by computation). However, this technique is not restricted to shape recognition. For example, suppose a camera is tracking a moving object from a fixed position, or suppose a camera attached to a mobile robot is aiming at an object fixed in the scene. Then, the above procedure assures that the object in question is always seen in the standard position.

The testing of equivalence is also viewed as the detection of *active motion*. We say that the motion of an object image is *passive* if the motion is induced by camera rotation alone, and *active* otherwise. When the camera orientation is changed, all object images move on the image plane, but at the same time the objects may also have moved in the scene independently of the camera. According to the procedure described in this section, we can detect active motion even if the angle and orientation of camera rotation are not known: if the two corresponding images are not equivalent, the object must have moved actively. This test can be done by computing the invariants of the object images, and *we need not know the point-to-point correspondence on the image plane.*[9]

Another possible application is camera orientation registration. Even if the camera is rotated by an unknown angle around an unknown axis, the camera orientation can be discovered as long as one particular region is identified on the image plane before and after the camera rotation. Again, we need not know the point-to-point correspondence.

---

[9] This idea can be extended to an arbitrary 3D object motion. In Chap. 7, we present a scheme of 3D motion detection without using the point-to-point correspondence.

Exercises

5.1 Using (5.2.3, 5), show that the camera rotation transformation in the image space $\mathcal{I}$ defined by (5.2.10) is written in the form of (5.2.11).

5.2 Prove that $L^{(l)}$ is an $(l+1)(l+2)/2$-dimensional space by showing that the set $\{m_{i_1} \ldots m_{i_l} | \; i_1, \ldots, i_l = 1, 2, 3\}$ has $(l+1)(l+2)/2$ distinct elements.

5.3 Prove that a scalar multiple of a tensor, (5.3.13) is also a tensor.

5.4 Prove that the sum of two tensors of the same degree is also a tensor.

5.5 Prove that the product of two tensors is also a tensor.

5.6 Prove that contraction of a tensor over a pair of indices is also a tensor.

5.7 Prove the quotient rule of (5.3.17).

5.8 (a) Prove that symmetrization of a tensor results in a tensor.
(b) Prove that a symmetric tensor is always transformed into a symmetric tensor.

5.9 (a) Prove that antisymmetrization of a tensor results in a tensor.
(b) Prove that an antisymmetric tensor is always transformed into an antisymmetric tensor.

5.10 Prove (5.3.22, 23).

5.11 Show that the inner product (5.4.3) satisfies (5.4.5–7).

5.12 Show that $c_1^{(2)} = -1/3$ if the value of $c_1^{(2)}$ is so determined that the tensor of (5.5.1) is a deviator.

5.13 Show that $c_1^{(3)} = -3/5$ if we determine the value of $c_1^{(3)}$ is so determined that the tensor of (5.5.3) is deviator.

5.14* Show that $c_s^{(r)}$ is given by (5.5.6) if the value of $c_s^{(r)}$ is so determined that the tensor of (5.5.5) is a deviator. *Hint*: Derive and solve a double recurrence equation of $c_s^{(r)}$ over s and r.

5.15 (a) Prove that the deviator part of a symmetric tensor is also a tensor.
(b) Prove that a deviator tensor is always transformed into a deviator tensor.

5.16* Prove

$$\int m_{i_1} \ldots m_{i_r} \, d\Omega = \frac{4\pi}{2r+1} \delta_{(i_1 i_2} \delta_{i_3 i_4} \ldots \delta_{i_{r/2-1} i_{r/2})} .$$

Hint: Argue from the viewpoint of the symmetry of indices that the left-hand side is a multiple of $\delta_{(i_1 i_2} \delta_{i_3 i_4} \ldots \delta_{i_{r/2-1} i_{r/2})}$, and determine the multiplier by contraction.

5.17 Prove that the subspace generated by $m_{\{i_1 \ldots i_l\}}$, $i_1, \ldots, i_l = 1, 2, 3$, is $2l+1$ dimensional by showing that there are $l(l-1)/2$ different ways to choose a pair of indices from among $i_1, \ldots, i_l$.

5.18 Derive (5.6.7) from (5.6.6) by Taylor expansion.

5.19 Show that the infinitesimal generators D_1, D_2, D_3 given by (5.6.8, 9) become (5.6.11) in spherical coordinates (θ, φ).

5.20 Show that the infinitesimal generators D_1, D_2, D_3 given by (5.6.11) satisfy the commutation relations (5.6.12).

5.21 Show that the infinitesimal generators D_1, D_2, D_3 given by (5.6.11) define the Casimir operator H in the form of (5.6.14).

5.22 Show that $s(\theta, \varphi)$ satisfies differential equation (5.6.16) if and only if $\tilde{s}(r, \theta, \varphi)$ satisfies differential equation (5.6.18).

5.23 Show that $1/r$ is a harmonic by direct differentiation.

5.24 Show that for $l = 1$, images $F_i(x, y)$, $i = 1, 2, 3$, are given by (5.7.6).

5.25 Show that for $l = 2$, images $F_{ij}(x, y)$, $i, j = 1, 2, 3$, are given by (5.7.7).

5.26 Show that for $l = 2$, images $F_{\{ij\}}(x, y)$, $i, j = 1, 2, 3$, are given by (5.7.11).

5.27 Show that the infinitesimal generators D_1, D_2, D_3 in (5.7.12) are given by (5.7.13).

5.28 Show that the infinitesimal generators D_1, D_2, D_3 given by (5.7.13) satisfy the commutation relations (5.7.15).

5.29 Show that the infinitesimal generators D_1, D_2, D_3 given by (5.7.13) define the Casimir operator H in the form of (5.7.16).

5.30 Show that images $F_{\{i_1 \ldots i_l\}}$, $i_1, \ldots, i_l = 1, 2, 3$, satisfy differential equation (5.7.18) for $l = 0, 1, 2$.

5.31 Differentiate (5.8.3) to obtain (5.8.4).

5.32 Show that the first fundamental form is given by (5.8.5) or equivalently (5.8.6, 7).

5.33 Substitute (5.8.11) into (5.8.12) to obtain (5.8.13).

5.34 Substituting (5.8.14, 15) into (5.8.16) and using (5.8.13) show that the infinitesimal area dS is given in terms of E, F, G in the form of (5.8.17).

5.35* Derive the equation of continuity (5.8.21) from the law of conservation of mass (5.8.20). *Hint*: Consider an instantaneous change and integrate the resulting equation by parts (the (*Gauss*) *divergence theorem*).

5.36 Show that the equation of continuity (5.8.20) becomes (5.8.22) for the flow of (5.8.18).

5.37 Show that for $l = 0$, function $w(x, y)$ is given by (5.9.13).

5.38 Show that for $l = 1$, functions $w_i(x, y)$, $i = 1, 2, 3$, are given by (5.9.14).

5.39 Show that for $l = 2$, functions $w_{ij}(x, y)$, $i, j = 1, 2, 3$, are given by (5.9.15).

5.40 Show that for $l = 2$, functions $w_{\{ij\}}(x, y)$, $i, j = 1, 2, 3$, are given by (5.9.19).

5.41 Show that the adjoint infinitesimal generators D_1^*, D_2^*, D_3^* in (5.9.20) are given by (5.9.21).

5.42 Show that the adjoint infinitesimal generators D_1^*, D_2^*, D_3^* given by (5.9.21) satisfy the commutation relations (5.9.23).

5.43 Show that the adjoint infinitesimal generators D_1^*, D_2^*, D_3^* given by (5.9.21) define the Casimir operator H^* in the form of (5.9.24).

5.44 Show that functions $w_{\{i_1 \ldots i_l\}}$, $i_1, \ldots, i_l = 1, 2, 3$, satisfy the differential equation (5.9.26) for $l = 0, 1, 2$.

5.45 Show that C defined by (5.10.3) is a scalar.

5.46 Show that a_i, $i = 1, 2, 3$, defined by (5.10.6) are transformed as a vector in the form of (5.10.7).
5.47 Show that b_{ij}, $i, j = 1, 2, 3$, defined by (5.10.9) are transformed as a tensor in the form of (5.10.10).
5.48 Prove (5.10.13).
5.49 Prove (5.10.14).
5.50 Prove (5.10.15).
5.51 Given the values of a_i, $i = 1, 2, 3$, and b_{ij}, $i, j = 1, 2, 3$, state the procedure to compute the camera rotation $\boldsymbol{R}$ that will bring the image into the standard position.

6. Representation of 3D Rotations

In this chapter, we study various aspects of 3D rotations. We first prove *Euler's theorem* and show how to parametrize a 3D rotation and how to compute compositions and inverses. For this purpose, we present many different representations of 3D rotations—by a rotation matrix, an angle and an axis, the *Euler angles*, the *Cayley–Klein parameters*, and a *quaternion*. We also discuss topological aspects of $SO(3)$. Finally, we derive *invariant measures* of $SO(3)$.

6.1 Representation of Object Orientations

In the previous chapters, we have studied various invariance properties of images under camera rotation. The importance of the study of 3D rotations is not limited to the study of camera rotations. Recall that the aim of image understanding is to obtain 3D descriptions of objects (Sect. 1.1). Rotations play a central role in object description: the orientation of an object is specified by a rotation from a fixed reference orientation, and the motion of an object is specified by the translation and rotation that took place between two frames.

In this chapter, we study various aspects of 3D rotations, in particular, how to parametrize a rotation and how to compute compositions and inverses. There are many different ways of parametrizing a rotation, all with their own advantages and disadvantages. We begin with the matrix representation of rotations and prove *Euler's theorem*, which enables us to represent a rotation by an axis and an angle. Then, we introduce the *Euler angles* and the *Cayley–Klein parameters*. The Cayley–Klein parameters are directly related to representations of $SO(3)$ by $SU(2)$, which is a *universal covering group* of $SO(3)$. We will prove that the matrix representation of $SO(3)$ is obtained as the *adjoint representation* of $SU(2)$.

Then, we introduce *quaternions*. We will also discuss the *topology* of $SO(3)$ and show that $SO(3)$ is not a simply connected group. Finally, we derive *invariant measures* that define the randomness of distributed 3D rotations.

6.2 Rotation Matrix

We start with a definition of a 3D rotation. Consider a Cartesian XYZ coordinate system in space, and let

$$\boldsymbol{i} = \begin{pmatrix} 1 \\ 0 \\ 0 \end{pmatrix}, \quad \boldsymbol{j} = \begin{pmatrix} 0 \\ 1 \\ 0 \end{pmatrix}, \quad \boldsymbol{k} = \begin{pmatrix} 0 \\ 0 \\ 1 \end{pmatrix} \tag{6.2.1}$$

be the respective unit vectors along the X-, Y-, Z-axes—they are called the *natural basis vectors.* Suppose a linear mapping maps the unit vectors $\{\boldsymbol{i}, \boldsymbol{j}, \boldsymbol{k}\}$ onto

$$\boldsymbol{r}_1 = \begin{pmatrix} r_{11} \\ r_{21} \\ r_{31} \end{pmatrix}, \quad \boldsymbol{r}_2 = \begin{pmatrix} r_{12} \\ r_{22} \\ r_{32} \end{pmatrix}, \quad \boldsymbol{r}_3 = \begin{pmatrix} r_{13} \\ r_{23} \\ r_{33} \end{pmatrix}, \tag{6.2.2}$$

respectively (Fig. 6.1). As is well known, we have

Lemma 6.1. *The linear mapping that maps the natural basis vectors* $\{\boldsymbol{i}, \boldsymbol{j}, \boldsymbol{k}\}$ *onto three vectors* $\{\boldsymbol{r}_1, \boldsymbol{r}_2, \boldsymbol{r}_3\}$ *is given by a matrix that has the vectors* $\{\boldsymbol{r}_1, \boldsymbol{r}_2, \boldsymbol{r}_3\}$ *as its three columns:*

$$\boldsymbol{R} = (\boldsymbol{r}_1 \boldsymbol{r}_2 \boldsymbol{r}_3) = \begin{pmatrix} r_{11} & r_{12} & r_{13} \\ r_{21} & r_{22} & r_{23} \\ r_{31} & r_{32} & r_{33} \end{pmatrix}. \tag{6.2.3}$$

Proof. Application of the matrix (6.2.3) to the vectors $\{\boldsymbol{i}, \boldsymbol{j}, \boldsymbol{k}\}$ of (6.2.1) results in the vectors $\{\boldsymbol{r}_1, \boldsymbol{r}_2, \boldsymbol{r}_3\}$ of (6.2.2).

The nine numbers r_{ij}, $i, j = 1, 2, 3$, are called the *matrix elements* of the linear mapping (Appendix, Sect. A.5). We say that this linear mapping is a *3D rotation* if

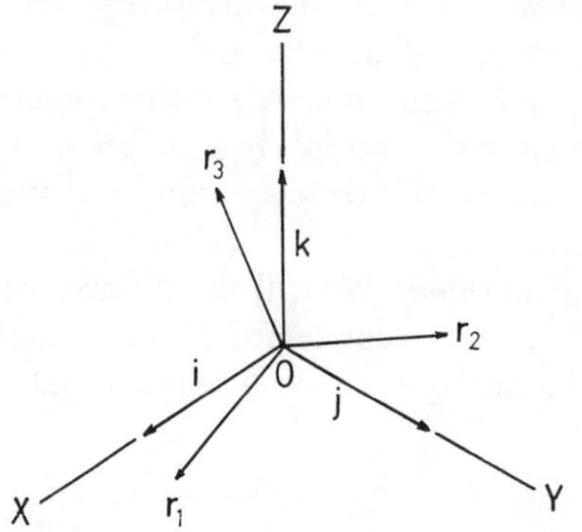

Fig. 6.1. The natural basis vectors $\{\boldsymbol{i}, \boldsymbol{j}, \boldsymbol{k}\}$ are respectively mapped onto vectors $\{\boldsymbol{r}_1, \boldsymbol{r}_2, \boldsymbol{r}_3\}$ by matrix $\boldsymbol{R}$. This mapping is a 3D rotation if the vectors are moved rigidly, maintaining their lengths and angles

(i) the lengths of the three vectors and the angles between them are preserved between $\{i, j, k\}$ and $\{r_1, r_2, r_3\}$, and
(ii) the *parity* is preserved, i.e., no reflection takes place.

Since the natural basis vectors $\{i, j, k\}$ are an orthonormal system, condition (i) is equivalent to

$$(r_i, r_j) = \delta_{ij}, \qquad i, j = 1, 2, 3, \tag{6.2.4}$$

where $(.,.)$ denotes the (Euclidean) inner product of vectors. From (6.2.3) we see that this condition is also equivalent to

$$RR^{\mathrm{T}} = R^{\mathrm{T}}R = I \tag{6.2.5}$$

(Exercise 6.1). A matrix that satisfies this condition is called an *orthogonal matrix*.

Since vectors $\{i, j, k\}$ are a right-hand system, condition (ii) means that the three vectors $\{r_1, r_2, r_3\}$ are also a right-hand system. Since vectors $\{i, j, k\}$ are an orthonormal system, this requirement is equivalent to saying that the volume of the cube defined by the three vectors $\{r_1, r_2, r_3\}$ is always 1:

$$|r_1 r_2 r_3| = (r_1 \times r_2, r_3) = (r_2 \times r_3, r_1) = (r_3 \times r_1, r_2) = 1 \ . \tag{6.2.6}$$

In other words, condition (iii) is equivalent to $|R| = 1$. Thus, we obtain the following observation:

Proposition 6.1. *A matrix R represents a 3D rotation if and only if R is an orthogonal matrix of determinant* 1:

$$RR^{\mathrm{T}} = R^{\mathrm{T}}R = I \ , \qquad |R| = 1 \ . \tag{6.2.7}$$

Corollary 6.1. *A 3D rotation R preserves the length of a vector and the inner product of two vectors:*

$$\|Ra\| = \|a\| \ , \qquad (Ra, Rb) = (a, b) \ . \tag{6.2.8}$$

Proof. The second equation is obtained from

$$(Ra, Rb) = (R^{\mathrm{T}} Ra, b) = (a, b) \ . \tag{6.2.9}$$

The first one is obtained by setting $a = b$ in (6.2.9).

We call an orthogonal matrix of determinant 1 a *rotation matrix*. A rotation matrix R has nine components, but these cannot be chosen independently because (6.2.7) must be satisfied. The following fact is essential to 3D rotations.

Proposition 6.2. *The degree of freedom of a 3D rotation is three.*

Proof. Equation (6.2.4) is symmetric over indices i, j, so there are six independent relations: $(i, j) = (1, 1), (2, 2), (3, 3), (1, 2), (2, 3), (3, 1)$. Hence, the degree of freedom is $9 - 6 = 3$.[1]

Proposition 6.3. *The set of all rotation matrices forms a group under matrix multiplication with the unit matrix $\boldsymbol{I}$ as the identity element.*

Proof. This is obvious from the geometrical interpretation of 3D rotations, but is also confirmed from (6.2.7): For two rotation matrices $\boldsymbol{R}$, $\boldsymbol{R}'$, we see that

$$(\boldsymbol{R}'\boldsymbol{R})(\boldsymbol{R}'\boldsymbol{R})^{\mathrm{T}} = \boldsymbol{R}'\boldsymbol{R}\boldsymbol{R}^{\mathrm{T}}\boldsymbol{R}'^{\mathrm{T}} = \boldsymbol{I} \,, \qquad (\boldsymbol{R}'\boldsymbol{R})^{\mathrm{T}} = \boldsymbol{R}^{\mathrm{T}}\boldsymbol{R}'^{\mathrm{T}}\boldsymbol{R}'\boldsymbol{R} = \boldsymbol{I} \,, \tag{6.2.10}$$

$$|\boldsymbol{R}'\boldsymbol{R}| = |\boldsymbol{R}'|\,|\boldsymbol{R}| = 1 \,. \tag{6.2.11}$$

Hence, the product $\boldsymbol{R}'\boldsymbol{R}$ is also a rotation matrix. It is easy to see that the inverse $\boldsymbol{R}^{-1}$ $(= \boldsymbol{R}^{\mathrm{T}})$ is also a rotation matrix.

This group is called the 3D *rotation group* and denoted by $SO(3)$. The letter "O" stands for "orthogonal", and the number "3" is the dimensionality, while the letter "S" stands for "special", meaning that the determinant is 1. Equation (6.2.8a) implies that a 3D rotation maps a sphere centered at the coordinate origin onto itself. We prove the converse in the following form. This result will play an important role in subsequent sections.

Proposition 6.4. *A linear mapping $\boldsymbol{T}$ is a* 3D *rotation if it maps a sphere centered at the origin onto itself and preserves parity.*

Proof. Let r be the radius of the sphere. The condition given above implies that for all vectors $\boldsymbol{a}$ such that $\|\boldsymbol{a}\| = r$ we have

$$\|\boldsymbol{T}\boldsymbol{a}\| = \|\boldsymbol{a}\| \,. \tag{6.2.12}$$

However, the requirement $\|\boldsymbol{a}\| = r$ can be dropped. Since $\boldsymbol{T}$ is a linear mapping, we can easily see, by multiplying (6.2.12) by an arbitrary number, that $\boldsymbol{T}$ maps a sphere of an arbitrary radius onto itself.

Since $\boldsymbol{T}$ is assumed to preserve parity and satisfies (6.2.12) for any vector $\boldsymbol{a}$, all we need to show is that for any two arbitrary vectors $\boldsymbol{a}, \boldsymbol{b}$,

$$(\boldsymbol{T}\boldsymbol{a}, \boldsymbol{T}\boldsymbol{b}) = (\boldsymbol{a}, \boldsymbol{b}) \,. \tag{6.2.13}$$

Replacing $\boldsymbol{a}$ by $\boldsymbol{a} - \boldsymbol{b}$ in (6.2.12), we see that

$$\|\boldsymbol{T}(\boldsymbol{a} - \boldsymbol{b})\|^2 = \|\boldsymbol{a} - \boldsymbol{b}\|^2 \,. \tag{6.2.14}$$

---

[1] The condition $|\boldsymbol{R}| = 1$ need not be counted. From $\boldsymbol{R}\boldsymbol{R}^{\mathrm{T}} = \boldsymbol{I}$, we see that $|\boldsymbol{R}|\,|\boldsymbol{R}^{\mathrm{T}}| = 1$. Since $|\boldsymbol{R}| = |\boldsymbol{R}^{\mathrm{T}}|$, we have $|\boldsymbol{R}| = \pm 1$. Hence, the constraint $|\boldsymbol{R}| = 1$ is equivalent to $|\boldsymbol{R}| > 0$, which does not affect the degree of freedom.

The left-hand side becomes

$$(T(a-b), T(a-b)) = (Ta - Tb, Ta - Tb)$$
$$= (Ta, Ta) - 2(Ta, Tb) + (Tb, Tb)$$
$$= \|a\|^2 - 2(Ta, Tb) + \|b\|^2 \,. \tag{6.2.15}$$

while the right-hand side becomes

$$\|a\|^2 - 2(a, b) + \|b\|^2 \,. \tag{6.2.16}$$

Hence, (6.2.13) follows.

This proof corresponds to the well-known theorem in elementary geometry that *two triangles whose corresponding sides have equal lengths are congruent to each other, and their corresponding angles are also the same* (Fig. 6.2).

6.2.1 The *n*D Rotation Group *SO*(*n*)

The argument in this section can be automatically extended to *n*D space. We call an n-dimensional matrix R an *nD rotation matrix* if (6.2.7) is satisfied. The degree of freedom of an *n*D rotation is $n(n-1)/2$. All *n*D rotations also form a group, which is denoted by $SO(n)$ and called the *nD rotation group*.

Since the proof of Proposition 6.4 does not make any use of the dimensionality, it also holds for $SO(n)$. Namely, *a linear mapping T of nD space is an nD rotation if it maps an* $(n-1)$*D sphere centered at the origin onto itself and preserves parity*.

All linear mappings that map a unit $(n-1)$D sphere onto itself form a group. This group is a subgroup of $O(n)$ (*n*D rotations and reflections), and the determinant is either 1 or -1. If all elements are reduced to the identity by continuously changing the parameters (that is, if the group is *connected*, see the Appendix, Sect. A.8, for details), this group is a subgroup of $SO(n)$. This is because *the value of the determinant cannot change smoothly between* 1 *and* -1. Namely, if all elements can be smoothly reduced to the identity, their determinant must be 1.

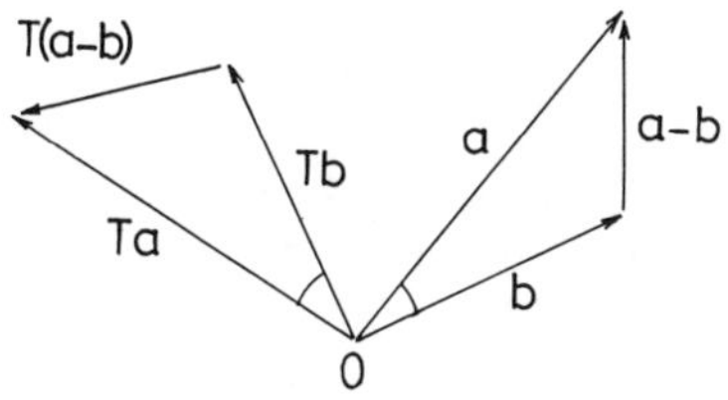

Fig. 6.2. Two triangles whose corresponding sides have equal lengths are congruent to each other. Hence, their corresponding angles are also equal

6.3 Rotation Axis and Rotation Angle

In Sect. 6.2 we defined a 3D rotation as a linear mapping that moves vectors rigidity, i.e., without changing lengths and angles. We now show that any 3D rotation so defined is realized as a rotation around a fixed axis. This fact is known as *Euler's theorem.*

Theorem 6.1 (*Euler's theorem*). *By a* 3D *rotation, all points move around a common axis, maintaining a fixed distance from it.*

Proof. Let $\boldsymbol{R}$ be a rotation matrix, σ be one of its eigenvalues, and $\boldsymbol{n} = (n_1, n_2, n_3)$ be the corresponding nonzero eigenvector:

$$\boldsymbol{R}\boldsymbol{n} = \sigma \boldsymbol{n} \ . \tag{6.3.1}$$

The eigenvalue σ may be a complex number. Consequently, $\boldsymbol{n}$ may be a vector consisting of complex components. Taking the complex conjugate of both sides of (6.3.1), we obtain

$$\boldsymbol{R}\boldsymbol{n}^* = \sigma^* \boldsymbol{n}^* \ , \tag{6.3.2}$$

where $\boldsymbol{n}^* = (n_1^*, n_2^*, n_3^*)$ is the vector obtained by taking the complex conjugate of each component. The inner product of the left-hand sides of (6.3.1, 2) becomes

$$(\boldsymbol{R}\boldsymbol{n}, \boldsymbol{R}\boldsymbol{n}^*) = (\boldsymbol{R}^{\mathrm{T}}\boldsymbol{R}\boldsymbol{n}, \boldsymbol{n}^*) = (\boldsymbol{n}, \boldsymbol{n}^*) \ , \tag{6.3.3}$$

while the inner product of the right-hand sides becomes

$$\sigma\sigma^*(\boldsymbol{n}, \boldsymbol{n}^*) = |\sigma|^2(\boldsymbol{n}, \boldsymbol{n}^*) \ . \tag{6.3.4}$$

Since $(\boldsymbol{n}, \boldsymbol{n}^*) = |n_1|^2 + |n_2|^2 + |n_3|^2 \neq 0$, we conclude that

$$|\sigma| = 1 \ . \tag{6.3.5}$$

Hence, all three eigenvalues $\sigma_1, \sigma_2, \sigma_3$ are of modulus 1. Since the determinant of a rotation matrix is 1, the product of these three eigenvalues is 1:

$$\sigma_1\sigma_2\sigma_3 = 1 \ . \tag{6.3.6}$$

On the other hand, the eigenvalues σ_1, σ_2, σ_3 are three roots of the characteristics equation

$$|\sigma \boldsymbol{I} - \boldsymbol{R}| = 0 \ , \tag{6.3.7}$$

which is a polynomial of degree three in σ with real coefficients. Hence, the three roots are either all real roots, or a pair of complex roots (mutually complex conjugate) and one real root. If (6.3.7) has three real roots, all are of absolute value 1 and their product is 1. This means that they are either 1, 1, 1 or 1, -1, -1. If (6.3.7) has one real root σ and a complex conjugate pair α, α^*, we see that $\alpha\alpha^* = |\alpha|^2 = 1$, and hence the remaining real root σ must be 1.

In any case, one of the three eigenvalues must be 1. The eigenvector for a real eigenvalue can be chosen as a real vector. In other words, there exists a real vector $\boldsymbol{n}$ such that

$$\boldsymbol{R}\boldsymbol{n} = \boldsymbol{n} \ . \tag{6.3.8}$$

This means that no points on the line of orientation $\boldsymbol{n}$ are moved by rotation $\boldsymbol{R}$. Hence, $\boldsymbol{n}$ defines an "axis" of this rotation. Consider a point represented by vector $\boldsymbol{r}$. The projected length of vector $\boldsymbol{r}$ onto this axis is $(\boldsymbol{r}, \boldsymbol{n})$, so the distance of the end point of $\boldsymbol{r}$ from the axis is $\|\boldsymbol{r} - (\boldsymbol{r}, \boldsymbol{n})\boldsymbol{n}\|$ (Fig. 6.3). From (6.2.8), we can readily prove that

$$\|\boldsymbol{R}\boldsymbol{r} - (\boldsymbol{R}\boldsymbol{r}, \boldsymbol{n})\boldsymbol{n}\| = \|\boldsymbol{R}[\boldsymbol{r} - (\boldsymbol{R}\boldsymbol{r}, \boldsymbol{R}\boldsymbol{n})\boldsymbol{n}]\| = \|\boldsymbol{r} - (\boldsymbol{r}, \boldsymbol{n})\boldsymbol{n}\| \ . \tag{6.3.9}$$

From this theorem, we can represent a 3D rotation by an axis specified by a unit vector $\boldsymbol{n}$ and an angle Ω of rotation around it. We take a convention that Ω is positive when rotation takes place screw-wise, i.e., in such a way that the axis would move in the direction of n if it were a screw.

Since $\boldsymbol{n}$ is a unit vector, only two of its three components are independent. Together with angle Ω, the degree of freedom is three, which confirms Proposition 6.2. The remaining question is how the rotation matrix $\boldsymbol{R}$ is given in terms of axis $\boldsymbol{n}$ and angle Ω. The result is as follows (we will give an alternative proof in terms of quaternions in Sect. 6.8).

Proposition 6.5. *The matrix representing a 3D rotation by angle Ω around unit vector $\boldsymbol{n}$ is given by*

$$\boldsymbol{R} = \begin{pmatrix} \cos\Omega + n_1^2(1-\cos\Omega) & n_1n_2(1-\cos\Omega) - n_3\sin\Omega & n_1n_3(1-\cos\Omega) + n_2\sin\Omega \\ n_2n_1(1-\cos\Omega) + n_3\sin\Omega & \cos\Omega + n_2^2(1-\cos\Omega) & n_2n_3(1-\cos\Omega) - n_1\sin\Omega \\ n_3n_1(1-\cos\Omega) - n_2\sin\Omega & n_3n_2(1-\cos\Omega) + n_1\sin\Omega & \cos\Omega + n_3^2(1-\cos\Omega) \end{pmatrix} . \tag{6.3.10}$$

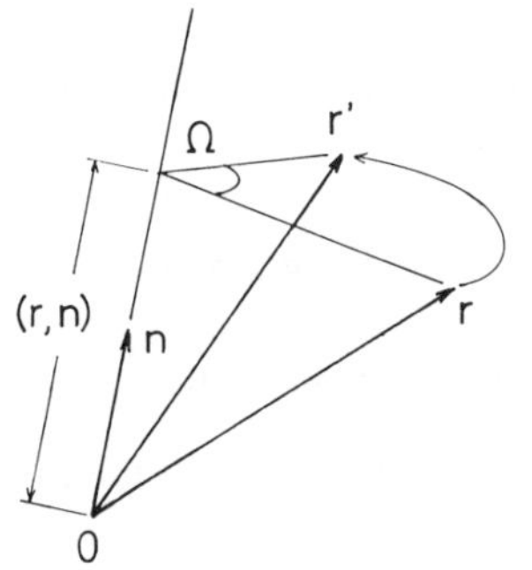

Fig. 6.3. A 3D rotation is represented by an angle Ω and an axis $\boldsymbol{n}$ of rotation

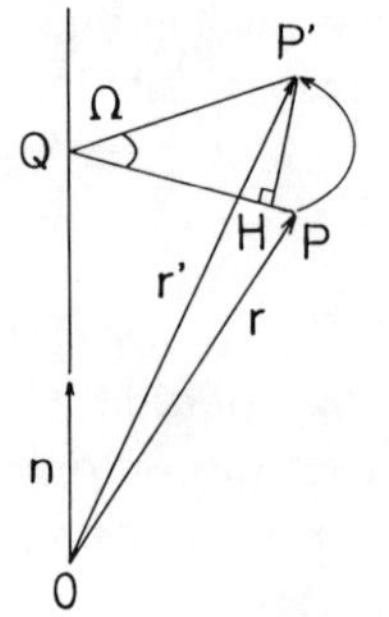

Fig. 6.4. Point P moves to point P' by a rotation around axis $\boldsymbol{n}$ through angle Ω

Proof. See Fig. 6.4. Point P rotates by angle Ω around the axis specified by unit vector $\boldsymbol{n}$. Let P' be the resulting new position. Let Q be the orthogonal projection of point P onto the axis. Then $|OP| = |OP'|$, $|QP| = |QP'|$, $\angle PQP' = \Omega$, $OQ \perp QP$, $OQ \perp OP'$. Let H be the orthogonal projection of point P' onto QP. If we put $\boldsymbol{r} = \overrightarrow{OP}$, $\boldsymbol{r}' = \overrightarrow{OP}'$, we see that

$$\boldsymbol{r}' = \overrightarrow{OQ} + \overrightarrow{QH} + \overrightarrow{HP}' . \tag{6.3.11}$$

Since $\overrightarrow{OQ}$ is the orthogonal projection of vector $\boldsymbol{r}$ onto the axis specified by unit vector $\boldsymbol{n}$, we have

$$\overrightarrow{OQ} = (\boldsymbol{r}, \boldsymbol{n})\boldsymbol{n} . \tag{6.3.12}$$

Similarly, $\overrightarrow{QH}$ is the orthogonal projection of vector $\overrightarrow{QP}'$ onto $\overrightarrow{QP}$. Noting that $|QP'| = |QP|$ and using (6.3.12), we have

$$\overrightarrow{QH} = \frac{\overrightarrow{QP}}{|QP|}|QP'|\cos\Omega = \overrightarrow{QP}\cos\Omega = (\overrightarrow{OP} - \overrightarrow{OQ})\cos\Omega$$

$$= [\boldsymbol{r} - (\boldsymbol{r}, \boldsymbol{n})\boldsymbol{n}]\cos\Omega . \tag{6.3.13}$$

Vector $\overrightarrow{HP}'$ is orthogonal to both $\boldsymbol{n}$ and $\boldsymbol{r}$, and has length $|QP'|\sin\Omega$. Noting that $|QP'| = |QP| = \|\boldsymbol{n} \times \boldsymbol{r}\|$, we have

$$\overrightarrow{HP}' = \frac{\boldsymbol{n} \times \boldsymbol{r}}{\|\boldsymbol{n} \times \boldsymbol{r}\|}|QP'|\sin\Omega = \boldsymbol{n} \times \boldsymbol{r}\sin\Omega . \tag{6.3.14}$$

If (6.3.12–14) are substituted, (6.3.11) becomes

$$\boldsymbol{r}' = \boldsymbol{r}\cos\Omega + \boldsymbol{n} \times \boldsymbol{r}\sin\Omega + (1 - \cos\Omega)(\boldsymbol{n}, \boldsymbol{r})\boldsymbol{n} . \tag{6.3.15}$$

This is called the *Rodrigues formula*. If this equation is rewritten in matrix form as $\boldsymbol{r}' = \boldsymbol{R}\boldsymbol{r}$, we obtain (6.3.10) (Exercise 6.5).

Now, let us determine the axis $\boldsymbol{n}$ and angle Ω from a rotation matrix

$$\boldsymbol{R} = \begin{pmatrix} r_{11} & r_{12} & r_{13} \\ r_{21} & r_{22} & r_{23} \\ r_{31} & r_{32} & r_{33} \end{pmatrix} . \tag{6.3.16}$$

Comparing the matrices of (6.3.10) and (6.3.16), we see that

$$r_{11} + r_{22} + r_{33} = 1 + 2\cos\Omega , \tag{6.3.17}$$

$$r_{23} - r_{32} = -2n_1 \sin\Omega, \quad r_{31} - r_{13} = -2n_2 \sin\Omega ,$$
$$r_{12} - r_{21} = -2n_3 \sin\Omega . \tag{6.3.18}$$

Hence,

$$\Omega = \cos^{-1}\frac{\mathrm{Tr}\{\boldsymbol{R}\} - 1}{2} ,$$
$$\boldsymbol{n} = -\frac{1}{\sin\Omega}\left(\frac{r_{23} - r_{32}}{2}, \frac{r_{31} - r_{13}}{2}, \frac{r_{12} - r_{21}}{2}\right) . \tag{6.3.19}$$

If the angle Ω is restricted to within the interval $-\pi < \Omega \leqq \pi$, equations (6.3.19a, b) give two values $\pm\Omega$ and the corresponding two vectors $\pm\boldsymbol{n}$. This is no surprise, because a rotation around axis $\boldsymbol{n}$ by angle Ω is equivalent to a rotation around axis $-\boldsymbol{n}$ by angle $-\Omega$. Equations (6.3.19a, b) are not defined for $\Omega = 0, \pm\pi$. Hence, when (6.3.19a, b) are used for numerical computation, special care is necessary if the angle of rotation is close to 0 or $\pm\pi$.

Thus, a 3D rotation is represented by a pair $\{\boldsymbol{n}, \Omega\}$ of its axis $\boldsymbol{n}$ (unit vector) and angle Ω. This representation is very convenient for many practical purposes. In manipulating a physical object by a robot arm, for example, it is very easy from the viewpoint of machine design to realize a new object orientation by rotating it around a given axis by a given angle. However, there is a very serious drawback. Suppose we want to apply a rotation specified by $\{\boldsymbol{n}, \Omega\}$ followed by another rotation specfied by $\{\boldsymbol{n}', \Omega'\}$. How can we determine the representation $\{\boldsymbol{n}'', \Omega''\}$ of the composite rotation? In other words, what is the *rule of composition* for $\boldsymbol{n}$ and Ω? Unfortunately, no simple expression is available.

6.4 Euler Angles

Another well-known representation of a 3D rotation are the *Euler angles* $\{\theta, \varphi, \psi\}$. Consider again the natural basis vectors $\{\boldsymbol{i}, \boldsymbol{j}, \boldsymbol{k}\}$ and the vectors $\{\boldsymbol{r}_1, \boldsymbol{r}_2, \boldsymbol{r}_3\}$ resulting from a 3D rotation $\boldsymbol{R}$ (Fig. 6.5). The Euler angles $\{\theta, \varphi, \psi\}$ are defined as follows. (In some books, they are defined somewhat differently.)

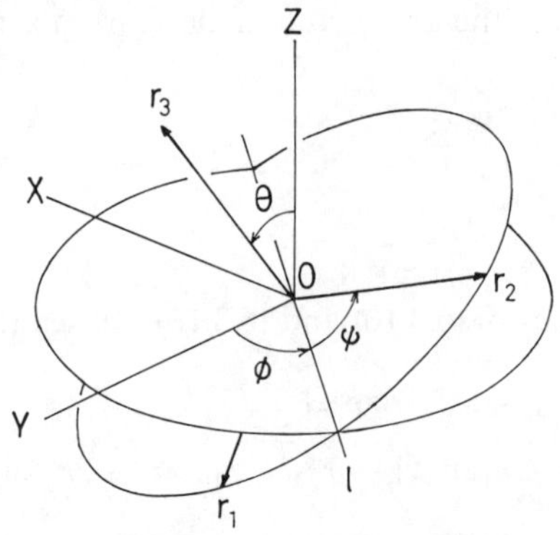

Fig. 6.5. The Euler angles $[\theta, \varphi, \psi\}$ are defined as follows. Angles (θ, φ) are the spherical coordinates of $\boldsymbol{r}_3$. Let l be the intersection of the XY-plane with the plane defined by $\boldsymbol{r}_1$ and $\boldsymbol{r}_2$. Angle ψ is theangle of $\boldsymbol{r}_2$ measured from l

First, angles (θ, φ), $0 \leqq \theta \leqq \pi$, $0 \leqq \varphi < 2\pi$, are the spherical coordinates of vector $\boldsymbol{r}_3$. Next, let $\boldsymbol{l}$ be the intersection of the XY-plane with the plane defined by two vectors $\boldsymbol{r}_1$, $\boldsymbol{r}_2$. Then, ψ, $0 \leqq \psi < 2\pi$, is the angle of vector $\boldsymbol{r}_2$ from line $\boldsymbol{l}$ measured screw-wise about vector $\boldsymbol{r}_3$.

Thus, we have again confirmed that a 3D rotation is prescribed by three parameters. The next question is how the rotation matrix $\boldsymbol{R}$ is given in terms of the Euler angles $\{\theta, \varphi, \psi\}$.

Proposition 6.6. *The matrix representing the 3D rotation specified by Euler angles* $\{\theta, \varphi, \psi\}$ *is given by*

$$\boldsymbol{R} = \begin{pmatrix} \cos\theta\cos\varphi\cos\psi - \sin\theta\sin\psi & -\cos\theta\cos\varphi\sin\psi - \sin\varphi\cos\psi & \sin\theta\cos\varphi \\ \cos\theta\sin\varphi\cos\psi + \cos\theta\sin\psi & -\cos\theta\sin\varphi\sin\psi + \cos\varphi\cos\psi & \sin\theta\sin\varphi \\ -\sin\theta\cos\psi & \sin\theta\sin\psi & \cos\theta \end{pmatrix}. \tag{6.4.1}$$

Proof. Let $\boldsymbol{R}_X(\Omega)$, $\boldsymbol{R}_Y(\Omega)$, $\boldsymbol{R}_Z(\Omega)$ be the rotation matrices representing rotations through angle Ω around the X-, Y-, and Z-axes, respectively. From (6.3.10), we have

$$\boldsymbol{R}_X(\Omega) = \begin{pmatrix} 1 & 0 & 0 \\ 0 & \cos\Omega & -\sin\Omega \\ 0 & \sin\Omega & \cos\Omega \end{pmatrix}, \qquad \boldsymbol{R}_Y(\Omega) = \begin{pmatrix} \cos\Omega & 0 & \sin\Omega \\ 0 & 1 & 0 \\ -\sin\Omega & 0 & \cos\Omega \end{pmatrix},$$

$$\boldsymbol{R}_Z(\Omega) = \begin{pmatrix} \cos\Omega & -\sin\Omega & 0 \\ \sin\Omega & \cos\Omega & 0 \\ 0 & 0 & 1 \end{pmatrix}, \tag{6.4.2}$$

From the definition of the Euler angles $\{\theta, \varphi, \psi\}$, we can write

$$\boldsymbol{R} = \boldsymbol{R}_Z(\varphi)\boldsymbol{R}_Y(\theta)\boldsymbol{R}_Z(\psi) \ . \tag{6.4.3}$$

(Consider carefully where the natural basis vectors $\{\boldsymbol{i}, \boldsymbol{j}, \boldsymbol{k}\}$ move to under the rotation of (6.4.3).) Equation (6.4.1) is obtained by substituting (6.4.2) into (6.4.3).

Now, consider how to determine the Euler angles $\{\theta, \varphi, \psi\}$ from a rotation matrix

$$\boldsymbol{R} = \begin{pmatrix} r_{11} & r_{12} & r_{13} \\ r_{21} & r_{22} & r_{23} \\ r_{31} & r_{32} & r_{33} \end{pmatrix} . \tag{6.4.4}$$

Since (θ, φ) are the spherical coordinates of the third column r_3, we first obtain

$$\theta = \cos^{-1} r_{33} \ , \qquad 0 \leqq \theta \leqq \pi \ . \tag{6.4.5}$$

If $\theta \neq 0$, angle φ is uniquely determined by

$$\cos\varphi = \frac{r_{13}}{\sin\theta} \ , \qquad \sin\varphi = \frac{r_{23}}{\sin\theta} \ , \qquad 0 \leqq \varphi < 2\pi \ . \tag{6.4.6}$$

The remaining angle ψ is determined from the third row of $\boldsymbol{R}$ by

$$\cos\psi = \frac{r_{31}}{\sin\theta} \ , \qquad \sin\psi = \frac{r_{32}}{\sin\theta} \ , \qquad 0 \leqq \psi \leqq 2\pi \ . \tag{6.4.7}$$

If $\theta = 0$, we can determine only $\varphi + \psi$, since the line $\boldsymbol{l}$ is not uniquely determined (Fig. 6.5). In fact, if $\theta = 0$, we see from (6.4.3) that

$$\boldsymbol{R} = \boldsymbol{R}_Z(\varphi)\boldsymbol{R}_Z(\psi) = \boldsymbol{R}_Z(\varphi + \psi) \ . \tag{6.4.8}$$

If a rotation is prescribed by the Euler angles $\{\theta, \varphi, \psi\}$, it can be easily realized by rotations around only the Z- and Y-axes. Again, there are many cases in which this realization is easy to implement. There arises, however, the same difficulty as in the axis and angle representation: no simple expression is available for the rule of composition. What are the Euler angles $\{\theta'', \varphi'', \psi''\}$ of the rotation composed from a rotation $\{\theta, \varphi, \psi\}$ followed by another rotation $\{\theta', \varphi', \psi'\}$?

6.5 Cayley–Klein Parameters

As we have pointed out in the previous sections, it is very desirable to parametrize a 3D rotation so that the rule of composition becomes simple. We now introduce such a parametrization. The first step is the *stereographic projection.*

Since 3D rotations do not change lengths between points, a point on a sphere centered at the coordinate origin O moves to another point on the same sphere.

This means that a 3D rotation can be regarded as a transformation of a sphere onto itself. Since the radius of the sphere is irrelevant, we consider the unit sphere

$$X^2 + Y^2 + Z^2 = 1 \ . \tag{6.5.1}$$

This sphere is denoted by S^2. (The superscript "2" indicates that the dimension of the sphere is two. It is a *two-dimensional manifold*. See the Appendix, Sect. A.8.) Thus, we can regard $SO(3)$ as a group of transformations acting on the topological space S^2.

Now, we introduce the following projection of S^2 onto the XY-plane: a point (X, Y, Z) on sphere S^2 is projected onto the intersection (x, y) of the XY-plane with the ray connecting the point (X, Y, Z) and the "south pole" $(0, 0, -1)$ (Fig. 6.6). This projection is called the *stereographic projection*. (Some books adopt the "north pole" $(0, 0, 1)$ rather than the "south pole" $(0, 0, -1)$. Then, subsequent equations must be modified accordingly.)

Under this stereographic projection, the "north pole" $(0, 0, 1)$ is projected onto the origin O. The "equator" $Z = 0$ is projected onto unit circle $x^2 + y^2 = 1$. (This unit circle is a "one-dimensional sphere" and denoted by S^1. We use lower-case (x, y) as the coordinates on the plane $Z = 0$, identifying the X- and Y-axes with the x- and y-axes, respectively.) The entire "northern hemisphere" $Z > 0$ is projected inside this circle, and the "southern hemisphere" $Z < 0$ is projected outside this circle. The "south pole" $(0, 0, -1)$ is projected onto the *point at infinity*, which is denoted by ∞. Let us denote the XY-plane by $\mathbb{R}^2$. The stereographic projection is a one-to-one mapping from S^2 onto $\mathbb{R}^2$ (with

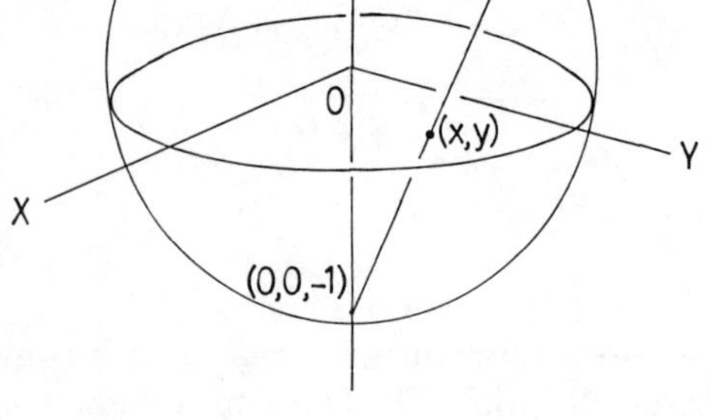

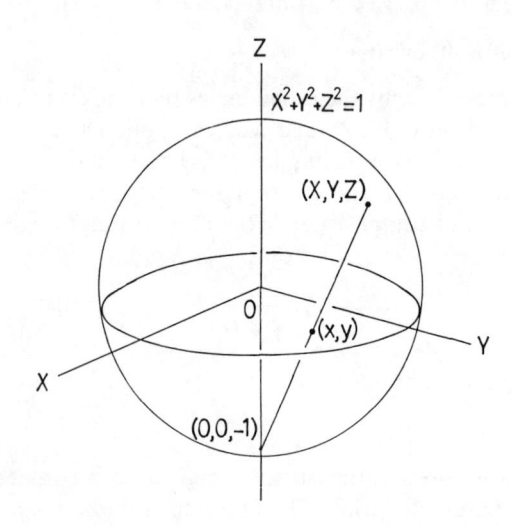

Fig. 6.6. Stereographic projection. A point (X, Y, Z) on the sphere $X^2 + Y^2 + Z^2 = 1$ is mapped onto the intersection (x, y) of the XY-plane with the ray connecting the point (X, Y, Z) and the "south pole" $(0, 0, -1)$

∞ added). It is easy to derive the following relationship between the point (X, Y, Z) on S^2 and the point (x, y) on $\mathbb{R}^2$ onto which it is projected:

$$x = \frac{X}{1+Z}, \qquad y = \frac{Y}{1+Z}. \tag{6.5.2}$$

$$X = \frac{2x}{1+x^2+y^2}, \quad Y = \frac{2y}{1+x^2+y^2}, \quad Z = \frac{1-x^2-y^2}{1+x^2+y^2}, \tag{6.5.3}$$

If the sphere S is rotated, a point (X, Y, Z) on it moves to another position (X', Y', Z'). As a result, the corresponding stereographic projection image (x, y) also moves to another position (x', y'). This means that a rotation of the sphere S^2 induces a transformation of $\mathbb{R}^2$. To put it differently, $SO(3)$ acts on $\mathbb{R}^2$ as a group of transformations via the stereographic projection. Evidently, this action is a homomorphism from $SO(3)$.

We now study how this transformation of $\mathbb{R}^2$ is expressed. To this end, we regard $\mathbb{R}^2$ as the complex number (or Gaussian) plane by identifying the X- and Y-axes with the real and the imaginary axes, respectively: a point (x, y) is identified with the complex number $z = x + \mathrm{i}y$, where i is the imaginary unit. We denote this complex number plane by $\mathbb{C}$.[2]

Theorem 6.2. *If $\mathbb{R}^2$ is identified with $\mathbb{C}$, the action of $SO(3)$ by the stereographic projection is given by*

$$z' = \frac{\gamma + \delta z}{\alpha + \beta z}, \tag{6.5.4}$$

where

$$\gamma = -\beta^*, \qquad \delta = \alpha^*, \qquad \alpha\delta - \beta\gamma = 1. \tag{6.5.5}$$

The proof will be given in the next section.[3] From this theorem, we find that the four complex parameters $\{\alpha, \beta, \gamma, \delta\}$ can be viewed as parameters of a 3D rotation. These four parameters are called the *Cayley–Klein parameters.*

The four Cayley–Klein parameters are not algebraically independent; they must satisfy the constraints (6.5.5). Four complex parameters are equivalent to eight real parameters. Equations (6.5.5a, b) are equivalent to four real equations.

---

[2] We assume that $\mathbb{C}$ also contains ∞. In this sense, $\mathbb{C}$ can be identified with S^2. The sphere S^2 equipped with the algebraic structure of $\mathbb{C}$ via the stereographic projection is called the *Riemann sphere.*

[3] If the constraints (6.5.5) are replaced by $\alpha\delta - \beta\gamma \neq 0$, the mappings of the complex number plane $\mathbb{C}$ in the form of (6.5.4) form a group known as the *linear fractional transformation group*. They are not only *conformal mappings* (hence angles are preserved) but also map a circle to a circle (a straight line is regarded as a circle of infinite radius). The action of $SO(3)$ on $\mathbb{C}$ is a subgroup of the linear fractional transformation group. Hence, it also induces conformal mappings that map a circle to a circle.

Equation (6.5.5c) is one real equation, since it can be rewritten as $|\alpha|^2 + |\beta|^2 = 1$. Hence, (6.5.5a–c) give five real constraining equations. Consequently, the degree of freedom (i.e., the number of independent parameters) is $8 - 5 = 3$, which again confirms Proposition 6.2.

A great advantage of the Cayley–Klein parameters is that the rule of composition is very simple. Suppose we apply a 3D rotation prescribed by $\{\alpha, \beta, \gamma, \delta\}$ followed by another rotation prescribed by $\{\alpha', \beta', \gamma', \delta'\}$. The Cayley–Klein parameters $\{\alpha'', \beta'', \gamma'', \delta''\}$ of the composite rotation are given as follows.

Proposition 6.7. *The rule of composition for the Cayley–Klein parameters is*

$$\begin{aligned} \alpha'' &= \alpha'\alpha + \beta'\gamma\ , \qquad & \beta'' &= \alpha'\beta + \beta'\delta\ , \\ \gamma'' &= \gamma'\alpha + \delta'\gamma\ , \qquad & \delta'' &= \gamma'\beta + \delta'\delta\ . \end{aligned} \tag{6.5.6}$$

Proof. Consider the following three transformations:

$$z' = \frac{\gamma + \delta z}{\alpha + \beta z}\ , \qquad z'' = \frac{\gamma' + \delta' z'}{\alpha' + \beta' z'}\ , \qquad z'' = \frac{\gamma'' + \delta'' z}{\alpha'' + \beta'' z}\ . \tag{6.5.7}$$

If the first equation is substituted into the second one, comparison of the resulting equation with the last one yields (6.5.6).

Corollary 6.2. *The identity is specified by Cayley–Klein parameters* $\{1, 0, 0, 1\}$, *and the inverse of the rotation specified by Cayley–Klein parameters* $\{\alpha, \beta, \gamma, \delta\}$ *is specified by Cayley–Klein parameters* $\{\delta, -\beta, -\gamma, \alpha\}$.

In the following sections, we derive explicit relationships between the Cayley–Klein parameters $\{\alpha, \beta, \gamma, \delta\}$ and the rotation matrix $\boldsymbol{R}$. To this end, we need some mathematical preliminaries.

6.6 Representation of $SO(3)$ by $SU(2)$

A point on the complex number plane $\mathbb{C}$ is specified by a single complex number z. However, this "coordinate" of $\mathbb{C}$, called the *inhomogeneous coordinate*, cannot specify the point at infinity ∞. It can be specified if we use a pair (z_0, z_1) of complex numbers as "coordinates" of $\mathbb{C}$: a pair (z_0, z_1) is identified with point z_1/z_0 if $z_0 \neq 0$, and ∞ if $z_0 = 0$. Hence, only the ratio $z_0 : z_1$ is important, and multiplication of (z_0, z_1) by any nonzero complex number does not change the designated point. The pair (z_0, z_1) is called the *homogeneous coordinates.*

In terms of homogeneous coordinates, it is easy to see that (6.5.4) is written as

$$\begin{pmatrix} z_0' \\ z_1' \end{pmatrix} = \begin{pmatrix} \alpha & \beta \\ \gamma & \delta \end{pmatrix} \begin{pmatrix} z_0 \\ z_1 \end{pmatrix}\ . \tag{6.6.1}$$

Let the coefficient matrix be U:

$$U = \begin{pmatrix} \alpha & \beta \\ \gamma & \delta \end{pmatrix} . \tag{6.6.2}$$

Equation (6.5.5c) means that $|U| = 1$. The inverse of U is

$$U^{-1} = \begin{pmatrix} \delta & -\beta \\ -\gamma & \alpha \end{pmatrix} . \tag{6.6.3}$$

The *Hermitian conjugate* $U^\dagger$ of U is defined as the element-wise complex conjugate of the transpose:

$$U^\dagger = (U^*)^{\mathrm{T}} . \tag{6.6.4}$$

In view of (6.6.3), equations (6.5.5a, b) state that $U^{-1} = U^\dagger$, or equivalently $U^\dagger U = I$, where I is the unit matrix. A matrix U that satisfies this condition is called a *unitary matrix*. Thus, (6.5.5) means that the matrix U of (6.6.1) is a unitary matrix of determinant 1:

$$U^\dagger U = I , \qquad |U| = 1 . \tag{6.6.5}$$

The set of two-dimensional (generally complex) matrices that satisfy (6.6.5) is denoted by $SU(2)$. (As in $SO(3)$, the letter "S" stands for "special", meaning determinant 1, and "U" stands for "unitary", "2" being the dimensionality.)

Lemma 6.2. *The set $SU(2)$ forms a group under matrix multiplication.*

Proof. For U, $U' \in SU(2)$,

$$(U'U)^\dagger(U'U) = U^\dagger U'^\dagger U' U = I , \qquad |U'U| = |U'||U| = 1 . \tag{6.6.6}$$

Hence, $U'U \in SU(2)$, too. It is easy to see that $I \in SU(2)$ and if $U \in SU(2)$, then $U^{-1}\ (= U^\dagger) \in SU(2)$, too.

Thus, we conclude the following.

Theorem 6.3. *The action of $SO(3)$ on the complex number plane $\mathbb{C}$ is given by*

$$\begin{pmatrix} z'_0 \\ z'_1 \end{pmatrix} = \begin{pmatrix} \alpha & \beta \\ \gamma & \delta \end{pmatrix} \begin{pmatrix} z_0 \\ z_1 \end{pmatrix} , \qquad \begin{pmatrix} \alpha & \beta \\ \gamma & \delta \end{pmatrix} \in SU(2) . \tag{6.6.7}$$

Corollary 6.3. *Equation (6.6.7) defines a representation of $SO(3)$ by $SU(2)$, i.e., the correspondence from $SO(3)$ to $SU(2)$ is a homomorphism, and the rule of composition is given by*

$$\begin{pmatrix} \alpha'' & \beta'' \\ \gamma'' & \delta'' \end{pmatrix} = \begin{pmatrix} \alpha' & \beta' \\ \gamma' & \delta' \end{pmatrix} \begin{pmatrix} \alpha & \beta \\ \gamma & \delta \end{pmatrix} . \tag{6.6.8}$$

Proof. Equation (6.6.8) is obvious. In fact, if we put

$$\begin{pmatrix} z_0' \\ z_1' \end{pmatrix} = \begin{pmatrix} \alpha & \beta \\ \gamma & \delta \end{pmatrix} \begin{pmatrix} z_0 \\ z_1 \end{pmatrix}, \quad \begin{pmatrix} z_0'' \\ z_1'' \end{pmatrix} = \begin{pmatrix} \alpha' & \beta' \\ \gamma' & \delta' \end{pmatrix} \begin{pmatrix} z_0' \\ z_1' \end{pmatrix},$$

$$\begin{pmatrix} z_0'' \\ z_1'' \end{pmatrix} = \begin{pmatrix} \alpha'' & \beta'' \\ \gamma'' & \delta'' \end{pmatrix} \begin{pmatrix} z_0 \\ z_1 \end{pmatrix}, \tag{6.6.9}$$

we obtain (6.6.8) by cmbining the first two and comparing the result with the third. (Equations (6.6.8, 9) are the matrix expressions of (6.5.6, 7).)

Now, we give the proof we skipped: we prove that $SO(3)$ acts on $\mathbb{C}$ in the form of (6.5.4, 5) or equivalently (6.6.7).

Suppose $\boldsymbol{U} \in SU(2)$ is infinitesimally close to the identity $\boldsymbol{I}$ and is expressed as

$$\boldsymbol{U} = \boldsymbol{I} + t\boldsymbol{W} + \mathrm{O}(t^2), \tag{6.6.10}$$

where t is a measure of "distance" of $\boldsymbol{U}$ from $\boldsymbol{I}$.[4] From the condition $\boldsymbol{U}^\dagger \boldsymbol{U} = \boldsymbol{I}$, we have

$$[\boldsymbol{I} + t\boldsymbol{W}^\dagger + \mathrm{O}(t^2)][\boldsymbol{I} + t\boldsymbol{W} + \mathrm{O}(t^2)] = \boldsymbol{I}, \tag{6.6.11}$$

hence

$$\boldsymbol{I} + t(\boldsymbol{W}^\dagger + \boldsymbol{W}) + \mathrm{O}(t^2) = \boldsymbol{I}. \tag{6.6.12}$$

Thus, $\boldsymbol{W}^\dagger + \boldsymbol{W} = \boldsymbol{O}$ ($\boldsymbol{O}$ is the two-dimensional zero matrix), or $\boldsymbol{W}^\dagger = -\boldsymbol{W}$. A matrix $\boldsymbol{W}$ that has this property is said to be *anti-Hermitian.*

From the condition $|\boldsymbol{U}| = 1$, we have

$$|\boldsymbol{I} + t\boldsymbol{W} + \mathrm{O}(t^2)| = 1, \tag{6.6.13}$$

hence

$$1 + t\mathrm{Tr}\{\boldsymbol{W}\} + \mathrm{O}(t^2) = 1 \tag{6.6.14}$$

(Exercise 6.11). Thus, $\mathrm{Tr}\{\boldsymbol{W}\} = 0$. Let $su(2)$ be the set of all two-dimensional matrices that are anti-Hermitian and of trace 0:[5]

$$\boldsymbol{W}^\dagger = -\boldsymbol{W}, \quad \mathrm{Tr}\{\boldsymbol{W}\} = 0. \tag{6.6.15}$$

The set $su(2)$ forms a real linear space. If $\boldsymbol{W}_1, \boldsymbol{W}_2 \in su(2)$, then $c_1\boldsymbol{W}_1 + c_2\boldsymbol{W}_2 \in su(2)$ for all real numbers c_1, c_2.

Lemma 6.3. *$SU(2)$ defines a representation of $SO(3)$.*

---

[4] To be precise, we are considering a *one-parameter subgroup* of $SU(2)$ passing through the identity $\boldsymbol{I}$, or, to put it differently, we are considering elements of the *tangent space* to $SU(2)$ at $\boldsymbol{I}$.

[5] This is the *Lie algebra* of $SU(2)$ (Sect. 3.2). We denote the Lie algebra of a Lie group by the lowercase letters of the corresponding uppercase letters that denote the Lie group. See the Appendix, Sect. A.9, for further details.

Proof. If $\boldsymbol{U}$ is infinitesimally close to $\boldsymbol{I}$ and has the form of (6.6.10), the *infinitesimal transformation* $\boldsymbol{W}$ is a member of $su(2)$. Since there are no other constraints, all members of $su(2)$ can be infinitesimal transformations of $SU(2)$.[6] If $\boldsymbol{W} \in su(2)$, then (6.6.15) implies that matrix $\boldsymbol{W}$ has the form

$$\boldsymbol{W} = \frac{1}{2}\begin{pmatrix} -ia_3 & -a_2 - ia_1 \\ a_2 - ia_1 & ia_3 \end{pmatrix}, \tag{6.6.16}$$

where a_1, a_2, a_3 are arbitrary real numbers. If we define $\boldsymbol{A}_1, \boldsymbol{A}_2, \boldsymbol{A}_3 \in su(2)$ by

$$\boldsymbol{A}_1 \equiv \frac{1}{2}\begin{pmatrix} & -i \\ -i & \end{pmatrix}, \quad \boldsymbol{A}_2 \equiv \frac{1}{2}\begin{pmatrix} & -1 \\ 1 & \end{pmatrix}, \quad \boldsymbol{A}_3 \equiv \frac{1}{2}\begin{pmatrix} -i & \\ & i \end{pmatrix}, \tag{6.6.17}$$

the matrix $\boldsymbol{W}$ of (6.6.16) is written as

$$\boldsymbol{W} = a_1\boldsymbol{A}_1 + a_2\boldsymbol{A}_2 + a_3\boldsymbol{A}_3 . \tag{6.6.18}$$

Hence, $su(2)$ is a three-dimensional real linear space, and the three matrices $\{\boldsymbol{A}_1, \boldsymbol{A}_2, \boldsymbol{A}_3\}$ are its basis. It is easy to confirm that $\{\boldsymbol{A}_1, \boldsymbol{A}_2, \boldsymbol{A}_3\}$ satisfy the following commutations relations:

$$[\boldsymbol{A}_1, \boldsymbol{A}_2] = \boldsymbol{A}_3 , \quad [\boldsymbol{A}_2, \boldsymbol{A}_3] = \boldsymbol{A}_1 , \quad [\boldsymbol{A}_3, \boldsymbol{A}_1] = \boldsymbol{A}_2 . \tag{6.6.19}$$

Hence, the linear space $su(2)$ is a Lie algebra isomorphic to that of $SO(3)$. (In this chapter, we consider rotations of the "space" relative to a fixed coordinate system—the *alibi* interpretation (Sect. 3.5.2). Hence, no minus signs appear on the right-hand sides of the commutation relations.) Thus, $SU(2)$ defines a representation of $SO(3)$ (Sect. 3.2). (However, since $SO(3)$ is not a simply connected Lie group, the correspondence between $SU(2)$ and $SO(3)$ is not globally one-to-one. But the correspondence is one-to-one if we restrict ourselves to a neighborhood of the identity (*local isomorphism*). See the Appendix for further details.)

As discussed in Sect. 3.2, (6.6.18, 19) imply a (local) isomorphism between $SO(3)$ and $SU(2)$. The correspondence between the infinitesimal transformation of $SU(2)$ by t, a_1, a_2, a_3 and the infinitesimal rotation by Ω around axis $\boldsymbol{n} = (n_1, n_2, n_3)$ is given by

$$t = \Omega , \quad a_1 = n_1 , \quad a_2 = n_2 , \quad a_3 = n_3 . \tag{6.6.20}$$

Integrating this relationship, we can establish the existence of a (local) isomorphism from $SO(3)$ into $SU(2)$ (its explicit expression will be shown later). What

---

[6] As we confirmed in the preceding section, the Cayley–Klein parameters have three degrees of freedom. Hence, $SU(2)$ is a three-dimensional Lie group. In general, the dimensionality, as a linear space, of the Lie algebra L of a Lie group G must be equal to the dimensionality of the Lie group G. Consequently, the entire $su(2)$ must be the Lie algebra of $SU(2)$, since $su(2)$ is a three-dimensional linear space, as will be subsequently shown.

remains to be shown is that this correspondence between $SO(3)$ and $SU(2)$ is given by (6.5.2, 3).

Lemma 6.4. *The action of $SO(3)$ on the complex number plane $\mathbb{C}$ coincides with the correspondence between $SO(3)$ and $SU(2)$ established by Lemma 6.3.*

Proof. As we have already observed, the action of $SO(3)$ on $\mathbb{C}$ via the stereographic projection is a homomorphism from $SO(3)$ into the induced group of transformations of $\mathbb{C}$. On the other hand, $SO(3)$ (locally) corresponds isomorphically to $SU(2)$ via the relationship implied in Lemma 6.3, and $SU(2)$ acts on $\mathbb{C}$ as the group of transformations defined by (6.6.7). Hence, what we need to show is that these two actions of $SO(3)$ on $\mathbb{C}$ are locally identical, i.e., they coincide near the identity transformation.

Consider an infinitesimal rotation by Ω (infinitesimal) around axis $\boldsymbol{n} = (n_1, n_2, n_3)$ (unit vector). As we saw in Chap. 3, (3.3.15), a point (X, Y, Z) moves to point $(X + \delta X, Y + \delta Y, Z + \delta Z)$, where $\delta X, \delta Y, \delta Z$ are given by

$$\delta \begin{pmatrix} X \\ Y \\ Z \end{pmatrix} = \Omega \begin{pmatrix} n_1 \\ n_2 \\ n_3 \end{pmatrix} \times \begin{pmatrix} X \\ Y \\ Z \end{pmatrix} + O(\Omega^2) = \Omega \begin{pmatrix} n_2 Z - n_3 Y \\ n_3 X - n_1 Z \\ n_1 Y - n_2 X \end{pmatrix} + O(\Omega^2) . \tag{6.6.21}$$

If we differentiate (6.5.2), substitute (6.6.21) into them, and eliminate X, Y, Z by using (6.5.3), we see that a point (x, y) on the XY-plane moves to point $(x + \delta x, y + \delta y)$, where

$$\delta x = \Omega \left[-n_1 xy + \frac{1}{2} n_2 (1 + x^2 - y^2) - n_3 y \right] + O(\Omega^2) ,$$

$$\delta y = \Omega \left[\frac{1}{2} n_1 (1 - x^2 + y^2) + n_2 xy + n_3 x \right] + O(\Omega^2) . \tag{6.6.22}$$

If we use the identification of (6.6.20), the Cayley–Klein parameters $\{\alpha, \beta, \gamma, \delta\}$ for an infinitesimal rotation are derived from $\boldsymbol{U} = \boldsymbol{I} + \Omega \boldsymbol{W} + O(\Omega^2)$, where $\boldsymbol{W}$ is given by (6.6.16). Namely,

$$\alpha = 1 - \frac{\mathrm{i}}{2} \Omega n_3 + O(\Omega^2) , \qquad \beta = -\frac{1}{2} \Omega (n_2 + \mathrm{i} n_1) + O(\Omega^2) ,$$

$$\gamma = \frac{1}{2} \Omega (n_2 - \mathrm{i} n_1) + O(\Omega^2) , \qquad \delta = 1 + \frac{\mathrm{i}}{2} \Omega n_3 + O(\Omega^2) . \tag{6.6.23}$$

If these are substituted into (6.5.4), we obtain

$$\delta z = \Omega \left[\frac{n_2 - \mathrm{i} n_1}{2} + \mathrm{i} n_3 z + \frac{n_2 + \mathrm{i} n_1}{2} z^2 \right] + O(\Omega^2) . \tag{6.6.24}$$

Taking the real and imaginary parts, we obtain

$$\delta x = \Omega\left[-n_1 xy + \frac{1}{2}n_2(1 + x^2 - y^2) - n_3 y \right] + O(\Omega^2) ,$$

$$\delta y = \Omega\left[-\frac{1}{2}n_1(1 - x^2 + y^2) + n_2 xy + n_3 x \right] + O(\Omega^2) . \qquad (6.6.25)$$

These are identical to (6.6.22). Since infinitesimal transformations are identical, both transformations must be locally identical.[7] This proves our assertion.

6.6.1 Spinors

Let us consider the irreducibility of the representation of $SO(3)$ defined by Lemma 6.3. The Casimir operator $H = -(A_1^2 + A_2^2 + A_3^2)$ computed from the infinitesimal generators (6.6.17) becomes

$$H = \begin{pmatrix} 3/4 & \\ & 3/4 \end{pmatrix} . \qquad (6.6.26)$$

This is a $2 \times (1/2) + 1 = 2$ dimensional unit matrix multiplied by $(1/2)(1/2 + 1) = 3/4$, so the representation is the irreducible representation $\mathscr{D}_{1/2}$ of weight 1/2 (Theorem 3.3). This means that the pair (z_0, z_1) of the homogeneous coordinates undergoes an irreducible representation of $SO(3)$ of a half-integer weight. Thus, it is a *spinor* (Sect. 3.3.3).

Since the weight is a half-integer, the spinor (z_0, z_1) changes its sign under a rotation by angle 2π (Sect. 3.3.3). This means that the representation of $SO(3)$ by $SU(2)$ is not globally one-to-one; it is only one-to-one near the identity (i.e., locally isomorphic). This is an inevitable consequence of the "topological difference" of $SO(3)$ and $SU(2)$: $SU(2)$ is a *simply connected* Lie group, while $SO(3)$ is not. We will discuss this in subsequent sections in more detail.

6.7 Adjoint Representation of *SU*(2)

In Sect. 6.6 we proved that an element of $SU(2)$ (locally) corresponds to an element of $SO(3)$. However, the correspondence was confirmed through the correspondence of infinitesimal generators only. An explicit correspondence for finite elements has yet to be established.

Suppose we have already established the correspondence between $SU(2)$ and $SO(3)$. Then, we can identify an element of $SU(2)$ with a 3D rotation, which is a transformation of a three-dimensional linear space. This suggests that we may

---

[7] The two transformations do coincide over the entire complex number plane $\mathbb{C}$. What we mean here is that we are considering transformations whose Cayley–Klein parameters $\{\alpha, \beta, \gamma, \delta\}$ are close to $\{1, 0, 0, 1\}$. We will study global aspects (of the "parameter space") later.

be able to establish the correspondence between $SU(2)$ and $SO(3)$ if we are able to find the action of $SU(2)$ on some "three-dimensional linear space".

Fortunately, we have already found a three-dimensional linear space on which $SU(2)$ acts; we proved in Sect. 6.6 that $su(2)$—the Lie algebra of $SU(2)$—is a three-dimensional linear space. In general, the Lie algebra L of an n-dimensional Lie group G is an n-dimensional linear space, and a Lie group G homomorphically acts on its Lie algebra L as linear transformations. Namely, a Lie group G can define a "representation" of itself over its Lie algebra L. This representation is called the *adjoint representation* (Appendix, Sect. A.9).

For $SU(2)$, the adjoint representation is defined as follows. Define the *adjoint transformation* Ad_U of $su(2)$ by

$$\mathrm{Ad}_U(\boldsymbol{W}) \equiv \boldsymbol{U}\boldsymbol{W}\boldsymbol{U}^\dagger , \qquad \boldsymbol{W} \in su(2) . \tag{6.7.1}$$

Proposition 6.8.

(i) *The adjoint transformation* Ad_U *is a linear transformation of* $su(2)$: *for all* $\boldsymbol{W}_1$, $\boldsymbol{W}_2 \in su(2)$ *and all real numbers* c_1, c_2,

$$\boldsymbol{W} \in su(2) \rightarrow \mathrm{Ad}_U(\boldsymbol{W}) \in su(2) , \tag{6.7.2}$$

$$\mathrm{Ad}_U(c_1\boldsymbol{W}_1 + c_2\boldsymbol{W}_2) = c_1\mathrm{Ad}_U(\boldsymbol{W}_1) + c_2\mathrm{Ad}_U(\boldsymbol{W}_2) . \tag{6.7.3}$$

(ii) *The set* $\mathrm{Ad}(SU(2))$ *of all the adjoint transformations* Ad_U: $su(2) \rightarrow su(2)$, $U \in SU(2)$ *is a group of transformations, and the correspondence from* $U \in SU(2)$ *to* Ad_U *is a homomorphism*:

$$\mathrm{Ad}_{U'} \circ \mathrm{Ad}_U = \mathrm{Ad}_{U'U} , \qquad \boldsymbol{U}', \boldsymbol{U} \in SU(2) . \tag{6.7.4}$$

$$\mathrm{Ad}_I = \boldsymbol{I} , \qquad (\mathrm{Ad}_U)^{-1} = \mathrm{Ad}_{U^{-1}} \quad (= \mathrm{Ad}_{U^*}) . \tag{6.7.5}$$

Proof. (i) If

$$\boldsymbol{W}^\dagger = -\boldsymbol{W} , \qquad \mathrm{Tr}\{\boldsymbol{W}\} = 0 . \tag{6.7.6}$$

then from (6.7.1)

$$[\mathrm{Ad}_U(\boldsymbol{W})]^\dagger = (\boldsymbol{U}\boldsymbol{W}\boldsymbol{U}^\dagger)^\dagger = \boldsymbol{U}\boldsymbol{W}^\dagger\boldsymbol{U}^\dagger = -\boldsymbol{U}\boldsymbol{W}\boldsymbol{U}^\dagger = -\mathrm{Ad}_U(\boldsymbol{W}) , \tag{6.7.7}$$

$$\mathrm{Tr}\{\mathrm{Ad}_U(\boldsymbol{W})\} = \mathrm{Tr}\{\boldsymbol{U}\boldsymbol{W}\boldsymbol{U}^\dagger\} = \mathrm{Tr}\{\boldsymbol{U}^\dagger\boldsymbol{U}\boldsymbol{W}\} = \mathrm{Tr}\{\boldsymbol{W}\} = 0 . \tag{6.7.8}$$

This means that if $\boldsymbol{W} \in su(2)$, then $\mathrm{Ad}_U(\boldsymbol{W}) \in su(2)$. Thus, Ad_U is a mapping from $su(2)$ into itself. From the definition (6.7.1), Ad_U is evidently a linear mapping. (ii) For $\boldsymbol{U}'$, $\boldsymbol{U} \in SU(2)$,

$$\begin{aligned}\mathrm{Ad}_{U'} \circ \mathrm{Ad}_U(\boldsymbol{W}) &= \mathrm{Ad}_{U'}(\boldsymbol{U}\boldsymbol{W}\boldsymbol{U}^\dagger) = \boldsymbol{U}'(\boldsymbol{U}\boldsymbol{W}\boldsymbol{U}^\dagger)\boldsymbol{U}'^\dagger \\ &= (\boldsymbol{U}'\boldsymbol{U})\boldsymbol{W}(\boldsymbol{U}'\boldsymbol{U})^\dagger = \mathrm{Ad}_{U'U}(\boldsymbol{W}) .\end{aligned} \tag{6.7.9}$$

From this follows (6.7.5). In other words, the set $\{\mathrm{Ad}_U | \boldsymbol{U} \in SU(2)\}$ is a group of linear transformations of $su(2)$, and its action is homomorphic to $SU(2)$.

We call the representation defined by the adjoint transformation Ad_U the *adjoint representation* of $SU(2)$ over its Lie algebra $su(2)$.

An element $\boldsymbol{W} \in su(2)$ must satisfy the condition $\boldsymbol{W}^\dagger = -\boldsymbol{W}$, $\mathrm{Tr}\{\boldsymbol{W}\} = 0$. Consequently, it must have the form

$$\boldsymbol{W} = \begin{pmatrix} -\mathrm{i}Z & -Y-\mathrm{i}X \\ Y-\mathrm{i}X & \mathrm{i}Z \end{pmatrix}, \tag{6.7.10}$$

where X, Y, Z are real numbers. We call these three real numbers the *natural coordinates* of $su(2)$.

Proposition 6.9. *The natural coordinates of* $su(2)$ *are transformed by the adjoint transformation* Ad_U *for*

$$\boldsymbol{U} = \begin{pmatrix} \alpha & \beta \\ \gamma & \delta \end{pmatrix} \tag{6.7.11}$$

in the form

$$\begin{pmatrix} X' \\ Y' \\ Z' \end{pmatrix} = \begin{pmatrix} (\alpha^2-\beta^2-\gamma^2+\delta^2)/2 & -\mathrm{i}(\alpha^2+\beta^2-\gamma^2-\delta^2)/2 & -\alpha\beta+\gamma\delta \\ \mathrm{i}(\alpha^2-\beta^2+\gamma^2-\delta^2)/2 & (\alpha^2+\beta^2+\gamma^2+\delta^2)/2 & -\mathrm{i}(\alpha\beta+\gamma\delta) \\ -\alpha\beta+\beta\delta & \mathrm{i}(\alpha\gamma+\beta\delta) & \alpha\delta+\beta\gamma \end{pmatrix} \times \begin{pmatrix} X \\ Y \\ Z \end{pmatrix}. \tag{6.7.12}$$

Proof. Equation (6.7.12) is obtained from (6.7.10, 11) by direct calculation:

$$\begin{pmatrix} -\mathrm{i}Z' & -Y'-\mathrm{i}X' \\ Y'-\mathrm{i}X' & \mathrm{i}Z' \end{pmatrix} = \begin{pmatrix} \alpha & \beta \\ \gamma & \delta \end{pmatrix} \begin{pmatrix} -\mathrm{i}Z & -Y-\mathrm{i}X \\ Y-\mathrm{i}X & \mathrm{i}Z \end{pmatrix} \begin{pmatrix} \delta & -\beta \\ -\gamma & \alpha \end{pmatrix}. \tag{6.7.13}$$

Lemma 6.5. *The linear mapping of* (6.7.12) *is a 3D rotation in the* XYZ*-space.*

Proof. First, note that, for $\boldsymbol{W} \in su(2)$,

$$|\mathrm{Ad}_U(\boldsymbol{W})| = |\boldsymbol{U}\boldsymbol{W}\boldsymbol{U}^\dagger| = |\boldsymbol{U}|\,|\boldsymbol{W}|\,|\boldsymbol{U}^\dagger| = |\boldsymbol{W}| . \tag{6.7.14}$$

In other words, the determinant $|\boldsymbol{W}|$ is preserved by the adjoint transformation Ad_U. From (6.7.10), we can see that

$$|\boldsymbol{W}| = X^2 + Y^2 + Z^2 . \tag{6.7.15}$$

This means that the XYZ-space of the natural coordinates is transformed by Ad_U in such a way that the distance from the origin O is fixed. Namely, a sphere centered at the origin O is mapped onto itself. Since Ad_U is a linear mapping in the XYZ-space, the transformation matrix of (6.7.12) is an orthogonal matrix.

Moreover, if we set $\alpha = 1$, $\beta = 0$, $\gamma = 0$, $\delta = 1$ in (6.7.12), the matrix reduces to the unit matrix $\boldsymbol{I}$. In other words, the mapping of (6.7.12) can be smoothly reduced to the identity mapping. Hence, the determinants of the transformation matrices are all 1 (Sect. 6.2.1). Thus, the matrix is an element of $SO(3)$.

We now show that the 3D rotation (6.7.12) exactly coincides with the 3D rotation specified by the Cayley–Klein parameters $\{\alpha, \beta, \gamma, \delta\}$ through the stereographic projection.

Proposition 6.10. The 3D *rotation specified by the Cayley–Klein parameters* $\{\alpha, \beta, \gamma, \delta\}$ *through the stereographic projection is given by* (6.7.12).

Proof. We have already proved the existence of a (local) one-to-one correspondence between a 3D rotation and an element of $SU(2)$ by identifying the infinitesimal generators of $su(2)$ with the infinitesimal generators of $SO(3)$ via (6.6.23). Hence, what remains to be shown is that the same identification holds for the rotation of (6.7.12).

Consider an infinitesimal rotation. If we substitute (6.6.23) into (6.7.12), we obtain

$$\delta \begin{pmatrix} X \\ Y \\ Z \end{pmatrix} = \Omega \begin{pmatrix} 0 & -n_3 & n_2 \\ n_3 & 0 & -n_1 \\ -n_2 & n_1 & 0 \end{pmatrix} + \mathrm{O}(\Omega^2)$$

$$= \Omega (n_1 \boldsymbol{A}_1 + n_2 \boldsymbol{A}_2 + n_3 \boldsymbol{A}_3) + \mathrm{O}(\Omega^2) \,, \tag{6.7.16}$$

where

$$\boldsymbol{A}_1 = \begin{pmatrix} & & \\ & & -1 \\ & 1 & \end{pmatrix}, \quad \boldsymbol{A}_2 = \begin{pmatrix} & & 1 \\ & & \\ -1 & & \end{pmatrix}, \quad \boldsymbol{A}_3 = \begin{pmatrix} & -1 & \\ 1 & & \\ & & \end{pmatrix}. \tag{6.7.17}$$

But these are exactly the infinitesimal generators of the 3D rotation. [In this chapter, we assume that the "space" is rotated relative to a fixed coordinate system—the *alibi* interpretation (Sect. 3.5.2). Hence, the infinitesimal generators of (6.7.17) have signs opposite to those considered in Chap. 3.] This completes our proof.

6.7.1 Differential Representation

We showed that $SU(2)$ acts on its Lie algebra $su(2)$ in the form of the adjoint transformation Ad_U. Now, let us consider infinitesimal adjoint transformations.

If $U \in SU(2)$ is infinitesimal, it is written in the form of (6.6.10), where $W \in su(2)$ has the form of (6.6.16). For such a $U \in SU(2)$, the adjoint transformation Ad_U becomes

$$\begin{aligned}\mathrm{Ad}_U(X) &= [I + tW + O(t^2)]X[I - tW + O(t^2)] \\ &= X + t(WX - XW) + O(t^2) = X + t[W, X] + O(t^2) \ , \end{aligned} \tag{6.7.18}$$

or we can write

$$\mathrm{Ad}_U(X) = X + t\mathrm{d}_W(X) + O(t^2) \ , \tag{6.7.19}$$

where

$$\mathrm{d}_W(X) \equiv [W, X] \ . \tag{6.7.20}$$

Equation (6.7.19) can be viewed as a kind of Taylor expansion; we can imagine that the second term on the right-hand side is some kind of "derivative" of X. Hence, the operation d_W is called the *derivation.* There exists another source of analogy: the *Jacobi identity* (3.2.12) can be rewritten as

$$\mathrm{d}_W([X, Y]) = [\mathrm{d}_W(X), Y] + [X, \mathrm{d}_W(Y)] \ . \tag{6.7.21}$$

Thus, the derivation d_W acts like the usual differentiation if the commutation is identified with multiplication. Put

$$\mathscr{D}(su(2)) = \{\mathrm{d}_W \,|\, W \in su(2)\} \ . \tag{6.7.22}$$

This is a set of linear transformations of $su(2)$. First, note that

$$\mathrm{d}_{c_1 W_1 + c_2 W_2} = c_1 \mathrm{d}_{W_1} + c_2 \mathrm{d}_{W_2} \tag{6.7.23}$$

for all $W_1, W_2 \in su(2)$ and all real numbers c_1, c_2. This means that the set $\mathscr{D}$ of transformations is a linear space (i.e., closed under scalar multiplication and addition) and the correspondence from $W \in su(2)$ to $\mathrm{d}_W \in \mathscr{D}$ is a *linear space homomorphism.* Furthermore, we can also prove that

$$\mathrm{d}_{[W_1, W_2]} = [\mathrm{d}_{W_1}, \mathrm{d}_{W_2}] \ . \tag{6.7.24}$$

This means that $\mathscr{D}$ is closed under the commutator. Hence, $\mathscr{D}$ is itself a Lie algebra. Moreover, (6.7.23, 24) imply that the correspondence from $W \in su(2)$ to $\mathrm{d}_W \in \mathscr{D}$ is a *Lie algebra homomorphism.* The Lie algebra $\mathscr{D}(su(2))$ is called the *differential representation* of $su(2)$ over itself (Appendix, Sect. A.9).

For $SU(2)$ and $su(2)$, we have proved the Lie group isomorphism

$$\mathrm{Ad}(SU(2)) \cong SO(3) \ .$$

We have also proved the Lie algebra isomorphism

$$\mathscr{D}(su(2)) \cong su(2) \ .$$

(In general, $\mathscr{D}(L) \cong L$ for all *semi-simple* Lie algebras L.)

6.8 Quaternions

The Cayley–Klein parameters $\{\alpha, \beta, \gamma, \delta\}$ are not mutually independent. From (6.5.5a, b), we can write

$$U = \begin{pmatrix} \alpha & \beta \\ \gamma & \delta \end{pmatrix} = \begin{pmatrix} q_0 - iq_3 & -q_2 - iq_1 \\ q_2 - iq_1 & q_0 + iq_3 \end{pmatrix}, \tag{6.8.1}$$

where q_0, q_1, q_2, q_3 are real parameters. From (6.5.5c), we have the constraint

$$q_0^2 + q_1^2 + q_2^2 + q_3^2 = 1 \ . \tag{6.8.2}$$

Equation (6.8.1) is rewritten as

$$U = q_0 I + q_1 S_1 + q_2 S_2 + q_3 S_3 \ , \tag{6.8.3}$$

where

$$I \equiv \begin{pmatrix} 1 & 0 \\ 0 & 1 \end{pmatrix}, \quad S_1 \equiv \begin{pmatrix} 0 & -i \\ -i & 0 \end{pmatrix},$$

$$S_2 \equiv \begin{pmatrix} 0 & -1 \\ 1 & 0 \end{pmatrix}, \quad S_3 \equiv \begin{pmatrix} -i & 0 \\ 0 & i \end{pmatrix}. \tag{6.8.4}$$

By direct calculation, we can observe the following relations:[8]

Lemma 6.6.

$$S_1^2 = S_2^2 = S_3^2 = -I \ ,$$

$$S_2 S_3 = -S_3 S_2 = S_1 \ , \qquad S_3 S_1 = -S_1 S_3 = S_2 \ ,$$

$$S_1 S_2 = -S_2 S_1 = S_3 \ . \tag{6.8.5}$$

In the preceding sections, we showed that the rule of composition of the Cayley–Klein parameters $\{\alpha, \beta, \gamma, \delta\}$ is given by multiplication of the Corresponding matrices in $SU(2)$: A rotation specified by $U \in SU(2)$ followed by another rotation specified by $U' \in SU(2)$ is the rotation specified by $U'U \in SU(2)$.

However, matrices of $SU(2)$ are expressed in the form of (6.8.3), and multiplications among I, S_1, S_2, S_3 are "closed", as shown by (6.8.5). This means that *we need not calculate matrix products*. In other words, the calculation can be done "symbolically", if (6.8.3) is replaced by the symbolic expression

$$q = q_0 + q_1 i + q_2 j + q_3 k \ , \tag{6.8.6}$$

[8] Matrices S_1, S_2, S_3, are all elements of $SU(2)$. If we put $P_1 = iS_1$, $P_2 = iS_2$, $P_3 = iS_3$, the matrices P_1, P_2, P_3 are all Hermitian matrices. They are called the *Pauli spin matrices* and play an important role in quantum mechanics.

where q_0, q_1, q_2, q_3 are real numbers satisfying

$$q_0^2 + q_1^2 + q_2^2 + q_3^2 = 1 \ , \tag{6.8.7}$$

while $\boldsymbol{i}, \boldsymbol{j}, \boldsymbol{k}$ are symbolic quantities. The rule of composition is given by multiplication just as for real numbers, except that $\boldsymbol{i}, \boldsymbol{j}, \boldsymbol{k}$ obey the rule corresponding to (6.8.5):

$$\begin{aligned} &\boldsymbol{i}^2 = \boldsymbol{j}^2 = \boldsymbol{k}^2 = -1 \ , \\ &\boldsymbol{ij} = -\boldsymbol{ji} = \boldsymbol{k} \ , \qquad \boldsymbol{jk} = -\boldsymbol{kj} = \boldsymbol{i} \ , \qquad \boldsymbol{ki} = -\boldsymbol{ik} = \boldsymbol{j} \ . \end{aligned} \tag{6.8.8}$$

An expression that has the form of (6.8.6) and obeys the rules (6.8.8) is called a (*Hamilton*) *quaternion*. Thus, we have shown the following result:

Proposition 6.11. *If rotations $\boldsymbol{R}$, $\boldsymbol{R}'$ are represented by quaternions $\boldsymbol{q}$, $\boldsymbol{q}'$, the composition $\boldsymbol{R}'' = \boldsymbol{R}'\boldsymbol{R}$ is represented by quaternion*

$$\boldsymbol{q}'' = \boldsymbol{q}'\boldsymbol{q} \ . \tag{6.8.9}$$

Corollary 6.4. *If a rotation $\boldsymbol{R}$ is represented by quaternion $\boldsymbol{q} = q_0 + q_1\boldsymbol{i} + q_2\boldsymbol{j} + q_3\boldsymbol{k}$, its inverse rotation $\boldsymbol{R}^{-1}$ ($= \boldsymbol{R}^{\mathrm{T}}$) is represented by quaternion*

$$\boldsymbol{q}^* = q_0 - q_1\boldsymbol{i} - q_2\boldsymbol{j} - q_3\boldsymbol{k} \ , \tag{6.8.10}$$

and

$$\boldsymbol{q}^*\boldsymbol{q} = \boldsymbol{q}\boldsymbol{q}^* = 1 \ , \tag{6.8.11}$$

where the real number 1 is identified with quaternion $1 + 0\boldsymbol{i} + 0\boldsymbol{j} + 0\boldsymbol{k}$, which represents the identity $\boldsymbol{I}$.

Equation (6.8.11) is obtained from (6.8.8) by direct calculation. The quaternion $\boldsymbol{q}^*$ of (6.8.10) is called the *conjugate* of quaternion $\boldsymbol{q}$.

The explicit form of the rotation matrix $\boldsymbol{R}$ represented by a quaternion $\boldsymbol{q}$ is given as follows:

Proposition 6.12. *The 3D rotation represented by quaternion $\boldsymbol{q} = q_0 + q_1\boldsymbol{i} + q_2\boldsymbol{j} + q_3\boldsymbol{k}$ is given by*

$$\begin{pmatrix} X' \\ Y' \\ Z' \end{pmatrix} = \begin{pmatrix} q_0^2 + q_1^2 - q_2^2 - q_3^2 & -2(q_0q_3 - q_1q_2) & 2(q_0q_2 + q_1q_3) \\ 2(q_0q_3 + q_1q_2) & q_0^2 - q_1^2 + q_2^2 - q_3^2 & -2(q_0q_1 - q_2q_3) \\ -2(q_0q_2 - q_1q_3) & 2(q_0q_1 + q_2q_3) & q_0^2 - q_1^2 - q_2^2 + q_3^2 \end{pmatrix} \begin{pmatrix} X \\ Y \\ Z \end{pmatrix} . \tag{6.8.12}$$

Proof. If we compare (6.7.10) and (6.8.3, 4), we see that (6.7.10) is written as

$$\boldsymbol{W} = X\boldsymbol{S}_1 + Y\boldsymbol{S}_2 + Z\boldsymbol{S}_3 \ . \tag{6.8.13}$$

This expression is identified with quaternion

$$\boldsymbol{r} = X\boldsymbol{i} + Y\boldsymbol{j} + Z\boldsymbol{k} \ . \tag{6.8.14}$$

Then, the relation $W' = \mathrm{Ad}_U(W) = UWU^+$ is written in terms of quaternions as

$$r' = qrq^* , \tag{6.8.15}$$

where q^* is the conjugate quaternion of q and r' is the quaternion

$$r' = X'i + Y'j + Z'k . \tag{6.8.16}$$

If we rewrite (6.8.15) in terms of X, Y, Z and X', Y', Z', the resulting mapping from (X, Y, Z) to (X', Y', Z') is the corresponding 3D rotation (see Lemma 6.4). Thus, we obtain (6.8.12).

Now, we give an alternative proof to Proposition 6.5. Let $q_n(\Omega)$ be the quaternion representing a rotation around axis $n = (n_1, n_2, n_3)$ (unit vector) by angle Ω. Let us identify the vector $n = (n_1, n_2, n_3)$ with quaternion

$$n = n_1 i + n_2 j + n_3 k . \tag{6.8.17}$$

Lemma 6.7.

$$\frac{dq_n}{d\Omega} = \frac{1}{2} n q_n . \tag{6.8.18}$$

Proof. First, note that quaternions commute with each other if the rotation axis is fixed:[9]

$$q_n(\Omega')q_n(\Omega) = q_n(\Omega' + \Omega) = q_n(\Omega)q_n(\Omega') . \tag{6.8.19}$$

If the rotation is infinitesimal, the Cayley–Klein parameters are given by (6.6.23). From (6.8.1), the corresponding quaternion becomes

$$q_n(\Omega) = 1 + \frac{\Omega}{2}(n_1 i + n_2 j + n_3 k) + O(\Omega^2) . \tag{6.8.20}$$

Dividing both sides by Ω, taking the limit $\Omega \to 0$, and noting that $q_n(0) = 1$, we have

$$\frac{dq_n}{d\Omega}(0) = \lim_{\Omega \to 0} \frac{q_n(\Omega) - 1}{\Omega} = \frac{1}{2} n . \tag{6.8.21}$$

Hence,

$$\begin{aligned} \frac{dq_n}{d\Omega} &= \lim_{\Omega \to 0} \frac{q_n(\Omega + \Omega') - q_n(\Omega)}{\Omega'} = \lim_{\Omega' \to 0} \frac{q_n(\Omega)q_n(\Omega') - q_n(\Omega)}{\Omega'} \\ &= \lim_{\Omega' \to 0} \frac{q_n(\Omega') - 1}{\Omega'} q_n(\Omega) = \frac{dq_n}{d\Omega}(0)\, q_n(\Omega) = \frac{1}{2} n q_n(\Omega) . \end{aligned} \tag{6.8.22}$$

---

[9] In general, quaternions do not commute, as seen in the rules (6.8.8), since the corresponding $SO(3)$ is not Abelian. However, if the axis is fixed, $SO(3)$ is restricted to $SO(2)$, which is Abelian (Sect. 3.6), so the corresponding quaternions also commute.

Theorem 6.4.

$$\boldsymbol{q}_n(\Omega) = \cos\frac{\Omega}{2} + \boldsymbol{n}\sin\frac{\Omega}{2} \ . \tag{6.8.23}$$

Proof. From (6.8.19), we can prove, by appropriately modifying (6.8.22), that $d\boldsymbol{q}_n(\Omega)/d\Omega = \boldsymbol{q}_n\boldsymbol{n}/2$. Hence, quaternion $\boldsymbol{n}$ commutes with quaternion $\boldsymbol{q}_n$:

$$\boldsymbol{n}\boldsymbol{q}_n = \boldsymbol{q}_n\boldsymbol{n} \ . \tag{6.8.24}$$

Thus, (6.8.18) can be treated just as if it were a usual differential equation; the initial condition is $\boldsymbol{q}_n(0) = 1$. The solution is given in the form of an infinite series

$$\boldsymbol{q}_n(\Omega) = \exp\frac{\Omega}{2}\boldsymbol{n}\left[\equiv \sum_{k=0}^{\infty}\frac{1}{k!}\left(\frac{\Omega}{2}\boldsymbol{n}\right)^k\right] . \tag{6.8.25}$$

It follows from the rules (6.8.8) that

$$\boldsymbol{n}^2 = -1 \tag{6.8.26}$$

as a quaternion. Hence, $\boldsymbol{n}$ can be treated just as if it were the imaginary unit, and if we recall *Euler's formula* $\exp(\mathrm{i}\theta) = \cos\theta + \mathrm{i}\sin\theta$, (6.8.25) is written as

$$\begin{aligned}\boldsymbol{q}_n(\Omega) &= \left[1 - \frac{1}{2!}\left(\frac{\Omega}{2}\right)^2 + \frac{1}{4!}\left(\frac{\Omega}{2}\right)^4 - \dots\right] \\ &\quad + \boldsymbol{n}\left[\frac{\Omega}{2} - \frac{1}{3!}\left(\frac{\Omega}{2}\right)^3 + \frac{1}{5!}\left(\frac{\Omega}{2}\right)^5 - \dots\right] \\ &= \cos\frac{\Omega}{2} + \boldsymbol{n}\sin\frac{\Omega}{2} \ . \end{aligned} \tag{6.8.27}$$

From Theorem 6.4, we immediately observe:

Corollary 6.5. *The angle Ω and axis $\boldsymbol{n}$ of the rotation represented by quaternion $\boldsymbol{q} = q_0 + \boldsymbol{i}q_1 + \boldsymbol{j}q_2 + \boldsymbol{k}q_3$ are*

$$\Omega = 2\cos^{-1} q_0 \ , \qquad \boldsymbol{n} = \frac{(q_1,\ q_2,\ q_3)}{\sqrt{q_1^2 + q_2^2 + q_3^2}} \ . \tag{6.8.28}$$

Proof of Proposition 6.5. Equation (6.3.10) is obtained if (6.8.23) is substituted into (6.8.12).

For the Euler angles $\{\theta, \varphi, \psi\}$, we obtain the following result:

Proposition 6.13. *The quaternion representing the rotation specified by Euler angles $\{\theta, \varphi, \psi\}$ is given by*

$$q(\theta, \varphi, \psi\} = \cos\frac{\theta}{2}\left(\cos\frac{\psi+\varphi}{2} + k\sin\frac{\psi+\varphi}{2}\right)$$
$$+ \sin\frac{\theta}{2}\left(i\sin\frac{\psi-\varphi}{2} + j\cos\frac{\psi-\varphi}{2}\right). \tag{6.8.29}$$

Proof. From (6.4.3),

$$q(\theta, \varphi, \psi) = q_k(\varphi)q_j(\theta)q_k(\psi)$$
$$= \left(\cos\frac{\theta}{2} + k\sin\frac{\varphi}{2}\right)\left(\cos\frac{\theta}{2} + j\sin\frac{\theta}{2}\right)\left(\cos\frac{\psi}{2} + k\sin\frac{\psi}{2}\right). \tag{6.8.30}$$

Equation (6.8.29) is obtained if (6.8.30) is computed according to the rules (6.8.8).

6.8.1 Quaternion Field

The *modulus* of a quaternion of (6.8.6) is defined by

$$|q| = \sqrt{q_0^2 + q_1^2 + q_2^2 + q_3^2}\ . \tag{6.8.31}$$

Hence, we can say that 3D *rotations are represented by quaternions of modulus* 1.
In terms of the conjugate q^*, (6.8.31) is also written as

$$|q|^2 = q^*q = qq^*\ . \tag{6.8.32}$$

This means that if we define the *inverse quaternion*

$$q^{-1} = \frac{q^*}{|q|^2} \tag{6.8.33}$$

for $q \neq 0$, then

$$q^{-1}q = qq^{-1} = 1\ . \tag{6.8.34}$$

In other words, *every nonzero quaternion has its inverse*. This means that the set $\mathbb{H}$ of quaternions is closed not only under addition, subtraction, and multiplication but also under division by a nonzero element. Thus, the set $\mathbb{H}$ of quaternions is a *field*. Since $q'q \neq qq'$ in general, as can be seen from the rules (6.8.8), the field H of quaternions is *not Abelian*.

As is well known, the smallest field that includes the set $\mathbb{N}$ of natural numbers (or the set $\mathbb{Z}$ of integers) is the set $\mathbb{Q}$ of rational numbers. This field is extended to the field $\mathbb{R}$ of real numbers by adding irrational numbers, which is then extended to the field $\mathbb{C}$ of complex numbers by adding imaginary numbers. The field $\mathbb{H}$ is viewed as an extension of $\mathbb{C}$ (hence of $\mathbb{R}$) with the identification

$$\mathbb{R} = \{q_0 + 0i + 0j + 0k \,|\, q_0 \in \mathbb{R}\},$$
$$\mathbb{C} = \{q_0 + q_1 i + 0j + 0k \,|\, q_0, q_1 \in \mathbb{R}\}.$$

In this sense, $\mathbb{R} \subset \mathbb{C} \subset \mathbb{H}$. It can be proved that the field $\mathbb{H}$ of quaternions is the "largest field" in the sense that it can no longer be extended to a larger field *as a finite-dimensional linear space over* $\mathbb{R}$.[10]

In Sect. 6.8, we often identified a vector $\boldsymbol{a} = (a_1, a_2, a_3)$ with quaternion $\boldsymbol{a} = a_1\boldsymbol{i} + a_2\boldsymbol{j} + a_3\boldsymbol{k}$. This is very natural, since the set

$$V = \{q_1\boldsymbol{i} + q_2\boldsymbol{j} + q_3\boldsymbol{k} \,|\, q_1, q_2, q_3 \in \mathbb{R}\}$$

is a three-dimensional linear space as far as addition, subtraction, and scalar multiplication are concerned. For this reason, the three components $\{q_1, q_2, q_3\}$ are called the *vector part* of $\boldsymbol{q}$, while q_0 is called the *scalar part*. (See the Exercises for some useful formulae.)

As was shown in Sect. 6.8, the set of quaternions of modulus 1 is a subgroup of $\mathbb{H}$ under multiplication, and is isomorphic to $SU(2)$. From (6.8.23, 28), we find that if a quaternion q represents a 3D rotation, *its scalar part indicates the angle of rotation, and its vector part indicates the axis of rotation.*

6.9 Topology of $SO(3)$[11]

We showed in Sect. 6.8 that an element of $SU(2)$ is uniquely specified by four real parameters q_0, q_1, q_2, q_3 satisfying

$$q_0^2 + q_1^2 + q_2^2 + q_3^2 = 1 \ . \tag{6.9.1}$$

In other words, there is a one-to-one correspondence between elements of $SU(2)$ and points on the "three-dimensional sphere" defined by (6.9.1) in the four-dimensional $q_0q_1q_2q_3$-space (Fig. 6.7). We state this fact by saying that the group $SU(2)$ is *homeomorphic* to (or more simply has the same *topology* as) a three-dimensional sphere. We write

$$SU(2) \approx S^3 \,(\text{homeomorphism}) \ ,$$

where S^3 denotes a three-dimensional sphere, which is usually referred to simply as a *3-sphere*. Since $SU(2)$ has the same topology as a 3-sphere S^3, the group $SU(2)$ is a *compact* Lie group, meaning that the parameter space is finitely bounded.

---

[10] The field H of quaternions can be extended by introducing *Cayley numbers* (or *octernions*) in such a way that the resulting set $\mathscr{C}$ is an eight-dimensional linear space over R and is closed under addition, subtraction, multiplication, and division by a nonzero element. However, the associative law $(ab)c = a(bc)$ does not hold, and hence $\mathscr{C}$ is not a field. The set $\mathscr{C}$ is called the *Cayley algebra*.

[11] This section contains many mathematical concepts that we have not fully discussed. Since the topics here are somewhat different from previous ones, this section can be skipped at the first reading. Here, we give only informal descriptions. See the Appendix, Sect. A.8, for basic definitions.

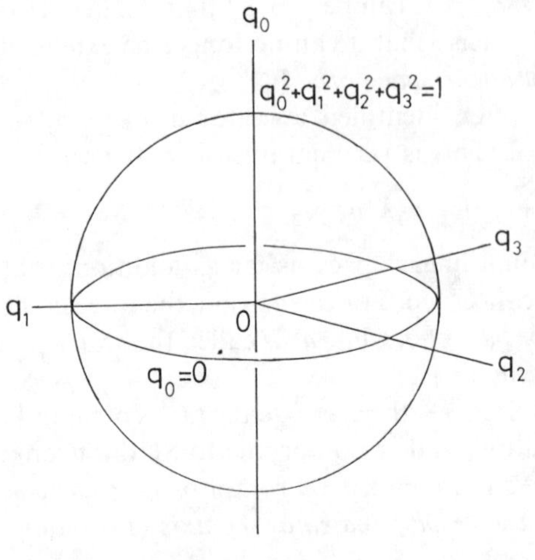

Fig. 6.7. A quaternion $\boldsymbol{q} = q_0 + q_1\boldsymbol{i} + q_2\boldsymbol{j} + q_3\boldsymbol{k}$ is identified with a point on the 3-sphere $q_0^2 + q_1^2 + q_2^2 + q_3^2 = 1$

We also showed in the previous sections that each element of $SU(2)$, or equivalently each quaternion of modulus 1, represents a 3D rotation. The correspondence is a homomorphism (Lemma 6.3). An explicit relationship is given by (6.7.10) or equivalently by (6.8.12). However, *the correspondence is not one-to-one*: the matrix elements in both (6.7.10) and (6.8.12) are quadratic in parameters so that both $\boldsymbol{q}$ and $-\boldsymbol{q}$ represent one and the same 3D rotation.

Since a point on a 3-sphere S^3 can be identified with a quaternion of modulus 1, the above consideration means that a point on a 3-sphere S^3 indicates a 3D rotation. In other words, a continuous mapping is defined from S^3 onto $SO(3)$. However, the mapping is not one-to-one: two antipodes (the "opposite" points with respect to the center) correspond to the same 3D rotation. This means that $SO(3)$ is homeomorphic to only a half of the 3-sphere S^3, say, the "northern hemisphere" $q_0 \geqq 0$. On the "equator" $q_0 = 0$, which is a 2-sphere $q_0^2 + q_1^2 + q_3^2 = 1$, two diametrically opposite points (antipodes of the 2-sphere) must be identified. We can imagine cutting out the northern hemisphere and pasting together the equator so that diametrically opposite points coincide. The compact space thus obtained is called the three-dimensional real *projective space* denoted by RP^3. Thus, $SO(3)$ is homeomorphic to RP^3:

$$SO(3) \approx RP^3 (\text{homeomorphism}) .$$

Hence, $SO(3)$ is also a compact Lie group. (A subset of a compact set is also compact.)

Since $\boldsymbol{q}$ and $-\boldsymbol{q}$ represent the same 3D rotation, one may tend to think that a unique description of 3D rotations could be obtained by quaternions such that $q_0 \geqq 0$. Why did we not do so from the beginning?

The observation we have made above is purely *topological*. We must also look at the *algebraic* structure. Consider the quaternion $\boldsymbol{q}_n(\Omega)$ representing rotations around a fixed axis $\boldsymbol{n}$, (6.8.23). We find that

$$\boldsymbol{q}_n(\Omega + 2\pi) = -\boldsymbol{q}_n(\Omega) \ . \tag{6.9.2}$$

In other words, the quaternion changes sign if the angle of rotation changes from 0 to 2π. If the angle is further increased from 2π to 4π, the quaternion returns to the original one.

Imagine this process on the 3-sphere S^3. Both the "north pole" (1, 0, 0, 0) and the "south pole" $(-1, 0, 0, 0)$ represent the identity $\boldsymbol{I}$. If we start from the north pole, gradually increasing the angle of rotation from 0 around a fixed axis, we go down through the northern hemisphere and arrive at the equator when the angle is π. As the angle further increases, we go on into the sourthern hemisphere and reach the south pole at angle 2π. Then, we go up again and come back to the north pole at angle 4π. Thus, we cannot simply cut out the northern hemisphere and work only on that; we would lose the fundamental rule of composition $\boldsymbol{q}'' = \boldsymbol{q}'\boldsymbol{q}$.

The north pole (1, 0, 0, 0) of S^3 corresponds to quaternion $\boldsymbol{q} = 1$, which corresponds to $\boldsymbol{I} \in SU(2)$ of (6.8.4). Similarly, the south pole $(-1, 0, 0, 0)$ corresponds to $-\boldsymbol{I} \in SU(2)$. We have already shown that the mapping from $SU(2)$ onto $SO(3)$ is a homomorphism. By this mapping, both $\boldsymbol{I}$ and $-\boldsymbol{I}$ (and these alone) are mapped onto the identity $\boldsymbol{I} \in SO(3)$. Namely, the set $\{\boldsymbol{I}, -\boldsymbol{I}\}$ is the kernel of this mapping. It is a subgroup of $SU(2)$. Since this subgroup consists of two elements (one is the identity, and the multiplication of the other element by itself is equal to the identity), it is isomorphic to $\mathbb{Z}_2$ (the group of integers under addition modulo 2). Hence, by the well-known *homomorphism theorem* of groups, $SO(3)$ is isomorphic to the quotient group of $SU(2)$ over this kernel. Thus, we conclude that

$$SO(3) \cong SU(2)/\mathbb{Z}_2 \ .$$

Any closed loop on the 3-sphere S^3 identified with the parameter space of $SU(2)$ can be shrunk into a point on S^3. This means that the Lie group $SU(2)$ is *simply connected* (Appendix, Sect. A.8). However, $SO(3)$ is not simply connected. As we have already pointed out, $SO(3)$ is homeomorphic to RP^3 (the northern hemisphere of S^3 with two diametrically opposite points on the equator identified with each other). Hence, a curve drawn on northern hemisphere starting and ending at two diametrically opposite points $\boldsymbol{q}$, $-\boldsymbol{q}$ on the equator is a "closed loop" in RP^3 (Fig. 6.8). This loop cannot be shrunk into a point, because if we move one endpoint $\boldsymbol{q}$ on the equator, the other endpoint $-\boldsymbol{q}$ moves in such a way that the two points $\boldsymbol{q}$, $-\boldsymbol{q}$ are always on opposite sides. Thus, RP^3 and hence $SO(3)$ are not simply connected.

Now, consider a curve starting from a point $\boldsymbol{q}$ on the equator, passing across the northern hemisphere and ending at another point $\boldsymbol{q}'$ on the equator, then starting from point $-\boldsymbol{q}'$ on the equator, passing through the northern hemi-

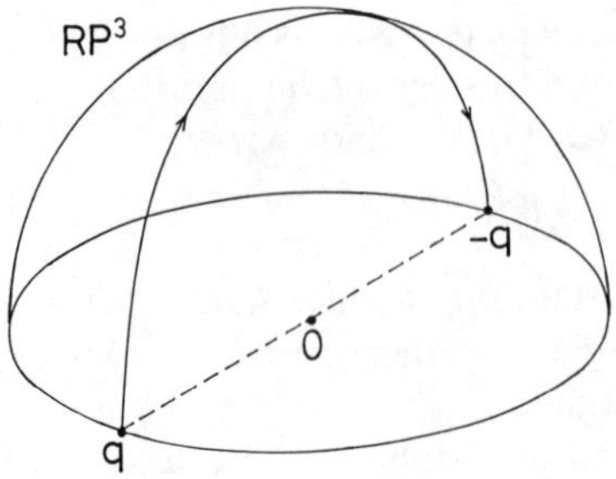

Fig. 6.8. A closed loop. The three-dimensional real projective space RP^3 is the "northern hemisphere" of S^3 with two diametrically opposite points on the "equator" identified with each other

sphere, and ending at point $-\boldsymbol{q}$ on the equator (Fig. 6.9). This is also a closed loop. This loop can be shrunk into one point; we only need to move $\boldsymbol{q}$ along the equator so that it coincides with $\boldsymbol{q}'$ (Fig. 6.10).

Similarly, a closed loop that has three pairs of endpoints on the equator can be deformed smoothly into a closed loop with a pair of endpoints (Fig. 6.11). Let us identify two closed curves when one is smoothly deformed into the other. Then, there are only two types of closed curves: those with $2n$ pairs of endpoints on the equator and those with $2n + 1$ pairs of endpoints, where $n = 0, 1, 2, \ldots$.

In general, two closed loops in a topological space X are said to be *homotopic* to each other if one is continuously deformed into the other (Appendix, Sect. A.8). We can define "multiplication" among all closed loops by identifying mutually homotopic loops, and define a group called the *fundamental group* (or the *first homotopy group*). This group is *topologically invariant*, and denoted by $\pi_1(X)$ (Appendix, Sect. A.8). We can prove that

$$\pi_1(SU(2)) \cong 0 , \qquad \pi_1(SO(3)) \cong \mathbb{Z}_2 .$$

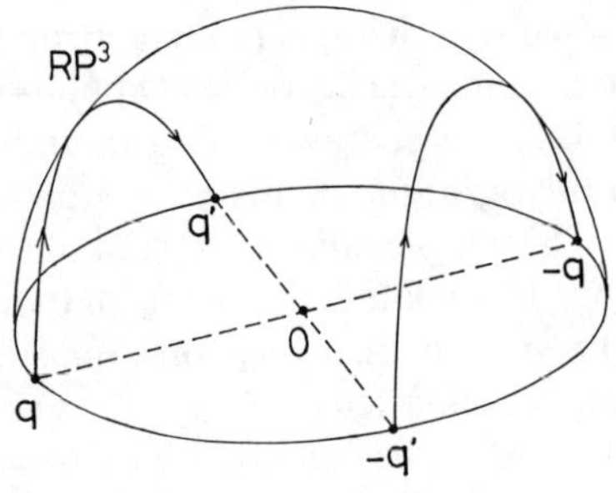

Fig. 6.9. A closed loop in RP^3 that has two pairs of endpoints on the equator

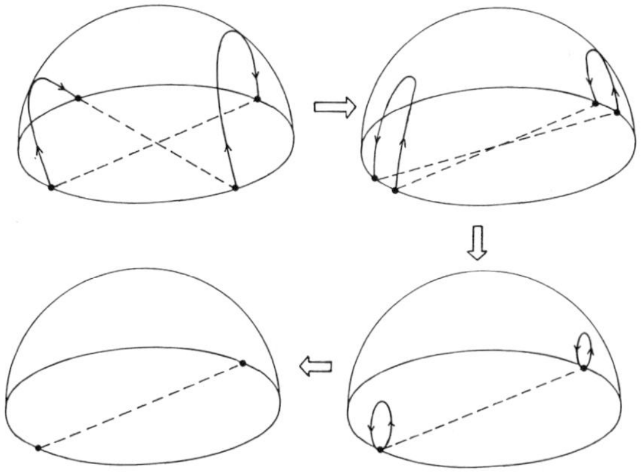

Fig. 6.10. A closed loop that has two pairs of endpoints on the equator can be shrunk into a point

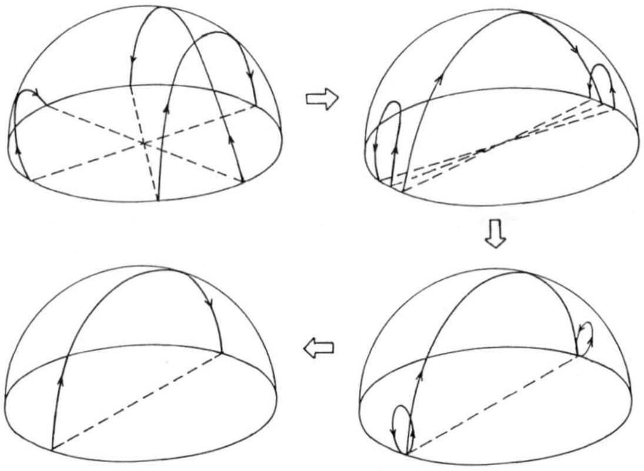

Fig. 6.11. A closed loop that has three pairs of endpoints on the equator can be deformed into a closed loop with a pair of endpoints

6.9.1 Universal Covering Group

The mapping from $SU(2)$ onto $SO(3)$ has the following properties:

(1) The mapping is a homomorphism from $SU(2)$ onto $SO(3)$.
(2) Two elements $\boldsymbol{U}$, $\boldsymbol{U}' \in SU(2)$ correspond to one element $\boldsymbol{R} \in SO(3)$, and for every open set $\mathcal{O}$ of $SO(3)$ that contains $\boldsymbol{R}$—a *neighborhood* of $\boldsymbol{R} \in SO(3)$, there exist neighborhoods of $\boldsymbol{U}$, $\boldsymbol{U}' \in SU(2)$ that are both homeomorphic to $\mathcal{O}$.[12]
(3) $SU(2)$ is simply connected.

[12] When we say an "open set" of $SU(2)$ or $SO(3)$, we are assuming that the topology has been introduced in a natural way, but we do not discuss the details. See the Appendix, Sect. A.8.

These conditions can be stated simply by saying that $SU(2)$ is the *universal covering group* (Appendix, Sect. A.8). Informally speaking, a universal covering group of a topological group that is not simply connected is a simply connected topological space constructed by "pasting together" as many of its copies as necessary, yet in such a way that the algebraic structure is preserved *globally* and the topological structure is preserved *locally*.

6.10 Invariant Measure of 3D Rotations

Suppose we have a set of many identical objects and want to test if the orientations, or postures, of these objects are randomly distributed. What should a precise definition of "randomness" be? Often, we are tempted to define randomness by a "uniform distribution" of the parameters. However, we must be careful.

The orientation of an object can be specified by a rotation from a fixed reference orientation. A rotation is specified, for example, by its angle and axis. The orientation of an axis can be specified in spherical coordinates (θ, φ), $0 \leqq \theta < \pi$, $0 \leqq \varphi < 2\pi$. The angle Ω of rotation can be restricted to interval $0 \leqq \Omega < 2\pi$. Then, does the "randomness" mean that parameters $\{\Omega, \theta, \varphi\}$ are distributed uniformly in intervals $0 \leqq \Omega \leqq 2\pi$, $0 \leqq \theta < \pi$, $0 \leqq \varphi < 2\pi$?

A rotation is also specified by the Euler angles $\{\theta, \varphi, \psi\}$. Does the "randomness" mean that these parameters are distributed uniformly in intervals $0 \leqq \theta < \pi$, $0 \leqq \varphi < 2\pi$, $0 \leqq \psi < 2\pi$? It can be proved easily that this distribution and the one given above are different. Which is more random than the other?

As can be seen from this argument, a uniform distribution in some parameters does not necessarily mean a uniform distribution in other parameters. Consequently, the "uniform distribution" does not have a physical meaning unless there is some reason, physical or mathematical, to say that uniformity in particular parameters has a special significance.

When we say that "orientations are random", our ituition is that one orientation does not have any preference over others, and if all the objects undergo the same rotation simultaneously, the resulting distribution is equally random. This fact is equivalent to saying that the description of the distribution *does not depend on the choice of the reference orientation* from which individual orientations are measured: rotating the reference orientation by $\boldsymbol{R}$ is equivalent to rotating all the objects by $\boldsymbol{R}^{-1}$ relative to the reference orientation. From this observation, a reasonable definition of randomness may be as follows.

Let $\{\theta_1, \theta_2, \theta_3\}$ be a set of parameters that specify a 3D rotation. Suppose the orientations are distributed according to a probability density $p_{\boldsymbol{I}}(\theta_1, \theta_2, \theta_3)$. Let $\{\theta'_1, \theta'_2, \theta'_3\}$ be the parameters of the orientation after all the objects are simultaneously rotated by $\boldsymbol{R}$. The probability density $p_{\boldsymbol{R}}(\theta'_1, \theta'_2, \theta'_3)$ of the new parameters $\{\theta'_1, \theta'_2, \theta'_3\}$ is related to the old distribution density $p_I(\theta_1, \theta_2, \theta_3)$ by

$$p_R(\theta'_1, \theta'_2, \theta'_3) d\theta'_1 \, d\theta'_2 \, d\theta'_3 = p_I(\theta_1, \theta_2, \theta_3) \, d\theta_1 \, d\theta_2 \, d\theta_3 \ . \tag{6.10.1}$$

Hence, the new density is

$$p_R(\theta'_1, \theta'_2, \theta'_3) = \frac{1}{|J|} p_I(\theta_1, \theta_2, \theta_3) \tag{6.10.2}$$

where J is the *Jacobian*

$$J = \begin{vmatrix} \partial\theta'_1/\partial\theta_1 & \partial\theta'_1/\partial\theta_2 & \partial\theta'_1/\partial\theta_3 \\ \partial\theta'_2/\partial\theta_1 & \partial\theta'_2/\partial\theta_2 & \partial\theta'_2/\partial\theta_3 \\ \partial\theta'_3/\partial\theta_1 & \partial\theta'_3/\partial\theta_2 & \partial\theta'_3/\partial\theta_3 \end{vmatrix} \ . \tag{6.10.3}$$

The randomness can be defined by requiring that the *probability densities are always identical*, i.e., for all $\boldsymbol{R} \in SU(3)$,

$$p_R(.,.,.) \equiv p_I(.,.,.) \ . \tag{6.10.4}$$

If this requirement is satisfied, we call the differential form

$$d\mu = p_I(\theta_1, \theta_2, \theta_3) \, d\theta_1 \, d\theta_2 \, d\theta_3 \tag{6.10.5}$$

the *invariant measure* of $SO(3)$ for the parameters $\{\theta_1, \theta_2, \theta_3\}$. We do not give the proof, but it can be proved that *a compact Lie group always has an invariant measure.*

As shown in Sect. 6.9, a 3D rotation is specified by a quaternion $\boldsymbol{q} = q_0 + q_1\boldsymbol{i} + q_2\boldsymbol{j} + q_3\boldsymbol{k}$, for which the parameter space is the 3-sphere S^3 given by $q_0^2 + q_1^2 + q_2^2 + q_3^2 = 1$. Hence, a probability density can be defined as a function over the 3-sphere S^3. Now, we show that application of a 3D rotation to all the points on S^3 can be regarded as a "4D rotation" of the 3-sphere S^3.

Lemma 6.8. *An element of $SO(3)$ acts on S^3 as a 4D rotation.*

Proof. Let $\boldsymbol{q} = q_0 + q_1\boldsymbol{i} + q_2\boldsymbol{j} + q_3\boldsymbol{k}$ be the quaternion representing the rotation to be applied. If points on S^3 are identified with quaternions, a "point" $q' = q'_0 + q_1\boldsymbol{i}' + q_2\boldsymbol{j}' + q_3\boldsymbol{k}'$ is mapped onto a "point" $\boldsymbol{q}'' = q''_0 + q_1\boldsymbol{i}'' + q_2\boldsymbol{j}'' + q_3\boldsymbol{k}''$ by the relation

$$\begin{aligned} & q''_0 + q''_1\boldsymbol{i} + q''_2\boldsymbol{j} + q''_3\boldsymbol{k} \\ & = (q_0 + q_1\boldsymbol{i} + q_2\boldsymbol{j} + q_3\boldsymbol{k})(q'_0 + q'_1\boldsymbol{i} + q'_2\boldsymbol{j} + q'_3\boldsymbol{k}) \ . \end{aligned} \tag{6.10.6}$$

This is a *linear mapping* in 4D space. In fact, if we calculate according to the rules (6.8.8), we find that

$$\begin{pmatrix} q_0'' \\ q_1'' \\ q_2'' \\ q_3'' \end{pmatrix} = \begin{pmatrix} q_0 & -q_1 & -q_2 & -q_3 \\ q_1 & q_0 & -q_3 & q_2 \\ q_2 & q_3 & q_0 & -q_1 \\ q_3 & -q_2 & q_1 & q_0 \end{pmatrix} \begin{pmatrix} q_0' \\ q_1' \\ q_2' \\ q_3' \end{pmatrix} . \tag{6.10.7}$$

This is a linear mapping that maps S^3 onto itself. Hence, the coefficient matrix is a 4D orthogonal matrix (see (6.2.1)). Moreover, the components of quaternion $\boldsymbol{q}$ can be smoothly changed into $q_0 = 1$, $q_1 = q_2 = q_3 = 0$, *reducing the mapping of* (6.10.7) *to the identity*. Hence, the determinant of the matrix must be 1. Consequently, the matrix is an element of $SO(4)$ (see Sect. 6.2.1).

If any point of a topological space is mapped onto any other point by some member of a group of transformations, the topological space is called a *homogeneous space* of the transformation group. We see:

Lemma 6.9. *The 3-sphere S^3 identified with $SU(2)$ is a homogeneous space of $SO(3)$.*

Proof. A point $\boldsymbol{q}$ is mapped onto a point $\boldsymbol{q}'$ by a 3D rotation represented by quaternion $\boldsymbol{q}'' = \boldsymbol{q}'\boldsymbol{q}^*$: $\boldsymbol{q}''\boldsymbol{q} = \boldsymbol{q}'\boldsymbol{q}^*\boldsymbol{q} = \boldsymbol{q}'$.

Thus, when a 3D rotation is applied, the 3-sphere S^3 "rotates rigidity" in the 4D space. Moreover, it can be rotated in any way, bringing any point on it to any position on it. From this, we can conclude:

Lemma 6.10. *The uniform distribution over the 3-sphere S^3 identified with $SU(2)$ defines an invariant measure of $SO(3)$.*

The position of a point P on a unit 3-sphere S^3 in the 4D $q_0q_1q_2q_3$-space centered at the origin can be specified by the 4D *spherical coordinates* $\{\theta_0, \theta_1, \theta_2\}$ defined as follows. First, let θ_0, $0 \leqq \theta_0 \leqq \pi$, be the angle from the q_0-axis of the line connecting the origin O and the point P. If we "cut" the 3-sphere S^3 by a 3D hyperplane perpendicular to the q_0-axis which passes through point P, the "cross section" $q_1^2 + q_2^2 + q_3^2 = \sin^2\theta_0$ is a 2-sphere of radius $\sin\theta_0$. Let (θ_1, θ_2) be the 3D spherical coordinates on this 2-sphere.[13] The 4D spherical coordinates $\{\theta_0, \theta_1, \theta_2\}$ are related to the $q_0q_1q_2q_3$-coordinates as follows:

$$\begin{aligned} q_0 &= \cos\theta_0 , \\ q_1 &= \sin\theta_0 \cos\theta_1 , \\ q_2 &= \sin\theta_0 \sin\theta_1 \cos\theta_2 , \\ q_3 &= \sin\theta_0 \sin\theta_1 \sin\theta_2 , \\ 0 \leqq \theta_0&, \theta_1 \leqq \pi, \qquad 0 \leqq \theta_2 < 2\pi . \end{aligned} \tag{6.10.8}$$

[13] By this recursive definition, we can similarly define the *nD spherical coordinates* $\{\theta_0, \ldots, \theta_{n-2}\}$. Let θ_0, $0 \leqq \theta_0 \leqq \pi$, be the angle from the first (or the last) coordinate axis, and let the remaining angles $\{\theta_1, \ldots, \theta_{n-2}\}$ be the $(n-1)$D spherical coordinates on the $(n-2)$-sphere of the cross-section $\theta_0 = \text{const}$.

In terms of the 4D spherical coordinates $\{\theta_0, \theta_1, \theta_3\}$, the invariant measure is obtained in the following form:

Proposition 6.14 (*Invariant measure for* $\{\theta_0, \theta_1, \theta_2\}$).

$$d\mu = \frac{1}{2\pi^2} \sin^2 \theta_0 \sin \theta_1 \, d\theta_0 \, d\theta_1 \, d\theta_2 \ , \quad 0 \leqq \theta_0, \theta_1 \leqq \pi \ , \quad 0 \leqq \theta_2 < 2\pi \ . \tag{6.10.9}$$

The coefficient in the above expression is chosen so that integration over S^3 is 1. We omit the details of calculation. The invariant measures in terms of other parameters are obtained from this by changes of variables. We list only the final results.

A 3D rotation is specified by the angle of rotation Ω around an axis, whose orientation can be specified in spherical coordinates (θ, φ). In terms of $\{\Omega, \theta, \varphi\}$, the invariant measure is given as follows:

Proposition 6.15 (*Invariant measure for* $\{\Omega, \theta, \varphi\}$).

$$d\mu = \frac{1}{2\pi^2} \sin^2 \frac{\Omega}{2} \sin \theta \, d\Omega \, d\theta \, d\varphi \ ,$$
$$0 \leqq \Omega < \pi \ , \quad 0 \leqq \theta \leqq \pi \ , \quad 0 \leqq \varphi < 2\pi \ . \tag{6.10.10}$$

The coefficient is chosen so that integration over the entire domain is 1. In terms of the Euler angles $\{\theta, \varphi, \psi\}$, the invariant measure is given as follows:

Proposition 6.16. *Invariant measure for* $\{\theta, \varphi, \psi\}$.

$$d\mu = \frac{1}{8\pi^2} \sin \theta \, d\theta \, d\varphi \, d\psi \ ,$$
$$0 \leqq \theta \leqq \pi \ , \quad 0 \leqq \varphi < 2\pi \ , \quad 0 \leqq \psi < 2\pi \ . \tag{6.10.11}$$

The coefficient is chosen so that integration over the entire domain is 1.

Exercises

6.1 Show that condition (6.2.4) is equivalent to (6.2.5).

6.2 For arbitrary n-dimensional vectors $\boldsymbol{a}$, $\boldsymbol{b}$ and matrix $\boldsymbol{A}$, prove that $(\boldsymbol{a}, \boldsymbol{A}\boldsymbol{b}) = (\boldsymbol{A}^{\mathrm{T}}\boldsymbol{a}, \boldsymbol{b})$.

6.3 Prove that for a real matrix, an eigenvector for a real eigenvalue can be chosen so that it consists of only real components.

6.4 Show that the length of the orthogonal projection of a vector $\boldsymbol{a}$ onto a line whose orientation is specified by unit vector $\boldsymbol{n}$ is given by $(\boldsymbol{a}, \boldsymbol{n})$ if $\boldsymbol{n}$ and the projection of $\boldsymbol{a}$ have the same orientation, and $-(\boldsymbol{a}, \boldsymbol{n})$ otherwise.

6.5 (a) Derive the Rodrigues formula (6.3.15) from (6.3.11–14). (b) From the Rodrigues formula (6.3.15), show that the rotation matrix $\boldsymbol{R}$ is given by (6.3.10).

6.6 Compute (6.4.3) by using (6.4.2), and derive (6.4.1).

6.7 Show that a point (X, Y, Z) on the sphere S^2 is projected by the stereographic projection onto the point (x, y) on R^2 given by (6.5.2). Also, show the inverse relation of (6.5.3).

6.8 (a) Derive the rule of composition of the Cayley–Klein parameters $\{\alpha, \beta, \gamma, \delta\}$ given by (6.5.6) by composing the first two transformations of (6.5.7) and comparing the result with the last one.
(b) Show that the inverse of the rotation specified by Cayley–Klein parameters $\{\alpha, \beta, \gamma, \delta\}$ is specified by Cayley–Klein parameters $\{\delta, -\beta, -\gamma, \alpha\}$.

6.9 Show that the linear fractional transformation (6.5.4) is rewritten in terms of the homogeneous coordinates of the complex number plane $\mathbb{C}$ in the form of (6.6.1).

6.10 For arbitrary square complex matrices $\boldsymbol{U}$, $\boldsymbol{V}$, prove that $(\boldsymbol{U}\boldsymbol{V})^\dagger = \boldsymbol{V}^\dagger \boldsymbol{V}^\dagger$.

6.11 Prove the approximation formula of the determinant

$$|\boldsymbol{I} + t\boldsymbol{A} + O(t^2)| = 1 + t\mathrm{Tr}\{\boldsymbol{A}\} + O(t^2) \ .$$

6.12 Show that the infinitesimal generators $\{\boldsymbol{A}_1, \boldsymbol{A}_2, \boldsymbol{A}_3\}$ of (6.6.17) satisfy the commutation relations (6.6.19).

6.13 Derive (6.6.23) by differentiating (6.5.2) and using (6.6.21).

6.14 (a) Derive (6.6.24) by taking an infinitesimal variation of the linear fractional transformation (6.5.4) together with (6.6.23).
(b) Show that the real and imaginary parts of (6.6.24) are given by (6.6.25).

6.15 Show that the infinitesimal generators $\{\boldsymbol{A}_1, \boldsymbol{A}_2, \boldsymbol{A}_3\}$ of (6.6.17) define the Casimir operator (6.6.26).

6.16 Derive the explicit 3D rotation matrix form of (6.7.12) by computing (6.7.13) element-wise.

6.17 Show from the relationship of (6.6.23) that (6.7.12) reduces to (6.7.16) if the rotation is infinitesimal.

6.18 (a) Prove (6.7.21) from the Jacobi identity (Sect. 3.2).
(b) Prove (6.7.23), which states that d_W is a linear space homomorphism from $su(2)$ into the group of linear transformations over $su(2)$.
(c) Prove (6.7.24), which, together with (6.7.23), states that d_W is a Lie algebra homomorphism from $su(2)$ into the group of linear transformations over $su(2)$.

6.19 Confirm that matrices $\boldsymbol{S}_1$, $\boldsymbol{S}_2$, $\boldsymbol{S}_3$ given by (6.8.4) satisfy the commutation relations (6.8.5).

6.20 Confirm that a quaternion $\boldsymbol{q}$ representing a 3D rotation and its conjugate $\boldsymbol{q}^*$ defined by (6.8.10) satisfy (6.8.11).

6.21 Derive the explicit 3D rotation matrix form of (6.8.12) by computing (6.8.15) with the rules given by (6.8.8).

6.22 From (6.6.23) and (6.8.1), show that the quaternion representing an infinitesimal rotation takes the form of (6.8.20).

6.23 Prove (6.8.24), i.e., show that quaternion $\boldsymbol{q}_n(\Omega)$ representing a rotation around axis $\boldsymbol{n}$ commutes with $\boldsymbol{n}$ if $\boldsymbol{n}$ is identified with a quaternion.

6.24 Show that the quaternion given by (6.8.25) satisfies the differential equation (6.8.11).

6.25 Prove (6.8.26) for a unit vector $\boldsymbol{n}$ by identifying it with a quaternion.

6.26 Prove Euler's formula $\exp(\mathrm{i}x) = \cos x + \mathrm{i}\sin x$ from the following definitions:

$$\mathrm{e}^{\mathrm{i}x} \equiv \sum_{k=0}^{\infty} \frac{1}{k!}(\mathrm{i}x)^k , \qquad \cos x \equiv \sum_{k=0}^{\infty} \frac{(-1)^k}{(2k)!} x^{2k} ,$$

$$\sin x \equiv \sum_{k=0}^{\infty} \frac{(-1)^k}{(2k+1)!} x^{2k+1} .$$

6.27 Derive (6.8.29) by computing (6.8.30) according to the rules of (6.8.8).

6.28 (a) Show that $(\boldsymbol{q}'\boldsymbol{q})^* = \boldsymbol{q}^*\boldsymbol{q}'^*$.

(b) Prove (6.8.32) relating the modulus and conjugate of the quaternion.

(c) Confirm (6.8.34), i.e., show that the inverse $\boldsymbol{q}^{-1}$ of a quaternion $\boldsymbol{q}$ is indeed given by (6.8.33).

6.29 (a) If two vectors $\boldsymbol{a} = (a_1, a_2, a_3)$, $\boldsymbol{b} = (b_1, b_2, b_3)$ are respectively identified with quaternions $\boldsymbol{a} = a_1\boldsymbol{i} + a_2\boldsymbol{j} + a_3\boldsymbol{k}$, $\boldsymbol{b} = b_1\boldsymbol{i} + b_2\boldsymbol{j} + b_3\boldsymbol{k}$, show that

$$\boldsymbol{ab} = -(\boldsymbol{a}, \boldsymbol{b}) + \boldsymbol{a} \times \boldsymbol{b} ,$$

where $(\boldsymbol{a}, \boldsymbol{b})$ is the inner product when $\boldsymbol{a}$, $\boldsymbol{b}$ are regarded as vectors, while $\boldsymbol{a} \times \boldsymbol{b}$ is obtained by first taking the vector product of $\boldsymbol{a}$, $\boldsymbol{b}$ regarded as vectors and then identifying the result with a quaternion.

(b) Let $\boldsymbol{q} = c + \boldsymbol{a}$, where c and $\boldsymbol{a}$ are respectively the scalar and vector parts of quaternion $\boldsymbol{q}$. Let $\boldsymbol{q}' = c' + \boldsymbol{a}'$ be another quaternion expressed similarly. Show that their product becomes

$$\boldsymbol{q}'\boldsymbol{q} = [c'c - (\boldsymbol{a}', \boldsymbol{a})] + (c'\boldsymbol{a} + c\boldsymbol{a}' + \boldsymbol{a}' \times \boldsymbol{a}) ,$$

where $(\boldsymbol{a}', \boldsymbol{a})$ and $\boldsymbol{a}' \times \boldsymbol{a}$ are interpreted as in (a).

(c) If the quaternion $\boldsymbol{q}$ in (6.8.15) is expressed as $\boldsymbol{q} = c + \boldsymbol{a}$, where c and $\boldsymbol{a}$ are respectively the scalar and vector parts of quaternion $\boldsymbol{q}$, show that (6.8.15) is written in the form

$$\boldsymbol{r}' = \boldsymbol{r} + 2c\boldsymbol{a} \times \boldsymbol{r} - 2(\boldsymbol{a} \times \boldsymbol{r}) \times \boldsymbol{a} ,$$

where $\boldsymbol{a}$, $\boldsymbol{r}$, $\boldsymbol{r}'$ are regarded as vectors.

6.30 Derive (6.10.7) by calculating (6.10.6) component-wise.

6.31 Show that integration of the invariant measure of (6.10.9) over the entire S^3 is 1.

6.32 Show that integration of the invariant measure of (6.10.10) over the entire domain of $\{\Omega, \theta, \varphi\}$ is 1.

6.33 Show that integration of the invariant measure of (6.10.11) over the entire domain of the Euler angles $\{\theta, \varphi, \psi\}$ is 1.

Part II
Principles of 3D Shape Recovery From Images

7. Shape from Motion

In Sect. 2.6, we analyzed the optical flow induced by orthographic projection of planar surface motion. We gave an analytical solution of the 3D recovery equations in terms of invariants constructed from image characteristics. These invariants correspond to irreducible representations of $SO(2)$—the group of rotations of the image coordinate system. We also discussed the geometrical meanings of these invariants. In the following, we give an analytical solution of the 3D recovery equations in terms of the invariants constructed in Sect. 2.6. We also study *adjacency conditions* of optical flow and their implications. Finally, we present a scheme of motion detection that does not require point-to-point correspondences between different image frames.

7.1 3D Recovery from Optical Flow for a Planar Surface

Reconstructing the 3D shape of an object moving in the scene from a sequence of its images is known as the problem of *shape from motion*. Since humans apparently have such an ability, this problem has long attracted psychologists in relation to human visual perception, and many psychological experiments have been conducted. Also, many attempts have been made to simulate this ability by means of a computer.

The first step is the *correspondence detection* to determine which point moves to which in different image frames. Since the object motion is usually small between consecutive image frames, detected pairs of corresponding points define a "flow" on the image plane. This flow is called *optical flow*, though many other terms are also used, and their precise meanings are different depending on the researchers (Sect. 2.6.1).

In this chapter, we analyze the optical flow induced by perspective projection of planar surface motion. The mathematical expression of the flow has already been given in Sect. 3.7. In the following, we will derive the 3D recovery equations and give an analytical solution. The analysis is almost parallel to the treatment in Sect. 2.6, where the 3D recovery equations were solved under orthographic projection. There, we first rewrote the 3D recovery equations in terms of *invariants* constructed from the flow parameters (the coefficients of the flow equations). These invariants were derived by *irreducible reduction* of the representation of $SO(2)$ (the group of rotations of the image coordinate system) defined by the flow parameters.

In the following, we will do exactly the same: we rewrite the 3D recovery equations in terms of invariants and then give an analytical solution. The solution is not unique. However, spurious solutions disappear if we observe two or more planar surfaces rigidly moving as a whole—for example, when we observe planar faces of the same rigid object in motion. We will derive a criterion to test whether two planar surfaces belong to the same object. Then, we show that the shape and motion of a rigid object of arbitrary shape is determined up to the absolute depth and a scale factor.

We also study how the formulation changes as the focal length f becomes larger. Taking the limit $f \to \infty$ means assuming orthographic projection, for which we studied the solution in detail in Sect. 2.6. In Sect. 7.5, we study the "transition" stage, which we call the *pseudo-orthographic approximation*, between the perspective and orthographic projections. One of the interesting results is that spurious solutions disappear in this approximation.

Finally, we present a simple scheme for detecting optical flow and object motion *without using point-to-point correspondences* between different image frames. The basic idea is the use of globally defined *image features* introduced in Chap. 5.

The analyses of this chapter are mathematically consistent, and from them we can derive many useful consequences. However, these analyses are not necessarily suitable as computer algorithms. For one thing, the shape-from-motion analysis has been known to be very sensitive to noise; if the detected optical flow is not very accurate, the computed solution may contain a large amount of error. This problem will be treated in Chap. 10, where we present an optimization scheme to cope with this noise sensitivity.

7.2 Flow Parameters and 3D Recovery Equations

In this chapter, we use the image-centered coordinate system (Sect. 1.2). Namely, the XY-plane is identified with the image plane, and a point in the scene is perspectively projected onto it from $(0, 0, -f)$ (Fig. 7.1). We use lowercase letters to indicate positions on the image plane. A point (X, Y, Z) in the scene is projected onto point (x, y) on the image plane according to the projection equations

$$x = \frac{fX}{f+Z}, \qquad y = \frac{fY}{f+Z}. \tag{7.2.1}$$

Consider a planar surface moving in the scene. Let

$$Z = pX + qY + r \tag{7.2.2}$$

be its equation. We call p, q, r the *surface parameters*. The pair (p, q) designates the *gradient* of the surface, while r designates the distance of the surface from the image plane along the Z-axis, which we call the *absolute depth*.

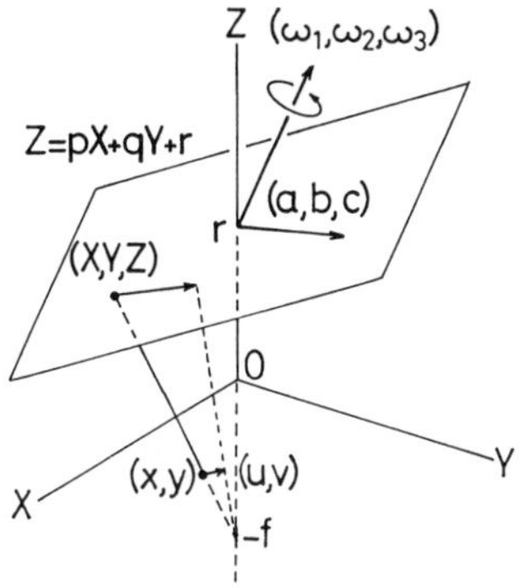

Fig. 7.1. A planar surface whose equation is $Z = pX + qY + r$ is moving with translation velocity (a, b, c) at $(0, 0, r)$ and rotation velocity $(\omega_1, \omega_2, \omega_3)$ around it. An optical flow is induced on the image plane by perspective projection from viewpoint $(0, 0, -f)$

An instantaneous rigid motion is specified by the velocity (the translation velocity) at an arbitrarily chosen reference point and the rotation velocity around it. As in Sect. 2.6, we take $(0, 0, r)$—the intersection of the Z-axis with the surface—as our reference point (Fig. 7.1). Let (a, b, c) be the translation velocity, and $(\omega_1, \omega_2, \omega_3)$ the rotation velocity. We call the six parameters $a, b, c, \omega_1, \omega_2, \omega_3$ the *motion parameters*.

The optical flow induced on the image plane by such a motion has already been given in Sect. 3.7 (Proposition 3.1), but there the viewer-centered coordinate system was used. In terms of the image-centered coordinate system, Proposition 3.1 and its proof are rewritten as follows.

Proposition 7.1 (*Flow equations*). *The optical flow induced by perspective projection of planar surface motion is given by*

$$u(x, y) = u_0 + Ax + By + (Ex + Fy)x \ ,$$

$$v(x, y) = v_0 + Cx + Dy + (Ex + Fy)y \ , \tag{7.2.3}$$

where

$$u_0 = \frac{fa}{f+r} \ , \qquad v_0 = \frac{fb}{f+r} \ ,$$

$$A = p\omega_2 - \frac{pa + c}{f+r} \ , \qquad B = q\omega_2 - \omega_3 - \frac{qa}{f+r} \ ,$$

$$C = -p\omega_1 + \omega_3 - \frac{pb}{f+r} \ , \quad D = -q\omega_1 - \frac{qb + c}{f+r} \ ,$$

$$E = \frac{1}{f}\left(\omega_2 + \frac{pc}{f+r}\right) \ , \quad F = \frac{1}{f}\left(-\omega_1 + \frac{qc}{f+r}\right) \ . \tag{7.2.4}$$

Proof. Combining the projection equations (7.2.1) and the surface equation (7.2.2), we can express the scene coordinates (X, Y, Z) of a point on the surface in terms of the image coordinates (x, y) of its projection as

$$X = \frac{(f+r)x}{f-px-qy}\ , \quad Y = \frac{(f+r)y}{f-px-qy}\ , \quad Z = \frac{f(px+qy+r)}{f-px-qy}\ . \tag{7.2.5}$$

According to our definition of the translation and rotation velocities, the velocity of point (X, Y, Z) in the scene is given by

$$\begin{pmatrix} \dot{X} \\ \dot{Y} \\ \dot{Z} \end{pmatrix} = \begin{pmatrix} a \\ b \\ c \end{pmatrix} + \begin{pmatrix} \omega_1 \\ \omega_2 \\ \omega_3 \end{pmatrix} \times \begin{pmatrix} X \\ Y \\ Z-r \end{pmatrix} . \tag{7.2.6}$$

Substituting the surface equation (7.2.2) into this, we obtain

$$\dot{X} = a + p\omega_2 X + (q\omega_2 - \omega_3)Y\ , \qquad \dot{Y} = b + (\omega_3 - p\omega_1)X - q\omega_1 Y\ ,$$
$$\dot{Z} = c - \omega_2 X + \omega_1 Y\ . \tag{7.2.7}$$

Differentiating both sides of the projection equation (7.2.1), we obtain the velocity of the image point as

$$\dot{x} = \frac{f\dot{X}}{f+Z} - \frac{fX\dot{Z}}{(f+Z)^2} = \frac{f\dot{X}}{f+Z} - \frac{x\dot{Z}}{f+Z}\ ,$$
$$\dot{y} = \frac{f\dot{Y}}{f+Z} - \frac{fY\dot{Z}}{(f+Z)^2} = \frac{f\dot{Y}}{f+Z} - \frac{x\dot{Y}}{f+Z}\ . \tag{7.2.8}$$

From (7.2.7) and the projection equations (7.2.1), we obtain

$$\frac{f\dot{X}}{f+Z} = \frac{fa}{f+Z} + p\omega_2 x + (q\omega_2 - \omega_3)y\ ,$$
$$\frac{f\dot{Y}}{f+Z} = \frac{fb}{f+Z} + (\omega_3 - p\omega_1)x - q\omega_1 y\ ,$$
$$\frac{f\dot{Z}}{f+Z} = \frac{fc}{f+Z} - \omega_2 x + \omega_1 y\ . \tag{7.2.9}$$

Substituting these into (7.2.8) and eliminating Z by (7.2.5c), we obtain the optical flow in the form of (7.2.3, 4).

We call the eight parameters u_0, v_0, A, B, C, D, E, F the *flow parameters*. Suppose the optical flow is detected at points (x_i, y_i), $i = 1, 2, \ldots$. Let us call these points *feature points*. Let (u_i, v_i), $i = 1, 2, \ldots$, be the velocities at these feature points. Then, the flow parameters u_0, v_0, A, B, C, D, E, F can be determined by fitting the flow equations (7.2.3). For example, we can use the least squares method, minimizing

$$M = \sum_i \{[u_0 + Ax_i + By_i + (Ex_i + Fy_i)x_i - u_i]^2$$
$$+ [v_0 + Cx_i + Dy_i + (Ex_i + Fy_i)y_i - v_i]^2\} , \quad (7.2.10)$$

where summation is taken over all the feature points. Since the flow parameters are determined from a given optical flow, they can be regarded as "image characteristics". Consequently, the left-hand sides of (7.2.4a–h) are all known quantities. Hence, they can be regarded as simultaneous equations to determine the "object parameters" (the surface parameters p, q, r and the motion parameters a, b, c, ω_1, ω_2, ω_3). Namely, (7.2.4a–h) are the "3D recovery equations".

Remark 7.1. By computing M of (7.2.10), which is called the *residual*, we obtain a *planarity condition*: If the resulting residual M is not less than a prescribed threshold value, the surface cannot be regarded as a plane. This also suggests the following procedure for segmenting the image of an object whose surface is not planar. Starting from a small number of feature points for which the residual M is very small, add feature points from their vicinity one by one, each time recomputing the flow parameters and checking the residual M, until it reaches a prescribed threshold value. Then, we end up with a region that is regarded as an image of a planar or almost planar part of the object. We call such a region a *planar patch*. If this procedure is repeated, the image domain is theoretically segmented into planar patches. Exact boundaries of these planar patches are not necessary. We will present the boundary reconstruction procedure in Sect. 7.6.

7.2.1 Least Squares Method

The *least squares method* is a well known and widely used scheme for estimating unknown parameters from observation. Suppose we observe m quantities $J_1, \ldots, J_m$, and suppose these quantities are known to be related to n parameters $\alpha_1, \ldots, \alpha_n$ in a known form. Let

$$J_j = F_j(\alpha_1, \ldots, \alpha_n) , \qquad j = 1, \ldots, m , \quad (7.2.11)$$

be those relations. The problem to be solved is: Given observed values β_j of quantities J_j, $j = 1, \ldots, m$, estimate the vaues of the unknown parameters α_i, $i = 1, \ldots, n$, for $n \leqq m$.

The least squares method computes the values $\alpha_1, \ldots, \alpha_n$ that minimize

$$M = \sum_{j=1}^{m} [F_j(\alpha_1, \ldots, \alpha_n) - \beta_j]^2 . \quad (7.2.12)$$

The solution is found by solving a set of simultaneous equations obtained by taking derivatives with respect to $\alpha_1, \ldots, \alpha_n$ and setting them to be zero:

$$\frac{\partial M}{\partial \alpha_1} = 0, \ldots, \frac{\partial M}{\partial \alpha_i} = 0 . \quad (7.2.13)$$

These equations are generally nonlinear, and computation of a numerical solution requires iterations. However, they become linear if (7.2.11) is linear in the unknown parameters. To see this, suppose

$$J_j = A_{j1}\alpha_1 + \ldots + A_{jn}\alpha_n , \qquad j = 1, \ldots, m . \tag{7.2.14}$$

Let us introduce the following matrix and vectors:

$$\boldsymbol{A} = \begin{pmatrix} A_{11} & \cdots\cdots & A_{1n} \\ & \cdots\cdots & \\ A_{m1} & \cdots\cdots & A_{mn} \end{pmatrix}, \quad \boldsymbol{a} = \begin{pmatrix} \alpha_1 \\ \vdots \\ \alpha_n \end{pmatrix}, \quad \boldsymbol{\beta} = \begin{pmatrix} \beta_1 \\ \vdots \\ \beta_m \end{pmatrix} . \tag{7.2.15}$$

Here, $\boldsymbol{A}$ is an $m \times n$ matrix, $\boldsymbol{a}$ is an n-dimensional vector, and $\boldsymbol{\beta}$ is an m-dimensional vector. Then, (7.2.12) is written as

$$M = \|\boldsymbol{A}\boldsymbol{a} - \boldsymbol{\beta}\|^2 , \tag{7.2.16}$$

where $\|.\|$ designates the (Euclidean) norm. Equations (7.2.13) are obtained by the *variational principle*: let $\boldsymbol{a} \to \boldsymbol{a} + \delta\boldsymbol{a}$ in (7.2.16) and set the *first variation* (or *derivative*) of M to be zero. The first variation δM of M is computed as

$$\delta M = \delta(\boldsymbol{A}\boldsymbol{a} - \boldsymbol{\beta}, \boldsymbol{A}\boldsymbol{a} - \boldsymbol{\beta}) = 2(\boldsymbol{A}\boldsymbol{a} - \boldsymbol{\beta}, \boldsymbol{A}\delta\boldsymbol{a}) = 2(\boldsymbol{A}^{\mathrm{T}}(\boldsymbol{A}\boldsymbol{a} - \boldsymbol{\beta}), \delta\boldsymbol{a}) . \tag{7.2.17}$$

Here, $(.,.)$ designates the (Euclidean) inner product. The first variation δM is zero for arbitrary variations $\delta\boldsymbol{a}$ of $\boldsymbol{a}$ if and only if

$$\boldsymbol{A}^{\mathrm{T}}\boldsymbol{A}\boldsymbol{a} = \boldsymbol{A}^{\mathrm{T}}\boldsymbol{\beta} . \tag{7.2.18}$$

This is called the *normal equation*. If the matrix $\boldsymbol{A}^{\mathrm{T}}\boldsymbol{A}$ is nonsingular, the solution is given by multiplying both sides by its inverse:

$$\boldsymbol{a} = \boldsymbol{A}^{-}\boldsymbol{\beta} , \qquad \boldsymbol{A}^{-} \equiv (\boldsymbol{A}^{\mathrm{T}}\boldsymbol{A})^{-1}\boldsymbol{A}^{\mathrm{T}} . \tag{7.2.19}$$

The $n \times m$ matrix $\boldsymbol{A}^{-}$ is called the *pseudo-inverse* (or *generalized inverse*) of $\boldsymbol{A}$. In more general terms, a pseudo-inverse of an $m \times n$ matrix $\boldsymbol{A}$ is an $n \times m$ matrix $\boldsymbol{A}^{-}$ that satisfies $\boldsymbol{A}\boldsymbol{A}^{-}\boldsymbol{A} = \boldsymbol{A}$. A pseudo-inverse of $\boldsymbol{A}$ can be defined even if matrix $\boldsymbol{A}^{\mathrm{T}}\boldsymbol{A}$ is singular. There are many ways to define a pseudo-inverse. However, if we want numerical values of $\alpha_1, \ldots, \alpha_n$, it is more practical to solve the normal equations (7.2.19) directly by some numerical scheme—the LU decomposition, Gaussian elimination, etc. (There also exist numerical schemes which do not require one to solve the normal equations. These methods are devised to avoid possible numerical instability when $\boldsymbol{A}^{\mathrm{T}}\boldsymbol{A}$ is ill-conditioned.)

Let $\hat{\boldsymbol{a}}$ be the solution of (7.2.18) and put $\hat{\boldsymbol{\beta}} = \boldsymbol{A}\,\hat{\boldsymbol{a}}$. Since $\boldsymbol{A}^{\mathrm{T}}\boldsymbol{A}\hat{\boldsymbol{a}} = \boldsymbol{A}^{\mathrm{T}}\boldsymbol{\beta}$,

$$\begin{aligned} M &= \|\boldsymbol{A}\hat{\boldsymbol{a}} - \boldsymbol{\beta}\|^2 = (\boldsymbol{A}\hat{\boldsymbol{a}} - \boldsymbol{\beta}, \boldsymbol{A}\hat{\boldsymbol{a}} - \boldsymbol{\beta}) = (\boldsymbol{A}\hat{\boldsymbol{a}} - \boldsymbol{\beta}, \boldsymbol{A}\hat{\boldsymbol{a}}) - (\boldsymbol{A}\hat{\boldsymbol{a}} - \boldsymbol{\beta}, \boldsymbol{\beta}) \\ &= (\boldsymbol{A}^{\mathrm{T}}(\boldsymbol{A}\hat{\boldsymbol{a}} - \boldsymbol{\beta}), \hat{\boldsymbol{a}}) - (\boldsymbol{A}\hat{\boldsymbol{a}}, \boldsymbol{\beta}) + (\boldsymbol{\beta}, \boldsymbol{\beta}) = \|\boldsymbol{\beta}\|^2 - (\hat{\boldsymbol{\beta}}, \boldsymbol{\beta}) . \end{aligned} \tag{7.2.20}$$

This quantity is called the *residual* and serves as a measure of how well the computed estimate fits.

Thus, the least squares method is very simple. However, there is one crucial problem. Suppose the quantities we observed are $W_j J_j$ instead of J_j, where W_j is a positive constant. Then, all observed values are simply scaled by positive weights. Let $W_j \beta_j$ be the observed values of quantities $W_j J_j$, $j = 1, \ldots, m$. According to the above procedure, we minimize

$$M = \sum_{j=1}^{m} W_j^2 [F_j(\alpha_1, \ldots, \alpha_n) - \beta_j]^2 \ , \tag{7.2.21}$$

but the solution is not necessarily the same as before. In other words, the least squares method is *not invariant to scaling*.

In real situations, the question of how to weigh individual equations is a very difficult problem. If errors are known to enter observations in the form

$$\beta_j = F_j(\alpha_1, \ldots, \alpha_n) + \varepsilon_j \ , \qquad j = 1, \ldots, m \ , \tag{7.2.22}$$

and errors ε_j can be regarded as mutually independent random variables with variances σ_j^2, the most reasonable choice is $W_j = 1/\sigma_j$. In particular, we need not assign any weights if all ε_j, $j = 1, \ldots, m$, can be regarded as identically and independently distributed random variables. For the flow equations (7.2.3), minimization of (7.2.10) is justified if errors in the image velocity components are distributed uniformly over the image plane and occur isotropically at each point.

7.3 Invariants of Optical Flow

Now, recall the argument we gave in Sect. 2.6. The flow equations (7.2.3) are obtained with reference to the image xy coordinate system, but the choice of the image coordinate system is arbitrary as long as the image origin corresponds to the camera optical axis. Suppose we use an $x'y'$ coordinate system obtained by rotating the original xy coordinate system around the image origin by angle θ clockwise. Since we are still observing the rigid motion of a planar surface, the flow equations must have the same form:

$$u' = u_0' + A'x' + B'y' + (E'x' + F'y')x' \ ,$$
$$v' = v_0' + C'x' + D'y' + (E'x' + F'y')y' \ , \tag{7.3.1}$$

Namely, the values of the coefficients may be different, but the *form* of the equation must be the same. From this fact, we can observe the following transformation rules of the flow parameters.

Proposition 7.2. *The pairs* $\{u_0, v_0\}$ *and* $\{E, F\}$ *are vectors, while parameters* $\{A, B, C, D\}$ *are a tensor:*

$$\begin{pmatrix} u_0' \\ v_0' \end{pmatrix} = \begin{pmatrix} \cos\theta & \sin\theta \\ -\sin\theta & \cos\theta \end{pmatrix} \begin{pmatrix} u_0 \\ v_0 \end{pmatrix}, \quad \begin{pmatrix} E' \\ F' \end{pmatrix} = \begin{pmatrix} \cos\theta & \sin\theta \\ -\sin\theta & \cos\theta \end{pmatrix} \begin{pmatrix} E \\ F \end{pmatrix}, \tag{7.3.2}$$

$$\begin{pmatrix} A' & B' \\ C' & D' \end{pmatrix} = \begin{pmatrix} \cos\theta & \sin\theta \\ -\sin\theta & \cos\theta \end{pmatrix} \begin{pmatrix} A & B \\ C & D \end{pmatrix} \begin{pmatrix} \cos\theta & -\sin\theta \\ \sin\theta & \cos\theta \end{pmatrix}. \tag{7.3.3}$$

Proof. Since the coordinates of a point are transformed as a (2D) vector under coordinate rotation, the original coordinates (x, y) and the new coordinates (x', y') are related by

$$\begin{pmatrix} x' \\ y' \end{pmatrix} = \begin{pmatrix} \cos\theta & \sin\theta \\ -\sin\theta & \cos\theta \end{pmatrix} \begin{pmatrix} x \\ y \end{pmatrix}. \tag{7.3.4}$$

Similarly, the original velocity components (u, v) and the new velocity components (u', v') are related by

$$\begin{pmatrix} u' \\ v' \end{pmatrix} = \begin{pmatrix} \cos\theta & \sin\theta \\ -\sin\theta & \cos\theta \end{pmatrix} \begin{pmatrix} u \\ v \end{pmatrix}. \tag{7.3.5}$$

Let us define vectors

$$\boldsymbol{u} = \begin{pmatrix} u \\ v \end{pmatrix}, \quad \boldsymbol{u}_0 = \begin{pmatrix} u_0 \\ v_0 \end{pmatrix}, \quad \boldsymbol{x} = \begin{pmatrix} x \\ y \end{pmatrix}, \quad \boldsymbol{k} = \begin{pmatrix} E \\ F \end{pmatrix}. \tag{7.3.6}$$

and matrices

$$\boldsymbol{A} = \begin{pmatrix} A & B \\ C & D \end{pmatrix}, \qquad \boldsymbol{R}(\theta) = \begin{pmatrix} \cos\theta & \sin\theta \\ -\sin\theta & \cos\theta \end{pmatrix}. \tag{7.3.7}$$

Then, the flow equations (7.2.3) and (7.3.1) are respectively rewritten as

$$\boldsymbol{u} = \boldsymbol{u}_0 + \boldsymbol{A}\boldsymbol{x} + (\boldsymbol{k}, \boldsymbol{x})\boldsymbol{x}, \qquad \boldsymbol{u}' = \boldsymbol{u}_0' + \boldsymbol{A}'\boldsymbol{x}' + (\boldsymbol{k}', \boldsymbol{x}')\boldsymbol{x}', \tag{7.3.8}$$

where (.,.) denotes the (Euclidean) inner product. Equations (7.3.4, 5) are rewritten as

$$\boldsymbol{x}' = \boldsymbol{R}(\theta)\boldsymbol{x}, \qquad \boldsymbol{u}' = \boldsymbol{R}(\theta)\boldsymbol{u}. \tag{7.3.9}$$

If we substitute (7.3.8a) into (7.3.9b) and use (7.3.9a) we obtain

$$\begin{aligned} \boldsymbol{u}' &= \boldsymbol{R}(\theta)[\boldsymbol{u}_0 + \boldsymbol{A}\boldsymbol{x} + (\boldsymbol{k}, \boldsymbol{x})\boldsymbol{x}] \\ &= \boldsymbol{R}(\theta)[\boldsymbol{u}_0 + \boldsymbol{A}\boldsymbol{R}(\theta)^{\mathrm{T}}\boldsymbol{x}' + (\boldsymbol{k}, \boldsymbol{R}(\theta)^{\mathrm{T}}\boldsymbol{x}')\boldsymbol{R}(\theta)^{\mathrm{T}}\boldsymbol{x}'] \\ &= \boldsymbol{R}(\theta)\boldsymbol{u}_0 + \boldsymbol{R}(\theta)\boldsymbol{A}\boldsymbol{R}(\theta)^{\mathrm{T}}\boldsymbol{x}' + (\boldsymbol{R}(\theta)\boldsymbol{k}, \boldsymbol{x}')\boldsymbol{x}', \end{aligned} \tag{7.3.10}$$

where we used the relations $\boldsymbol{R}(\theta)^{-1} = \boldsymbol{R}(\theta)^{\mathrm{T}}$ and $(\boldsymbol{R}(\theta)\boldsymbol{k}, \boldsymbol{x}') = (\boldsymbol{k}, \boldsymbol{R}(\theta)^{\mathrm{T}}\boldsymbol{x}')$. Comparison of (7.3.10) with the second of the flow equations (7.3.8) yields (7.3.2, 3).

Hence, as in Chap. 2, we can construct the following invariants:

$$T = A + D \ , \quad R = C - B \ , \quad U_0 = u_0 + \mathrm{i}v_0 \ , \quad K = E + \mathrm{i}F \ ,$$
$$S = (A - D) + \mathrm{i}(B + C) \ . \tag{7.3.11}$$

These quantities define one-dimensional irreducible representations of $SO(2)$. Namely, T and R are absolute invariants, U_0 and K are relative invariants of weight 1, and S is a relative invariant of weight 2:

$$T' = T \ , \quad R' = R \ , \quad U_0' = \mathrm{e}^{-\mathrm{i}\theta} U_0 \ ,$$
$$K' = \mathrm{e}^{-\mathrm{i}\theta} K \ , \ S' = \mathrm{e}^{-2\mathrm{i}\theta} S \ . \tag{7.3.12}$$

Since these invariants define irreducible representations, each of these is regarded as indicating a distinctive geometrical meaning (*Weyl's thesis*). As we saw in Sect. 2.6, U_0 represents *translation* (Fig. 7.2a), T *divergence* (Fig. 7.2b), R *rotation* (Fig. 7.2c), and S *shearing* (Fig. 7.2d). The new invariant K is characteristic of perspective projection; it represents *fanning* (or *foreshortening*) (Fig. 7.2e).

The transformation rules of parameters p, q, r, ω_1, ω_2, ω_3 have already been given in Sect. 2.6. Recall that whenever the image xy coordinate system is rotated around the image origin, the scene XYZ coordinate system is accordingly rotated around the Z-axis so as to preserve the projection equations (7.2.1). We reiterate Proposition 2.7 and its proof:

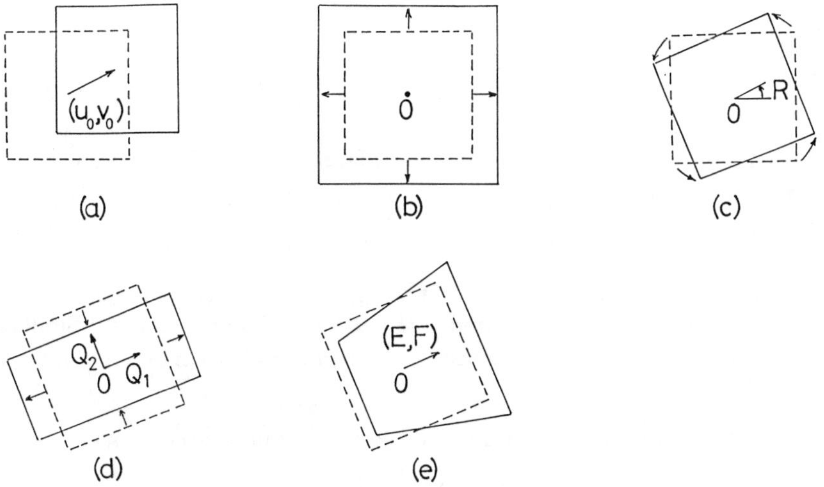

Fig. 7.2a. Translation by (u_0, v_0). **b** Divergence by $\boldsymbol{T}$. **c** Rotation by $\boldsymbol{R}$. **d** Shearing; Q_1 and Q_2 respectively indicate the axes of maximum expansion and maximum compression. **e** Fanning along (E, F)

Proposition 7.3.

(i) *The surface gradient* (p, q) *is a vector, while the absolute depth* r *is a scalar*:

$$\begin{pmatrix} p' \\ q' \end{pmatrix} = \begin{pmatrix} \cos\theta & \sin\theta \\ -\sin\theta & \cos\theta \end{pmatrix} \begin{pmatrix} p \\ q \end{pmatrix}, \qquad r' = r . \tag{7.3.13}$$

(ii) *The translation velocity components* $\{a, b\}$ *are a vector, while* c *is a scalar*:

$$\begin{pmatrix} a' \\ b' \end{pmatrix} = \begin{pmatrix} \cos\theta & \sin\theta \\ -\sin\theta & \cos\theta \end{pmatrix} \begin{pmatrix} a \\ b \end{pmatrix}, \qquad c' = c . \tag{7.3.14}$$

(iii) *The rotation velocity components* $\{\omega_1, \omega_2\}$ *are a vector, while* ω_3 *is a scalar*:

$$\begin{pmatrix} \omega'_1 \\ \omega'_2 \end{pmatrix} = \begin{pmatrix} \cos\theta & \sin\theta \\ -\sin\theta & \cos\theta \end{pmatrix} \begin{pmatrix} \omega_1 \\ \omega_2 \end{pmatrix}, \qquad \omega'_3 = \omega_3 . \tag{7.3.15}$$

Proof. (i) The surface equation (7.2.2) is written in vector form as

$$Z = r + (p\ q) \begin{pmatrix} X \\ Y \end{pmatrix} . \tag{7.3.16}$$

If the XYZ coordinate system is rotated around the Z-axis, the equation still has the same form:

$$Z' = r' + (p'\ q') \begin{pmatrix} X' \\ Y' \end{pmatrix} . \tag{7.3.17}$$

The old XYZ coordinate system and the new $X'Y'Z'$ coordinate system are related by

$$\begin{pmatrix} X \\ Y \end{pmatrix} = \begin{pmatrix} \cos\theta & -\sin\theta \\ \sin\theta & \cos\theta \end{pmatrix} \begin{pmatrix} X' \\ Y' \end{pmatrix}, \qquad Z = Z' . \tag{7.3.18}$$

On substitution of these equations, the surface equation (7.3.16) becomes

$$Z' = r + (p\ q) \begin{pmatrix} \cos\theta & -\sin\theta \\ \sin\theta & \cos\theta \end{pmatrix} \begin{pmatrix} X' \\ Y' \end{pmatrix} . \tag{7.3.19}$$

Comparison of this with (7.3.17) gives (7.3.13). (ii) Since $\{a, b, c\}$ are a 3D vector, the pair $\{a, b\}$ is a 2D vector, and c is a 2D scalar (Sect. 3.6). (iii) Similarly, $\{\omega_1, \omega_2, \omega_3\}$ are a 3D vector. Hence, the pair $\{\omega_1, \omega_2\}$ is a 2D vector, and ω_3 is a 2D scalar.

From this result, we find that r, c, ω_3 are absolute invariants, while

$$P = p + iq , \qquad V = a + ib , \qquad W = \omega_1 + i\omega_2 , \tag{7.3.20}$$

are relative invariants of weight 1:

$$P' = e^{-i\theta} P , \qquad V' = e^{-i\theta} V , \qquad W' = e^{-i\theta} W . \tag{7.3.21}$$

7.4 Analytical Solution of the 3D Recovery Equations

Now, we give the solution of the 3D recovery equations (7.2.4). First, we rewrite these equations in terms of the invariants. (The asterisk $*$ denotes the complex conjugate.)

Proposition 7.4 (3D *recovery equations*).

$$V = \frac{f+r}{f} U_0 \ , \quad PW'^* = (2\omega_3 - R) - \mathrm{i}(2c' + T) \ ,$$

$$PW' = \mathrm{i}S \ , \quad c'P - \mathrm{i}W' = L \ , \tag{7.4.1}$$

where

$$c' \equiv \frac{c}{f+r} \ , \quad W' \equiv W - \frac{\mathrm{i}}{f} U_0 \ , \quad L \equiv fK - \frac{1}{f} U_0 \ . \tag{7.4.2}$$

Proof. If (7.2.4) are substituted into (7.3.11), we obtain

$$T = p\omega_2 - q\omega_1 - \frac{pa + qb + 2c}{f+r} \ ,$$

$$R = -p\omega_1 - q\omega_2 + 2\omega_3 - \frac{pb - qa}{f+r} \ , \quad U_0 = \frac{f(a + \mathrm{i}b)}{f+r} \ ,$$

$$K = \frac{1}{f}\omega_2 + \frac{cp}{f(f+r)} + \mathrm{i}\left(-\frac{1}{f}\omega_1 \frac{pb + qa}{f+r} \right) \ ,$$

$$S = p\omega_2 + q\omega_1 - \frac{pa - qb}{f+r} + \mathrm{i}\left(q\omega_2 - p\omega_1 - \frac{pb + qa}{f+r} \right) \ . \tag{7.4.3}$$

If these equations are rewritten in terms of the complex expressions P, V, W of (7.3.20), then (7.4.3c) yields (7.4.1a), and (7.4.3a, b) are combined into one complex equation

$$R + \mathrm{i}T = -PW^* + 2\omega_3 - \frac{\mathrm{i}(PV^* + 2c)}{f+r} \ . \tag{7.4.4}$$

If (7.4.1a) is used, this equation becomes

$$P\left(W^* + \frac{\mathrm{i}}{f} U_0^* \right) = (2\omega_3 - R) - \mathrm{i}\left(\frac{2c}{f+r} + T \right) \ , \tag{7.4.5}$$

from which follows (7.4.1b). In terms of P, V, W, (7.4.3e) becomes

$$S = -\mathrm{i}PW - \frac{PV}{f+r} \ . \tag{7.4.6}$$

If (7.4.1a) is used, this equation becomes

$$P\left(W - \frac{\mathrm{i}}{f}U_0\right) = \mathrm{i}S \ , \tag{7.4.7}$$

from which follows (7.4.1c). Finally, (7.4.3d) is rewritten as

$$K = -\frac{\mathrm{i}}{f}W + \frac{cP}{f(f+r)} \ , \tag{7.4.8}$$

from which follows (7.4.1d).

Remark 7.2. The 3D recovery equations (7.2.4) consist of eight real equations, which have now been equivalently reduced to four complex equations in unknowns P, r, V, c', W', ω_3. The weight of the product of two invariants is the sum of the weights of the invariants, and the weight of the complex conjugate of an invariant is the negative of its weight. In (7.4.1), we can confirm that *only terms of the same weight are added or subtracted.* In particular, *the weights of both sides of each equation are the same.* Namely, if the image xy coordinate system is rotated clockwise by angle θ, both sides are multiplied by $\exp(-\mathrm{i}n\theta)$ with the same weight n.

In order to solve the 3D recovery equations (7.4.1), we must first check whether or not $c' = 0$.

Lemma 7.1. *The case* $c' = 0$ *occurs if and only if*

$$\mathrm{Re}\{Se^{-2\mathrm{i}\alpha}\} = T \ , \qquad \alpha \equiv \arg(L) \ . \tag{7.4.9}$$

In this case, the solution of the 3D recovery equations (7.4.1) *is given by*

$$V = \frac{f+r}{f}U_0 \ , \qquad P = \frac{S}{L} \ , \qquad W = \mathrm{i}fK \ ,$$

$$\omega_3 = \frac{1}{2}(R + \mathrm{Im}\{Se^{-2\mathrm{i}\alpha}\}) \ . \tag{7.4.10}$$

Proof. Equation (7.4.10a) is the same as (7.4.1a). If $c' = 0$, we have $W' = \mathrm{i}L$ from (7.4.1d). Then, we obtain (7.4.10b, c) from (7.4.1c) and (7.4.2b, c). By taking the real and imaginary parts of both sides, we rewrite (7.4.1b) equivalently as

$$\omega_3 = \frac{1}{2}(R + \mathrm{Re}\{PW'^*\}) \ , \qquad \mathrm{Im}\{PW'^*\} = -\mathrm{i}T \ . \tag{7.4.11}$$

Since $W' = \mathrm{i}L$, we see that

$$PW'^* = \frac{S}{L}(-\mathrm{i}L^*) = -\mathrm{i}Se^{-2\mathrm{i}\alpha} \ . \tag{7.4.12}$$

Hence, (7.4.11a) yields (7.4.10d). The only remaining condition to be satisfied is (7.4.11b), namely (7.4.9).

Remark 7.3. Note that the solution given by (7.4.10) is indeterminate when $T=0,\ S=0,\ L=0$. We can see that the condition $T=0,\ S=0,\ L=0$ is equivalent to the vanishing of $B=(b_{ij})$, $i, j = 1, 2, 3$, defined by (3.7.28). This means that the *tensor part* of the optical flow vanishes. Thus, this ambiguity occurs when and only when the surface is "orbiting" around the viewpoint with the relative configuration fixed (Sect. 3.7.1).

If $c' \neq 0$,[1] the solution of the 3D recovery equations (7.4.1) is given as follows:

Theorem 7.1 (*Solution of the 3D recovery equations*). *If $c' \neq 0$, the cubic equation*

$$X^3 + TX^2 + \frac{1}{4}(T^2 - |S|^2 - |L|^2)X + \frac{1}{8}(\mathrm{Re}\{L^2 S\} - T|L|^2) = 0 \tag{7.4.13}$$

has three real roots. Let c' be the middle one. The object parameters are given as[2]

$$V = \frac{f+r}{f}U_0\ , \quad c = (f+r)c'\ , \quad P = \frac{1}{2c'}(L \pm \sqrt{L^2 - 4c'S})\ ,$$

$$W = \frac{\mathrm{i}}{2}(L \mp \sqrt{L^2 - 4c'S}) + \frac{\mathrm{i}}{f}U_0\ , \quad \omega_3 = \frac{1}{2}R \pm \mathrm{Im}\,\{L^* \sqrt{L^2 - 4c'S}\}\ . \tag{7.4.14}$$

Hence, (i) *the absolute depth r is indeterminate,* (ii) $a/(f+r)$, $b/(f+r)$, $c/(f+r)$ *are uniquely determined,*[3] *and* (iii) *two sets of solutions exist for $p, q, \omega_1, \omega_2, \omega_3$.*

Proof. Equations (7.4.14a, b) are the same as (7.4.1a, 2a). If $c' \neq 0$, then (7.4.1c) is rewritten as $(c'P)(-\mathrm{i}W') = c'S$. Consequently, (7.4.1c, d) imply that $c'P$ and $-\mathrm{i}W'$ are the two roots of the quadratic equation

$$X^2 - LX + c'S = 0\ . \tag{7.4.15}$$

Hence, P and W' are given as functions of c' in the form

$$P(c') = \frac{1}{2c'}(L \pm \sqrt{L^2 - 4c'S})\ ,$$

$$W'(c') = \frac{\mathrm{i}}{2}(L \mp \sqrt{L^2 - 4c'S})\ . \tag{7.4.16}$$

---

[1] Since measurements in real circumstances always contain errors, the equality of the criterion of (7.4.9) must be given some allowance for error.

[2] One particular branch is chosen for the complex square root; the square root of a complex argument has two values, and we can choose either one arbitrarily, but we must use the same one for all three square roots in (7.4.14).

[3] To be very strict, *one* of $a, b, c, \omega_1, \omega_2, \omega_3$ is indeterminate. For example, we could alternatively assert that a is indeterminate while $r = f(a/u_0 - 1)$, b/a, c/a are uniquely determined.

Taking the real and imaginary parts of both sides of (7.4.1b), we obtain

$$\omega_3 = \frac{1}{2}[R + \mathrm{Re}\,\{P(c')W'(c')^*\}] \ , \ c' = -\frac{1}{2}[T + \mathrm{Im}\,\{P(c')W'(c')^*\}] \ . \tag{7.4.17}$$

Equation (7.4.14e) is obtained by substituting (7.4.16) into (7.4.17a). The equation to determine c' is obtained by substituting (7.4.16) into (7.4.17b):

$$\sqrt{16|S|^2c'^2 - 8\mathrm{Re}\,(L^2S^*)c' + |L|^4} = 8c'^2 - 4Tc' + |L|^2 \ . \tag{7.4.18}$$

The left-hand side of (7.4.18) is a smooth concave function (or constant if $S = 0$) passing through $(0, |L|^2)$ (Fig. 7.3). In view of the assumption $c' \neq 0$, we can see from Fig. 7.3 that there exists a single unique nonzero solution c'. Taking the squares of both sides and dropping off c' from both sides, we obtain (7.4.13). From Fig. 7.3, we can easily see that (7.4.13) has three real roots and that the middle one is the desired root; the other roots were introduced when both sides were squared.

Corollary 7.1 (*Bounds of c' and ω_3*).

$$-\frac{1}{2}(|S| + T) \leqq c' \leqq \frac{1}{2}(|S| - T) \ , \quad \frac{1}{2}(R - |S|) \leqq \omega_3 \leqq \frac{1}{2}(R + |S|) \ . \tag{7.4.19}$$

Proof. Since $|PW'^*| = |PW'|$, we see from (7.4.1b, c) that

$$(2\omega_3 - R)^2 + (2c' + T)^2 = |S|^2 \ . \tag{7.4.20}$$

This means that point (ω_3, c') lies on the circle of center $(R/2, -T/2)$ and of radius $|S|/2$ in the ω_3c'-plane.

Example 7.1. Consider the optical flow of Fig. 7.4. The focal length is $f = 2$. The flow parameters are: $u_0 = -0.04$, $v_0 = 0.04$, $A = -0.068$, $B = -0.196$, $C = 0.142$, $D = -0.079$, $E = 0.059$, $F = -0.054$. The invariants are: $U_0 = -0.04 + 0.04\mathrm{i}$, $T = -0.146$, $R = 0.338$, $S = 0.011 - 0.054\mathrm{i}$, $K = 0.059 - 0.054\mathrm{i}$. Since $|T| > |S| = 0.055$, the flow cannot be regarded as an ortho-

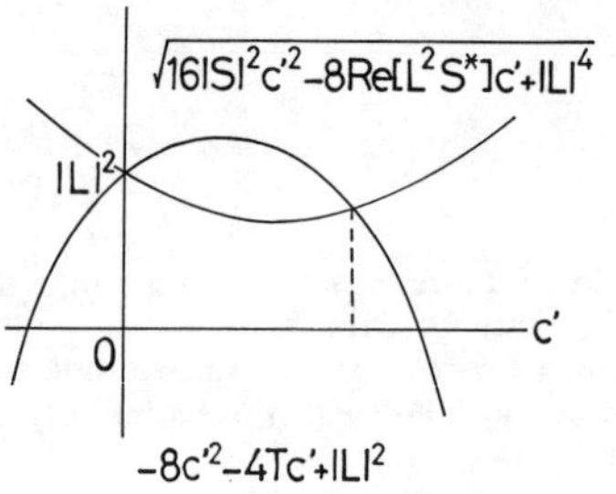

Fig. 7.3. Existence and uniqueness of nonzero c'

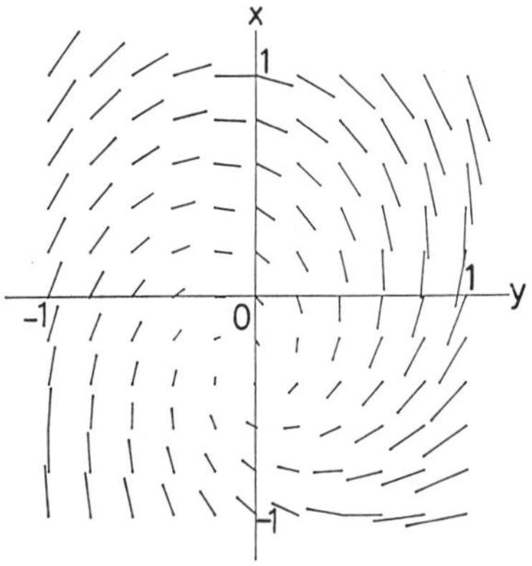

Fig. 7.4. An example of optical flow

graphic projection of a moving plane (Corollary 2.1). Applying Theorem 7.1, we obtain $a/(r+2) = -0.02$, $b/(r+2) = 0.02$, $c/(r+2) = 0.10$, and two sets of solutions: (i) $(p, q) = (0.300, -0.200)$, $(\omega_1, \omega_2, \omega_3) = (5.00, 5.00, 10.00)$ deg/s, (ii) $(p, q) = (1.073, -1.073)$, $(\omega_1, \omega_2, \omega_3) = (0.00, 0.57, 9.39)$ deg/s. The absolute depth r is indeterminate.

7.5 Pseudo-orthographic Approximation

Since the image-centered XYZ coordinate system is used, orthographic projection is realized in the limit of $f \to \infty$. If we take this orthographic limit, the 3D recovery equations (7.2.4) reduce to

$$u_0 = a\,, \quad v_0 = b\,, \qquad A = p\omega_2\,, \qquad B = q\omega_2 - \omega_3\,,$$
$$C = -p\omega_1 + \omega_3\,, \qquad D = -q\omega_1\,. \tag{7.5.1}$$

In terms of the invariants, the 3D recovery equations (7.4.1) reduce to

$$U_0 = V\,, \quad pW^* = 2\omega_3 - (R + \mathrm{i}T)\,, \quad PW = \mathrm{i}S\,. \tag{7.5.2}$$

These equations are exactly those we studied in Sect. 2.6. There, we obtained the solution in analytical form (Theorem 2.3). We reiterate the result.

Theorem 7.2 (*Orthographic solution*).

$$\omega_3 = \frac{1}{2}(R \pm \sqrt{SS^* - T^2})\,,$$
$$W = k\exp\left[\mathrm{i}\left(\frac{\pi}{4} + \frac{1}{2}\arg(S) - \frac{1}{2}\arg((2\omega_3 - R) - \mathrm{i}T)\right)\right],$$
$$P = \frac{S}{k}\exp\left[\mathrm{i}\left(\frac{\pi}{4} - \frac{1}{2}\arg(S) + \frac{1}{2}\arg((2\omega_3 - R) - \mathrm{i}T)\right)\right], \tag{7.5.3}$$

where k is an indeterminate parameter. For each k, there exist two solutions of p, q, $\omega_1, \omega_2, \omega_3$.

Remark 7.4. If we compare this solution with Theorem 7.1, we notice that *the two solutions are not analytically connected.* In other words, the perspective solution of Theorem 7.1 does not reduce to the orthographic solution of Theorem 7.2 in the orthographic limit $f \to \infty$; expressions in (7.4.13) take the form of $\infty - \infty$.

Suppose the focal length f is very large, but not large enough for the projection to be regarded as orthographic. To describe this transient stage, we omit, in the 3D recovery equations, terms of $O(1/f^2)$ but retain terms of $O(1/f)$. We call this the *pseudo-orthographic approximation.* From (7.2.4), we obtain

$$u_0 = \frac{fa}{f+r}, \quad v_0 = \frac{fb}{f+r}, \quad A = p\omega_2 - \frac{pa+c}{f+r},$$

$$B = q\omega_2 - \omega_3 - \frac{qa}{f+r}, \quad C = -p\omega_1 + \omega_3 - \frac{pb}{f+r},$$

$$D = -q\omega_1 - \frac{qb+c}{f+r}, \quad E = \frac{1}{f}\omega_2, \quad F = -\frac{1}{f}\omega_1, \tag{7.5.4}$$

or in terms of the invariants

$$V = \frac{f+r}{f}U_0, \quad PW'^* = (2\omega_3 - R) - \mathrm{i}(2c' + T),$$

$$PW' = \mathrm{i}S, \quad K = -\frac{\mathrm{i}}{f}W. \tag{7.5.5}$$

The analytical solution is given as follows:

Theorem 7.3 (*Pseudo-orthographic solution*).

$$V = \frac{f+r}{f}U_0, \quad P = \frac{S}{L}, \quad W = \mathrm{i}fK,$$

$$\omega_3 = \frac{1}{2}(R + \mathrm{Im}\,\{S\mathrm{e}^{-2\mathrm{i}\alpha}\}), \quad c = \frac{f+r}{2}(T - \mathrm{Re}\,\{S\mathrm{e}^{-2\mathrm{i}\alpha}\}). \tag{7.5.6}$$

Hence, (i) *the absolute depth r is indeterminate,* (ii) $a/(f+r)$, $b/(f+r)$, $c/(f+r)$ *are uniquely determined, and* (iii) p, q, ω_1, ω_2, ω_3 *are uniquely determined. In particular, no spurious solution exists.*

Proof. Equation (7.5.6a) is the same as (7.5.5a). Equation (7.5.6c) is immediately obtained by (7.5.5d). If we recall the definitions of W' and L (7.4.2b, c), we obtain

$$W' = W - \frac{\mathrm{i}}{f}U_0 = \mathrm{i}fK - \frac{\mathrm{i}}{f}U_0 = \mathrm{i}\left(fK - \frac{1}{f}U_0\right) = \mathrm{i}L. \tag{7.5.7}$$

From this and (7.5.5c) follows (7.5.6b). Taking the real and imaginary parts of both sides of (7.5.6b), we obtain

$$\omega_3 = \frac{1}{2}(R + \mathrm{Re}\,\{PW'^*\})\;, \qquad c' = -\frac{1}{2}(T + \mathrm{Im}\,\{PW'^*\})\;, \qquad (7.5.8)$$

but from (7.5.7 and 6b), we see that

$$PW'^* = \frac{S}{L}(-\mathrm{i}L^*) = -\mathrm{i}Se^{-2\mathrm{i}\alpha}\;. \qquad (7.5.9)$$

(Recall the definition $\alpha \equiv \arg(L)$.) From this and (7.5.8) follow (7.5.6d, e).

Example 7.2. Consider the optical flow of Fig. 7.4 again. If we apply the pseudo-orthographic approximation, the solution is unique: $a/(r+2) = -0.02$, $b/(r+2) = 0.02$, $c/(r+2) = 0.10$, $(p, q) = (0.238, -0.171)$, $(\omega_1, \omega_2, \omega_3) = (6.19, 6.76, 9.88)$ deg/s.

Remark 7.5. Let us consider the geometrical interpretation of this pseudo-orthographic approximation. From (7.4.8), we see that the fanning (Fig. 7.2e) is primarily caused by rotation W around an axis parallel to the image plane. However, it is also caused by the velocity c along the Z-axis when the surface gradient is large or the surface is very close to the viewer. Intuitively speaking, the translation along the line of sight mimics the visual effect of rotation.[4] The pseudo-orthographic approximation removes the spurious solution by assuming that the surface is located far away and the projective distortion is not very large.

As we saw in the preceding section, there exist two solutions if the projection is perspective: one is true and the other spurious. It seems that the two solutions cannot be distinguished, because the two solutions predict exactly the same optical flow. However, the fact that the pseudo-orthographic approximation predicts a unique solution provides us with a way to distinguish them. For example, compare the results of Example 7.1 and Example 7.2. The pseudo-orthographic solution of Example 7.2 is close to the first of the perspective solutions of Example 7.1. Hence, we can conclude that the first solution of Example 7.1 is the true one if the surface is located not very close to the viewer and the projective distortion is not very large.

The *visual ambiguity* that two different planar surface motions are capable of producing an identical optical flow is well known among psychologists. Usually, however, humans do not suffer from this kind of illusion. From this fact, it can be conjectured that human visual perception employs some recognition mech-

---

[4] This does not happen if $c = 0$. In this case, the perspective solution (Lemma 7.1) coincides with the pseudo-orthographic solution (Theorem 7.3). Compare (7.4.10) and (7.5.6).

anism close to the pseudo-orthographic approximation rather than the full perspective solution. However, nothing definite can be said at present.

7.5.1 Robustness of Computation

In Remark 7.5 we noted the possibility that human visual perception may not correspond to the perspective solution of the optical flow. More support for this conjecture comes from the fact that the perspective solution given by Theorem 7.1 is known to be computationally vulnerable to noise. Although we do not give a precise error sensitivity analysis, let us try a simple estimation of error behavior.

We assumed that the flow parameters u_0, v_0, A, B, C, D, E, F are obtained by applying a fitting scheme to the observed optical flow. In the flow equations (7.2.3), parameters u_0, v_0 are the constant terms and parameters A, B, C, D are coefficients of the first-order terms in x, y, while parameters E, F are coefficients of the second-order terms. Hence, estimation of parameters E, F may be more sensitive to image noise than the other flow parameters.

Suppose the estimates of E, F contain numerical error, or in terms of the invariants, K contains error. From the 3D recovery equations (7.4.1, 2), we see that numerical error enters through the value of L. If we look at the solution of (7.4.13, 14), we find that L enters the solution in a very complicated manner. In particular, the value of c' may be greatly influenced by the error in L.

On the other hand, consider the pseudo-orthographic solution (7.5.6). The third equation states that the error in W is proportional to the error in K. The second equation contains L in its denominator. Hence, the error in P is not great unless L is close to zero or S is very large. The last two equations show that error enters through the value of $\alpha = \arg(L)$, which is stable unless L is close to zero. Since $\exp(-2i\alpha)$ is multiplied by S, the errors in ω_3 and c are not great unless S is very large or L is close to zero. Thus, we conclude that the pseudo-orthographic solution is robust against the errors in E, F unless S is very large or L is close to zero. From the second of (7.5.6), we see that this undesirable case corresponds to a surface with a very large surface gradient P, for which instability is intuitively very easy to understand. Thus, the pseudo-orthographic solution is robust against noise unless the surface gradient is very large.

7.6 Adjacency Condition of Optical Flow

So far, we have considered a single planar surface. Now, let us consider a rigid object that has multiple planar faces or whose surface can be approximated by multiple planar faces. This means that the image consists of multiple planar patches.

First, note that if we put

$$U = u + iv\ , \qquad Z = x + iy\ , \tag{7.6.1}$$

the x- and y-components of the flow equations (7.2.3) can be combined into a single complex equation in complex variables in the following form:

Lemma 7.2 (*Computer flow equations*).

$$U = U_0 + \frac{1}{2}[(T + iR)Z + SZ^* + KZZ^* + K^*Z^2] \ . \tag{7.6.2}$$

Suppose the flow equation (7.6.2) holds over one planar patch. Let

$$U' = U'_0 + \frac{1}{2}[(T' + iR')Z + S'Z^* + K'ZZ^* + K'^*Z^2] \tag{7.6.3}$$

be the optical flow over another planar patch. Here, we introduce a new notation: if J is a quantity that characterizes the first patch and J' is the corresponding quantity for the second, their difference is expressed by double square brackets:

$$[\![J]\!] \equiv J' - J \ . \tag{7.6.4}$$

If we use this notation, the difference of (7.6.2) and (7.6.3) is

$$[\![U]\!] = [\![U_0]\!] + \frac{1}{2}([\![T + iR]\!]Z + [\![S]\!]Z^* + [\![K]\!]ZZ^* + [\![K]\!]^*Z^2) \ . \tag{7.6.5}$$

Suppose the two planar patches belong to the same object. Then, the two optical flows must be continuous over the intersection line (Fig. 7.5). To be precise, what we mean is the "image" of the intersection line of the two corresponding planar faces of the object. The two faces need not actually meet, nor need the intersection line be visible on the image plane. Hence, at any point (x, y) on the intersection line, which may or may not appear on the image plane, the relation $[\![U]\!] = 0$ must be satisfied. From (7.6.5), we obtain

$$([\![K]\!]Z + [\![S]\!])Z^* + ([\![K]\!]^*Z^2 + [\![T + iR]\!]Z + 2[\![U_0]\!]) = 0 \ . \tag{7.6.6}$$

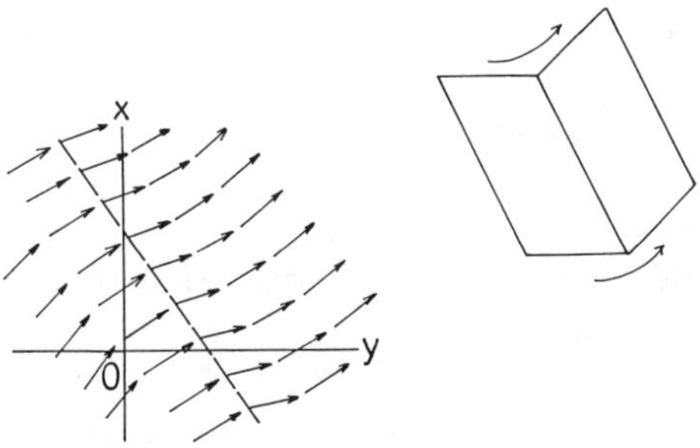

Fig. 7.5. If two adjacent planar surfaces are in motion, the resulting optical flow must be continuous across the straight line that separates the two regions

This equation holds at all points (x, y) on the intersection line. Hence, this is the *equation of the intersection line*, if it exists.

We say that two optical flows over different regions are *linearly adjacent* (or simply *adjacent*) if there exists a straight line across which the two flows, extended if necessary, are continuous. Two optical flows must be linearly adjacent if they result from two planar faces of the same object. However, these two planar faces need not be rigidly connected; if two planar faces are flexibly "hinged together" and the angle between them changes in the course of motion, they still produce two adjacent optical flows on the image plane.

Let us consider the cases of $[\![K]\!] \neq 0$ and $[\![K]\!] = 0$ separately.

Proposition 7.5. *If* $[\![K]\!] \neq 0$, *two optical flows are linearly adjacent if and only if*

$$[\![K]\!]^* [\![S]\!]^2 - [\![T + iR]\!][\![S]\!][\![K]\!] + 2[\![U_0]\!][\![K]\!]^2 = 0 \ ,$$
$$\arg([\![U_0]\!]) + \arg([\![K]\!]) = \arg([\![S]\!]) (\bmod \pi) \ . \tag{7.6.7}$$

If this condition is satisfied, their intersection line is given by

$$[\![E]\!]x + [\![F]\!]y + \frac{[\![U_0]\!][\![K]\!]}{[\![S]\!]} = 0 \ . \tag{7.6.8}$$

Proof. By definition, two optical flows are linearly adjacent if and only if (7.6.6) defines a straight line. Since (7.6.6) is quadratic in Z, it must be factored into two linear equations if it defines a line. This is possible if and only if the quadratic polynomial $[\![K]\!]^* Z^2 + [\![T + iR]\!] Z + 2[\![U_0]\!]$ in Z is divisible by $[\![K]\!] Z + [\![S]\!]$. By the *remainder theorem* (Exercise 7.11), the necessary and sufficient condition for that is

$$[\![K]\!]^* [\![S]\!]^2 - [\![T + iR]\!][\![S]\!][\![K]\!] + 2[\![U_0]\!][\![K]\!]^2 = 0 \ . \tag{7.6.9}$$

If this is satisfied, (7.6.6) is factored into

$$([\![K]\!]Z + [\![S]\!])\left(Z^* + \frac{[\![K]\!]^*}{[\![K]\!]} Z + 2\frac{[\![U_0]\!]}{[\![S]\!]}\right) = 0 \ . \tag{7.6.10}$$

Since $[\![K]\!]Z + [\![S]\!] = 0$ describes one point, the intersection line must be

$$Z^* + \frac{[\![K]\!]^*}{[\![K]\!]} Z + 2\frac{[\![U_0]\!]}{[\![S]\!]} = 0 \ . \tag{7.6.11}$$

In terms of x and y, this equation becomes (7.6.8), which is an equation of a line if and only if $[\![U_0]\!][\![K]\!]/[\![S]\!]$ is a real number. The necessary and sufficient condition for it is given by (7.6.7).

Proposition 7.6. *If* $[\![K]\!] = 0$, *two optical flows are linearly adjacent if and only if*

$$[\![u_0]\!] : [\![v_0]\!] = [\![A]\!] : [\![C]\!] = [\![B]\!] : [\![D]\!] \ . \tag{7.6.12}$$

If this condition is satisfied, their intersection line is given by

$$[\![u_0]\!] + [\![A]\!]\,x + [\![B]\!]\,y = 0 \ . \tag{7.6.13}$$

Proof. If $[\![K]\!] = 0$, (7.6.6) is linear in Z. In terms of x and y, it is rewritten as

$$([\![u_0]\!] + \mathrm{i}\,[\![v_0]\!]) + ([\![A]\!] + \mathrm{i}[\![C]\!])x + ([\![B]\!] + \mathrm{i}\,[\![D]\!])y = 0 \ . \tag{7.6.14}$$

This is an equation of a line if and only if the three coefficients have a real ratio, which is equivalent to requiring that their arguments be equal modulo π. Hence, we obtain (7.6.12). If (7.6.12) holds, (7.6.14) is equivalent to (7.6.13).[5]

Example 7.3. Consider the flow of Fig. 7.6. The focal length is $f = 2$. The flow as a whole does not satisfy the planarity condition stated in Remark 7.1. Hence, it cannot be regarded as a single flow. Suppose we estimate the flow parameters for the upper right part and the lower left part separately. For the upper right part, $u_0 = -0.061$, $v_0 = 0.126$, $A = 0.003$, $B = -0.134$, $C = 0.056$, $D = -0.148$, $E = 0.112$, $F = -0.077$, and for the lower left part $u_0' = -0.097$, $v_0' = 0.167$, $A' = -0.176$, $B' = -0.264$, $C' = 0.252$, $D' = -0.006$, $E' = 0.071$, $F' = -0.109$. In this case, $[\![K]\!] \neq 0$, and the adjacency condition (7.6.7) is satisfied within the rounding error. The intersection line given by (7.6.8) is $y = -1.30x - 0.27$, which is indicated by a broken line in Fig. 7.6.

Remark 7.6. Two adjacent optical flows need not be "physically" adjacent to each other; the domains over which these flows are defined can be far apart on the image plane. The intersection line also need not physically appear on the

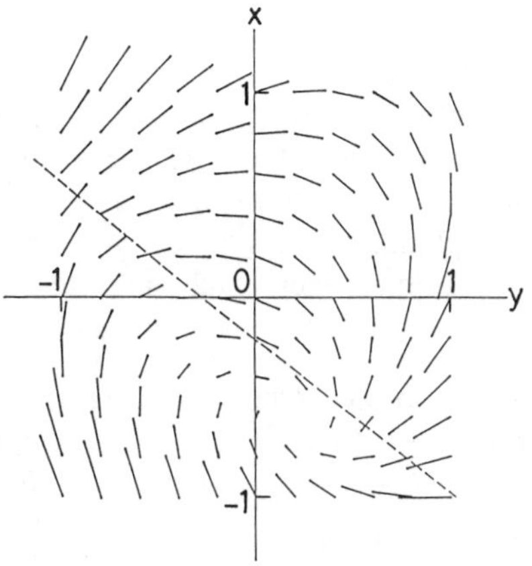

Fig. 7.6. This flow cannot be regarded as a single flow of a planar patch. There exist two planar patches, and the dashed line indicates the intersection line computed from the flow

[5] If (7.6.12) is satisfied, (7.6.13) can be replaced equivalently by $[\![v_0]\!] + [\![C]\!]\,x + [\![D]\!]\,y = 0$.

image plane; it can be computed from the flow parameters of the two flows. Thus, once two planar patches (which may be located far apart) are detected (e.g., by the scheme mentioned in Remark 7.1), we can immediately test whether or not the two flows are linearly adjacent. If they are, we can compute their intersection line; there is no need for edge detection. In particular, if the optical flow we observe results from a motion of a polyhedron, we can segment the image and obtain a line drawing from the optical flow alone, at least in principle.

Remark 7.7. As we pointed out earlier, the adjacency of two optical flows alone does not assure that the two surfaces belong to the same rigid object. To assure this, we must solve the 3D recovery equations for each planar patch by using the formulae of Theorem 7.1. As we have already seen, two solutions exist for p, q, ω_1, ω_2, ω_3 at each patch. However, *the values of* ω_1, ω_2, ω_3 *must be common to all the patches if the two surfaces are rigidly connected.* Hence, we can easily pick out the true solution for each patch. If common ω_1, ω_2, ω_3 are not found, we can conclude that these patches are not rigidly connected together.

7.7 3D Recovery of a Polyhedron

Suppose an optical flow resulting from a motion of a polyhedron is given. The procedures suggested in the previous sections are as follows.

1. Applying the planarity condition stated in Remark 7.1, identify planar patches on the image plane, and compute the flow parameters for each patch by means of the least squares method.
2. Applying the adjacency condition (Propositions 7.5, 7.6), test whether all the patches are linearly adjacent to each other. If so, compute their intersection lines and determine the exact boundary of each planar patch.
3. For each planar patch, solve the 3D recovery equations by applying Theorem 7.1 and compute two sets of solutions for p, q, ω_1, ω_2, ω_3.
4. Check whether all the planar patches have common values of ω_1, ω_2, ω_3. If so, choose the true solution for each patch and discard the spurious solution. The next step is given by the following lemma:

Lemma 7.3 (*Relative depth*). *If the intersection line between two planar patches is* $y = mx + n$ *and the patches correspond to planar surfaces* $Z = pX + qY + r$ *and* $Z = p'X + q'Y + r'$, *the relative depth* $[\![r]\!] = r' - r$ *is given by*

$$[\![r]\!] = \frac{(f+r)n}{fm+pn}[\![p]\!] = \frac{(f-r)n}{nq-f}[\![q]\!] \ . \tag{7.7.1}$$

Proof. A point (X, Y, Z) on the intersection line of two planar surfaces $Z = pX + qY + r$ and $Z = p'X + q'Y + r'$ satisfies

$$[\![p]\!]X + [\![q]\!]Y + [\![r]\!] = 0 \ . \tag{7.7.2}$$

Substituting (7.2.5) into this, we obtain

$$[(f+r)[\![p]\!] - p[\![r]\!]]x + [(f+r)[\![q]\!] - q[\![r]\!]]y + f[\![r]\!] = 0 \ . \qquad (7.7.3)$$

This is a linear equation in x and y, and all points (x, y) on the intersection line must satisfy this equation. This means that (7.7.3) itself is the equation of the intersection line. Comparing this with $y = mx + ny$, we obtain (7.7.1).

Thus, the *relative depth* $[\![r]\!]$ can be computed from the intersection lines (either observed or computed). Hence, if the absolute depth r is known or assumed for one planar patch, the absolute depths of all other patches are uniquely determined. We summarize this observation in general terms as Theorem 7.4. (Strictly speaking, the reconstruction may not be unique if the optical flows of *all* patches happen to have two identical sets of solution for $\omega_1, \omega_2, \omega_3$. However, it can be proved that this does not happen for polyhedron images. It can be shown that non-uniqueness arises only for a special type of quadric surface.)

Theorem 7.4. *The structure and motion of an object whose surface consists of, or is approximated by, planar faces are uniquely determined from its optical flow up to a single indeterminate absolute depth r.*

Example 7.4. Consider the optical flow of Fig. 7.6 in Example 7.3 again. If the formulae of Theorem 7.1 are applied, the upper right part gives $(\omega_1, \omega_2, \omega_2) = (10.0, 10.0, 10.0), (-4.7, 1.1, 0.9)$ deg/s, and the lower left part gives $(\omega_1, \omega_2, \omega_2) = (-2.3, -4.6, 19.5)$ $(10.0, 10.0, 10.0)$ deg/s. Hence, the true solution is $(\omega_1, \omega_2, \omega_3) = (10.0, 10.0, 10.0)$ deg/s. The surface gradient (p, q) is uniquely determined accordingly for both parts. For the upper right part $(p, q) = (0.5, 0.2)$, and for the lower left part $(p, q) = (-0.3, -0.4)$. If the absolute depth of the upper right surface is r, the intersection computed in Example 7.3 and (7.7.3) predict the relative depth $[\![r]\!]$ to be $-0.08r - 0.16$. Hence, the equations of the surfaces are $Z = 0.5X + 0.2Y + r$ for the upper right part and $Z = -0.3X - 0.4Y + (0.92r - 0.16)$ for the lower left part.

If edges are detected on the image plane, or a line drawing is obtained, the 3D recovery of a polyhedron becomes much easier. Suppose the shape of the polyhedron is such that all the vertices are incident to at least one face that has four or more corners. Consider a face which has four corners. If image velocities are observed at these four corners, the optical flow is completely determined:

Proposition 7.7. *The optical flow resulting from a planar surface in motion is uniquely determined by image velocities at four points if no three of them are collinear.*

Proof. If velocities (u_i, v_i) are observed at four points (x_i, y_i) $i = 1, \ldots, 4$, the flow parameters $u_0, v_0, A, B, C, D, E, F$ are determined from the flow equations

(7.2.3) by solving the following simultaneous linear equations:

$$\begin{pmatrix} 1 & & x_1 & y_1 & & & x_1^2 & x_1y_1 \\ 1 & & x_2 & y_2 & & & x_2^2 & x_2y_2 \\ 1 & & x_3 & y_3 & & & x_3^2 & x_3y_3 \\ 1 & & x_4 & y_4 & & & x_4^2 & x_4y_4 \\ & 1 & & & x_1 & y_1 & x_1y_1 & y_1^2 \\ & 1 & & & x_2 & y_2 & x_2y_2 & y_2^2 \\ & 1 & & & x_3 & y_3 & x_3y_3 & y_3^2 \\ & 1 & & & x_4 & y_4 & x_4y_4 & y_4^2 \end{pmatrix} \begin{pmatrix} u_0 \\ v_0 \\ A \\ B \\ C \\ D \\ E \\ F \end{pmatrix} = \begin{pmatrix} u_1 \\ u_2 \\ u_3 \\ u_4 \\ v_1 \\ v_2 \\ v_3 \\ v_4 \end{pmatrix}. \tag{7.7.4}$$

It can be shown (we skip the proof) that the determinant of the above matrix is equal to

$$\begin{vmatrix} 1 & x_1 & y_1 \\ 1 & x_2 & y_2 \\ 1 & x_3 & y_3 \end{vmatrix} \cdot \begin{vmatrix} 1 & x_2 & y_2 \\ 1 & x_3 & y_3 \\ 1 & x_4 & y_4 \end{vmatrix} \cdot \begin{vmatrix} 1 & x_3 & y_3 \\ 1 & x_4 & y_4 \\ 1 & x_1 & y_1 \end{vmatrix} \cdot \begin{vmatrix} 1 & x_4 & y_4 \\ 1 & x_1 & y_1 \\ 1 & x_2 & y_2 \end{vmatrix}, \tag{7.7.5}$$

which does not vanish unless three of the four points are collinear (Exercise 7.14). Hence, if no three points are collinear, the flow parameters are uniquely determined from the velocities at the four points.

We can then construct invariants from the computed flow parameters and solve the 3D recovery equations. The spurious solution can be eliminated if two or more faces of the same polyhedron are observed. The relative depths of the faces are computed from their intersection lines (Lemma 7.3). Thus, the 3D shape of the object is determined uniquely up to a single absolute depth.

Example 7.5. Suppose image velocities are observed at four vertices (0.4, 0.2), (– 0.4, 0.4), (– 0.2, – 0.4), (0.6, – 0.2) of the quadrilateral face shown in Fig. 7.7a. The focal length is $f = 2$. Let (– 0.105, – 0.015), (0.036, 0.067), (– 0.085, 0.112), (– 0.196, – 0.011) be their respective velocities. By solving (7.7.4), we obtain $u_0 = -0.037$, $v_0 = 0.023$, $A = -0.162$, $B = 0.124$, $C = -0.120$, $D = -0.080$, $E = 0.019$, $F = 0.059$. The corresponding flow is given in Fig. 7.7b. By solving the 3D recovery equations, we obtain $a/(r+2) = -0.14$, $b/(r+2) = 0.09$, $c/(r+2) = 0.40$, $(p, q) = (1.24, 0.65)$ or $(-0.50, 0.30)$, $(\omega_1, \omega_2, \omega_3) = (-3, -5, -9)$ or $(-5, 5, -5)$ deg/s. For each of the two solutions, the XYZ-coordinates of the four vertices in the scene are given by (7.2.5), containing r as a single indeterminate parameter.

7.7.1 Noise Sensitivity of Computation

The results in Sect. 7.7 are all "theoretical", and the suggested procedures are not necessarily suitable for actual computation of real data.

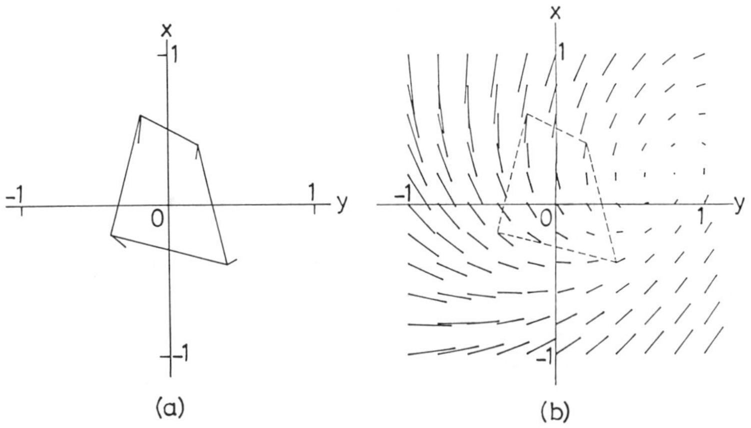

Fig. 7.7a. Image velocities observed at four vertices of a quadrilateral planar face. **b** The computed optical flow

First, identification of planar patches may be unstable due to noise. If the observed optical flow is not accurate, the resulting planar patches may depend on the order of the search if we employ the method suggested in Remark 7.1. The same applies to the adjacency condition and detection of intersection lines. Also, the solution of the 3D recovery equations is sensitive to the error contained in the input data. In order to discard spurious solutions, we must find common values of $\omega_1, \omega_2, \omega_3$ for all planar patches, but the use of real data may result in different values for all patches. Hence, some appropriate clustering is necessary in order to group slightly different solutions together.

The relative depth is determined by Lemma 7.3, but (7.7.1) has two equalities. "Theoretically" both must hold at the same time, but "computationally" the second and the third terms may have different values. Which should we adopt?

If the (observed or computed) intersection line is not accurate, the two separately reconstructed planar surfaces may be inconsistent; a "crevice" may appear along the expected intersection line.

If a line drawing of the object is given by some other means, computation may be more reliable. Still, the solution of (7.7.4) may be greatly affected if the velocity measurement at the corner vertices is not accurate. In Chap. 10, we will present a scheme of optimization to cope with noise sensitivity of 3D recovery from real data.

7.8 Motion Detection Without Correspondence

So far, we have assumed that optical flows are available. It is true that many effective techniques for optical flow detection have been proposed and tested

(Sect. 2.6.1). These techniques aim at establishing point-to-point correspondences between one image and another image taken after a very short time. However, very complicated computations are involved, and long computation time is consumed. The difficulty arises from the fact that we have no prior knowledge about the scene: we do not know how many objects there are, where they are located, and what shapes they have.

If, on the other hand, the number of objects and their shapes are known beforehand (but not their locations and motions), the detection of optical flow becomes much easier. The unknown object locations and motions can be specified by a small number of parameters, so the problem reduces to parameter estimation. In particular, *we need not find the point-to-point correspondence between image frames*. We now discuss this approach.

The principle is stated in general terms as follows. If n unknown parameters $\alpha_1, \ldots, \alpha_n$ are involved, we measure n quantities $J_1, \ldots, J_n$ for one image and the same quantities $J'_1, \ldots, J'_n$ for the other. We call these quantities *observables*.[6] Since all the changes that occur in the scene can be specified by the n parameters $\alpha_1, \ldots, \alpha_n$, we can derive, at least in principle, equations

$$\Delta J_j = F_j(\alpha_1, \ldots, \alpha_n; J_1, \ldots, J_n, \Delta t) \, , \qquad j = 1, \ldots, n \, , \tag{7.8.1}$$

that predict how those observables should change, where Δt is the lapse of time between the two images and $\Delta J_j \equiv J'_j - J_j$, $j = 1, \ldots, n$. The values of the unknown parameters $\alpha_1, \ldots, \alpha_n$ are determined by solving these equations. (This is a typical *2D non-Euclidean approach* (see Sect. 1.5).)

Thus, the principle is very simple. However, finding a relationship like (7.8.1) in a simple form is very difficult. Let us consider, for simplicity, the motion of a planar surface, and assume that the time interval between two image frames is very short so that the difference ΔJ_j of observable J_j can be identified with the instantaneous rate of change dJ_j/dt. (The time interval Δt is taken to be unit time.)

Since the flow equations (7.2.3) are linear in the flow parameters u_0, v_0, A, B, C, D, E, F, the change rate of such observables must also be linear in these flow parameters:[7]

$$\frac{dJ_j}{dt} = C_{u0}^{(j)} u_0 + C_{v0}^{(j)} v_0 + C_A^{(j)} A + C_B^{(j)} B + C_C^{(j)} C + C_D^{(j)} D + C_E^{(j)} E + C_F^{(j)} F \, . \tag{7.8.2}$$

[6] We used the term "image features" in Chap. 5 with the implication that they are used for identification and classification of object images; here we use the term "observable" to emphasize the fact that this is what is actually observed on the image plane.

[7] The flow parameters describe the deformation of the image *to a first-order approximation*. Hence, the first-order approximation of the difference of an observable is also given in terms of these flow parameters.

The coefficients $C_{u_0}^{(j)}$, $C_{v_0}^{(j)}$, $C_A^{(j)}$, $C_B^{(j)}$, $C_C^{(j)}$, $C_D^{(j)}$, $C_E^{(j)}$, $C_F^{(j)}$ are determined by the current image. Hence, they are also "observables"; we can compute their values from the given image. Thus, (7.8.2) assigns a *linear constraint* on the flow parameters u_0, v_0, A, B, C, D, E, F. Consequently, if eight observables J_i, $i = 1, \ldots, 8$, are observed, the flow parameters are computed by solving simultaneous linear equations

$$\begin{pmatrix} C_{u_0}^{(1)} & C_{v_0}^{(1)} & C_A^{(1)} & C_B^{(1)} & C_C^{(1)} & C_D^{(1)} & C_E^{(1)} & C_F^{(1)} \\ C_{u_0}^{(2)} & C_{v_0}^{(2)} & C_A^{(2)} & C_B^{(2)} & C_C^{(2)} & C_D^{(2)} & C_E^{(2)} & C_F^{(2)} \\ \cdots & \cdots & \cdots & \cdots & \cdots & \cdots & \cdots & \cdots \\ C_{u_0}^{(8)} & C_{v_0}^{(8)} & C_A^{(8)} & C_B^{(8)} & C_C^{(8)} & C_D^{(8)} & C_E^{(8)} & C_F^{(8)} \end{pmatrix} \begin{pmatrix} u_0 \\ v_0 \\ \vdots \\ F \end{pmatrix}$$

$$= \begin{pmatrix} dJ_1/dt \\ dJ_2/dt \\ \vdots \\ dJ_8/dt \end{pmatrix} . \tag{7.8.3}$$

(If more than eight observables are available, we use the least squares method (Sect. 7.2.1)).

Example 7.6. Suppose specific feature points (e.g., surface markings) can be identified on the image plane. Then, we can define observables in the form

$$J_j = \sum_i w_j(x_i, y_i) , \qquad j = 1, \ldots, 8 , \tag{7.8.4}$$

where $w_j(x, y)$, $j = 1, \ldots, 8$, are fixed functions, and the summation is taken over all the feature points.[8] If a flow field (u, v) exists on the image plane, their rates of change are given by

$$\frac{dJ_j}{dt} = \sum_i \left[\frac{\partial w_j}{\partial x}(x_i, y_i) u(x_i, y_i) + \frac{\partial w_j}{dy}(x_i, y_i) v(x_i, y_i) \right] . \tag{7.8.5}$$

Substituting the flow equations (7.2.3), we obtain the linear constraint (7.8.2) with the following coefficients:

$$C_{u_0}^{(j)} = \sum_i \frac{\partial w_j}{\partial x}(x_i, y_i) , \qquad C_{v_0}^{(j)} = \sum_i \frac{\partial w_j}{\partial y}(x_i, y_i) ,$$

$$C_A^{(j)} = \sum_i x_i \frac{\partial w_j}{\partial x}(x_i, y_i) , \qquad C_B^{(j)} = \sum_i y_i \frac{\partial w_j}{\partial x}(x_i, y_i) ,$$

[8] Although we must identify feature points on the image plane, we *need not find correspondences* of these features points between two image frames, because we do not compare two images; we only compare the values of the observables of (7.8.4).

$$C_C^{(j)} = \sum_i x_i \frac{\partial w_j}{\partial y}(x_i, y_i) , \qquad C_D^{(j)} = \sum_i y_i \frac{\partial w_j}{\partial y}(x_i, y_i) ,$$

$$C_E^{(j)} = \sum_i \left[x_i^2 \frac{\partial w_j}{\partial x}(x_i, y_i) + x_i y_i \frac{\partial w_j}{\partial y}(x_i, y_i) \right] ,$$

$$C_F^{(j)} = \sum_i \left[x_i y_i \frac{\partial w_j}{\partial x}(x_i, y_i) + y_i^2 \frac{\partial w_j}{\partial y}(x_i, y_i) \right] , \tag{7.8.6}$$

All these quantities are "observables" computable from the image.

Example 7.7. Suppose specific (straight or curved) line segments can be identified on the image plane. They may be cracks of the object surface, but a typical one is the boundary of the surface. Consider the sum of the curvilinear integrals along individual line segments or around the surface boundary:

$$J_j = \sum_i \int_{L_i} w_j(x, y)\, ds , \qquad j = 1, \ldots, 8 , \tag{7.8.7}$$

where $w_j(x, y)$, $j = 1, \ldots, 8$, are fixed functions, and s is the arc length along individual line segments. The summation is taken over all the line segments. (Again, we need not find correspondences of these line segments, because we only compare the values of the observables of (7.8.7)). If a flow field (u, v) exists on the image plane, their rates of change are given as follows (Exercise 7.16):

$$\frac{dJ_j}{dt} = \sum_i \int_{L_i} \left[\frac{\partial w_j}{\partial x} u + \frac{\partial w_j}{\partial y} v + \left\{ \frac{\partial u}{\partial x} t_1^2 + \left(\frac{\partial u}{\partial y} + \frac{\partial v}{\partial x} \right) t_1 t_2 + \frac{\partial v}{\partial y} t_2^2 \right\} w_j \right] ds . \tag{7.8.8.}$$

Here, (t_1, t_2) is the unit tangent vector to individual line segments oriented in the direction of increasing arc length s. Substituting the flow equations (7.2.3), we obtain the linear constraint (7.8.2) with the following coefficients:

$$C_{u_0}^{(j)} = \sum_i \int_{L_i} \frac{\partial w_j}{\partial x} ds , \qquad C_{v_0}^{(j)} = \sum_i \int_{L_i} \frac{\partial w_j}{\partial y} ds ,$$

$$C_A^{(j)} = \sum_i \int_{L_i} \left[x \frac{\partial w_j}{\partial x} + t_1^2 w_j \right] ds, \qquad C_B^{(j)} = \sum_i \int_{L_i} \left[y \frac{\partial w_j}{\partial x} + t_1 t_2 w_j \right] ds ,$$

$$C_C^{(j)} = \sum_i \int_{L_i} \left[x \frac{\partial w_j}{\partial y} + t_1 t_2 w_j \right] ds, \qquad C_D^{(j)} = \sum_i \int_{L_i} \left[y \frac{\partial w_j}{\partial y} + t_2^2 w_j \right] ds ,$$

$$C_E^{(j)} = \sum_i \int_{L_i} \left[x^2 \frac{\partial w_j}{\partial x} + xy \frac{\partial w_j}{\partial y} + (2xt_1^2 + yt_1t_2 + xt_2^2) w_j \right] ds \ ,$$

$$C_F^{(j)} = \sum_i \int_{L_i} \left[xy \frac{\partial w_j}{\partial x} + y^2 \frac{\partial w_j}{\partial y} + (yt_1^2 + xt_1t_2 + 2yt_2^2) w_j \right] ds \ . \qquad (7.8.9)$$

All these quantities are "observables" computable from the image.

Example 7.8. If the boundary of the planar surface is detected, we can use, as we did in Chap. 5, the weighted integral of the gray level $F(x, y)$ over the region S corresponding to the surface:[9]

$$J_j = \int_S w_j(x, y) F(x, y)\, dx\, dy \ , \qquad j = 1, \ldots, 8 \ . \qquad (7.8.10)$$

Again, $w_j(x, y)$, $j = 1, \ldots, 8$, are fixed functions. In this case, we must assume that the gray level $F(x, y)$ is characteristic to the object surface itself and does not change in the course of its motion. If a flow field (u, v) exists on the image plane, their change rates are given as follows (Exercise 7.17):

$$\frac{dJ_j}{dt} = \int_S \left[\frac{\partial w_j}{\partial x} u + \frac{\partial w_j}{\partial y} v + \left(\frac{\partial u}{\partial x} + \frac{\partial v}{\partial y} \right) w_j \right] F(x, y)\, dx\, dy \ . \qquad (7.8.11)$$

Substituting the flow equations (7.2.3), we obtain the linear constraint (7.8.2) with the following coefficients:

$$C_{u_0}^{(j)} = \int_S \frac{\partial w_j}{\partial x} F(x, y)\, dx\, dy \ , \quad C_{v_0}^{(j)} = \int_S \frac{\partial w_j}{\partial y} F(x, y)\, dx\, dy \ ,$$

$$C_A^{(j)} = \int_S \left[w_j + x \frac{\partial w_j}{\partial x} \right] F(x, y)\, dx\, dy \ , \quad C_B^{(j)} = \int_S y \frac{\partial w_j}{\partial x} F(x, y)\, dx\, dy \ ,$$

$$C_C^{(j)} = \int_S x \frac{\partial w_j}{\partial y} F(x, y)\, dx\, dy \ , \quad C_D^{(j)} = \int_S \left[w_j + y \frac{\partial w_j}{\partial x} \right] F(x, y)\, dx\, dy \ ,$$

$$C_E^{(j)} = \int_S \left[3xw_j + x^2 \frac{\partial w_j}{\partial x} + xy \frac{\partial w_j}{\partial y} \right] F(x, y)\, dx\, dy \ ,$$

$$C_F^{(j)} = \int_S \left[3yw_j + xy \frac{\partial w_j}{\partial x} + y^2 \frac{\partial w_j}{\partial y} \right] F(x, y)\, dx\, dy \ . \qquad (7.8.12)$$

All these quantities are "observables" computable from the image.

---

[9] We need not necessarily use the gray level; we could also simply put $F(x, y) = 1$ in all the subsequent equations (Exercise 7.18).

Once the flow parameters u_0, v_0, A, B, C, D, E, F are computed, the position and motion of the surface are computed by first constructing invariants and then solving the 3D recovery equations. As we have seen in the previous sections, the absolute depth r is indeterminate, and two solutions exist. However, if the 3D position of the surface is known for one image frame, its 3D positions for the subsequent frames are uniquely determined. This is shown as follows.

If we substitute (7.2.4) into (7.8.2), we obtain

$$\frac{dJ_j}{dt} = C_a^{(j)}a + C_b^{(j)}b + C_c^{(j)}c + C_{\omega_1}^{(j)}\omega_1 + C_{\omega_2}^{(j)}\omega_2 + C_{\omega_3}^{(j)}\omega_3 \ , \qquad (7.8.13)$$

where

$$C_a^{(j)} = \frac{1}{f+r}(fC_{u_0}^{(j)} - pC_A^{(j)} - qC_B^{(j)}) \ ,$$

$$C_b^{(j)} = \frac{1}{f+r}(fC_{v_0}^{(j)} - pC_C^{(j)} - qC_D^{(j)}) \ ,$$

$$C_c^{(j)} = \frac{1}{f+r}\left(C_A^{(j)} + C_D^{(j)} - \frac{1}{f}(pC_E^{(j)} + qC_F^{(j)})\right) ,$$

$$C_{\omega_1}^{(j)} = -\left(pC_C^{(j)} + qC_D^{(j)} + \frac{1}{f}C_F^{(j)}\right) , \qquad C_{\omega_2}^{(j)} = pC_A^{(j)} + qC_B^{(j)} + \frac{1}{f}C_E^{(j)} \ ,$$

$$C_{\omega_3}^{(j)} = C_D^{(j)} - C_B^{(j)} \ . \qquad (7.8.14)$$

Suppose the surface parameters p, q, r are known for one image frame, but the motion parameters a, b, c, ω_1, ω_2, ω_3 are unknown. If the surface parameters p, q, r are known, all $C_a^{(j)}$, $C_b^{(j)}$, $C_c^{(j)}$, $C_{\omega_1}^{(j)}$, $C_{\omega_2}^{(j)}$, $C_{\omega_3}^{(j)}$ can be computed from the image, and (7.8.13) gives a linear constraint on the motion parameters a, b, c, ω_1, ω_2, ω_3. Hence, if six observables J_j, $j = 1, \ldots, 6$, are available, the six motion parameters a, b, c, ω_1, ω_2, ω_3 are determined by solving the simultaneous linear equation

$$\begin{pmatrix} C_a^{(1)} & C_b^{(1)} & C_c^{(1)} & C_{\omega_1}^{(1)} & C_{\omega_2}^{(1)} & C_{\omega_3}^{(1)} \\ C_a^{(2)} & C_b^{(2)} & C_c^{(2)} & C_{\omega_1}^{(2)} & C_{\omega_2}^{(2)} & C_{\omega_3}^{(2)} \\ \cdot & \cdot & \cdot & \cdot & \cdot & \cdot \\ C_a^{(6)} & C_b^{(6)} & C_c^{(6)} & C_{\omega_1}^{(6)} & C_{\omega_2}^{(6)} & C_{\omega_3}^{(6)} \end{pmatrix} \begin{pmatrix} a \\ b \\ \vdots \\ \omega_3 \end{pmatrix}$$

$$= \begin{pmatrix} dJ_1/dt \\ dJ_2/dt \\ \vdots \\ dJ_6/dt \end{pmatrix} . \qquad (7.8.15)$$

(If more than six observables are available, we use the least squares method (Sect. 7.2.1)).

If the motion parameters $a, b, c, \omega_1, \omega_2, \omega_3$ are obtained, the change rates of the surface parameters p, q, r are given by the following surface evolution equations.

Proposition 7.8 (*Surface evolution equations*).

$$\frac{dp}{dt} = pq\omega_1 - (p^2 + 1)\omega_2 - q\omega_3 \ , \qquad \frac{dq}{dt} = (q^2 + 1)\omega_1 - pq\omega_2 + p\omega_3 \ ,$$

$$\frac{dr}{dt} = c - pa - qb \ . \tag{7.8.16}$$

Proof. Let $X' = X + \dot{X}\Delta t$, $Y' = Y + \dot{Y}\Delta t$, $Z' = Z + \dot{Z}\Delta t$. If point (X, Y, Z) satisfies the surface equation $Z = pX + qY + r$, point (X', Y', Z') satisfies equation

$$Z' - \dot{Z}\,\Delta t = p(X' - \dot{X}\,\Delta t) + q\,(Y' - \dot{Y}\Delta t) + r \ . \tag{7.8.17}$$

Substituting (7.2.7) into this, and comparing the result with $Z' = (p + \Delta p)X' + (q + \Delta q)\,Y' + (r + \Delta r)$, we obtain

$$\Delta p = [pq\omega_1 - (p^2 + 1)\omega_2 - q\omega_3]\Delta t \ ,$$

$$\Delta q = [(q^2 + 1)\omega_1 - pq\omega_2 + p\omega_3]\Delta t \ ,$$

$$\Delta r = (c - pa - qb)\Delta t \ . \tag{7.8.18}$$

In the limit of $\Delta t \to 0$, the ratios $\Delta p/\Delta t$, $\Delta q/\Delta t$, $\Delta r/\Delta t$ reduce to (7.8.16).

From (7.8.16), we can estimate the surface parameters p', q', r' of the next frame by

$$p' \approx p + [pq\omega_1 - (p^2 + 1)\omega_2 - q\omega_3]\Delta t \ ,$$

$$q' \approx q + [(q^2 + 1)\omega_1 - pq\omega_2 + p\omega_3]\Delta t \ ,$$

$$r' \approx r + (c - pa - qb)\Delta t \ . \tag{7.8.19}$$

(A higher-order formula can be used if the surface parameters are available for multiple image frames.) After the surface parameters p', q', r' are known for the second frame, the same procedure can be repeated to estimate the surface parameters p'', q'', r'' for the third frame, and so on.

Example 7.9. Fig. 7.8a shows a motion of the contour of a planar surface. The surface parameters are $(p, q, r) = (-0.3, -1.71, 1.00)$ for C. (Length is scaled by the focal length f.) If we put

$$w_1(x, y) = x \ , \quad w_2(x, y) = y \ , \quad w_3 = (x, y) = x^2 \ ,$$

$$w_4(x, y) = xy \ , \quad w_5(x, y) = y^2 \ , \quad w_6(x, y) = x^2y^2 \ , \tag{7.8.20}$$

and define observables by curvilinear integration along the contour

$$J_j = \int_C w_j(x, y)\, ds\ , \qquad j = 1, \ldots, 6\ , \tag{7.8.21}$$

the motion parameters are estimated to be $(a,\ b,\ c) = (0.27,\ 0.27,\ 0.46)$, $(\omega_1, \omega_2, \omega_3) = (0.031, 0.031, 0.027)$. (The time interval is taken as unit time, and the unit of angle is the radian.) The true motion parameters are $(a, b, c) = (0.03, 0.03, 0.05)$, $(\omega_1, \omega_2, \omega_3) = (0.03, 0.03, 0.03)$. The estimated surface parameters for C' are $(p', q', r') = (-0.27,\ -1.61,\ 1.10)$, while the true values are $(-0.17,\ -1.61,\ 1.11)$. Figure 7.8b shows the detected optical flow.

Remark 7.8. As we can observe from the above example, the solution is only an approximation because time derivatives are approximated by finite differences. If this procedure is iterated without any correction, errors will grow rapidly. One way to avoid this is to "move" the initial image according to the estimated motion parameters. Namely, we generate the object image we would observe if the object were moved according to the estimated motion parameters. If this generated image completely coincides with the target image, our estimation is correct. Otherwise, we apply the same procedure to the generated image and the target image. This process is repeated until the two images sufficiently coincide. (The test for coincidence can be done either by matching the two images or by comparing the values of the observables.)

Remark 7.9. The principle stated so far also works if the object is not a planar surface but a solid object. If the 3D shape and position of the object is known in one image frame, we can compute how the image would look if the object

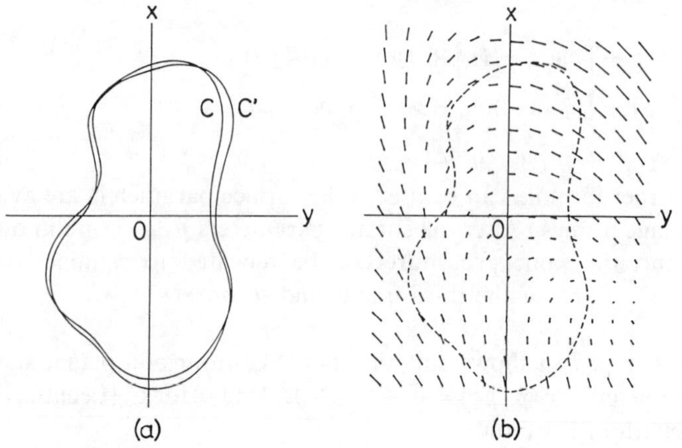

Fig. 7.8a. A moving contour surrounding a planar surface. **b** The computed optical flow

underwent a motion specified by motion parameters we arbitrarily prescribe. Hence, we can compute how given observables, such as those defined by (7.8.4, 8, 10), change for prescribed motion parameters a, b, c, ω_1, ω_2, ω_3. This means that we can compute the observables $C_a^{(j)}$, $C_b^{(j)}$, $C_c^{(j)}$, $C_{\omega_1}^{(j)}$, $C_{\omega_2}^{(j)}$, $C_{\omega_3}^{(j)}$ for the object image and obtain linear constraints in the form of (7.8.13) (rather than (7.8.2)). Hence, we can apply the iteration procedure without detecting point-to-point correspondences.

This method can be applied to the *model-based object recognition* problem, i.e., to determine the 3D position and orientation of an object when its 3D shape is stored in a database and its projection image is given. We first "guess" its 3D position, and generate its graphic image we would observe if it were placed in that position. Then, the 3D position of the observed object can be determined by comparing the given image and the generated image. For example, suppose the image of Fig. 7.9a is given, and the 3D shape of the object is known. If we generate an appropriate projection image of it and apply the iterative process discussed in Remark 7.8, we obtain a sequence of object motion shown in Fig. 7.9b. Here, no point-to-point correspondence is assumed.

In order to produce a good guess, we must register, beforehand, *typical* images of the object viewed from various positions, and then choose the one that best matches the observed image.[10]

7.8.1 Stereo Without Correspondence

The above iterative procedure requires knowledge of the object position for the initial image. *Stereo* is a well-known technique for such a measurement: The absolute depth of a point in the scene is determined by triangulation if the point is identified on two (or more) images. However, identifying corresponding points between two images is a time-consuming process.

On the other hand, a slight modification to our motion detection technique leads to a stereo algorithm that does not require correspondence detection. Suppose the camera is moved relative to a stationary planar surface by a small distance $\Delta\alpha$ in the negative direction of the X-axis. This means that the surface moves relative to the camera by distance $\Delta\alpha$ in the positive direction of the X-axis. Let ΔJ be the change of observable J caused by this camera translation.[11] Then, the value of C_α can be approximated by $\Delta J/\Delta a$. (If the camera is

---

[10] The choice is not easy. On way is to try all the candidates and see if their generated images finally coincide with the observed one. The test for coincidence can be done either directly by their images or indirectly by their observables.

[11] In this case, we need only one observable J, and hence, superscript (j) is dropped from $C_a^{(j)}$, etc. We can also use multiple observables. For example, if we use three observables, the camera needs to be moved, theoretically, in only one direction, say along the X-axis. However, increasing the number of camera translation directions rather than the number of observables is expected to yield more stable results.

(a)

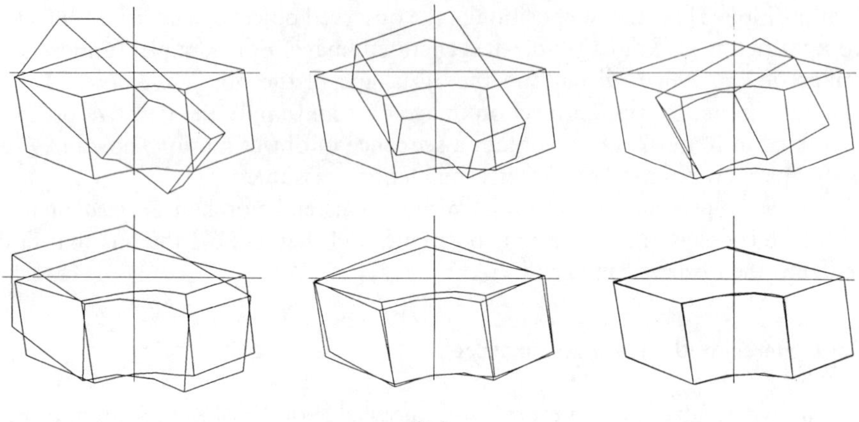

(b)

Fig. 7.9a. An object image. The 3D shape of the object is assumed to be stored in a database. **b** Iteration steps to move a generated object image into the position of the observed image. Here, no point-to-point correspondence is assumed

allowed to move continuously, and multiple images are available for each direction, a higher-order formula can be used.) Similarly, the values of C_b and C_c can be estimated by translating the camera along the Y- and Z-axes. Then, the surface parameters p, q, r are obtained by regarding (7.8.14a–c) as simultaneous equations in unknowns p, q, r and solving them. Eliminating r, we see that the gradient (p, q) is given as a solution of

$$\begin{pmatrix} C_c C_A + C_a C_E/f & C_c C_B + C_a C_F/f \\ C_c C_C + C_b C_E/f & C_c C_D + C_b C_F/f \end{pmatrix} \begin{pmatrix} p \\ q \end{pmatrix} = \begin{pmatrix} f C_c C_{u_0} + C_a(C_A + C_D) \\ f C_c C_{u_0} + C_a(C_A + C_D) \end{pmatrix}, \tag{7.8.22}$$

and the absolute depth r is given by

$$r = \frac{fC_{u_0} - pC_A - qC_B}{C_a} - f = \frac{fC_{v_0} - pC_C - qC_D}{C_b} - f\,. \qquad (7.8.23)$$

Figure 7.10 shows images of a planar surface when the camera is moved leftward (Fig. 7.10a), downward (Fig. 7.10b), and away from the surface (Fig. 7.10c). Using the contour length $J = \int_C ds$ as the observable, we obtain the surface parameters $(p, q, r) = (-0.36, -1.93, 1.22)$, while the true values are $(-0.30, -1.71, 1.00)$ (length is scaled by the focal length f).

Here, we assumed that the surface boundary contour was detected. If the surface boundary is known, it is not difficult to find point-to-point correspondences, because the corresponding point on the image plane must be somewhere on a line—called the *epipolar line*—determined by the camera displacement. For example, if the camera is displaced horizontally, the corresponding point on the image plane must be on a horizontal line passing through the original point.[12] Hence, only points on the epipolar line need to be searched. This fact is known as the *epipolar constraint*.

A comparison of the above method and ordinary stereo provides us with a good example of two contrasting principles of 3D recovery (Sects 1.4, 1.5).

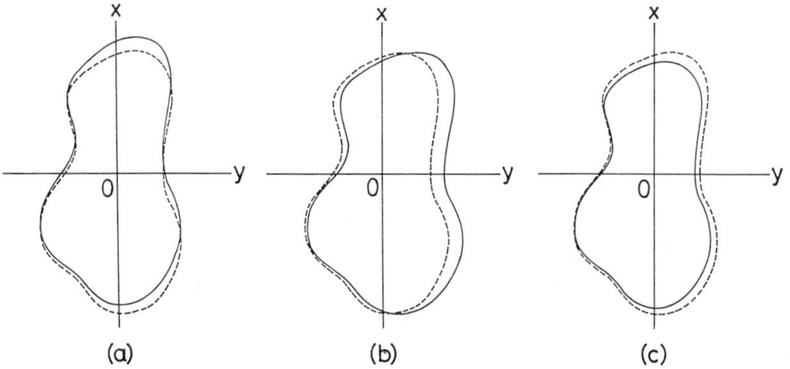

Fig. 7.10 a–c. Images of a contour surrounding a planar surface obtained by displacing the camera. The camera is moved **a** along the X-axis, **b** along the Y-axis, **c** along the Z-axis

[12] In general terms, the *epipolar line* is the perspective projection image of the "ray" starting from the center of the lens of the other camera and passing through the corresponding point on the image plane of that camera. Epipolar lines can be easily generated if the relative location and orientation of the other camera are given.

Note that although point-to-point correspondences may be detected by the epipolar constraint, what is obtained is the *depth map* of individual points, not the surface itself. Even if the true surface is planar, the computed boundary points may not be coplanar. Hence, we must fit a plane by some criterion (least squares, etc.). In general, stereo does not provide information about the object shape and position: it is obtained only after an appropriate *model* is fitted to the computed depth map. If a model must be fitted eventually, we can use this knowledge from the beginning and estimate the model parameters directly from observed images. All the methods suggested in this section are typical examples of this model-based approach or the 2D *non-Euclidean approach* (Sect. 1.5).

Exercises

7.1 Confirm the proof of Proposition 7.1 by deriving (7.2.5–9).

7.2 (a) Derive a set of simultaneous linear equations to determine the flow parameters u_0, v_0, A, B, C, D, E, F that minimize M of (7.2.10).
(b) Let $\hat{u}_0$, $\hat{v}_0$, $\hat{A}$, $\hat{B}$, $\hat{C}$, $\hat{D}$, $\hat{E}$, $\hat{F}$ be the flow parameters that minimize M of (7.2.10). Show that the residual M for these flow parameters is given by

$$M = -\sum_i \{[\hat{u}_0 + \hat{A}x_i + \hat{B}y_i + (\hat{E}x_i + \hat{F}y_i)x_i - u_i]u_i$$
$$+ [\hat{v}_0 + \hat{C}x_i + \hat{D}y_i + (\hat{E}x_i + \hat{F}y_i)y_i - v_i]v_i\} .$$

7.3 (a) Show that an n-dimensional vector $\boldsymbol{u}$ satisfies $(\boldsymbol{u}, \boldsymbol{v}) = 0$ for an arbitrary n-dimensional vector $\boldsymbol{v}$ if and only if $\boldsymbol{u} = 0$. *Hint*: What happens if $\boldsymbol{u} = \boldsymbol{v}$?
(b) Let $J(\boldsymbol{x}) = (\boldsymbol{A}\boldsymbol{x}, \boldsymbol{x}) + (\boldsymbol{b}, \boldsymbol{x}) + \boldsymbol{c}$ be an (inhomogeneous) quadratic form in n-dimensional vector $\boldsymbol{x}$, where $\boldsymbol{A}$ is an n-dimensional symmetric matrix, and $\boldsymbol{b}$ and $\boldsymbol{c}$ are n-dimensional vectors. Show that J takes its extremum value if and only if $2\boldsymbol{A}\boldsymbol{x} + \boldsymbol{b} = 0$. *Hint*: $(\boldsymbol{A}\boldsymbol{x}, \boldsymbol{x}) = (\boldsymbol{x}, \boldsymbol{A}^{\mathrm{T}}\boldsymbol{x})$.

7.4 Confirm (7.4.3–8).

7.5 For a nonzero complex number z, prove that $z^*/z = \exp[-2i\arg(z)]$.

7.6 Confirm (7.4.15–18).

7.7 (a) Derive (7.4.19) from (7.4.16, 18).
(b) Why do two additional roots enter if both sides of (7.4.19) are squared? Where are these additional roots located?

7.8 Give an algorithm to compute the complex square root $\sqrt{z}$ for a complex number $z = a + ib$.

7.9 Confirm (7.5.7–9).

7.10 Derive (7.6.2).

7.11 (a) Prove the *remainder theorem*: Polynomial $f(x)$ is divisible by $(x - a)$ if and only if $f(a) = 0$.

(b) Confirm (7.6.9–11).

(c) Prove that if z, z' and z'' are complex numbers, zz'/z'' is a real number if and only if $\arg(z) + \arg(z') - \arg(z'') = 0 \pmod{\pi}$.

7.12 (a) Derive (7.6.14) from (7.6.6) when $[\![K]\!] = 0$.

(b) Prove that (7.6.4) expresses a line on the xy-plane if and only if (7.6.12) is satisfied.

7.13 Confirm (7.7.3), and derive (7.7.1).

7.14 (a) Derive (7.7.4).

(b)* Prove that the determinant of the matrix in (7.7.4) is given by the expression (7.7.5).

(c) Show that the equation of a line passing through two points (a, b) and (a', b') on the xy-plane is given by

$$\begin{vmatrix} 1 & x & y \\ 1 & a & b \\ 1 & a' & b' \end{vmatrix} = 0 \ .$$

(c) Give the equation of a plane passing through three points (a, b, c), (a', b', c'), and (a'', b'', c'') in XYZ-space.

7.15 Prove (7.8.5), and confirm (7.8.6).

7.16* (a) Consider two points (x, y) and $(x + t_1 ds, y + t_2 ds)$ infinitesimally far apart on the xy-plane, where (t_1, t_2) is a unit vector. What is the distance between these two points?

(b) If a flow field $(u(x, y), v(x, y))$ is present, what is the distance between these two points Δt seconds later? Compute it to first order in Δt. *Hint*: Point (x, y) moves to point $(x + u\Delta t, y + v\Delta t)$, while point $(x + dx, y + dy)$ moves to point $(x + dx + [u + (\partial u/\partial x)dx + (\partial u/\partial y)dy]\Delta t,\ y + dy + [v + (\partial v/\partial x)dx + (\partial v/\partial y))dy]\Delta t)$. Also, use the approximation $\sqrt{1 + x} \approx 1 + x/2$ when x is close to zero.

(c) Prove (7.8.8), and confirm (7.8.9).

7.17* (a) Consider the square defined by four points (x, y), $(x + dx, y)$, $(x + dx, y + dy)$, $(x, y + dy)$ infinitesimally far apart on the xy-plane. Show that if a flow field $(u(x, y), v(x, y))$ is present, the area of the parallelogram defined by these four points becomes $[1 + (\partial u/\partial x + \partial v/\partial y)\Delta t]\, dx\, dy$ to first order in Δt after Δt seconds.

(b) Prove (7.8.11), and confirm (7.8.12).

7.18* (a) Prove *Green's theorem*: For a function $F(x, y)$ of variables x, y, there exist two functions $P(x, y)$, $Q(x, y)$ such that

$$F(x, y) = \frac{\partial Q}{\partial x} - \frac{\partial P}{\partial y} \ ,$$

and the double integral of $F(x, y)$ over a region S is converted into a curvilinear integral along the boundary contour C of S in the form

$$\int_S F(x, y)dx\,dy = \int_C [P(x, y)t_1 + Q(x, y)t_2]\,ds \ ,$$

where (t_1, t_2) is the unit tangent vector to the contour C along the positive orientation (counterclockwise for the ordinary xy-plane, but clockwise for our image xy-plane).

(b) Show that if $F(x, y) = 1$ (i.e., if the gray levels are ignored and integration is performed over the domain S), the double integral of (7.8.11) is converted into a curvilinear integral

$$\frac{dJ_j}{dt} = \int_C (ut_2 - vt_1)m_j\,ds \ .$$

(c) Show that if $F(x, y) = 1$, the double integrals of (7.8.12) are converted into the following curvilinear integrals.

$$C_{u_0}^{(j)} = \int_C t_2 m_j\,ds \ , \quad C_{v_0}^{(j)} = -\int_C t_1 m_j\,ds \ ,$$

$$C_A^{(j)} = \int_C xt_2 m_j\,ds \ , \quad C_B^{(j)} = \int_C yt_2 m_j\,ds \ ,$$

$$C_C^{(j)} = -\int_C xt_1 m_j\,ds \ , \quad C_D^{(j)} = -\int_C yt_1 m_j\,ds \ ,$$

$$C_E^{(j)} = -\int_C (xyt_1 - x^2t_2)m_j\,ds \ , \quad C_F^{(j)} = -\int_C (y^2t_1 - xyt_2)m_j\,ds \ .$$

7.19 Derive (7.8.14) by substituting (7.2.4) into (7.8.2).

7.20* (a) For an ordinary differential equation $dx/dt = f(x, t)$, give a formula to estimate the value of $x(t)$ when the values of $x(t - \Delta t)$ and $x(t - 2\Delta t)$ are given.
(b) Similarly, give the formula when the values of $x(t - \Delta t)$, $x(t - 2\Delta t)$, and $x(t - 3\Delta t)$ are given.

7.21 (a) For the functions $m_1(x, y), \ldots, m_6(x, y)$ of (7.8.20), give explicit forms of observables $C_a^{(j)}$, $C_b^{(j)}$, $C_c^{(j)}$, $C_{\omega_1}^{(j)}$, $C_{\omega_2}^{(j)}$, $C_{\omega_3}^{(j)}$ associated with the observable of (7.8.21).
(b) Do the same for observables of (7.8.4).
(c) Do the same for observables of (7.8.10).

7.22 Confirm (7.8.22, 23).

7.23 (a) Take the orthographic limit $f \to \infty$ of (7.8.14). How is (7.8.15) modified? Discuss what is determinate and what is undeterminate.
(b) Make a similar argument for (7.8.20, 21) in the orthographic limit $f \to \infty$.

7.24* (a) Give a formula to estimate the derivative $f'(0)$ at $x = 0$ when the values of $f(\Delta x)$ and $f(2\Delta x)$ are given.
(b) Similarly, give the formula when the values of $f(\Delta x)$, $f(2\Delta x)$, and $f(3\Delta x)$ are given.
(c) By Taylor expansion, discuss the accuracy of the formulae derived in (a) and (b). Also, analyze the statistical behavior when the given data contain errors that can be regarded as identically and independently distributed random variables.

8. Shape from Angle

In this chapter, we study the problem of 3D recovery of polyhedra on the assumption that the corners are rectangular (*rectangularity hypothesis*). First, we study how the 3D orientations of edges defining a rectangular corner are computed from their orthographic projection. Then, we study the constraints on the image of a rectangular polyhedron as a whole. Our analysis is based on the observation that two rectangular corners are either identical or mirror images of each other with respect to an appropriately placed mirror, and all the mappings between different types of rectangular corners form a group of transformations. Next, we turn to perspective projection. We make use of the fact that as far as the 3D interpretation of a corner is concerned, the distinction between orthographic and perspective projections disappears if the camera is rotated in such a way that the vertex is moved to the center of the image plane. Then, we study the constraints imposed by *vanishing points* and *vanishing lines*, which are typical of perspective projection. The *duality* of points and lines introduced in Chap. 4 plays an essential role.

8.1 Rectangularity Hypothesis

In this chapter, we study the problem of 3D recovery from a "single" image. This is impossible without some knowledge of, or *constraints* on, the object to be reconstructed, since depth information is completely lost in the process of projection. One typical assumption is that the object is a polyhedron. This is a reasonable assumption, since many man-made objects are polyhedra. From this assumption, we can deduce many kinds of strong constraints on the object shape. In the past, various types of qualitative inference have been proposed on the basis of the consistency of a polyhedron as a whole. Such work is known as the study of the *blocks world*. One of the best known techniques is *Huffman–Clowes edge labeling* (Sect. 8.3.1). A variety of other types of information can be incorporated into this technique: shading and shadows, hidden lines, paper-like thin surfaces, etc. (see the Bibliography).

However, all of these are qualitative in the sense that we can either justify an input polyhedral image or reject it as an "impossible object", but we cannot uniquely reconstruct the 3D shape. For unique reconstruction, we need stronger hypotheses. In this respect, this study has been closely associated with psychology, since humans can often interpret a line drawing as a unique 3D object when mathematically no other constraints exist. Apparently, humans are invoking some hypotheses. The following three have been most widely studied, though many other hypotheses have also been proposed. (For example, various hypotheses about closed contours have been proposed—humans tend to interpret closed contours to be planar curves with maximum (area)/(perimeter)^2, etc.)

The first one is the *rectangularity hypothesis*. The 3D orientation of a corner can be computed if its three edges are assumed to meet perpendicularly. Since many man-made objects—buildings, machine parts, furniture, etc.—have rectangular corners, this hypothesis is also very natural.

The second is the *parallelism hypothesis*. If lines in the image converge to a single point, they are interpreted to be projections of parallel lines. Such a point is called a *vanishing point*, which determines the 3D orientation of these lines. Since foreshortening greatly helps humans perceive 3D depth, this reasoning is also very natural for humans.

The third is the *skewed-symmetry hypothesis*. If the shape of a region on the image plane can be obtained by distorting an axially symmetric shape, the region is interpreted to be a projection of an axially symmetric shape. This hypothesis is also very natural, and gives a partial condition for the surface gradient.

In the following, we first focus on the rectangularity hypothesis and study how the 3D edge orientations of a rectangular corner are computed from its 2D projection under orthographic projection. Our analysis also involves Euler angles and quaternions introduced in Chap. 6.

Then, we go on to the constraints on the image of a rectangular polyhedron. Our analysis is based on the observation that two rectagnular corners are either identical or mirror images of each other with respect to an appropriately placed mirror. We can see that all the transformations that map one rectangular corner into another form a group of transformations. This fact assigns strong constraints on two rectangular corners linked by an edge. We will give a complete list of such constraints and present a shape interpretation scheme. We will also discuss the uniqueness issues and compare our scheme with Huffman–Clowes edge labeling.

Next, we turn to perspective projection. Here, the camera rotation transformation plays an essential role. Our method is based on the observation that as far as the 3D interpretation of a corner is concerned, *the distinction between orthographic and perspective projections disappears if the camera is rotated in such a way that the vertex is brought to the center of the image plane.*

Finally, we study the constraints imposed by vanishing points and vanishing lines, which are typical of perspective projection. In order to make use of this type of constraint, we must make adequate allowance for the noise and error contained in the image data. For example, when projected, parallel lines in the scene should intersect at their common vanishing point on the image plane, but they may not meet at a single point in the presence of noise. We present averaging schemes that do not cause computational overflow even when the resulting point is located very far apart from the image origin.

8.2 Spatial Orientation of a Rectangular Corner

Consider an orthographic projection of a *rectangular corner*—a corner at which three edges meet perpendicularly. Since the projection is orthographic, we can arbitrarily translate the xy coordinate system across the image plane, and we do not lose generality by assuming that the vertex is projected onto the image origin (i.e., the vertex is on the Z-axis). Let the three edges be arbitrarily numbered from 1 to 3 and call them the 1-, 2-, 3-edges respectively. Let $\boldsymbol{n}_i$ be the unit vector starting from the vertex and extending along the i-edge, and (θ_i, φ_i) be the spherical coordinates of $\boldsymbol{n}_i$. Then, the image of the i-edge makes, upon orthographic projection, angle φ_i with the x-axis (Fig. 8.1). The angles $\varphi_1, \varphi_2, \varphi_3$ can be observed on the image plane, while the angles $\theta_1, \theta_2, \theta_3$ are unknowns. By definition, we can write

$$\boldsymbol{n}_i = (\sin\theta_i \cos\varphi_i, \sin\theta_i \sin\varphi_i, \cos\theta_i) \ , \qquad i = 1, 2, 3 \ . \tag{8.2.1}$$

If the corner is rectangular, the angles $\theta_1, \theta_2, \theta_3$ are determined from the observed angles $\varphi_1, \varphi_2, \varphi_3$ as follows:[1]

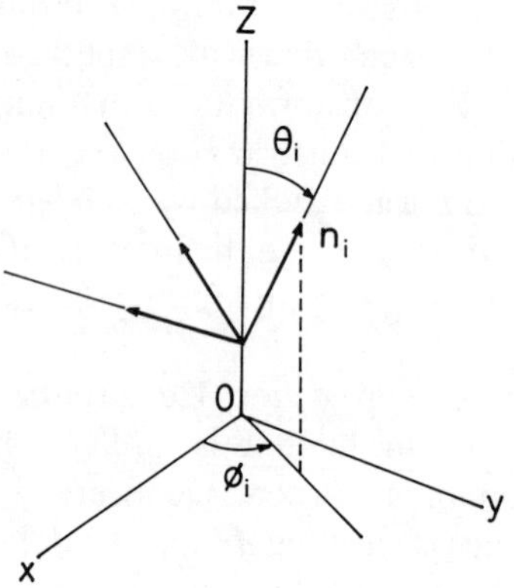

Fig. 8.1. The angles (θ_i, φ_i) are the spherical coordinates of the unit vector $\boldsymbol{n}_i$ along the i-edge

[1] We assume that the corner is in a *general position*. In particular, we assume that no two edges are projected exactly onto one straight line, and no edge is projected exactly onto the vertex. In other words, $\varphi_i - \varphi_j \neq \pi (\mathrm{mod}\ \pi)$ for $i \neq j$, and $\theta_i \neq 0 \ (\mathrm{mod}\ \pi)$ for $i = 1, 2, 3$.

Proposition 8.1. *If all the three edges are directed away from the viewer* ($0 < \theta_1$, θ_2, $\theta_3 < \pi/2$), *then*

$$\theta_1 = \tan^{-1}\sqrt{-\frac{\cos(\varphi_2 - \varphi_3)}{\cos(\varphi_1 - \varphi_2)\cos(\varphi_3 - \varphi_1)}} ,$$

$$\theta_2 = \tan^{-1}\sqrt{-\frac{\cos(\varphi_3 - \varphi_1)}{\cos(\varphi_2 - \varphi_3)\cos(\varphi_1 - \varphi_2)}} ,$$

$$\theta_3 = \tan^{-1}\sqrt{-\frac{\cos(\varphi_1 - \varphi_2)}{\cos(\varphi_3 - \varphi_1)\cos(\varphi_2 - \varphi_3)}} . \qquad (8.2.2)$$

If the i-edge is directed toward the viewer ($\pi/2 < \theta_i < \pi$), *its angle* θ_i *computed above is replaced by* $\pi - \theta_i$.

Proof. The requirement that these three vectors be mutually orthogonal is $(\boldsymbol{n}_i, \boldsymbol{n}_j) = 0$, $i, j = 1, 2, 3$, or

$$\sin\theta_i \sin\theta_j \cos(\varphi_i - \varphi_j) + \cos\theta_i \cos\theta_j = 0 , \quad i, j = 1, 2, 3, \quad i \neq j , \qquad (8.2.3)$$

which can be rewritten as

$$\tan\theta_i \tan\theta_j = -\frac{1}{\cos(\varphi_i - \varphi_j)} , \quad i, j = 1, 2, 3, \quad i \neq j . \qquad (8.2.4)$$

Multiplying both sides of the three equations (8.2.4), we obtain

$$\tan^2\theta_1 \tan^2\theta_2 \tan^2\theta_3 = \frac{1}{\cos(\varphi_1 - \varphi_2)\cos(\varphi_2 - \varphi_3)\cos(\varphi_3 - \varphi_1)} . \qquad (8.2.5)$$

If $0 \leqq \theta_1, \theta_2, \theta_3 < \pi/2$, we get

$$\tan\theta_1 \tan\theta_2 \tan\theta_3 = \sqrt{-\frac{1}{\cos(\varphi_1 - \varphi_2)\cos(\varphi_2 - \varphi_3)\cos(\varphi_3 - \varphi_1)}} . \qquad (8.2.6)$$

Equations (8.2.2a–c) are obtained from (8.2.4, 6). If $\pi/2 < \theta_i < \pi$, the above computed θ_i must be replaced by $\pi - \theta_i$.

In deciding which edges go away from or toward the viewer, we must distinguish two configurations. One is the *fork* (or 'Y'): all pairs of edges make angles of larger than $\pi/2$ but less than π (Fig. 8.2a). The other configuration is the *arrow*: one pair of edges makes an angle of larger than π but less than $3\pi/2$, and the other pairs make angles of less than $\pi/2$ (Fig. 8.2b). All other cases—all pairs of edges make angles of less than $\pi/2$, etc.—cannot be images of rectangular corners. Namely:

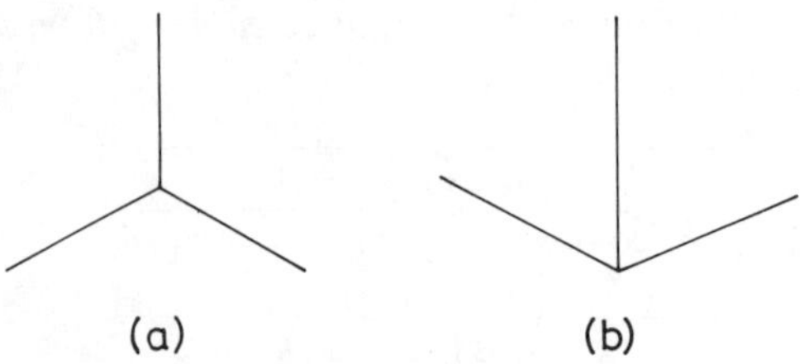

Fig. 8.2a. Fork. All pairs of edges make angles of larger than $\pi/2$ but less than π. **b** Arrow. One pair of edges makes an angle of larger than π but less than $3\pi/2$, and the other pairs make angles of less than $\pi/2$

Lemma 8.1. *The necessary and sufficient condition for a given image corner to be interpreted as a rectangular corner is that the configuration is either a fork or an arrow.*

Proof. The condition is given by the requirement that all the expressions in the square roots of (8.2.2, 6) be positive, namely,

$$\cos(\varphi_1 - \varphi_2)\cos(\varphi_2 - \varphi_3)\cos(\varphi_3 - \varphi_1) < 0 \ . \tag{8.2.7}$$

By checking exhaustively, we can easily confirm that the configuration must be either a fork or an arrow as defined above. (Due to our assumption of general position, we ignore the possibility that the configuration is "L" or "T".)

By checking the signs of both sides of (8.2.4) for all cases, we can easily find:

Lemma 8.2. *If the configuration is a fork, either all the edges are directed away from the viewer, or all toward the viewer. If the configuration is an arrow, either the side edges are directed toward the viewer and the central edge away from the viewer, or the side edges are directed away from the viewer and the central edge toward the viewer.*

Thus, we obtain the following conclusion:

Theorem 8.1. *The 3D orientation of a rectangular corner is determined uniquely from its orthographic projection except for the mirror image.*

Example 8.1. Consider the corner images of Fig. 8.3a. The observed angles are

$$\varphi_1 = 254° \ , \qquad \varphi_2 = 143° \ , \qquad \varphi_3 = 47° \ . \tag{8.2.8}$$

Hence, the configuration is a fork. Assuming that all the edges are directed toward the viewer, we obtain one solution

$$\theta_1 = 150.2° \ , \qquad \theta_2 = 101.6°, \qquad \theta_3 = 117.0° \ . \tag{8.2.9}$$

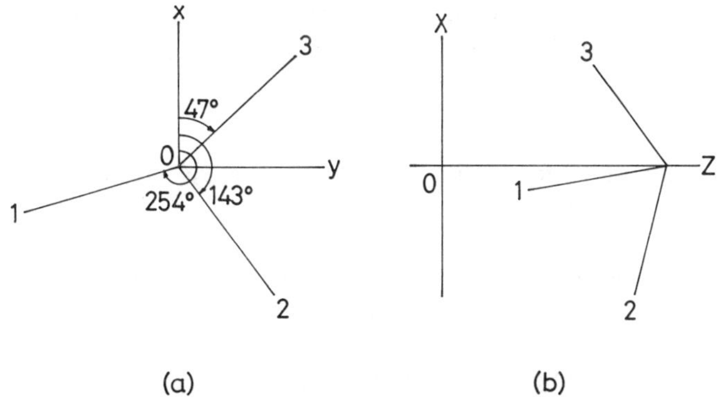

Fig. 8.3a. A projected image of a rectangular corner. This is a fork. **b** The orthographic projection onto the ZX-plane of the three edges reconstructed from **a**

From (8.2.1), the unit vectors along the three edges are given by

$$\boldsymbol{n}_1 = \begin{pmatrix} -0.1369 \\ -0.4774 \\ -0.8680 \end{pmatrix}, \quad \boldsymbol{n}_2 = \begin{pmatrix} -0.7824 \\ 0.5896 \\ -0.2009 \end{pmatrix}, \quad \boldsymbol{n}_3 = \begin{pmatrix} 0.6076 \\ 0.6516 \\ -0.4542 \end{pmatrix}. \tag{8.2.10}$$

Figure 8.3b is the orthographic projection of these vectors onto the ZX-plane. The other solution is given as the mirror image of this solution.

Example 8.2. Consider the corner images of Fig. 8.4a. The observed angles are

$$\varphi_1 = 355^\circ, \quad \varphi_2 = 133^\circ, \quad \varphi_3 = 36^\circ. \tag{8.2.11}$$

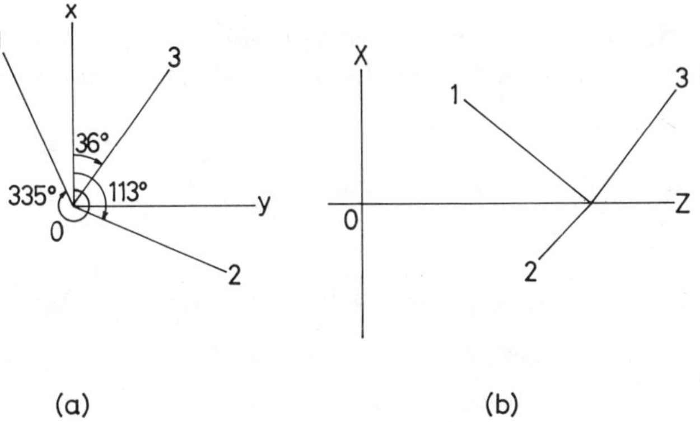

Fig. 8.4a. A projected image of a rectangular corner. This is an arrow. **b** The orthographic projection onto the ZX-plane of the three edges reconstructed from **a**

Hence, the configuration is an arrow. Assuming that all the edges are directed toward the viewer, we obtain one solution

$$\theta_1 = 141.5^\circ \ , \qquad \theta_2 = 110.5^\circ \ , \qquad \theta_3 = 59.0^\circ \ . \tag{8.2.12}$$

From (8.2.1), the unit vectors along the three edges are given by

$$\boldsymbol{n}_1 = \begin{pmatrix} 0.6208 \\ -0.0543 \\ -0.7821 \end{pmatrix} , \ \boldsymbol{n}_2 = \begin{pmatrix} -0.3660 \\ 0.8622 \\ -0.3504 \end{pmatrix} , \ \boldsymbol{n}_3 = \begin{pmatrix} 0.6933 \\ 0.5037 \\ 0.5153 \end{pmatrix} . \tag{8.2.13}$$

Figure 8.4b is the orthographic projection of these vectors onto the ZX-plane. The other solution is given as the mirror image of this solution.

Suppose the three edges are numbered in such a way that the 1-, 2-, 3-edges form a right-hand system in this order. Then, the 3D orientation of the corner is represented by the rotation that maps the X-, Y-, Z-axes onto the 1-, 2-, 3-edges, respectively. As we studied in Chap. 6, the rotation matrix $\boldsymbol{R}$ is given by $\boldsymbol{R} = (\boldsymbol{n}_1 \boldsymbol{n}_2 \boldsymbol{n}_3)$ (Sect. 6.2). From (3.2.1), the rotation matrix $\boldsymbol{R}$ is given in terms of the spherical coordinates (θ_i, φ_i), $i = 1, 2, 3$, as

$$\boldsymbol{R} = \begin{pmatrix} \sin\theta_1 \cos\varphi_1 & \sin\theta_2 \cos\varphi_2 & \sin\theta_3 \cos\varphi_3 \\ \sin\theta_1 \sin\varphi_1 & \sin\theta_2 \sin\varphi_2 & \sin\theta_3 \sin\varphi_3 \\ \cos\theta_1 & \cos\theta_2 & \cos\theta_3 \end{pmatrix} . \tag{8.2.14}$$

A 3D rotation is also specified by three Euler angles $\{\theta, \varphi, \psi\}$ (Sect. 6.3). Their values can be determined from the rotation matrix $\boldsymbol{R}$, but there exists a direct method of determination; we can compute the Euler angles $\{\theta, \varphi, \psi\}$ directly from the given projection image as follows.

Proposition 8.2. *If φ_1, φ_2, φ_3 are respectively the polar coordinates of the projections of the* 1-, 2-, 3-*edges, the Euler angles $\{\theta, \varphi, \psi\}$ are given by*

$$\theta = \theta_0 \ \textit{or} \ \pi - \theta_0, \ \varphi = \varphi_3,$$
$$\psi = \psi_0, \ -\psi_0, \ \pi - \psi_0, \ \textit{or} \ \pi + \psi_0, \tag{8.2.15}$$

where

$$\theta_0 = \cos^{-1} \sqrt{\cot(\varphi_3 - \varphi_1)\cot(\varphi_2 - \varphi_3)} \ ,$$
$$\psi_0 = \tan^{-1} \sqrt{\frac{\tan(\varphi_3 - \varphi_1)}{\tan(\varphi_2 - \varphi_3)}} \ . \tag{8.2.16}$$

The values of (8.2.15) *are chosen according to the configuration of the corner image as shown in Fig.* 8.5.

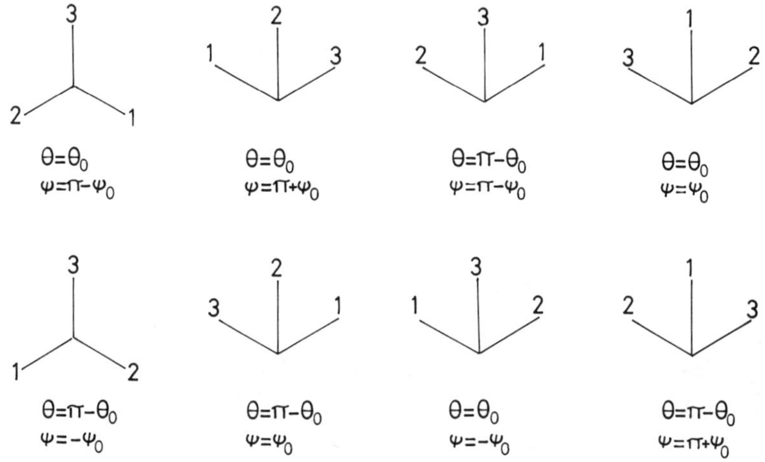

Fig. 8.5. The choice of the Euler angles $\{\theta, \varphi, \psi\}$ depends on the configuration of the corner image as shown here

Proof. The unit vectors $\boldsymbol{n}_i$, $i = 1, 2, 3$, along the three edges are given in terms of the Euler angles $\{\theta, \varphi, \psi\}$ as

$$\boldsymbol{n}_1 = \begin{pmatrix} \cos\theta\cos\varphi\cos\theta - \sin\varphi\sin\psi \\ \cos\theta\sin\varphi\cos\theta + \cos\varphi\sin\psi \\ -\sin\theta\cos\psi \end{pmatrix},$$

$$\boldsymbol{n}_2 = \begin{pmatrix} -\cos\theta\cos\varphi\sin\theta - \sin\varphi\cos\psi \\ -\cos\theta\sin\varphi\sin\theta + \cos\varphi\cos\psi \\ \sin\theta\sin\psi \end{pmatrix},$$

$$\boldsymbol{n}_3 = \begin{pmatrix} \sin\theta\cos\varphi \\ \sin\theta\sin\varphi \\ \cos\theta \end{pmatrix} \tag{8.2.17}$$

(Sect. 6.4). We immediately see that $\varphi_3 = \varphi$. From (8.2.1), $\tan\varphi_1$ and $\tan\varphi_2$ are given by the ratios of the X- and Y-components of $\boldsymbol{n}_1$ and $\boldsymbol{n}_2$:

$$\tan\varphi_1 = \frac{\cos\theta\sin\varphi\cos\psi + \cos\varphi\sin\psi}{\cos\theta\cos\varphi\cos\psi - \sin\varphi\sin\psi},$$

$$\tan\varphi_2 = \frac{\cos\theta\sin\varphi\sin\psi - \cos\varphi\cos\psi}{\cos\theta\cos\varphi\sin\psi + \sin\varphi\cos\psi}, \tag{8.2.18}$$

Rearranging these equations, we obtain

$$\tan(\varphi_3 - \varphi_1) = -\frac{\tan\psi}{\cos\theta}\,, \quad \tan(\varphi_2 - \varphi_3) = -\frac{1}{\cos\theta\tan\psi}\,. \tag{8.2.19}$$

Hence,

$$\cos^2\theta = \cot(\varphi_3 - \varphi_1)\cot(\varphi_2 - \varphi_3)\,, \quad \tan^2\psi = \frac{\tan(\varphi_3 - \varphi_1)}{\tan(\varphi_2 - \varphi_3)}\,. \tag{8.2.20}$$

The solution is given by (8.2.15, 16)[2]. The validity of Fig. 8.5 is confirmed by checking the signs of the Z-components of vectors $\boldsymbol{n}_1, \boldsymbol{n}_2, \boldsymbol{n}_3$ of (8.2.17).

Example 8.3. Consider Figs. 8.3a and 8.4a again. If the 1-, 2-, 3-edges form a right-hand system in this order, the Euler angles are

$$\begin{aligned} &\theta = 117.0°\,, \quad \varphi = 47.0°\,, \quad \psi = -13.0°\,, \\ &\theta = 59.0°\,, \quad \varphi = 36.0°\,, \quad \psi = -24.1° \end{aligned} \tag{8.2.21}$$

for Figs. 8.3a and 8.4a, respectively. If the unit vectors $\boldsymbol{n}_1, \boldsymbol{n}_2, \boldsymbol{n}_3$ are computed by (8.2.17), vectors (8.2.10) are obtained again.

A 3D rotation is also represented by a quaternion (Sect. 6.8). The use of quaternions is convenient for computing composition and inverse. The quaternion that represents a 3D rotation is obtained by first computing the rotation matrix $\boldsymbol{R}$, (8.2.14), and then converting it into its axis $\boldsymbol{n}$ (unit vector) and angle Ω of rotation (Sects. 6.3, 6.8); the quaternion is then given by

$$\boldsymbol{q} = \cos\frac{\Omega}{2} + \boldsymbol{n}\sin\frac{\Omega}{2}\,, \tag{8.2.22}$$

where the unit vector $\boldsymbol{n}$ is identified as a quaternion. However, it is more straightforward first to compute the Euler angles $\{\theta, \varphi, \psi\}$ from the image by Proposition 8.2. Then, the quaternion is given by

$$\begin{aligned} \boldsymbol{q} = {} & \cos\frac{\theta}{2}\left(\cos\frac{\psi+\varphi}{2} + \boldsymbol{k}\sin\frac{\psi+\varphi}{2}\right) \\ & + \sin\frac{\theta}{2}\left(\boldsymbol{i}\sin\frac{\psi-\varphi}{2} + \boldsymbol{j}\cos\frac{\psi-\varphi}{2}\right)\,. \end{aligned} \tag{8.2.23}$$

If quaternion $\boldsymbol{q} = q_0 + q_1\boldsymbol{i} + q_2\boldsymbol{j} + q_3\boldsymbol{k}$ represents the 3D orientation of a rectangular corner, the unit vectors $\boldsymbol{n}_1, \boldsymbol{n}_2, \boldsymbol{n}_3$ along the three edges are given as

[2] Since $\theta = \theta_3$ by definition, (8.2.2c) must be equivalent to (8.2.16a). We can easily check this (Exercise 8.3).

follows (Sect. 6.8):

$$n_1 = \begin{pmatrix} q_0^2 + q_1^2 - q_2^2 - q_3^2 \\ 2(q_0 q_3 + q_1 q_2) \\ -2(q_0 q_2 - q_1 q_3) \end{pmatrix}, \quad n_2 = \begin{pmatrix} -2(q_0 q_3 - q_1 q_2) \\ q_0^2 - q_1^2 + q_2^2 + q_3^2 \\ 2(q_0 q_1 + q_2 q_3) \end{pmatrix},$$

$$n_3 = \begin{pmatrix} 2(q_0 q_2 + q_1 q_3) \\ -2(q_0 q_1 - q_2 q_3) \\ q_0^2 - q_1^2 - q_2^2 + q_3^2 \end{pmatrix}. \tag{8.2.24}$$

Let $\boldsymbol{q}$ and $\boldsymbol{q}'$ be the quaternions representing the 3D orientations of two rectangular corners. Then, the *relative rotation* that brings corner $\boldsymbol{q}$ to corner $\boldsymbol{q}'$ is represented by the quaternion $\boldsymbol{q}'' = \boldsymbol{q}'\boldsymbol{q}^*$, where $\boldsymbol{q}^* = q_0 - q_1\boldsymbol{i} - q_2\boldsymbol{j} - q_3\boldsymbol{k}$ is the conjugate of $\boldsymbol{q}$ (Sect. 6.8). It is very easy to convert a quaternion $\boldsymbol{q} = q_0 + q_1\boldsymbol{i} + q_2\boldsymbol{j} + q_3\boldsymbol{k}$ into the axis $\boldsymbol{n}$ (unit vector) and angle Ω of the 3D rotation it represents (Sect. 6.8):

$$\Omega = 2\cos^{-1} q_0, \qquad \boldsymbol{n} = \frac{(q_1, q_2, q_3)}{\sqrt{q_1^2 + q_2^2 + q_3^2}}. \tag{8.2.25}$$

Using this fact, we can easily compute from the projection image alone how a rectangular corner has moved from its original orientation.

Example 8.4. Consider Figs. 8.3a and 8.4a again. By (8.2.23), the Euler angles (8.2.21) are converted into quaternions

$$\boldsymbol{q} = 0.4996 - 0.4265\boldsymbol{i} + 0.7383\boldsymbol{j} + 0.1526\boldsymbol{k},$$

$$\boldsymbol{q}' = 0.8658 - 0.2466\boldsymbol{i} + 0.4260\boldsymbol{j} + 0.0900\boldsymbol{k}. \tag{8.2.26}$$

If these are substituted into (8.2.24), the unit vectors $\boldsymbol{n}_1$, $\boldsymbol{n}_2$, $\boldsymbol{n}_3$ (8.2.10) are obtained again.

From (8.2.26), we see that the 3D orientation of the corner of Fig. 8.4 relative to the corner of Fig. 8.3 is represented by quaternion

$$\boldsymbol{q}'\boldsymbol{q}^* = 0.8660 + 0.2475\boldsymbol{i} - 0.4256\boldsymbol{j} - 0.0868\boldsymbol{k}. \tag{8.2.27}$$

From (8.2.25), the corresponding axis $\boldsymbol{n}$ and angle Ω of rotation are

$$\Omega = 60.0^\circ, \qquad \boldsymbol{n} = \begin{pmatrix} 0.4950 \\ -0.8514 \\ -0.1736 \end{pmatrix}. \tag{8.2.28}$$

In other words, the corner of Fig. 8.4 results from the corner of Fig. 8.3 by a 3D rotation around the axis $\boldsymbol{n}$ by the angle Ω given above. (An alternative method is first to compute the rotation matrices $\boldsymbol{R}$ and $\boldsymbol{R}'$ corresponding to the two corners, then to compute the product $\boldsymbol{R}'\boldsymbol{R}^{\mathrm{T}}$, and finally to convert it into the axis $\boldsymbol{n}$ and angle Ω by the method described in Sect. 6.3. However, the use of quaternions is more straightforward.)

8.2.1 Corners with Two Right Angles

The analysis of Proposition 8.1 can be extended to an arbitrary corner that has three edges. Suppose we know that the 1- and 2-edges make angle α, the 2- and 3-edges make angle β, and the 3- and 1-edges make angle γ. Then, the equations to be solved are $(\boldsymbol{n}_1, \boldsymbol{n}_2) = \cos\alpha$, $(\boldsymbol{n}_2, \boldsymbol{n}_3) = \cos\beta$, $(\boldsymbol{n}_3, \boldsymbol{n}_1) = \cos\gamma$. Namely,

$$\begin{aligned} &\sin\theta_1 \sin\theta_2 \cos(\varphi_1 - \varphi_2) + \cos\theta_1 \cos\theta_2 = \cos\alpha \ , \\ &\sin\theta_2 \sin\theta_3 \cos(\varphi_2 - \varphi_3) + \cos\theta_2 \cos\theta_3 = \cos\beta \ , \\ &\sin\theta_3 \sin\theta_1 \cos(\varphi_3 - \varphi_1) + \cos\theta_3 \cos\theta_1 = \cos\gamma \ . \end{aligned} \tag{8.2.29}$$

We can always solve these equations by a numerical scheme, say the Newton iterations (Sect. 9.6.2), but it is very difficult to obtain a general solution in an analytical form. However, these equations can be solved analytically if *two of the three angles are* $\pi/2$.

Suppose the 1- and 2-edges make angle α $(0 < \alpha < \pi)$ and the other pairs make right angles. We must solve

$$\begin{aligned} &\sin\theta_1 \sin\theta_2 \cos(\varphi_1 - \varphi_2) + \cos\theta_1 \cos\theta_2 = \cos\alpha \ , \\ &\sin\theta_2 \sin\theta_3 \cos(\varphi_2 - \varphi_3) + \cos\theta_2 \cos\theta_3 = 0 \ , \\ &\sin\theta_3 \sin\theta_1 \cos(\varphi_3 - \varphi_1) + \cos\theta_3 \cos\theta_1 = 0 \ . \end{aligned} \tag{8.2.30}$$

As for a rectangular corner, the second and the third equations are rewritten as

$$\begin{aligned} \tan\theta_1 \tan\theta_3 &= -\frac{1}{\cos(\varphi_1 - \varphi_3)} \ , \\ \tan\theta_2 \tan\theta_3 &= -\frac{1}{\cos(\varphi_2 - \varphi_3)} \ , \end{aligned} \tag{8.2.31}$$

Taking squares of both sides of (8.2.30a), we obtain

$$\begin{aligned} &\tan^2\theta_1 \tan^2\theta_2 \cos^2(\varphi_1 - \varphi_2) + 2\tan\theta_1 \tan\theta_2 \cos(\varphi_1 - \varphi_2) + 1 \\ &\quad = (1 + \tan^2\theta_1)(1 + \tan^2\theta_2)\cos^2\alpha \ . \end{aligned} \tag{8.2.32}$$

From (8.2.31), we obtain

$$\tan\theta_2 = \frac{\cos(\varphi_1 - \varphi_3)}{\cos(\varphi_2 - \varphi_3)} \tan\theta_1 \ . \tag{8.2.33}$$

Substituting this in (8.2.32), we obtain

$$A \tan^4 \theta_1 + B \tan^2 \theta_1 + C = 0 \ , \tag{8.2.34}$$

where

$$A = \frac{\cos^2 (\varphi_1 - \varphi_3)}{\cos^2 (\varphi_2 - \varphi_3)} [\cos^2 (\varphi_1 - \varphi_2) - \cos^2 \alpha] \ ,$$

$$B = 2\frac{\cos (\varphi_1 - \varphi_2) \cos (\varphi_1 - \varphi_3)}{\cos (\varphi_2 - \varphi_3)} - \left(1 + \frac{\cos^2 (\varphi_1 - \varphi_3)}{\cos^2 (\varphi_2 - \varphi_3)}\right) \cos^2 \alpha \ ,$$

$$C = \sin^2 \alpha \ . \tag{8.2.35}$$

Hence, if we put $X \equiv \tan^2 \theta_1$, then (8.2.34) gives

$$X = \frac{-B \pm \sqrt{B^2 - 4AC}}{2A} \ . \tag{8.2.36}$$

We choose only positive X. From (8.2.31), the solutions θ_i, $i = 1, 2, 3$, are given as follows:

$$\theta_1 = \tan^{-1} \sqrt{X} \quad \text{or} \quad \pi - \tan^{-1} \sqrt{X} \ ,$$

$$\theta_2 = \tan^{-1} \left(\frac{\cos (\varphi_1 - \varphi_3)}{\cos (\varphi_2 - \varphi_3)} \tan \theta_1 \right) \ ,$$

$$\theta_3 = \tan^{-1} \frac{-1}{\cos (\varphi_3 - \varphi_1) \tan \theta_1} \ . \tag{8.2.37}$$

Since (8.2.30a) was squared to derive (8.2.32), the above solutions contain those that satisfy (8.2.30a) with the right-hand side replaced by $-\cos \alpha$ ($= \cos (\pi - \alpha)$). Hence, we must choose the solutions that really satisfy (8.2.30). From (8.2.30), it is easy to see that if $\theta_1, \theta_2, \theta_3$ are one solution, so are $\pi - \theta_1$, $\pi - \theta_3$, (the mirror image).

8.3 Interpretation of a Rectangular Polyhedron

Now, we consider shape interpretation, exploiting the fact that the object is a rectangular polyhedron. Here, we assume that the polyhedron is *trihedral*, i.e., all its corners have exactly three edges (Fig. 8.6)[3]. In contrast to a general

[3] A non-trihedral rectangular corner can be regarded as a degenerate case of vertices of two or more trihedral corners accidentally coinciding (Fig. 8.6). Hence, we can also perturb it, at least in principle, into two nearby rectangular corners before interpreting it.

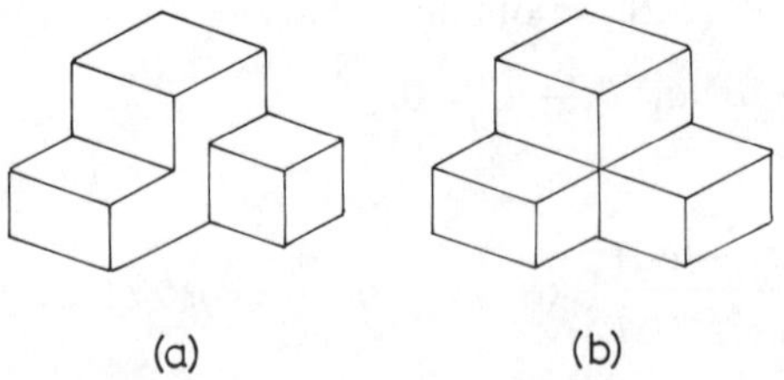

Fig. 8.6a. A trihedral polyhedron. b A non-trihedral corner can be regarded as two or more degenerate trihedral corners

polyhedron, a trihedral rectangular polyhedron has the following distinctive characteristics:

Fact 1: There exist only three different edge orientations if edges are regarded as undirected axes.
Fact 2: For any two corners, either their configurations are identical or one is obtained from the other by reversing one or more edges.
Fact 3: There exist only three face orientations in the scene if the distinction between inside and outside is disregarded.
Fact 4: The angle between two edges that define a face at a corner is either $\pi/2$ or $3\pi/2$ in the scene.

Fact 1 enables us to fix reference orientations and specify the edge orientation in reference to them. We number the three possible orientations arbitrarily as 1, 2, 3, and assign their directions arbitrarily. These three directed orientations play the role of a "coordinate system", and we refer to them as the 1-, 2-, and 3-axes, respectively.

At this stage, we do not consider "visibility"; no distinction is made between visible edges and hidden edges. This is because we are not considering faces or opaque substances. Let us call the edge parallel to the i-axis the i-edge. We define the *type* of a corner image by a triplet $c = (c_1 c_2 c_3)$, where $c_i = 0$ if the *i-edge* has the same direction as the i-axis, and $c_i = 1$ if it has the opposite direction (Fig. 8.7).

Fact 2 implies the existence of a *group of transformations* T generated by the identity and the reversals of each edge direction. There are eight elements in T because each of the three edges is either preserved or reversed. Let us write an element of T as a triplet $\boldsymbol{t} = (t_1 t_2 t_3)$, where $t_i = 0$ if it preserves the orientation of the i-edge, and $t_i = 1$ if it reverses it. The identity is given by $0 = (000)$. The group operation is given by $\boldsymbol{t} \oplus \boldsymbol{t}'$, where $\oplus$ denotes component-wise addition modulo 2 ("exclusive or").

It is immediately seen that T is a group. Since applying edge reversal twice means applying the identity, each component defines the group $\mathbb{Z}_2$ of addition modulo 2. Hence, T is isomorphic to the direct product:

$$T \cong \mathbb{Z}_2 \times \mathbb{Z}_2 \times \mathbb{Z}_2 \ . \tag{8.3.1}$$

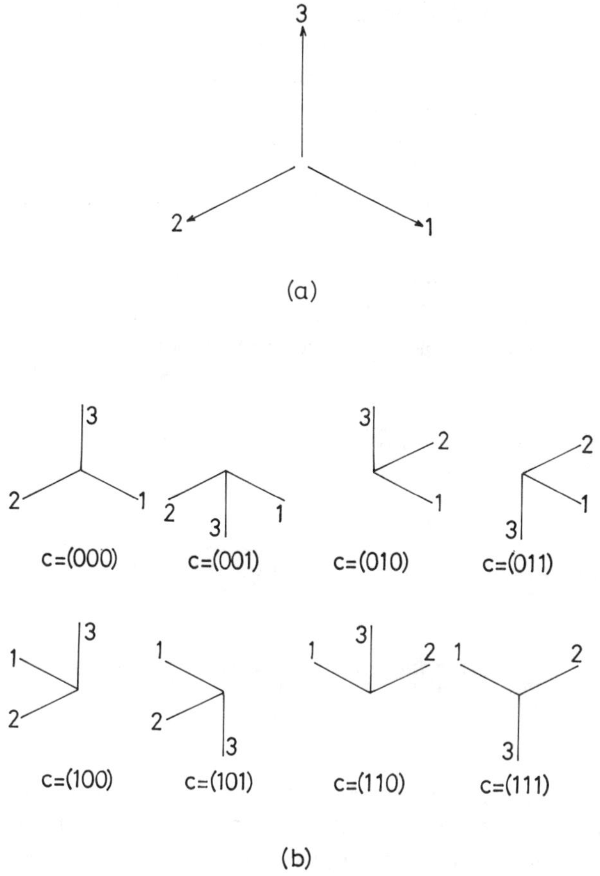

Fig. 8.7a. Reference axes orientations. **b** The eight possible corner types with respect to the reference axes of **a**

Evidently, T is associative (as a group should be) and Abelian, and each element is the inverse (or negative) of itself:

$$(\boldsymbol{t} \oplus \boldsymbol{t}') \oplus \boldsymbol{t}'' = \boldsymbol{t} \oplus (\boldsymbol{t}' \oplus \boldsymbol{t}'') \,, \tag{8.3.2}$$

$$\boldsymbol{t} \oplus \boldsymbol{t}' = \boldsymbol{t}' \oplus \boldsymbol{t} \,, \qquad \boldsymbol{t} \oplus \boldsymbol{t} = 0 \,. \tag{8.3.3}$$

By definition, we observe the following fact:

Proposition 8.3. *If transformation* $\boldsymbol{t}$ *maps a corner of type* $\mathbf{c}$ *to a corner of type* $\mathbf{c}'$, *then*

$$\boldsymbol{c}' = \boldsymbol{c} \oplus \boldsymbol{t} \,, \quad or \quad \boldsymbol{t} = \boldsymbol{c} \oplus \boldsymbol{c}' \,. \tag{8.3.4}$$

Now, we consider the 3D configuration of a corner and define its *orientation* by a triplet $\boldsymbol{p} = (p_1 p_2 p_3)$, where $p_i = 0$ if its i-edge is directed away from the viewer, and $p_i = 1$ if its i-edge is directed toward the viewer. For a given corner image, two 3D configurations are possible, one being the mirror image of the other (Theorem 8.1). We rewrite this result in terms of this new notation:

Proposition 8.4. *Two orientations can be assigned to a corner image, one being the complement of the other.*

In other words, if the value of $\boldsymbol{p}$ is assigned to one corner image, another possible assignment is obtained by interchanging 0 and 1 for all components. However, the assignment cannot be done independently for each corner. It is easy to see:

Proposition 8.5. *If transformation $\boldsymbol{t}$ maps a corner of orientation $\boldsymbol{p}$ to a corner of orientation $\boldsymbol{p}'$, then*

$$\boldsymbol{p}' = \boldsymbol{p} \oplus \boldsymbol{t} \ , \quad or \quad \boldsymbol{t} = \boldsymbol{p} \oplus \boldsymbol{p}' \ . \tag{8.3.5}$$

Combining this with Proposition 8.3, we obtain the following important consequence:

Corollary 8.1. *If one corner has type $\boldsymbol{c}$ and orientation $\boldsymbol{p}$ while another has type $\boldsymbol{c}'$ and orientation $\boldsymbol{p}'$, then*

$$\boldsymbol{p}' = \boldsymbol{p} \oplus \boldsymbol{c} \oplus \boldsymbol{c}' \ . \tag{8.3.6}$$

From this, we can see that, once an orientation $\boldsymbol{p}$ is assigned to an arbitrarily chosen corner, the orientations of all the other corners are uniquely determined by (8.3.6). Combining this with Proposition 8.5, we conclude:

Proposition 8.6. *Two 3D configurations are possible for the skeleton image of a rectangular polyhedron, one being the mirror image of the other.*[4]

This is a familiar fact known as the *Necker cube phenomenon* (Fig. 8.8).

Up to now, we have considered only skeletons. Now, we consider faces. As pointed out in Fact 3, there exist only three types of face. Let us call the face defined by the i- and j-edges the ij-face. At this stage, we are not considering on which side the opaque substance exists, and hence we do not make any distinction between the outside and inside. Thus, the ij-face and the ji-face are identified with one another, and there are only three types: the 12-face, the 23-face, and the 31-face. Fact 4 enables us to characterize a corner by a *state* defined as a triplet $\boldsymbol{s} = (s_{23} s_{31} s_{12})$, where $s_{ij} = 0$ if the ij-face is at angle $\pi/2$, and $s_{ij} = 1$ if it is at angle $3\pi/2$.

---

[4] So far, we have considered edges only. Faces and opaque substances will be considered subsequently.

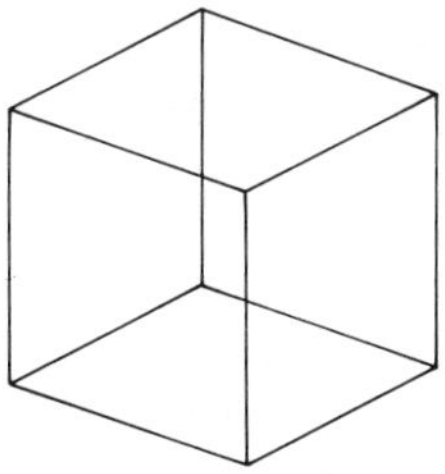

Fig. 8.8. Necker cube phenomenon. Two interpretations are possible for the 3D shape, one being the mirror image of the other

By considering all the possibilities exhaustively, it is easy to see that there are two constraints on the state. First, *no two faces at a corner can be at angle* $3\pi/2$. The other is on the connection of two corners: *If two corners are linked by an i-edge, the ij-face has the same angle at the two corners if and only if the two j-edges have the same orientation; it has different angles otherwise* (Fig. 8.9). These two constraints are written in terms of our notation as follows:

Proposition 8.7. *No two components of any state* $\mathbf{s}$ *are* 1, *i.e.*,

$$s_{23}s_{31} + s_{31}s_{12} + s_{12}s_{23} = 0 \ . \tag{8.3.7}$$

Proposition 8.8. *If two corners of states* $\mathbf{s}$ *and* $\mathbf{s}'$ *are linked by an i-edge, and if* $\mathbf{t}$ *is the transformation that maps one corner to the other, then*

$$s_{ij} = s'_{ij} \oplus t_j \ , \qquad j \neq i \ . \tag{8.3.8}$$

Combining this with Proposition 8.3, we obtain:

Corollary 8.2. *If two adjacent corners linked by an i-edge have types* $\mathbf{c}$ *and* $\mathbf{c}'$ *and states* $\mathbf{s}$ *and* $\mathbf{s}'$, *respectively, then*

$$s_{ij} = s'_{ij} \oplus c_j \oplus c'_j \ , \qquad j \neq i \ . \tag{8.3.9}$$

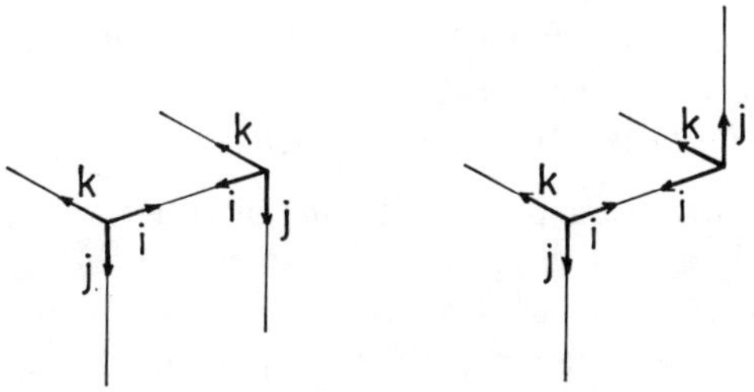

Fig. 8.9. If two corners are linked by an *i*-edge, the *ij*-face has the same angle at the two corners if and only if the two *j*-edges have the same orientation; it has different angles otherwise

Up to now, we have treated the faces of a polyhedron as "transparent membranes". Now, we introduce "visibility conditions" by regarding the polyhedron as filled with an opaque substance. If a corner image has three visible edges, these form either a fork or an arrow. If a corner image has only two visible edges, these make either an obtuse angle or an acute angle, and for each case there are two possibilities for the direction of the remaining hidden edge.

Hence, we first need a criterion which tells us whether two edges make an obtuse angle or an acute angle. This is done by fixing a special coordinate system. So far, the three reference orientations have been numbered arbitrarily, and directions have been assigned to them arbitrarily. Let us assign directions to them in such a way that these three directed lines form a fork. We call such a set a *fork coordnate system.* The coordinate system of Fig. 8.7 is a fork coordinate system. As we can easily see from Fig. 8.7, we find:

Lemma 8.3. *The i- and j-edges of a corner of type* **c** *make an obtuse angle if* $c_i \oplus c_j = 0$ *and an acute angle if* $c_i \oplus c_j = 1$.

Lemma 8.4. *A corner is a fork if its type is* (000) *or* (111). *It is an arrow otherwise.*

In detecting hidden edges, the following obvious facts play a fundamental role.

Lemma 8.5. *If a corner is a fork, its orientation is either* (000) *or* (111). *If it is an arrow in which the i-edge is the central edge and the j- and k-edges are the side edges, then either* $p_i = 0$, $p_j = p_k = 1$ *or* $p_i = 1$, $p_j = p_k = 0$.

Lemma 8.6. *For a corner with two visible edges, the hidden edge is always directed away from the viewer.*

Lemma 8.6 is obtained by exhausting all the cases in which one edge is hidden by the face defined by the other two edges. From these lemmas and exhaustive consideration of all possibilities, we obtain the following theorem:

Theorem 8.2 (*Visibility conditions*). *For a corner of a given type* **c**, *its orientation* **ap** *and state* **s** *must satisfy the constraints shown in Fig.* 8.10 *and those derived from them by permutations of the* 1-, 2-, *and* 3-*components.*

An actual interpretation procedure runs as follows. Suppose a portion of a rectangular polyhedron image is given. First, we choose a fork coordinate system and define the types of all visible corners. For a corner with two visible edges, only the corresponding components are determined. Then, we can freely talk of obtuse angles, acute angles, forks, and arrows (Lemmas 8.3, 8.4). Next, find a corner that has a hidden edge, and hypothesize one of the two

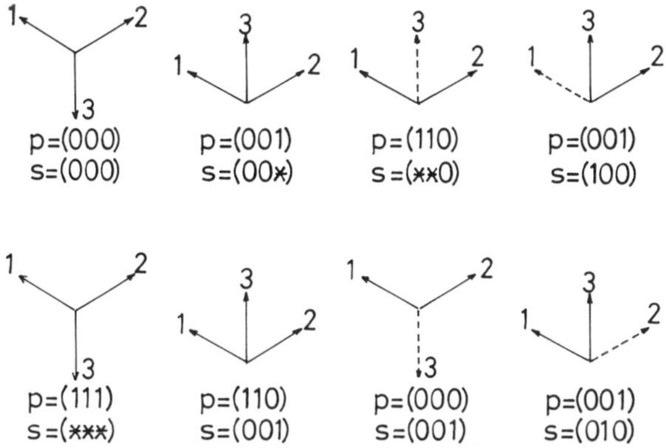

Fig. 8.10. For a corner of a given type ***c***, its orientation ***p*** and state ***s*** must satisfy the constraints shown here and those derived by permutations of the 1-, 2- and 3-components. Dashed lines indicate hidden edges, and * denotes an indeterminate bit

possibilities for its direction. The orientation ***p*** of the corner is uniquely determined (Lemmas 8.5, 8.6). Consequently, the 3D orientations of all the visible corners are uniquely determined (Corollary 8.2).

Next, hypothesize a state ***s*** for each corner (some components may be missing) in such a way that the visibility conditions of Fig. 8.10 are satisfied. Then, check each pair of corners linked by an edge to see if Propositions 8.7 and 8.8 are satisfied. As soon as any inconsistency is found, the previous hypothesis is replaced by another. If we proceed in this way and obtain an assignment of types, orientations, and states without any inconsistency (i.e., Proposition 8.7 and 8.8 and the visibility conditions of Theorem 8.2 are all satisfied), we say that an *interpretation* has been obtained. The interpretation may not be unique.

Example 8.5. Consider the image of Fig. 8.11a. Using the fork coordinate system of Fig. 8.11b, we find that corner A has type (111), while corner B has type (00*) with the 3-edge hidden. If we hypothesize its type to be (001) as shown in Fig. 8.11c, the visibility conditions of Fig. 8.10 demand that the orientation and state of corner B be (110) and (**1), respectively. Then, Corollary 8.2 demands that the orientation of corner A be (000). Hence, the visibility conditions of Fig. 8.10 require that its state be (000).

However, this contradicts Proposition 8.8, because $s_{12} = 0$ at corner B would imply $s_{12} = 1$ at corner A. Hence, the hidden edge is drawn as in Fig. 8.11d, and consequently corner B is given type (000). From the visibility conditions of Fig. 8.10, its orientation and state are respectively (000) and (000). From Corollary 8.2, corner A has orientation (111). From Propositions 8.7 and 8.8, its state is (010). Thus, the interpretation is unique.

Example 8.6. Consider the image of Fig. 8.12a. Using the fork coordinate system of Fig. 8.12b, we find that corner A has type (010), while corner B has

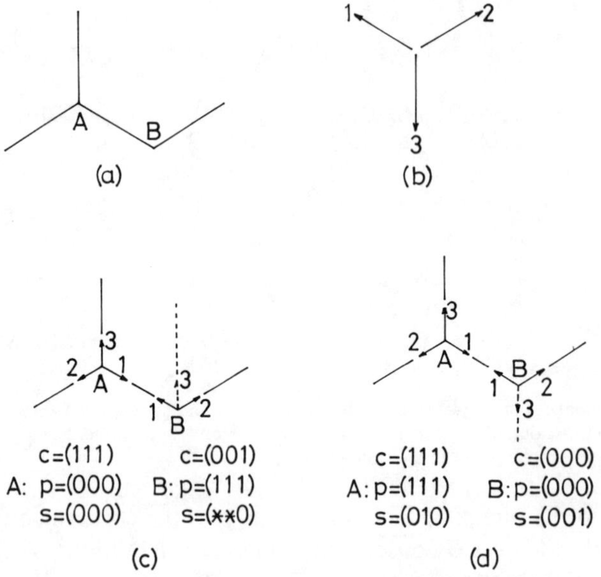

Fig. 8.11a. A corner image. **b** A fork coordinate system. **c** An assignment of orientations and states that yields inconsistency. **d** A uniquely determined consistent interpretation

type (*11) with the 1-edge hidden. If we hypothesize its type to be (111) as shown in Fig. 8.12d, the visibility conditions of Fig. 8.10 demand that the orientation and state of corner *B* be (000) and (100), respectively. Then, Corollary 8.2 demands that the orientation of corner *A* be (101). Hence, the visibility conditions of Fig. 8.10 require that its state be (010).

However, this contradicts Proposition 8.8, because $s_{23} = 1$ at corner *B* would imply $s_{23} = 1$ at corner *A*. Hence, the hidden edge is drawn as in Fig. 8.12c, and consequently corner *B* is given type (011). From Corollary 8.2, its orientation is (011). However, the states of corners *A* and *B* are not uniquely determined. From the visibility conditions of Fig. 8.10, possible states of corners *A* and *B* are $s(A) = s(B) = (000)$ (part of an isolated box), $s(A) = s(B) = (010)$ (part of a box sticking out of a vertical wall) or $s(A) = (000)$ and $s(B) = (001)$ (part of a box on a horizontal plane).

Remark 8.1. As we saw in Example 8.6, the interpretation may not be unique. Are all the resulting interpretations physically admissible? The answer is no. We call this fact the *wall paradox*.

In our scheme, an edge that belongs to the outermost boundary yields two interpretations—one as an occluding edge and the other as a concave edge adjacent to an outside "wall". In our scheme, we took no account of the lengths

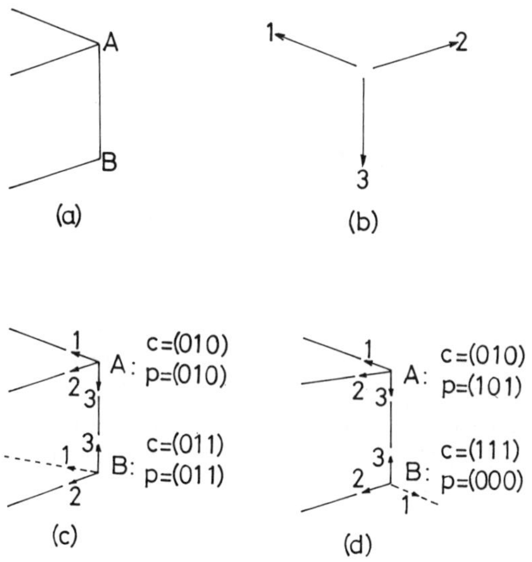

Fig. 8.12a. A corner image. **b** A fork coordinate system. **c** The states are not uniquely determined: part of an isolated box, part of a box sticking out of a vertical wall, or part of a box on a horizontal plane. **d** An assignment of orientations and states that yields inconsistency

of edges; we did not use any *metric properties* (xy-coordinates of vertices, lengths of edges, angles between edges, etc.). Hence, the interpretations of Fig. 8.13, for example, include one which tells us that edges a and b are both adjacent to an outside "wall". This wall paradox does not necessarily occur at edges of the truly "outermost boundary. The "wall" adjacent to edges a and b of Fig. 8.13 can be a face of another big polyhedron, and thus the paradox can occur inside such a face as well.

In order to avoid this wall paradox, we must incorporate metric properties. The wall paradox cannot be removed by Huffman–Clowes edge labeling (Sect. 8.3.1), either. This is because it, too, does not incorporate metric properties. The 3D shape recovery of polyhedra based on metric properties will be discussed in Chap. 10.

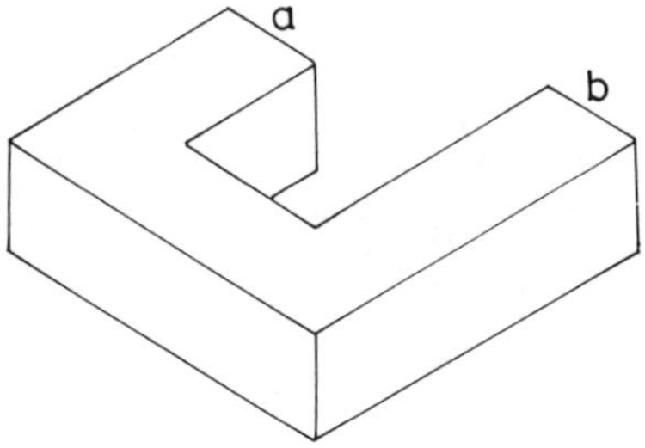

Fig. 8.13. Wall paradox. Edges a and b are both adjacent to an outside "wall"!?

Remark 8.2. What happens if our interpretation scheme is applied to a false image purporting to be a projection of a real rectangular polyhedron such as those in Fig. 8.14a and Fig. 8.15? These *impossible objects* can be classified into two categories.

One consists of those images that can be projections of real objects if curved edges and surfaces are allowed in such a way that Facts 1–4 are all satisfied. Figure 8.14a is one such example, which might look like Fig. 8.14b from a slightly different viewpoint. As we can see, each edge has the same orientation at both ends (but is curved elsewhere), and a face has the same orientation at its corners (but is curved elsewhere). Besides, there exist only three edge orientations and three face orientations at all corners. Thus, all the requirements on which our interpretation scheme is based are satisfied.

The other category consists of those images that cannot be interpreted in this way. Figure 8.15 is an example. Our scheme produces no interpretations for such images, rejecting them as impossible. In other words, only the former type (such as Fig. 8.14a) cannot be rejected, because all the requirements used in our scheme are satisfied. However, our scheme is "stronger" in rejecting impossible

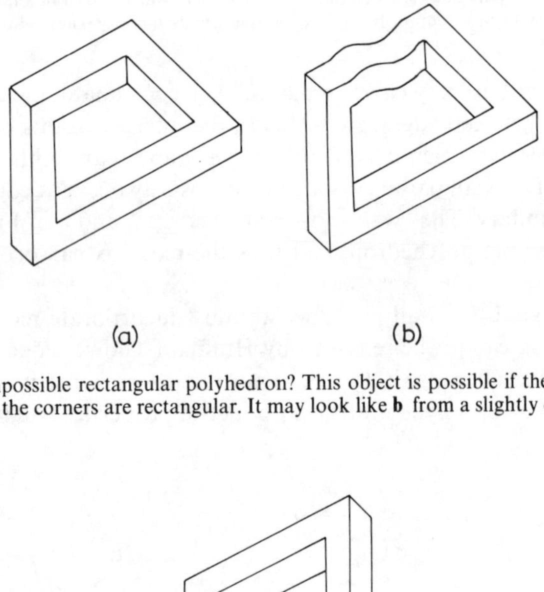

Fig. 8.14a. An impossible rectangular polyhedron? This object is possible if the edges and surface are curved but all the corners are rectangular. It may look like **b** from a slightly different viewpoint

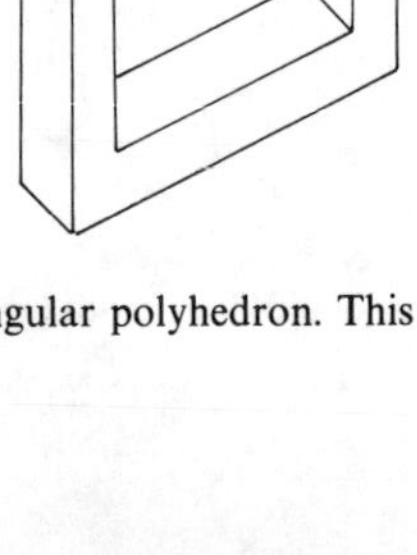

Fig. 8.15. An impossible rectangular polyhedron. This object is rejected as impossible

objects than the Huffman–Clowes edge labeling (see Sect. 8.3.1), which can reject neither Fig. 8.14a nor Fig. 8.15.

8.3.1 Huffman–Clowes Edge Labeling

In our interpretation scheme, each "corner" is labeled, and consistency conditions are checked at "edges that link corners". In the case of a general polyhedron, the best known interpretation scheme is the *Huffman–Clowes edge labeling* due to David A. Huffman and Maxwell B. Clowes. In contrast to our scheme, each "edge" is labeled, and consistency conditions are checked at "corners where edges meet".

The edge labels consist of four symbols $+$, $-$, $\rightarrow$, $\leftarrow$. The symbol $+$ means that the edge is convex, while the symbol $-$ means that the edge is concave. The arrows $\rightarrow$, $\leftarrow$ mean that the edge is an occluding edge constituting the boundary of the polyhedron; its direction is assigned in such a way that a face is adjacent on the right side and the background is on the left side (Fig. 8.16). The consistency conditions for a trihedral polyhedron are listed in Fig. 8.17, where T-junctions are also included. Let us collectively call corner images and T-junctions simply *junctions*. Then, we can easily confirm that all possible labelings of a junction must be those listed there. Figure 8.17 is called the *Huffman–Clowes junction dictionary*. The junctions are classified into forks, arrows and T's according to their 2D configurations. (However, the definition of forks and arrows is different from ours given in Sect. 8.2, because strong conditions are required on the three angles between edges. Here, the configuration is an arrow, a T, or a fork as the largest angle between edges is larger than, equal to, or less than π, respectively).

While Huffman–Clowes edge labeling can be applied to a general polyhedron, the constraints it gives are weaker than ours. For example, if we apply Huffman–Clowes edge labeling to Fig. 8.11a, we end up with the seven different interpretations shown in Fig. 8.18. In other words, the knowledge of rectangularity reduces these seven to only one, namely Fig. 8.18e, as we have shown in Example 8.5. Moreover, while our corner labeling works locally, Huffman–Clowes edge labeling requires a large number of junctions—usually an entire polyhedron image—to yield a unique interpretation (but still the interpretation may not be unique).

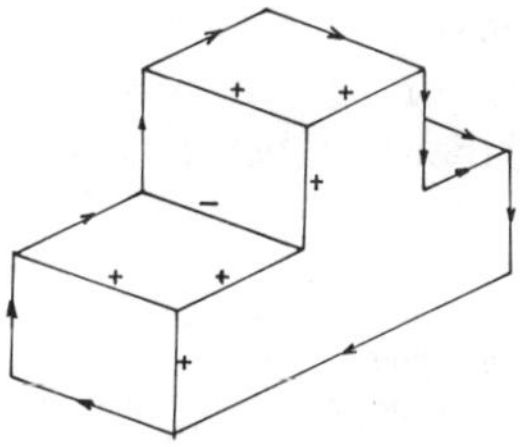

Fig. 8.16. Huffman–Clowes edge labeling

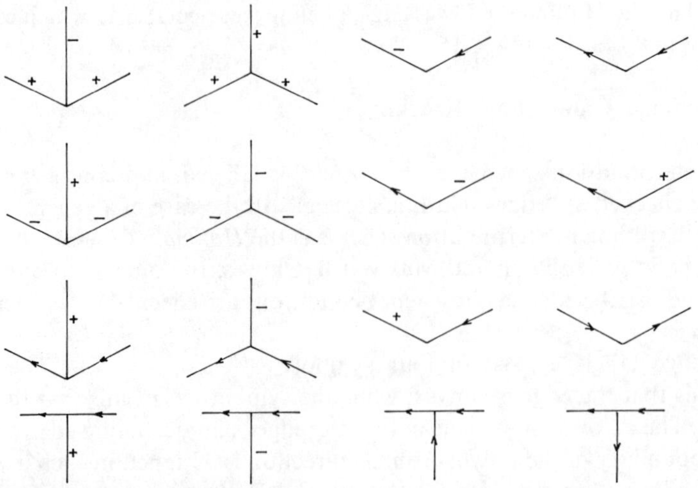

Fig. 8.17. Huffman–Clowes junction dictionary. All possible labelings of three meeting edges are listed

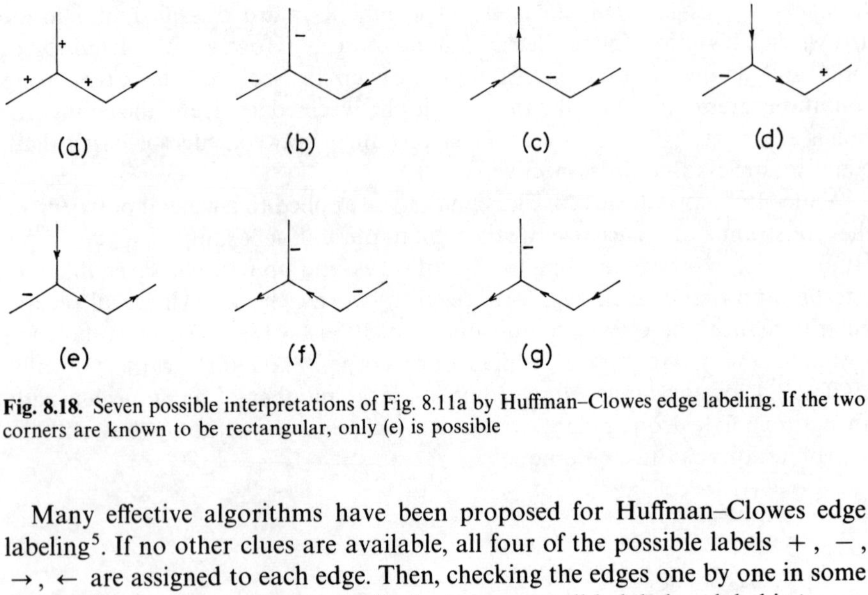

Fig. 8.18. Seven possible interpretations of Fig. 8.11a by Huffman–Clowes edge labeling. If the two corners are known to be rectangular, only (e) is possible

Many effective algorithms have been proposed for Huffman–Clowes edge labeling[5]. If no other clues are available, all four of the possible labels $+$, $-$, $\rightarrow$, $\leftarrow$ are assigned to each edge. Then, checking the edges one by one in some appropriate order, we iteratively eliminate incompatible labels: a label is *incom-*

[5] Readers who are not familiar with computer algorithms—"combinatorial algorithms" in particular—may find the following descriptions difficult to visualize. If so, skip the rest of this section.

patible if no labels can be chosen from among the labels of the adjacent edges to form a labeled junction in the Huffman–Clowes junction dictionary. Alternatively, we can first assign each junction all possible labelings according to the Huffman–Clowes junction dictionary, and then iteratively eliminate incompatible labelings: two labeled junctions are *incompatible* if the labels assigned to the edge connecting them are not identical. Whichever is used, this type of iterative elimination of incompatible labels is called *constraint propagation.*

If multiple labels remain for an edge or a junction after the above procedure, we hypothesize one of them. From this hypothesis, the constraints are again propogated. If we come across an inconsistency (i.e., no label has survived) or end up with a unique labeling, we backtrack to the most recent hypothesis and replace it with an alternative, if any exist. In this way, we can exhaust all possible interpretations.

The process becomes simpler if the object can be distinguished from the background, because the outermost edges can be assumed to be occluding edges. (Then, no wall paradox arises at the outermost boundary.) If appropriately directed arrows are assigned to all the edges constituting the boundary, the constraints are propagated from these boundary edges. In general, however, Huffman–Clowes edge labeling is known to be a "hard" problem. (It is known to be an *NP-complete problem.* However, if the polyhedron image is not so very complicated, it is usually performed very efficiently.)

8.4 Standard Transformation of Corner Images

So far, we have considered only orthographic projection. Now, we study how to compute the 3D orientation of a corner from its perspectively projected image. There is one crucial distinction between orthographic and perspective projections.

Under orthographic projection, we can freely translate a projected image to an arbitrary position across the image plane. The process of 3D recovery is not affected except for the corresponding translation in the scene. This is because the relationship between the 3D shape and its 2D projection is the same all over the image plane. In this sense, the image plane is geometrically homogeneous.

On the other hand, the image distortion due to perspective projection depends on the position on the image plane. Hence, the 3D recovery formulae necessarily involve the position where the object image is observed. This means that the relationship between the 3D shape and its 2D projection differs from position to position. In this sense, the image plane is geometrically inhomogeneous.

But must we always analyze a perspectively projected image in that position? Can we not move it in some way to another position on the image plane so that analysis becomes easy? These questions lead us to the following observations on human perception. When a human finds a familiar object in his field of vision, he first rotates his eye or head so that the object image comes to the center of the field. Then, he estimates the 3D shape and orientation of the object. Finally, recalling the angle of eye or head rotation, he interprets the 3D information in reference to his body.

This human reaction can be simulated by the camera rotation transformation we introduced in Sect. 1.3. We showed that no information is lost by applying the camera rotation transformation, because the original image can be recovered by applying its inverse.[6] Hence, no knowledge about the 3D scene is required. Moreover, many properties are obtained by computation; we need not actually rotate the camera or generate the entire transformed image.

These considerations lead to the following observation: an object image can be moved by the camera rotation transformed into a particular position where analysis becomes easy. We call such a position a *canonical position*. Evidently, the image origin is a prime candidate. Note that as far as 3D corner shape is concerned, *the distinction between orthographic and perspective projections disappears if the vertex is at the image origin*. This is because perspective distortion occurs "radially", and their projected orientations are not affected.

In the following, we construct a camera rotation transformation that maps a given point onto the image origin. We call it the *standard transformation* of the point. Then, we derive a formula that predicts the 2D configuration of a corner image after its vertex is mapped onto the image origin. By this formula, we can reduce the problem to the one under orthographic projection. Hence, the results of Sect. 8.2 can be applied without modification, and the computed 3D information is transformed back into the original configuration.

Let us take the viewer-centered coordinate system; the viewpoint is at the origin O of the XYZ-coordinate system, and the image plane is $Z = f$, where f is the focal length. Consider a point (a, b) on the image plane. The camera rotation transformation that maps point (a, b) onto the image origin is given as follows:

Proposition 8.9 (*Standard rotation and standard transformation*). *If the camera is rotated by*

$$\boldsymbol{R}(a, b) = \begin{pmatrix} E & F & l_1 \\ F & G & l_2 \\ -l_1 & -l_2 & l_3 \end{pmatrix}, \tag{8.4.1}$$

[6] Of course, we need the proviso pointed out in Chaps. 4 and 5. Namely, we must assume that the object image is localized and remains in the finite part of the image plane when the camera is rotated.

where

$$l_1 \equiv \frac{a}{\sqrt{a^2+b^2+f^2}}, \quad l_2 \equiv \frac{b}{\sqrt{a^2+b^2+f^2}}, \quad l_3 \equiv \frac{f}{\sqrt{a^2+b^2+f^2}}, \tag{8.4.2}$$

$$E \equiv \frac{a^2 l_3 + b^2}{a^2+b^2}, \quad F \equiv \frac{ab(l_3-1)}{a^2+b^2}, \quad G \equiv \frac{b^2 l_3 + a^2}{a^2+b^2}, \tag{8.4.3}$$

point (a, b) *is mapped onto the image origin. The induced image transformation takes the form*

$$x' = f\frac{Ex+Fy-l_1 f}{l_1 x + l_2 y + l_3 f}, \qquad y' = f\frac{Fx+Gy-l_2 f}{l_1 x + l_2 y + l_3 f}. \tag{8.4.4}$$

Proof. The 3D unit vector starting from the viewpoint O pointing toward the point (a, b) on the image plane $Z = f$ is given by

$$\boldsymbol{m} = \left(\frac{a}{\sqrt{a^2+b^2+f^2}}, \quad \frac{b}{\sqrt{a^2+b^2+f^2}}, \quad \frac{f}{\sqrt{a^2+b^2+f^2}}\right). \tag{8.4.5}$$

(Fig. 8.19). This vector makes angle

$$\Omega = \tan^{-1}\frac{\sqrt{a^2+b^2}}{f} \tag{8.4.6}$$

with the unit vector $\boldsymbol{k} = (0, 0, 1)$ along the Z-axis. The unit vector perpendicular to both $\boldsymbol{m}$ and $\boldsymbol{k}$ is

$$\boldsymbol{n} = \frac{\boldsymbol{k}\times\boldsymbol{m}}{\|\boldsymbol{k}\times\boldsymbol{m}\|} = \left(-\frac{b}{\sqrt{a^2+b^2}}, \quad \frac{a}{\sqrt{a^2+b^2}}, \quad 0\right). \tag{8.4.7}$$

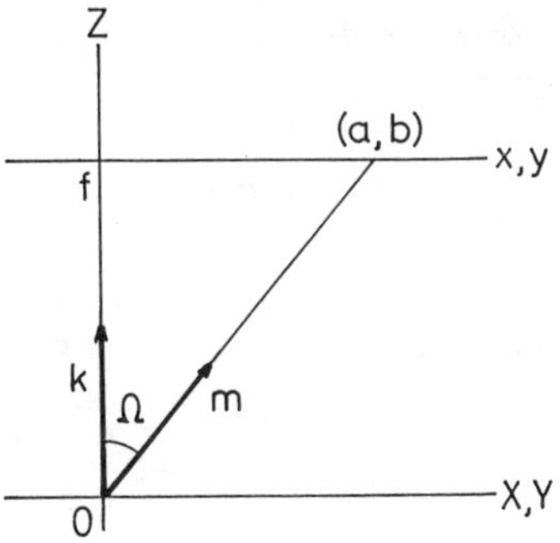

Fig. 8.19. Point (a, b) on the image plane $Z = f$ defines unit vector $\boldsymbol{m}$, which starts from the viewpoint O and points toward (a, b)

If the camera is rotated around vector $\boldsymbol{n} = (n_1, n_2, n_3)$ screw-wise by angle Ω, the point (a, b) is mapped onto the image origin. The corresponding rotation matrix is

$$\boldsymbol{R} = \begin{pmatrix} \cos\Omega + n_1^2(1-\cos\Omega) & n_1 n_2(1-\cos\Omega) - n_3\sin\Omega & n_1 n_3(1-\cos\Omega) + n_2\sin\Omega \\ n_2 n_1(1-\cos\Omega) + n_3\sin\Omega & \cos\Omega + n_2^2(1-\cos\Omega) & n_2 n_3(1-\cos\Omega) + n_1\sin\Omega \\ n_3 n_1(1-\cos\Omega) - n_2\sin\Omega & n_3 n_2(1-\cos\Omega) + n_1\sin\Omega & \cos\Omega + n_3^2(1-\cos\Omega) \end{pmatrix} \tag{8.4.8}$$

(Proposition 6.5). Substituting (8.4.6, 7) into this, we obtain (8.4.1) (this is a special case of Theorem 4.3). Since the image transformation induced by camera rotation $\boldsymbol{R} = (r_{ij})$, $i = 1, 2, 3$, is

$$x' = f\frac{r_{11}x + r_{21}y + r_{31}f}{r_{13}x + r_{23}y + r_{33}f}, \quad y' = f\frac{r_{12}x + r_{22}y + r_{33}f}{r_{13}x + r_{23}y + r_{33}f} \tag{8.4.9}$$

(Theorem 1.1), we obtain (8.4.4).

We call the rotation $\boldsymbol{R}(a, b)$ the *standard rotation* to map point (a, b) onto the image origin, and the transformation (8.4.4), which we denote by $T_{(a,b)}$, the *standard transformation* with respect to point (a, b). From Corollary 1.1, we obtain:

Corollary 8.3. *The inverse* $T_{(a,b)}^{-1}$ *of the standard transformation* $T_{(a,b)}$ *is given by*

$$x = f\frac{Ex' + Fy' + l_1 f}{l_1 x' + l_2 y' - l_3 f} \quad y = f\frac{Fx' + Gy' + l_2 f}{l_1 x' + l_2 y' - l_2 f}. \tag{8.4.10}$$

Remark 8.3. As we pointed out in Sect. 4.8, the standard rotation $\boldsymbol{R}(a, b)$ is not the only camera rotation that maps the point (a, b) onto the image origin: after applying $\boldsymbol{R}(a, b)$, we can add an arbitrary rotation around the Z-axis. In this sense, the standard rotation $\boldsymbol{R}(a, b)$ is regarded as a rotation *that does not contain rotations around the Z-axis.*[7] This is similar to the rotations of the eye or the head; they rotate up, down, right, and left, but not around the line of sight.

Consider the orthographic limit $f \to \infty$. From (8.4.2, 3), we see that $(l_1, l_2, l_3) \to (0, 0, 1)$, $E \to 1$, $F \to 0$, $G \to 1$ as $f \to \infty$. If we note that $l_1 f \to a$ and $l_2 f \to b$ from (8.4.2), the standard transformation $T_{(a,b)}$ reduces to translation

$$x' = x - a\,, \qquad y' = y - b\,. \tag{8.4.11}$$

Thus, the standard transformation $T_{(a,b)}$ is a natural extension of image translation.

[7] The rotation of a camera around its optical axis is called *swing*, while the rotations around the other two axes perpendicular to it are called *tilt* and *pan*—which is which depends on convention. Thus, the standard rotation defined here is a rotation with *zero swing*.

Suppose we observe a half-line starting from point (a, b) on the image plane. Let us call the angle φ it makes from the x-axis measured clockwise the *angle* of the half-line. If we apply the standard transformation $T_{(a,b)}$, the transformed half-line now starts from the image origin. Let us call the angle $\bar{\varphi}$ of the transformed half-line the *canonical angle* (Fig. 8.20).

Theorem 8.3 (*Canonical angle*). *The canonical angle $\bar{\varphi}$ of a half-line of angle φ starting from point (a, b) is given by*

$$\bar{\varphi} = -\tan^{-1}\frac{(fE + al_1)\tan\varphi - (fF + bl_1)}{(fF + al_2)\tan\varphi - (fG + bl_2)} \,. \tag{8.4.12}$$

From among the two values resulting from $\tan^{-1}$, *the one closer to φ is chosen.*

Proof. A line on the image plane is written in the form

$$Ax + By + C = 0 \ , \tag{8.4.13}$$

or $A\!:\!B\!:\!C$ (Sect. 4.2). This line is mapped by the camera rotation $\boldsymbol{R} = (r_{ij})$, $i = 1, 2, 3$, onto line

$$A'x + B'y + C' = 0 \ , \tag{8.4.14}$$

or $A'\!:\!B'\!:\!C'$, where

$$A'\!:\!B'\!:\!C' = \left(r_{11}A + r_{21}B + \frac{1}{f}r_{31}C\right)\!:\!\left(r_{12}A + r_{22}B + \frac{1}{f}r_{32}C\right)$$
$$:\!(f(r_{13}A + r_{23}B) + r_{33}C) \tag{8.4.15}$$

(Lemma 4.1). A line passing through point (a, b) is written as

$$A(x - a) + B(y - b) = 0 \ , \tag{8.4.16}$$

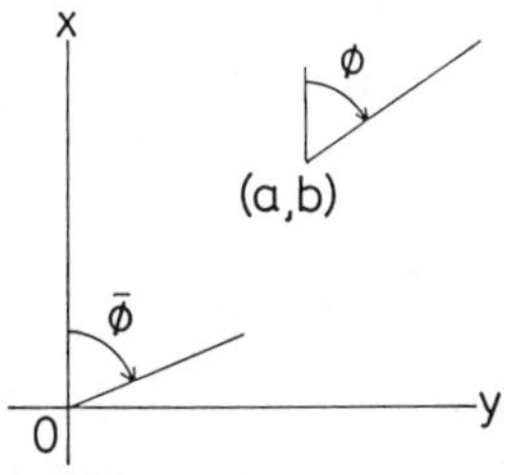

Fig. 8.20. Canonical angle $\bar{\varphi}$. A half-line starting from point (a, b) at an angle φ from the x-axis is mapped by the standard transformation $T_{(a,b)}$ onto a half-line starting from the image origin with angle $\bar{\varphi}$

or $A:B:-(Aa+Bb)$. Applying (8.4.15), we see that line (8.4.16) is mapped by the standard transformation $T_{(a,b)}$ onto line

$$\bar{A}x+\bar{B}y=0 \ , \tag{8.4.17}$$

or $\bar{A}:\bar{B}:0$, where the ratio $\bar{A}:\bar{B}$ is given by

$$\frac{\bar{A}}{\bar{B}}=\frac{(fE+al_1)A+(fF+bl_1)B}{(fF+al_2)A+(fG+bl_2)B} \ . \tag{8.4.18}$$

If the original angle is φ and the canonical angle is $\bar{\varphi}$, we have

$$\frac{A}{B}=-\tan\varphi \ , \qquad \frac{\bar{A}}{\bar{B}}=-\tan\bar{\varphi} \ . \tag{8.4.19}$$

From these follows (8.4.12), where one values of $\tan^{-1}$ is chosen by assuming that the camera rotation is not very large.

Corollary 8.4. *If a half-line has canonical angle $\bar{\varphi}$, its angle φ at (a, b) is given by*

$$\varphi=\tan^{-1}\frac{(fG+bl_2)\tan\bar{\varphi}+(fF+bl_1)}{(fF+al_2)\ \tan\bar{\varphi}+(fE+al_1)} \ . \tag{8.4.20}$$

Again, from among the two values of φ, the one closer to $\bar{\varphi}$ is chosen.

Remark 8.4. Although (8.4.19) is sufficient for theoretical purposes, it is not desirable for computational purposes because $\tan\varphi\to\infty$ as $\varphi\to\pi/2$. One way to avoid this is to use (8.4.19) for $0\leqq\varphi<\pi/4$, $3\pi/4\leqq\varphi<5\pi/4$, $7\pi/4\leqq\varphi<2\pi$, and otherwise use

$$\bar{\varphi}=-\cot^{-1}\frac{(fF+al_2)-(fG+bl_2)\cot\varphi}{(fF+al_1)-(fF+bl_1)\ \cot\varphi} \ , \tag{8.4.21}$$

which is equivalent to (8.4.19). Similarly, we can use, instead of (8.4.10),

$$\varphi=\cot^{-1}\frac{(fF+al_2{}^1)+(fE+al_1)\cot\bar{\varphi}}{(fG+bl_2)+(fF+bl_1)\cot\bar{\varphi}} \ , \tag{8.4.22}$$

depending upon the angle $\bar{\varphi}$.

Thus, when analyzing the 3D shape of a corner, we need not apply the standard transformation $T_{(a,b)}$ to the image itself: *the canonical angles can be computed.* Then, the 3D edge orientations can be computed by Proposition 8.1. Once the unit vector $\bar{\boldsymbol{n}}$ along an edge is computed in this canonical position, its 3D orientation in the original configuration is computed as follows:

Proposition 8.10. *If an edge has* 3D *orientation $\bar{\boldsymbol{n}}$ in the canonical position, its* 3D *orientation in the original configuration where the vertex appeared at point (a, b) is given by unit vector*

$$\boldsymbol{n}=\boldsymbol{R}(a,b)\bar{\boldsymbol{n}} \ . \tag{8.4.23}$$

Proof. Application of the standard transformation $T_{(a,b)}$ means that the camera has been rotated by $\boldsymbol{R}(a, b)$ relative to the scene. This means that the scene has been rotated by $\boldsymbol{R}(a, b)^{-1}$ relative to the camera. Hence, vector $\bar{\boldsymbol{n}}$ is transformed back into its original configuration if it is multiplied by $\boldsymbol{R}(a, b)$.

Example 8.7. Consider the building of Fig. 8.21a. The focal length is $f = 28$mm. The image coordinates of the upper-right vertex are (10.0 mm, 7.9 mm), and the angles of the three edges are

$$\varphi_1 = 110^\circ \ , \quad \varphi_2 = 168^\circ \ , \quad \varphi_3 = 224^\circ \ . \tag{8.4.24}$$

From Theorem 8.3, their canonical angles at the image origin become

$$\bar{\varphi}_1 = 111.5^\circ \ , \quad \bar{\varphi}_2 = 165.4^\circ \ , \quad \bar{\varphi}_3 \doteq 224.6^\circ \tag{8.4.25}$$

(Fig. 8.21b). Suppose we know that the corner is rectangular. The configuration

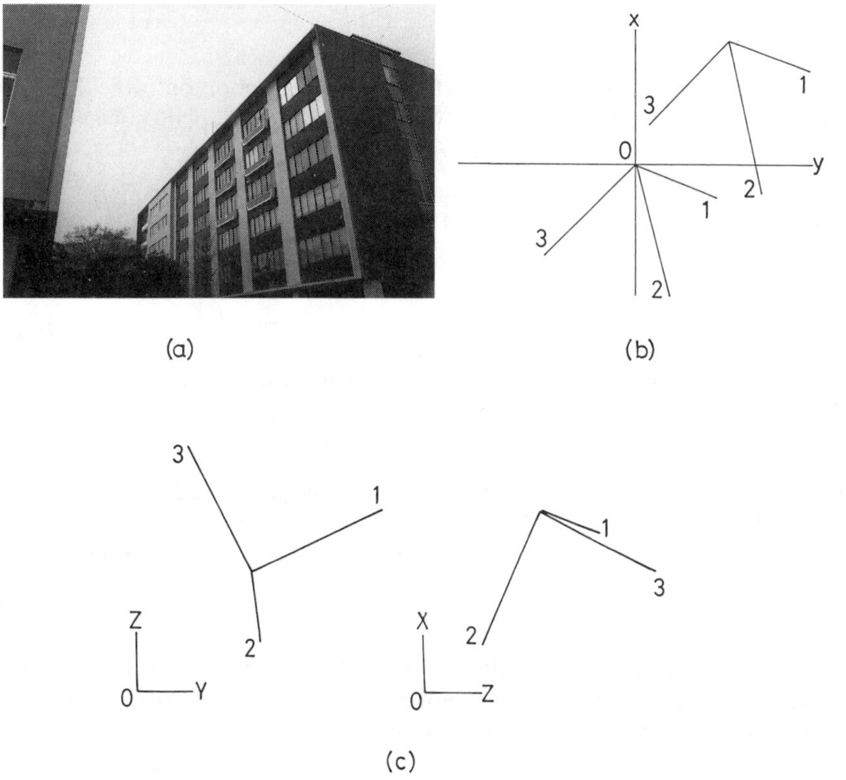

Fig. 8.21a. An image of a building. The upper-right corner is a rectangular corner. **b** The standard transformation applied to the corner of **a**. **c** The top view and the side view of the corner reconstructed from **a**

of (8.4.25) is an arrow. By Proposition 8.1, the angles of these edges from the Z-axis are computed to be

$$\theta_1 = 56.1^\circ\ , \quad \theta_2 = 131.3^\circ\ , \quad \theta_3 = 59.8^\circ\ , \tag{8.4.26}$$

if we assume that the 2-edge extends away from the viewer and the 1- and 3-edges toward the viewer. (If we do not assume this, the mirror image is also obtained.) From (8.2.1), the corresponding unit vectors are

$$\bar{\boldsymbol{n}}_1 = \begin{pmatrix} -0.305 \\ 0.772 \\ 0.557 \end{pmatrix}, \quad \bar{\boldsymbol{n}}_2 = \begin{pmatrix} -0.727 \\ 0.190 \\ -0.660 \end{pmatrix}, \quad \bar{\boldsymbol{n}}_3 = \begin{pmatrix} -0.616 \\ -0.606 \\ 0.503 \end{pmatrix}. \tag{8.4.27}$$

From (8.4.23), their 3D orientations in the original configuration are

$$\boldsymbol{n}_1 = \begin{pmatrix} -0.141 \\ 0.902 \\ 0.408 \end{pmatrix}, \quad \boldsymbol{n}_2 = \begin{pmatrix} -0.909 \\ 0.045 \\ -0.414 \end{pmatrix}, \quad \boldsymbol{n}_3 = \begin{pmatrix} -0.392 \\ -0.429 \\ 0.814 \end{pmatrix}. \tag{8.4.28}$$

Figure 8.21c shows the top view (orthographic projection onto the YZ-plane) and the side view (orthographic projection onto the ZX-plane), where the 3D position of the vertex is taken arbitrarily.

Example 8.8. Consider the object in Fig. 8.22a. The focal length is $f = 28$ mm. The image coordinates of the upper-right vertex are (9.0 mm, 11.1 mm), and the angles of the three edges are

$$\varphi_1 = 163^\circ\ , \quad \varphi_2 = 193^\circ\ , \quad \varphi_3 = 257^\circ\ . \tag{8.4.29}$$

From Theorem 8.3, their canonical angles at the image origin become

$$\bar{\varphi}_1 = 160.8^\circ\ , \quad \bar{\varphi}_2 = 189.7^\circ\ , \quad \bar{\varphi}_3 = 259.7^\circ \tag{8.4.30}$$

(Fig. 8.22b). Suppose we know that the 1- and 2-edges make an angle of 60°, the 2- and 3-edges make 90°, and the 3- and 1-edges also make 90°. By the analysis of Sect. 8.2.1, the angles between these edges and the Z-axis are computed to be

$$\theta_1 = 72.1^\circ\ , \quad \theta_2 = 125.5^\circ\ , \quad \theta_3 = 64.3^\circ, \tag{8.4.31}$$

if we assume that the 2-edge extends toward the viewer and the 1- and 3-edges away from the viewer. (If we do not assume this, the mirror image is also obtained.) The corresponding unit vectors are

$$\bar{\boldsymbol{n}}_1 = \begin{pmatrix} -0.899 \\ 0.313 \\ 0.307 \end{pmatrix}, \quad \bar{\boldsymbol{n}}_2 = \begin{pmatrix} -0.803 \\ -0.138 \\ -0.580 \end{pmatrix}, \quad \bar{\boldsymbol{n}}_3 = \begin{pmatrix} -0.161 \\ -0.887 \\ 0.433 \end{pmatrix}. \tag{8.4.32}$$

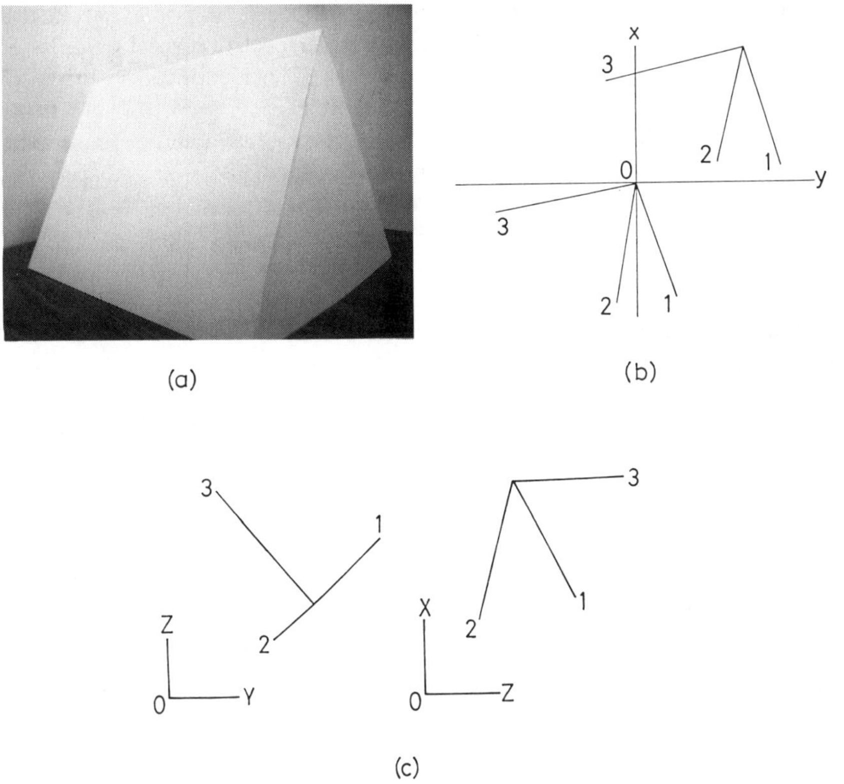

Fig. 8.22a. An object image. The three edges of the upper-right corner make angles of 60°, 90°, and 90°. **b** The standard transformation applied to the three edges of **a**. **c** The top view and the side view of the corner reconstructed from **a**

From (8.4.23), their 3D orientations in the original configuration are

$$\boldsymbol{n}_1 = \begin{pmatrix} -0.789 \\ 0.449 \\ 0.420 \end{pmatrix}, \quad \boldsymbol{n}_2 = \begin{pmatrix} -0.927 \\ -0.291 \\ -0.239 \end{pmatrix}, \quad \boldsymbol{n}_3 = \begin{pmatrix} 0.017 \\ -0.667 \\ 0.745 \end{pmatrix}. \qquad (8.4.33)$$

Figure 8.22c shows the top view (orthographic projection onto the YZ-plane) and the side view (orthographic projection onto the ZX-plane), where the 3D position of the vertex is taken arbitrarily.

8.5 Vanishing Points and Vanishing Lines

Recall the interpretation scheme for rectangular polyhedra constructed in Sect. 8.3. We immediately notice that the entire formulation is unaffected by whether or not the projection is orthographic, *because no metric properties are involved.* Hence, all the results are also applicable under perspective projection *as long as parallel edges are identified on the image plane.* The identification of parallel edges is not difficult in principle,[8] because there exists a powerful clue: projections of parallel lines in the scene must be *concurrent* on the image plane. Namely, they must intersect at a common point—called the *vanishing point.* In order to prove this well-known fact, we only need to show that the vanishing point of a line is determined by its 3D orientation alone, irrespective of its location in the scene:

Theorem 8.4. *The vanishing point of a line whose 3D orientation is* $\boldsymbol{m} = (m_1, m_2, m_3)$ *is given by* $(fm_1/m_3, fm_2/m_3)$.

Proof. A line passing through point (X_0, Y_0, Z_0) and extending in the direction of vector $\boldsymbol{m} = (m_1, m_2, m_3)$ is given by

$$X = X_0 + tm_1 \ , \qquad Y = Y_0 + tm_2 \ , \qquad Z = Z_0 + tm_3 \ , \tag{8.5.1}$$

where t is a real parameter. The perspective projection of this line is given by

$$x = f\frac{X}{Z} = f\frac{X_0 + tm_1}{Z_0 + tm_3} \ , \qquad y = f\frac{Y}{Z} = f\frac{Y_0 + tm_2}{Z_0 + tm_3} \ . \tag{8.5.2}$$

The vanishing point (a, b) of this line is obtained by taking the limit $t \to \pm\infty$:

$$a = f\frac{m_1}{m_3} \ , \qquad b = f\frac{m_2}{m_3} \ . \tag{8.5.3}$$

This result holds irrespective of the position (X_0, Y_0, Z_0).

Corollary 8.5. *The* 3D *orientation of a line whose vanishing point is* (a, b) *is given by the unit vector*

$$\boldsymbol{m} = \pm\left(\frac{a}{\sqrt{a^2 + b^2 + f^2}}, \frac{b}{\sqrt{a^2 + b^2 + f^2}}, \frac{f}{\sqrt{a^2 + b^2 + f^2}}\right) . \tag{8.5.4}$$

Thus, once parallel edges are detected, their vanishing point determines their 3D orientation. There are, however, three issues to be considered. One is the presence of noise. Since the edges are detected by image processing techniques, they may not be accurate. As a result, parallel lines may not necessarily meet at a single point when projected onto the image plane (Fig. 8.23). How can we estimate their common vanishing point?

---

[8] However, many intricate technicalities will be involved if we are to construct an actual algorithm. We will discuss this in Chap. 10 in detail.

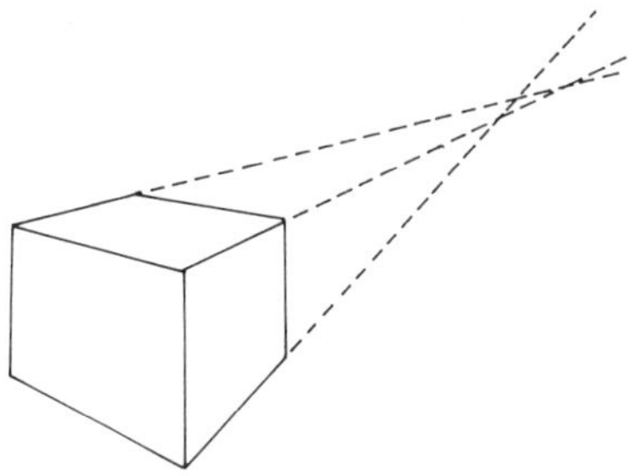

Fig. 8.23. The vanishing point uniquely determines the 3D orientation of the edges. However, how can we estimate it if the edges are, when extended, not exactly concurrent on the image plane?

The second issue is the possibility of overflow. Two edges that are almost parallel on the image plane may meet, when extended, at a point very far apart from the image origin. It follows that the computation may break down or result in poor accuracy. It is desirable that all intermediate numerical data be kept within a reasonable finite range. In other words, computation must be done *in the finite domain.*

The third issue is the question of how to detect parallel edges on the image plane, but in this section, we specifically focus on the first two issues. (The parallel edge detection algorithm will be discussed in Chap. 10.) To this end, we make use of the *duality* of points and lines. Most of the necessary concepts have already been introduced in Chap. 4. Since the study in Chap. 4 was done from a purely algebraic point of view, let us reiterate these results from the viewpoint of 3D recovery. We begin with the following definitions.

Consider a point $P:(a, b)$ on the image plane. The unit vector m starting from the viewpoint O and pointing toward the point P on the image plane $Z=f$ is given by

$$\boldsymbol{m} = \pm\left(\frac{a}{\sqrt{a^2+b^2+f^2}}, \frac{b}{\sqrt{a^2+b^2+f^2}}, \frac{f}{\sqrt{a^2+b^2+f^2}}\right) \tag{8.5.5}$$

(Fig. 8.24a). We call this $\boldsymbol{m}$ the unit vector *associated with* point P.[9] In terms of this definition, Corollary 8.5 is rephrased as follows:

Corollary 8.6. *The vector associated with the vanishing point indicates the 3D orientation of the line.*

Next, consider a line $l: Ax + By + C = 0$ on the image plane. The plane passing through the viewpoint O and intersecting the image plane $Z = f$ along

[9] The use of unit vectors associated with points and lines on the image plane is, from the viewpoint of projective geometry, the use of (normalized) homogeneous coordinates. See Sect. 8.5.1.

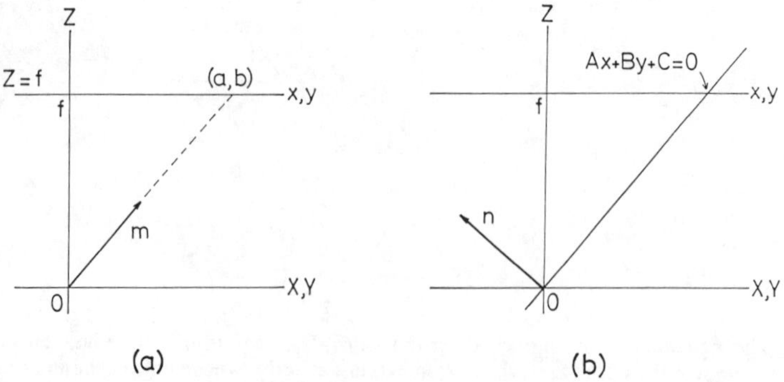

Fig. 8.24. a The unit vector $\boldsymbol{m}$ starting from the viewpoint O and pointing toward point (a, b) on the image plane $Z = f$ is associated with the point (a, b). **b** The unit vector $\boldsymbol{n}$ starting from the viewpoint O and normal to the plane passing through the viewpoint and intersecting the image plane $Z = f$ along line $Ax + By + C = 0$ is associated with the line $Ax + By + C = 0$

this line is $AX + BY + CZ/f = 0$. The unit vector normal to this plane is given by

$$\boldsymbol{n} = \pm\left(\frac{A}{\sqrt{A^2 + B^2 + C^2/f^2}}, \frac{B}{\sqrt{A^2 + B^2 + C^2/f^2}}, \frac{C/f}{\sqrt{A^2 + B^2 + C^2/f^2}}\right) \tag{8.5.6}$$

(Fig. 8.24b). We call this $\boldsymbol{n}$ the unit vector *associated with* the line l.[9]

In order to make use of Corollary 8.5 or 8.6 in the presence of noise, we must first identify a common intersection of not necessarily concurrent lines. Evidently, we must take some average. The direct average of all the intersections of N lines requires the computation of $N(N-1)/2$ intersections, which is obviously undesirable. Moreover, in order to keep the computation within the finite domain, we must avoid directly computing the coordinates of intersections. A simple principle is the least squares method, but there are many alternatives depending on the quantities to be minimized.

Example 8.9. Let $A_i x + B_i y + C_i = 0$, $i = 1, \ldots, N$, be not necessarily concurrent lines on the image plane. Since these equations can be multiplied by an arbitrary nonzero constant, we can assume that the coefficients A_i, B_i, C_i are so chosen that $A_i^2 + B_i^2 = 1$, $i = 1, \ldots, m$. Then, as shown in Fig. 8.25, the distance of line $A_i x + B_i y + C_i = 0$ from point (a, b) is given by $|A_i a + B_i b + C_i|$ (Exercise 8.22). Hence, we can determine the common vanishing point (a, b) by minimizing

$$\sum_{i=1}^{N} (A_i a + B_i b + C_i)^2 \ . \tag{8.5.7}$$

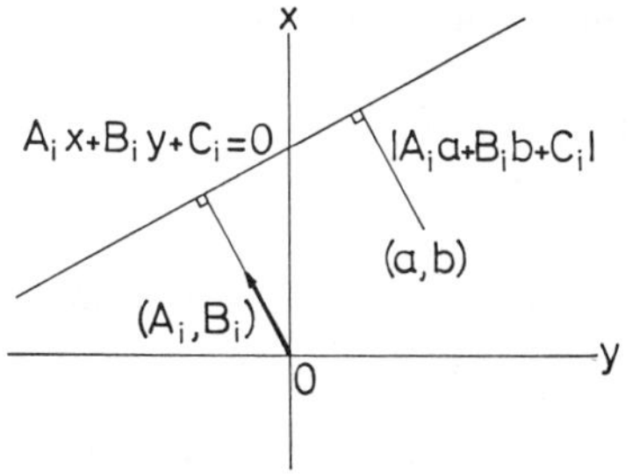

Fig. 8.25. The distance of line $A_i x + B_i y + C_i = 0$ from point (a, b) is $|A_i a + B_i b + C_i|$ if the equation is normalized so that $A_i^2 + B_i^2 = 1$

Taking derivatives with respect to a and b, and setting the respective results to be zero, we obtain the following normal equation (Sect. 7.2):

$$\begin{pmatrix} \sum_{i=1}^{N} A_i^2 & \sum_{i=1}^{N} A_i B_i \\ \sum_{i=1}^{N} A_i B_i & \sum_{i=1}^{N} B_i^2 \end{pmatrix} \begin{pmatrix} a \\ b \end{pmatrix} = - \begin{pmatrix} \sum_{i=1}^{N} A_i C_i \\ \sum_{i=1}^{N} B_i C_i \end{pmatrix} . \tag{8.5.8}$$

The solution is given by

$$a = - \frac{\sum_{i=1}^{N} A_i C_i \sum_{i=1}^{N} B_i^2 - \sum_{i=1}^{N} B_i C_i \sum_{i=1}^{N} A_i B_i}{\sum_{i=1}^{N} A_i^2 \sum_{i=1}^{N} B_i^2 - \left(\sum_{i=1}^{N} A_i B_i \right)^2} ,$$

$$b = - \frac{\sum_{i=1}^{N} A_i^2 \sum_{i=1}^{N} B_i C_i - \sum_{i=1}^{N} A_i B_i \sum_{i=1}^{N} A_i C_i}{\sum_{i=1}^{N} A_i^2 \sum_{i=1}^{N} B_i^2 - \left(\sum_{i=1}^{N} A_i B_i \right)^2} . \tag{8.5.9}$$

If we put

$$m_1' = \sum_{i=1}^{N} A_i C_i \sum_{i=1}^{N} B_i^2 - \sum_{i=1}^{N} B_i C_i \sum_{i=1}^{N} A_i B_i ,$$

$$m_2' = \sum_{i=1}^{N} A_i^2 \sum_{i=1}^{N} B_i C_i - \sum_{i=1}^{N} A_i B_i \sum_{i=1}^{N} A_i C_i ,$$

$$m_3' = -f \left[\sum_{i=1}^{N} A_i^2 \sum_{i=1}^{N} B_i^2 - \left(\sum_{i=1}^{N} A_i B_i \right)^2 \right] , \tag{8.5.10}$$

the vector $\boldsymbol{m}' = (m_1', m_2', m_3')$ indicates the 3D orientation of these parallel lines. Thus, the computation can be confined within the finite domain. The unit vector

$\boldsymbol{m}$ associated with the common intersection is obtained by normalizing $\boldsymbol{m}'$ into $\boldsymbol{m} = \boldsymbol{m}'/\|\boldsymbol{m}'\|$.

Figure 8.26a is an image of a rectangular polyhedron. Suppose the line drawing of Fig. 8.26b is obtained. Let the edges be labeled as indicated in the figure. If the parallel edge detection algorithm presented in Chap. 10 is applied, the edges are grouped together into the following three sets of parallel edges:

$$\{e_1, e_5, e_8, e_{11}, e_{14}\}\ ,\quad \{e_2, e_4, e_9, e_{10}, e_{15}\}\ ,\quad \{e_3, e_6, e_7, e_{12}, e_{13}\}\ .$$

If the above algorithm is applied, their respective 3D orientations are given by the following unit vectors:

$$\begin{pmatrix} 0.4348 \\ 0.8453 \\ 0.3106 \end{pmatrix}, \quad \begin{pmatrix} 0.5967 \\ -0.5584 \\ 0.5763 \end{pmatrix}, \quad \begin{pmatrix} -0.6837 \\ 0.2356 \\ 0.7294 \end{pmatrix}. \tag{8.5.11}$$

Example 8.10. The algorithm in Example 8.9 may break down or result in poor accuracy if the original lines are obtained *by computation* and are not necessarily located in the actual image; if these lines lie very far away from the image origin, some of the coefficients may become very large. This difficulty is overcome if we use, instead of the equations of the lines, the unit vectors associated with them.

Let $\boldsymbol{n}_i = (n_{i(1)}, n_{i(2)}, n_{i(3)})$ be the unit vectors associated with lines $A_i x + B_i y + C_i = 0$, $i = 1, \ldots, N$. As shown in Fig. 8.27, the distance from point (a, b) on the image plane $Z = f$ to the plane passing through the viewpoint O and intersecting the image plane along the line $A_i x + B_i y + C_i = 0$ is $|n_{i(1)}a + n_{i(2)}b + n_{i(3)}f|$ (Exercise 8.23). It follows that we can determine the common vanishing point (a, b) by minimizing

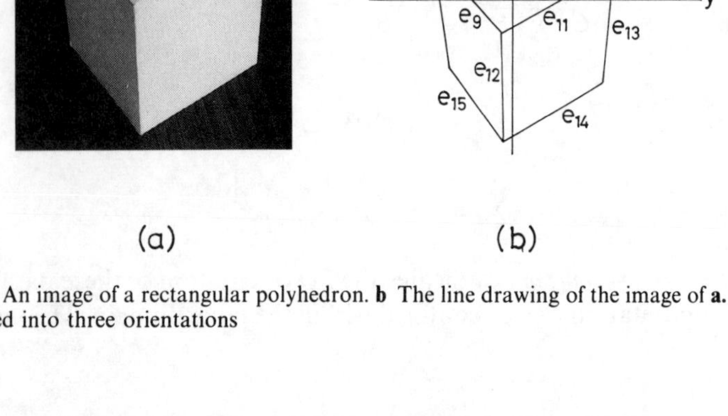

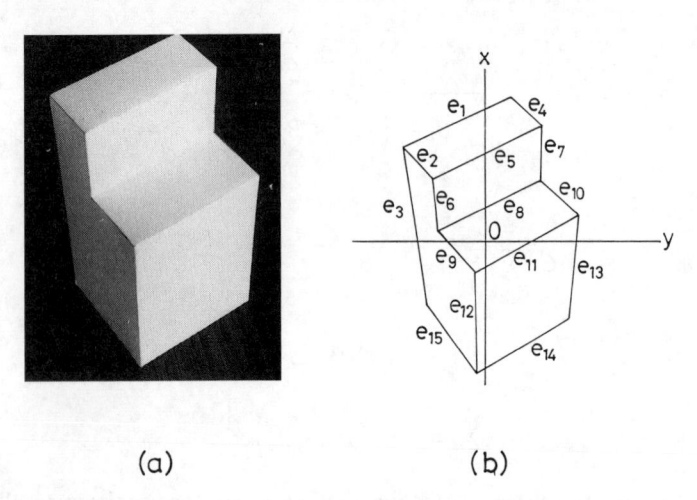

Fig. 8.26a. An image of a rectangular polyhedron. **b** The line drawing of the image of **a.** The edges are classified into three orientations

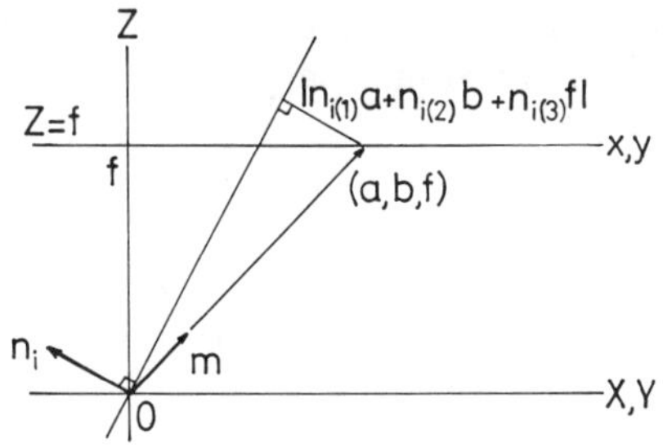

Fig. 8.27. The distance from point (a, b) on the image plane $Z = f$ to the plane passing through the viewpoint and intersecting the image plane along line $A_ix + B_iy + C_i = 0$ is $|n_{i(1)}a + n_{i(2)}b + n_{i(3)}f|$, where $\boldsymbol{n}_i = (n_{i(1)}, n_{i(2)}, n_{i(3)})$ is the unit vector associated with the line $A_ix + B_iy + C_i = 0$

$$\sum_{i=1}^{N} (n_{i(1)}a + n_{i(2)}b + n_{i(3)}f)^2 \ . \tag{8.5.12}$$

Taking derivatives with respect to a and b, and setting the respective results to be zero, we obtain the following normal equations:

$$\begin{pmatrix} \sum_{i=1}^{N} n_{i(1)}^2 & \sum_{i=1}^{N} n_{i(1)}n_{i(2)} \\ \sum_{i=1}^{N} n_{i(1)}n_{i(2)} & \sum_{i=1}^{N} n_{i(2)}^2 \end{pmatrix} \begin{pmatrix} a \\ b \end{pmatrix} = -f \begin{pmatrix} \sum_{i=1}^{N} n_{i(1)}n_{i(3)} \\ \sum_{i=1}^{N} n_{i(2)}n_{i(3)} \end{pmatrix} . \tag{8.5.13}$$

The solution is given by

$$a = -f \frac{\sum_{i=1}^{N} n_{i(1)}n_{i(3)} \sum_{i=1}^{N} n_{i(2)}^2 - \sum_{i=1}^{N} n_{i(2)}n_{i(3)} \sum_{i=1}^{N} n_{i(1)}n_{i(2)}}{\sum_{i=1}^{N} n_{i(1)}^2 \sum_{i=1}^{N} n_{i(2)}^2 - \left(\sum_{i=1}^{N} n_{i(1)}n_{i(2)}\right)^2} ,$$

$$b = -f \frac{\sum_{i=1}^{N} n_{i(1)}^2 \sum_{i=1}^{N} n_{i(2)}n_{i(3)} - \sum_{i=1}^{N} n_{i(1)}n_{i(2)} \sum_{i=1}^{N} n_{i(1)}n_{i(3)}}{\sum_{i=1}^{N} n_{i(1)}^2 \sum_{i=1}^{N} n_{i(2)}^2 - \left(\sum_{i=1}^{N} n_{i(1)}n_{i(2)}\right)^2} . \tag{8.5.14}$$

If we put

$$m'_1 = \sum_{i=1}^{N} n_{i(1)}n_{i(3)} \sum_{i=1}^{N} n_{i(2)}^2 - \sum_{i=1}^{N} n_{i(2)}n_{i(3)} \sum_{i=1}^{N} n_{i(1)}n_{i(2)} ,$$

$$m'_2 = \sum_{i=1}^{N} n_{i(1)}^2 \sum_{i=1}^{N} n_{i(2)}n_{i(3)} - \sum_{i=1}^{N} n_{i(1)}n_{i(2)} \sum_{i=1}^{N} n_{i(1)}n_{i(3)} ,$$

$$m'_3 = \left[\sum_{i=1}^{N} n_{i(1)}^2 \sum_{i=1}^{N} n_{i(2)}^2 - \left(\sum_{i=1}^{N} n_{i(1)}n_{i(2)}\right)^2\right] , \tag{8.5.15}$$

the vector $\boldsymbol{m}' = (m'_1, m'_2, m'_3)$ indicates the 3D orientation of these parallel lines, and the computation is done within the finite domain. The unit vector $\boldsymbol{m}$ is obtained by normalizing $\boldsymbol{m}'$ into $\boldsymbol{m} = \boldsymbol{m}'/\|\boldsymbol{m}'\|$.

For example, if we use this algorithm on the image of Fig. 8.26, we obtain three unit vectors

$$\begin{pmatrix} 0.4348 \\ 0.8453 \\ 0.3107 \end{pmatrix}, \quad \begin{pmatrix} 0.5964 \\ -0.5583 \\ 0.5766 \end{pmatrix}, \quad \begin{pmatrix} -0.6837 \\ 0.2353 \\ 0.7294 \end{pmatrix}. \tag{8.5.16}$$

Example 8.11. The algorithms shown in the above two examples are *not invariant to the camera rotation*: if we apply the camera rotation and then execute these algorithms, the results do not necessarily coincide with the camera rotation of the original results. Also, both fail when the original lines happen to be parallel to each other on the image plane (the vector $\boldsymbol{m}'$ becomes zero). These difficulties can be overcome by the following least squares method. Since the unit vector $\boldsymbol{m}$ associated with the common intersection is supposed to be orthogonal to all the unit vectors $\boldsymbol{n}_i$, $i = 1, \ldots, N$, associated with the concurrent lines, we can estimate the vector $\boldsymbol{m}$ by minimizing

$$\sum_{i=1}^{N} (\boldsymbol{n}_i, \boldsymbol{m})^2 \tag{8.5.17}$$

on the condition that $\boldsymbol{m}$ is a unit vector. It is easy to see that (8.5.17) can be rewritten as the quadratic form

$$\sum_{j,k=1}^{3} N_{jk} m_j m_k \tag{8.5.18}$$

in m_1, m_2, m_3, where $\boldsymbol{N} = (N_{jk})$, $j, k = 1, 2, 3$, is a symmetric matrix given by

$$N_{jk} = \sum_{i=1}^{N} n_{i(j)} n_{i(k)}, \qquad j, k = 1, 2, 3 \tag{8.5.19}$$

(Exercise 8.24). As is well known, minimization of the quadratic form (8.5.18) for unit vector $\boldsymbol{m}$ is attained by choosing the eigenvector of the matrix $\boldsymbol{N}$ for the minimum eigenvalue (Exercise 8.24).

For example, if we compute these eigenvectors for the image shown in Fig. 8.26, we obtain three unit vectors

$$\begin{pmatrix} 0.4351 \\ 0.8461 \\ 0.3080 \end{pmatrix}, \quad \begin{pmatrix} 0.6025 \\ -0.5647 \\ 0.5640 \end{pmatrix}, \quad \begin{pmatrix} -0.6883 \\ 0.2363 \\ 0.7251 \end{pmatrix}. \tag{8.5.20}$$

However, since computation of eigenvalues and eigenvectors is involved, this method is computationally more costly than the previous methods of Examples

8.9 and 8.10. In contrast, the solutions of Examples 8.9 and 8.10 are given as explicit expressions. As we observe from these three examples, the computed values are not so very different. Hence, although this method may be theoretically desirable, the methods of the previous two examples are better suited in practice.

In Chapter 4, we introduced the notion of the *duality of points and lines.* In the present context, this can be stated as follows. A point P and a line l are *dual* to each other *if their associated unit vectors are identical.* Then, we say that the point P is the *pole* of the line l, and the line l is the *polar* of the point P. From this definition, we can immediately prove the following facts (Exercise 8.25):

Lemma 8.7. *The pole of line* $Ax + By + C = 0$ *is* $(f^2A/C, f^2B/C)$.[10]

Lemma 8.8. *The polar of point* (a, b) *is* $ax + by + f^2 = 0$.

Now, we prove the following two essential theorems about this duality (or *polarity*).

Theorem 8.5. *The poles of concurrent lines are collinear, and the common line passing through them is the polar of the common intersection of the concurrent lines.*

Proof. Let $A_ix + B_iy + C_i = 0$, $i = 1, \ldots, N$, be concurrent lines with (a, b) as their the common intersection: $A_ia + B_ib + C_i = 0$, $i = 1, \ldots, N$. Then, their poles $(f^2A_i/C_i, f^2B_i/C_i)$, $i = 1, \ldots, N$, are all on line $ax + by + f^2 = 0$.

Theorem 8.6. *The polars of collinear points are concurrent, and the common intersection is the pole of the common line passing through the collinear points.*

Proof. Let (a_i, b_i), $i = 1, \ldots, N$, be collinear points with $Ax + By + C = 0$ as the common line passing through all of them: $Aa_i + Bb_i + C = 0$, $i = 1, \ldots, N$. Then, their polars $a_ix + b_iy + f^2 = 0$, $i = 1, \ldots, N$, all pass through point $(f^2A/C, f^2B/C)$.

In Examples 8.9–11, we presented algorithms to find a common intersection of not necessarily concurrent lines. Taking advantage of the duality shown above, we can convert these algorithms to ones for *line fitting*, i.e., finding a common line passing through not necessarily collinear points. Instead of finding such a line, we only need to find a common intersection of their polars. The desired line is obtained as the polar of the common intersection (Fig. 8.28).

---

[10] If $C = 0$, the pole is regarded as being located at infinity.

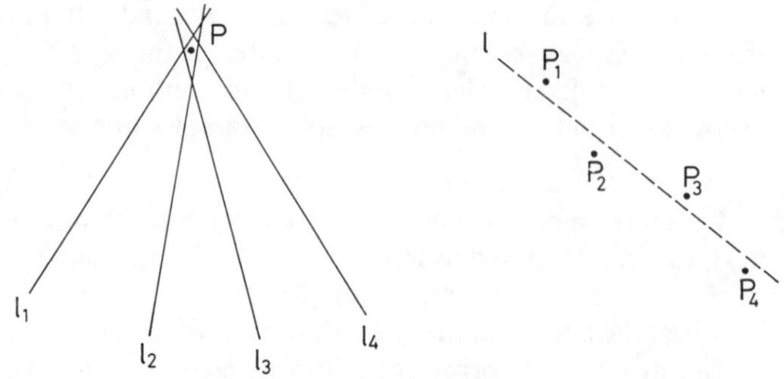

Fig. 8.28. The Hough transform. The poles of concurrent lines are collinear, and the polars of collinear lines are concurrent. The common intersection and the common line are dual to each other. Hence, line fitting is converted into intersection estimation, and vice versa

Example 8.12. Let (a_i, b_i), $i = 1, \ldots, N$, be the data points on the image plane. Their polars are $a_i x + b_i y + f^2 = 0$, $i = 1, \ldots, N$. If we apply the algorithm of Example 8.9, we first multiply these equations by appropriate constants and convert them into $A_i x + B_i y + C_i = 0$ such that $A_i^2 + B_i^2 = 1$, $i = 1, \ldots, N$. Then, the common intersection (a, b) is estimated by (8.5.9). The vector $\boldsymbol{m}' = (m'_1, m'_2, m'_3)$ associated with this point is given by (8.5.10), and the equation of the desired line is given by $m'_1 x + m'_2 y + m'_3 f = 0$.

Example 8.13. Let $\boldsymbol{n}_i = (n_{i(1)}, n_{i(2)}, n_{i(3)})$ be the unit vectors associated with the data points. If we use the algorithm of Example 8.10, the common intersection (a, b) of their polars is estimated by (8.5.14). The vector $\boldsymbol{m}' = (m'_1, m'_2, m'_3)$ associated with this point is given by (8.5.15), and the equation of the desired line is given by $m'_1 x + m'_2 y + m'_3 f = 0$. Like Example 8.10, this algorithm is useful when the original points are given *by computation* and are not necessarily located in the actual image. Computation can be done within the finite domain even if these points are located very far away from the image origin.

Example 8.14. The above two algorithms fail if the data points happen to be on a line passing through the image origin (the vector $\boldsymbol{m}'$ becomes zero). If we use the camera rotation invariant algorithm of Example 8.11, the unit vector $\boldsymbol{m} = (m_1, m_2, m_3)$ associated with the common intersection of the polars is given by the eigenvector of the matrix $\boldsymbol{N}$ of (8.5.19) for the minimum eigenvalue, where $\boldsymbol{n}_i$, $i = 1, \ldots, N$, are the unit vectors associated with the data points. The desired line is given by $m_1 x + m_2 y + m_3 f = 0$.

Remark 8.5. The fitting of a line to not necessarily collinear points by finding a common intersection of their polars is known as the *Hough transform*. This technique is frequently used to detect lines from image data. In practice, points (a_i, b_i) are usually associated with curves $a_i \cos\theta + b_i \sin\theta = r$ on the $r\theta$-plane.

The $r\theta$-plane is digitized into cells, and a cell through which such a curve passes accumulates one "vote". The cell (r_0, θ_0) that gets the maximum number of votes wins, and the line is estimated to be $x \cos\theta_0 + y \sin\theta_0 = r_0$.

Next, we study perspective projection of a planar surface. The starting point is the following well-known fact about the *vanishing line*. Note the "duality" between Theorems 8.7 and 8.4, between Corollaries 8.7 and 8.5, and between Corollaries 8.8 and 8.6.

Theorem 8.7. *The vanishing line of a plane whose surface gradient is* (p, q) *is* $px + qy = f$.

Proof. Consider a point (X, Y, Z) on plane $Z = pX + qY + r$ whose surface gradient is (p, q). Its image coordinates are given by $x = fX/Z$ and $y = fY/Z$. Eliminating X and Y from these three equations, we obtain

$$Z = \frac{fr}{f - px - qy}. \tag{8.5.21}$$

The depth Z is infinite along the line $px + qy = f$ on the image plane irrespective of the value of r. (If $r = 0$, the plane passes through the viewpoint and intersects the image plane along line $px + qy = f$.)

Corollary 8.7. *The surface gradient* (p, q) *of a plane whose vanishing line is* $Ax + By + C = 0$ *is*

$$p = -\frac{fA}{C}, \qquad q = -\frac{fB}{C}. \tag{8.5.22}$$

Corollary 8.8. *The unit vector* $\boldsymbol{n}$ *associated with the vanishing line is the unit surface normal to the plane.*

Proof. The unit surface normal to a plane whose surface gradient is (p, q) is

$$\boldsymbol{n} = \left(\frac{p}{\sqrt{p^2 + q^2 + 1}}, \frac{q}{\sqrt{p^2 + q^2 + 1}}, \frac{-1}{\sqrt{p^2 + q^2 + 1}} \right) \tag{8.5.23}$$

(Exercise 8.26), which is the unit vector associated with the vanishing line $px + qy = f$.

Remark 8.6. From Theorem 8.7, the surface gradient is uniquely determined once two pairs of parallel edges lying on the surface are detected; each pair determines its vanishing point, and the two vanishing points determine the vanishing line uniquely (Fig. 8.29a). If more than two sets of parallel lines belong to the same surface, their vanishing points must be all collinear. In the presence of noise, however, the vanishing points may not be strictly collinear (Fig. 8.29b).

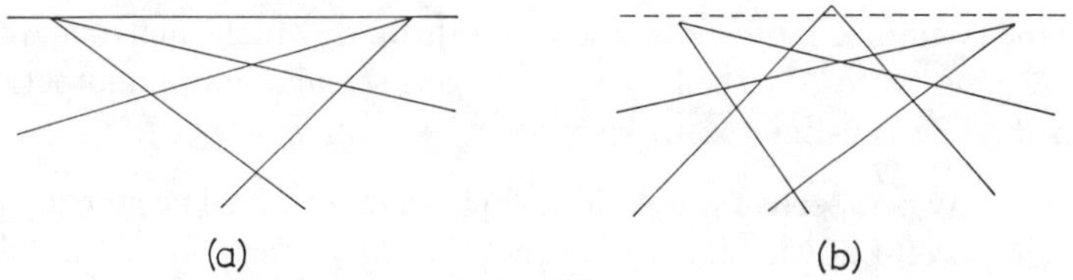

Fig. 8.29a. The vanishing line of a plane is uniquely determined by two pairs of parallel lines on it; their vanishing points must be on the vanishing line. **b** If there are there or more parallel line pairs, their vanishing points may not be exactly collinear in the presence of noise

If three or more vanishing points are not collinear, we must fit an appropriate line. Since the locations of these vanishing points can be very far apart from the image origin, usual line fitting schemes like the least squares method may fail due to possible overflow. This difficulty can be avoided if we use the Hough transform technique of Example 8.13 or 8.14. Then, the computation directly yields the (not necessarily normalized) vector $\boldsymbol{m}'$ associated with the vanishing line, which is the surface normal to the plane (Corollary 8.8).

Finally, recall the definition of *conjugacy* introduced in Chap. 4. Let P and Q be two points on the image plane, and let l be the line passing through them. Let H be the foot of the perpendicular line drawn from the image origin O to line l (Fig. 8.30). We say that points P and Q on line l are *conjugate* to each other if they are on the opposite sides of H and

$$PH \cdot QH = OH^2 + f^2 \ . \tag{8.5.24}$$

We also say that two lines are *conjugate* to each other if their poles are conjugate to each other. We can easily prove the following facts:

Theorem 8.8. *Two points on the image plane are conjugate to each other if and only if their associated vectors are orthogonal to each other.*

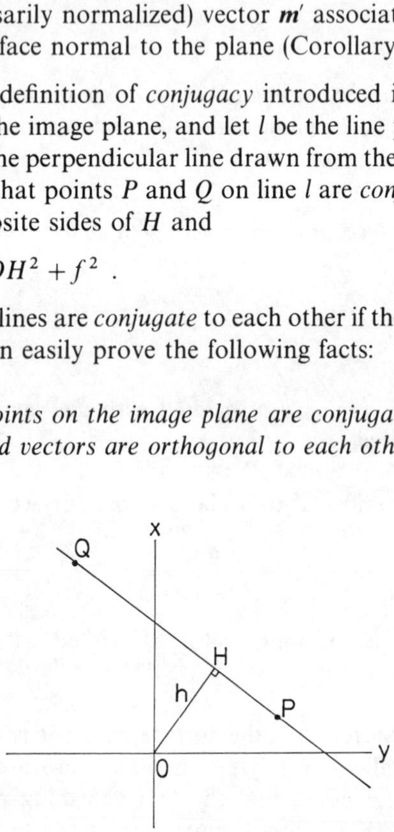

Fig. 8.30. Let H be the foot of the perpendicular line drawn from the image origin O to line l, and let h be the distance of line l from the image origin O. Two points P and Q on line l are conjugate to each other if they are on the opposite sides of H and $PH \cdot QH = f^2 + h^2$

Corollary 8.9. *Two lines are orthogonal to each other in the scene if and only if their vanishing points are conjugate to each other on the image plane.*

Corollary 8.10. *Two planes are orthogonal to each other in the scene if and only if their vanishing lines are conjugate to each other on the image plane.*

Remark 8.7. Making use of these facts, we can theoretically test the orthogonality of two sets of parallel lines in the scene by checking if their vanishing points are conjugate to each other. However, a direct computation may break down or result in poor accuracy if their vanishing points are located very far away from the image origin. Hence, it is more practical to compute the unit vectors associated with these vanishing points, as we showed earlier, and then check their orthogonality.

Remark 8.8. If two sets of parallel lines are known to be orthogonal to each other in the scene, we can determine the focal length f. Namely, if P and Q are their vanishing points, and if H is the foot of the perpendicular line drawn from the image origin O to the line passing through points P and Q, we obtain

$$f = \sqrt{PH \cdot QH - OH^2} \ . \tag{8.5.25}$$

However, the computation may result in poor accuracy if one of these points is near H (and hence the other is very far away).

Let us briefly summarize some theoretically interesting results concerning three mutually orthogonal lines in the scene in terms of the notions of duality and conjugacy, although these are not necessarily useful from the computational viewpoint. The proofs of the following propositions are left as exercises.

Consider a triangle $\triangle ABC$ on the image plane. We say that triangle $\triangle A'B'C'$ is the *polar triangle* of triangle $\triangle ABC$ if line $B'C'$ is the polar of A, line $C'A'$ is the polar of B, and line $A'B'$ is the polar of C. It is easy to prove that if $\triangle A'B'C'$ is the polar triangle of $\triangle ABC$, then $\triangle ABC$ is also the polar triangle of $\triangle A'B'C'$. A triangle is *self-polar* if it is the polar triangle of itself.

Proposition 8.11. *A triangle is self-polar if and only if each pair of vertices is conjugate to each other.*

Proposition 8.12. *A triangle is self-polar if and only if each pair of edges is conjugate to each other.*

Proposition 8.13. *Three lines in the scene are mutually orthogonal if and only if their vanishing points define a self-polar triangle on the image plane.*

Proposition 8.14. *Three planes in the scene are mutually orthogonal if and only if their vanishing lines define a self-polar triangle on the image plane.*

8.5.1 Spherical Geometry and Projective Geometry

Our technique for confining computation within the finite domain can be viewed as the use of the *image sphere* (Sect. 4.6): a point is projected onto the sphere of radius f centered at the viewpoint O, and a point on the image sphere can be specified by the unit vector starting from the center and pointing toward it. This vector is precisely the unit vector associated with the point (Fig. 8.31a). A line in the scene, on the other hand, is projected onto part of a great circle on the image sphere. A great circle is the intersection of the image sphere with a plane passing through its center. The unit surface normal to that plane is precisely the unit vector associated with the line defined as the intersection of this line and the image plane (Fig. 8.31b). We will make use of this interpretation in Chap. 10 when we construct an algorithm for detecting parallelism.

Note that although the image sphere interpretation is useful to "visualize" various practical algorithms, we need not generate images on the image sphere. The image sphere is a completely auxiliary concept, and computation is always done from the original data.

The use of the image sphere can be viewed also as the use of (normalized) homogeneous coordinates of points and lines from the standpoint of projective geometry. If expressed in terms of homogeneous coordinates, points and lines at infinity do not require different treatment. In fact, the formulation in Sect. 8.5 is closely related to projective geometry.

As was pointed out in Sect. 1.3, a *2D projective transformation* of the image plane (with *points at infinity* and the *line at infinity* added) is a *collineation*—a mapping that maps collinear points onto collinear points and concurrent lines

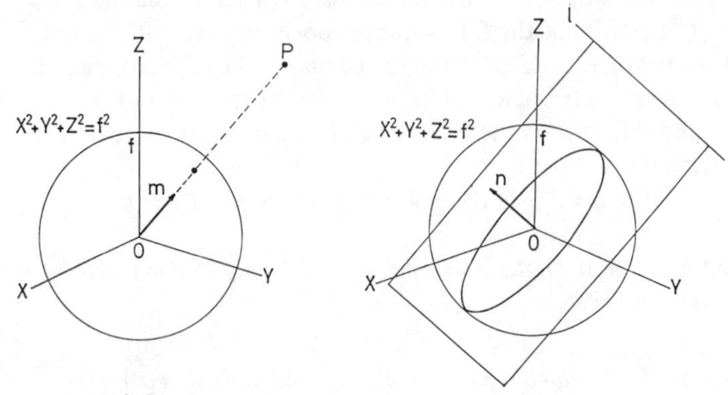

Fig. 8.31a, b. The image sphere is a sphere of radius f centered at the viewpoint O. **a** A point P in the scene is projected onto the intersection of the image sphere with the ray connecting the viewpoint O and the point P. The unit vector $\boldsymbol{m}$ along this ray is associated with the point P. **b** A line l in the scene is projected onto the great circle defined as the intersection of the image sphere and the plane passing through the viewpoint O and the line l. The unit $\boldsymbol{n}$ vector normal to this plane is associated with the line l

onto concurrent lines, and preserves *incidence* (i.e., if a point is on a line, the image point is also on the image line). Another fundamental notion is a *correlation*—a mapping that maps collinear points onto concurrent lines and concurrent lines onto collinear points, and preserves incidence (i.e., if a point is on a line, the image line passes through the image point). A correlation is said to be a *polarity* if, whenever point P is mapped onto line l, line l is also mapped onto point P. Then, the point P is said to be the *pole* of the line l, and the line l is said to be the *polar* of the point P. If three lines l, l', l'' are the respective polars of points P, P', P'', the triangle defined by the three lines l, l', l'' is called the *polar triangle* of triangle $\triangle PP'P''$. A triangle that is a polar triangle of itself is called a *self-polar triangle*.

Two points P, P' are said to be *conjugate* if the polar of point P passes through point P', and the polar of point P' passes through point P. Similarly, two lines l, l' are said to be *conjugate* if the pole of line l is on line l', and the pole of line l' is on line l. Hence, the three vertices (or edges) of a self-polar triangle are conjugate to each other. A point that is conjugate to itself is said to be *self-conjugate*. For a given polarity, the set of all self-conjugate points is called the *conic* (or *quadric*) of the polarity.

In terms of homogeneous coordinates, the above definitions are expressed in a simple form. A collineation is a linear mapping from the homogeneous coordinates of a point (or line) to the homogeneous coordinates of another point (or line), while a correlation is a linear mapping from the homogeneous coordinates of a point (or line) to the homogeneous coordinates of a line (or point). A polarity is a correlation whose transformation matrix is symmetric. The conic (or quadric) of a polarity is a quadratic curve constructed from the matrix of the polarity (i.e., if A_{ij}, $i, j = 1, 2, 3$, is the matrix of the polarity, its conic is $\sum_{i,j=1}^{3} A_{ij}u_iu_j = 0$ in homogeneous coordinates u_i, $i = 1, 2, 3$). Conversely, a conic expressed as a homogeneous quadratic equation in homogeneous coordinates uniquely defines a polarity.

In view of these definitions, it is easy to confirm that our "duality" is simply the "polarity" with respect to "conic" $x^2 + y^2 + f^2 = 0$ (an imaginary circle centered at the image origin with imaginary radius if). In fact, our "camera rotation transformation" can be characterized as the projective transformation that maps the conic $x^2 + y^2 + f^2 = 0$ onto itself. Our "conjugacy" is also the conjugacy with respect to this conic.

Exercises

8.1 (a) Show that (8.2.3) can be rewritten as (8.2.4).
(b) Confirm (8.2.2).

8.2 (a) Show that (8.2.18) can be written as (8.2.19).
(b) Derive (8.2.20) and prove (8.2.16).
(c) Confirm the validity of Fig. 8.15.

8.3 Show that (8.2.2c) is equivalent to (8.2.16a).
Hint: $\varphi_1 - \varphi_2 = -(\varphi_3 - \varphi_1) - (\varphi_2 - \varphi_3)$.

8.4 (a) Confirm (8.2.32).
(b) Confirm (8.2.33) and derive (8.2.34, 35).
(c) Prove (8.2.36, 37).

8.5 Confirm (8.3.2, 3), and show that T is indeed an Abelian group.

8.6 Confirm Proposition 8.3.

8.7 Confirm Proposition 8.5 and Corollary 8.1, and deduce Proposition 8.6.

8.8 (a) Show that the only constraint on the state of one corner when no other information is available is given by Proposition 8.7.
(b) Show that the only constraint on the states of two corners linked by an edge when no other information is available is given by Proposition 8.8 if the constraint of Proposition 8.7 is satisfied at both of the corners.
(c) Confirm Corollary 8.2.

8.9 Prove Lemmas 8.3 and 8.4 for a fork coordinate system.

8.10 Prove Lemmas 8.5 and 8.6 by considering all the possibilities exhaustively.

8.11 By checking all the possibilities exhaustively, prove that the constraints on the state and orientation of a corner when visibility is considered are given by Fig. 8.10.

8.12 Confirm the deduction of Example 8.5.

8.13 Confirm the deduction of Example 8.6.

8.14 Give one image of a rectangular polyhedron for which the wall paradox given in Remark 8.1 occurs at one of its own faces.

8.15 Besides those shown in Remark 8.2, show two impossible objects such that our interpretation scheme accepts one and rejects the other.

8.16 Confirm that the Huffman–Clowes junction dictionary given in Fig. 8.17 exhausts all possible labelings of a trihedral corner.

8.17 Show that the seven different interpretations shown in Fig. 8.18 result from the image in Fig. 8.11a if Huffman–Clowes edge labeling is applied.

8.18 (a) Prove (8.4.20).
(b) Show that (8.4.21, 22) are respectively equivalent to (4.12, 20).

8.19 Prove Corollary 8.5.

8.20 (a) Show that the unit vector $\boldsymbol{m}$ starting from the viewpoint O and pointing toward point (a, b) on the image plane $Z = f$ is given by (8.5.5).
(b) Show that the unit vector $\boldsymbol{n}$ starting from the viewpoint O and normal to the plane passing through the viewpoint O and intersecting the image plane $Z = f$ along line $Ax + By + C = 0$ is given by (8.5.6).

8.21 Show that the line passing through two points (x_1, y_1), (x_2, y_2) is given by

$$\begin{vmatrix} x & y & 1 \\ x_1 & y_1 & 1 \\ x_2 & y_2 & 1 \end{vmatrix} = 0 ,$$

namely

$$(y_1 - y_2)x - (x_1 - x_2)y + (x_1 y_2 - x_2 y_1) = 0 \ .$$

8.22 (a) If $A^2 + B^2 = 1$, show that the distance of line $Ax + By + C = 0$ from point (a, b) is given by $|Aa + Bb + C|$, *Hint*: 2D vector (A, B) is a unit vector perpendicular to the line $Ax + By + C = 0$. The distance of the line from (a, b) is obtained by the orthogonal projection of vector $(x - a, y - b)$ onto the direction defined by unit vector (A, B), where (x, y) is a point on the line.
(b) Derive (8.5.8) by differentiating (8.5.7) with respect to a and b and setting the respective results to zero.
(c) Show that the solution of (8.5.8) is given by (8.5.9).
(d) Show that the vector $\boldsymbol{m}'$ associated with the common vanishing point is given by (8.5.10).

8.23 (a) If $\boldsymbol{n} = (n_1, n_2, n_3)$ is the unit vector associated with line $Ax + By + C = 0$, show that the distance from point (a, b) on the image plane $Z = f$ to the plane passing through the viewpoint and intersecting the image plane along the line $Ax + By + C = 0$ is $|n_1 a + n_2 b + n_3 f|$. *Hint*: The distance is given by the orthogonal projection of vector $(X - a, Y - b, Z - f)$ onto the direction defined by the unit vector $\boldsymbol{n}$, where (X, Y, Z) is a point on the plane.
(b) Derive (8.5.13) by differentiating (8.5.12) with respect to a and b and setting the respective results to zero.
(c) Show that the solution of (8.5.13) is given by (8.5.14).
(d) Show that the vector $\boldsymbol{m}'$ associated with the common vanishing point is given by (8.5.14).

8.24 (a) Show that (8.5.17) is rewritten as (8.5.18) if we define the matrix $\boldsymbol{N}$ by (8.5.19).
(b) Show that the matrix $\boldsymbol{N}$ defined by (8.5.19) is symmetric and positive definite if vectors $\boldsymbol{n}_i$, $i = 1, \ldots, N$, include at least three linearly independent ones.
(c) Prove that the eigenvalues of an n-dimensional symmetric matrix are all real.
(d) Prove that the eigenvalues of an n-dimensional symmetric positive definite matrix are all positive.
(e) Prove that the quadratic form $A[\boldsymbol{x}] = \sum_{i,j=1}^{n} A_{ij} x_i x_j$ in n-dimensional unit vector $\boldsymbol{x} = (x_i)$ for an n-dimensional symmetric matrix $\boldsymbol{A} = (A_{ij})$ takes its minimum when $\boldsymbol{x}$ is the eigenvector of matrix $\boldsymbol{A}$ for the minimum eigenvalue.

8.25 (a) Prove Lemma 8.7.
(b) Prove Lemma 8.8.

8.26 Show that the unit surface normal $\boldsymbol{n}$ to a plane whose surface gradient is (p, q) is given by (8.5.23).

8.27 (a) Prove Theorem 8.8.
(b) Prove Corollary 8.9.
(c) Prove Corollary 8.10.

8.28 (a) Show that if triangle $\triangle A'B'C'$ is the polar triangle of $\triangle ABC$, then triangle $\triangle ABC$ is also the polar triangle of $\triangle A'B'C'$.
(b) Prove Proposition 8.11.
(c) Prove Proposition 8.12.
(d) Prove Proposition 8.13.

9. Shape from Texture

In Sect. 2.7, we gave an analysis of the *shape-from-texture* problem. There, we assumed that the "texture density" was somehow observed on the image plane. Then, we showed that the surface parameters were computed in an analytically closed form in terms of invariants. We also studied the ambiguity in interpretation of the surface shape. In this chapter, we give a rigorous mathematical treatment for characterizing discrete textures. In particular, we give precise definitions of *texture density* and *homogeneity* in terms of the theory of distributions and differential geometry. Then, we present numerical schemes for solving the resulting 3D recovery equations. We also give some numerical examples for synthetic data.

9.1 Shape from Texture from Homogeneity

Seeing a pattern that has some kind of regularity, or *texture*, and converges on a receding plane, humans can easily perceive the 3D depth of the scene. This fact has long since interested many researchers, and efforts have been made to simulate, by a computer, this seemingly "highly intelligent" human perception. This problem is now widely known as 3D recovery of *shape from texture.*

In general, 3D recovery from texture is possible if we have some prior knowledge about the true texture; if the observed texture has properties different from those of the true texture, the 3D shape is computed in such a way that the discrepancy is accounted for. For example, if the true texture is known to be an array of elements with a known shape, say circular, the surface gradient can be inferred from the observed distorted shape, say elliptical, of the elements. If the true texture elements are known to be distributed periodically with constant intervals, the surface gradient can be computed from the ratio of the converging interval lengths. If the true texture elements are aligned on parallel lines, or if individual texture elements have parallel line segments, the surface gradient is inferred from the vanishing points of such lines (Sect. 8.5). Similar reasoning is possible if the true texture elements or their alignments possess orthogonality or symmetry.

One major issue of these approaches is that we must first recognize the "structure" of the true texture—regularity, periodicity, collinearity, parallelism, orthogonality, symmetry, and so on. This is very difficult to automate by a computer, because the observed texture does not exhibit regularity, periodicity, etc. (That is why the 3D shape can be recovered.) Despite this difficulty, these *structure-based approaches* were attempted by many researchers by heuristic means. Perhaps this is because humans seem to employ this type of inference; humans can recognize such "structures" very easily.

Then, a new approach, which does not require such structures appeared. It is based on *statistical assumptions* about the true texture distribution. For example, if the true texture is distributed *isotropically*, meaning that the line segments constituting the texture have no preferred orientations, the 3D surface shape can be computed from observed preferred orientations. Another possible statistical assumption is *homogeneity*: when observed, the texture looks dense on the part of the surface apart from the observer and sparse on the part near the observer.

In this chapter, we give a mathematically consistent treatment based on differential geometry and the theory of distributions—in particular, we show how to relate the discreteness of textures to the description of smooth surfaces in differential terms. As it turns out, we need not recognize the "structure" of the texture at all: the problem is easily solved by a *computational principle.*

We first give a precise definition of *homogeneity* of a texture. If a texture consists of dots or line segments, the *texture density*, in its literal sense, is a singular function taking the value ∞ at the texture dots or line segments and 0 elsewhere. How can we say that the density is uniform? How can we tell that a given texture is homogeneously distributed? As we will see, how to define texture homogeneity is the core of the theory. We define the texture density in a formal way based on the *theory of distributions.*

We next give a mathematical analysis of the texture distortion due to perspective projection. We describe the perspective distortion in terms of the *first fundamental form* expressed as a function of the *image* coordinates. Our formulation consists of two stages. First, we present the 3D recovery equations for determining the surface gradient. Then, we propose numerical schemes for solving these equations. There, the camera rotation transformation plays an important role. Finally, we give some numerical examples for synthetic images.

9.2 Texture Density and Homogeneity

In this section, we study how to define the *density* of a discrete texture. Consider textures composed of dots and line segments. If we are to seek a function $\rho(x, y)$ describing the amount of texture divided by the area it occupies, we are forced to consider delta-function-like singularities; the value of $\rho(x, y)$ becomes ∞ at the texture elements and 0 elsewhere, because the area of a dot or a line segment is 0. In order to avoid such singularities, we define the texture density in a formal way as a *linear functional* or *distribution.*

Let us fix a domain W, called a *window*, on the image plane, and consider the texture image inside W. First, consider textures consisting of distributed dots. The "trick" to avoid singularities is that we do not define the *value* of $\rho(x, y)$. Instead, we define only the *rule of integration* of $\rho(x, y)$ multiplied by some function $w(x, y)$—we call it a *test function*—over the window W. To be specific, we define

$$\int_W \rho(x, y) w(x, y) dx\, dy = \sum_{P_i \in W} w(x_i, y_i) , \tag{9.2.1}$$

where $P_i(x_i, y_i)$ are the positions of the dot texture elements on the image plane.

Note that the left-hand side of (9.2.1) is merely a symbolic notation of its right-hand side. Since the left-hand side is not an ordinary integration, we might as well write it as, say, $\rho[w(x, y)]$, which would make clear the meaning that the texture density ρ is merely a *linear functional*, mapping a test function $w(x, y)$ to a real number. However, the symbol of integration appeals more to our intuition.

Since the texture density $\rho(x, y)$ is defined formally, we need not worry about its singularities. We can simply "imagine" that $\rho(x, y)$ takes the value ∞ at texture elements and 0 elsewhere.

The texture density $\rho(x, y)$ for a line segment texture is similarly defined by the following rule of integration:

$$\int_W \rho(x, y) w(x, y) dx\, dy = \sum_{L_i \subset W} \int_{L_i} w(x(l), y(l))\, dl . \tag{9.2.2}$$

Here, L_i are the line segments on the image plane, and the right-hand side is the sum of curvilinear integrals along all line segments parametrized by arc length l. Again, we can "imagine" that $\rho(x, y)$ takes the value ∞ along texture line segments and 0 elsewhere.

Now, we define *homogeneity* of the texture density. Let $\rho(x, y)$ be the texture density formally defined above. We would like to say that the texture is homogeneous if the texture density is approximately equal to a constant: $\rho(x, y) \approx c$. Since the texture density is defined as a functional, this requirement must be interpreted *in the weak sense* or *in the sense of a distribution*. Namely, we say that a texture density $\rho(x, y)$ is *homogeneous* for the class $\mathscr{W}$ of test functions if

$$\int_W \rho(x, y) w(x, y) dx\, dy \approx c \int_W w(x, y) dx\, dy , \tag{9.2.3}$$

for all test functions $w(x, y) \in \mathscr{W}$, where c is a constant independent of the test functions $w(x, y) \in \mathscr{W}$.

The constant c can be interpreted as the *texture density* in the intuitive sense—the "total number of dots per unit area" or the "total length of line segments per unit area". Combined with the definitions (9.2.1, 2), the homogeneity is characterized as follows:

Lemma 9.1. *The texture density is homogeneous if*

$$\int_W w(x, y)dx\,dy \approx \begin{cases} \dfrac{1}{c} \displaystyle\sum_{P_i \in W} w(x_i, y_i) & \text{for dot textures ,} \\ \dfrac{1}{c} \displaystyle\sum_{L_i \subset W} \int_{L_i} w(x(l), y(l))dl & \text{for line segment textures .} \end{cases} \tag{9.2.4}$$

Remark 9.1. Equation (9.2.4) can be viewed as the *Monte Carlo method* to integrate a test function $w(x, y)$, where $1/c$ is the "area per dot" or the "area per unit length line segment". This type of Monte Carlo method is known to give a good approximation only when the distribution of the sampling points is "homogeneous". Here, we use this very fact to "define" the homogeneity. Namely, a texture is homogeneous *if* the Monte Carlo method of integration over the texture elements yields a good approximation.

Remark 9.2. If we choose, as a test function $w(x, y)$, the characteristic function

$$\chi_S(x, y) = \begin{cases} 1 & (x, y) \in S \\ 0 & \text{otherwise} \end{cases} \tag{9.2.5}$$

of a region S, then (9.2.4) states that the number of dots or the total length of line segments in the region S is approximately proportional to the area of S, the constant c being the number or the length of texture elements in the region S divided by its area. This is the interpretation which most people informally think of.

Remark 9.3. The definition of homogeneity depends on the choice of the class $\mathscr{W}$ of test functions $w(x, y)$. Even if the texture is sparse, it can be homogeneous for very smooth test functions $w(x, y)$ (i.e., viewed *coarsely* or viewed with low resolution), while it may not be homogeneous for rapidly varying test functions $w(x, y)$ (i.e., viewed *finely* or viewed with high resolution). Figuratively, we are looking at a discrete texture through "filters" $w(x, y)$, and the homogeneity depends on the coarseness of the filter. If, for example, we take $\mathscr{W} = \{\exp[i\pi(kx/a + ly/b)]\}$, assuming that the window W is a rectangle of size $2a \times 2b$ and setting a certain threshold for the approximation of (9.2.3), we can define the *degree of homogeneity* by those integer pairs (k, l) that satisfy the approximation. However, what we have described so far is sufficient for the subsequent discussions.

Since the integration over the texture is defined as a functional (9.2.1, 2), we must be careful when we change the variables of integration; the rule for usual integration does not apply here. Consider two smooth functions $x'(x, y)$, $y'(x, y)$ such that the correspondence between (x, y) and (x', y') is one-to-one. Let $x(x', y')$, $y(x', y')$ be the inverse relationship. Suppose we use (x', y') as new coordinates. Let W' be the domain in the $x'y'$-plane that corresponds to the window W in the xy-plane. Define the transformed texture density $\rho'(x', y')$ also as a functional by

$$\int_{W'} \rho'(x', y') w'(x', y') dx' dy' = \int_{W} \rho(x, y) w(x, y) dx\, dy \ , \tag{9.2.6}$$

where $w'(x', y') \equiv w(x(x', y'), y(x', y'))$. Then, the action, as a functional, of the new density $\rho'(x', y')$ on a given test function $w'(x', y')$ is given as follows:

Proposition 9.1 (*Change of variables of texture densities*).

$$\rho'(x', y') = \begin{cases} \rho(x(x', y'), y(x', y')) & \text{for dot textures ,} \\ \rho(x(x', y'), y(x', y'))\Gamma(x', y', \boldsymbol{t}') & \text{for line segment textures ,} \end{cases} \tag{9.2.7}$$

where

$\Gamma(x', y', \boldsymbol{t}') \equiv$

$$\sqrt{\left[\left(\frac{\partial x}{\partial x'}\right)^2 + \left(\frac{\partial x}{\partial y'}\right)^2\right] t_1'^2 + 2\left(\frac{\partial x}{\partial x'}\frac{\partial x}{\partial y'} + \frac{\partial y}{\partial x'}\frac{\partial y}{\partial y'}\right) t_1' t_2' + \left[\left(\frac{\partial x}{\partial y'}\right)^2 + \left(\frac{\partial y}{\partial y'}\right)^2\right] t_2'^2} \ , \tag{9.2.8}$$

and $\boldsymbol{t}' = (t_1', t_2')$ is the unit tangent vector to the line segment at point (x', y') in the $x'y'$-domain.

Proof. First, consider a dot texture. Let P_i' be the point in the $x'y'$-plane that corresponds to point P_i in the xy-plane. Then, we see that

$$\begin{aligned} \int_W \rho(x, y) w(x, y) dx\, dy &= \sum_{P_i \in W} w(x_i, y_i) \\ &= \sum_{P_i' \in W'} w(x(x_i', y_i'), y(x_i', y_i')) \\ &= \sum_{P_i' \in W'} w'(x_i', y_i') \ . \end{aligned} \tag{9.2.9}$$

Since this relation defines the action, as a functional, of density $\rho'(x', y')$ on the test function $w'(x', y')$ in the $x'y'$-plane, we obtain (9.2.7) for dot textures.

Next, consider a line segment texture. Let L_i' be the line segment (parametrized by arc length l') in the $x'y'$-plane that corresponds to the line segment

L_i (parametrized by arc length l) in the xy-plane. We have the following differential relationship:

$$dl = \sqrt{dx^2 + dy^2}$$

$$= \sqrt{\left(\frac{\partial x}{\partial x'} dx' + \frac{\partial x}{\partial y'} dy'\right)^2 + \left(\frac{\partial y}{\partial x'} dx' + \frac{\partial y}{\partial y'} dy'\right)^2}$$

$$= \sqrt{\left[\left(\frac{\partial x}{\partial x'}\right)^2 + \left(\frac{\partial y}{\partial x'}\right)^2\right] dx'^2 + 2\left(\frac{\partial x}{\partial x'}\frac{\partial x}{\partial y'} + \frac{\partial y}{\partial x'}\frac{\partial y}{\partial y'}\right) dx'\,dy' + \left[\left(\frac{\partial x}{\partial y'}\right)^2 + \left(\frac{\partial y}{\partial y'}\right)^2\right] dy'^2}$$

$$= \sqrt{\left[\left(\frac{\partial x}{\partial x'}\right)^2 + \left(\frac{\partial y}{\partial x'}\right)^2\right]\left(\frac{dx'}{dl'}\right)^2 + 2\left(\frac{\partial x}{\partial x'}\frac{\partial x}{\partial y'} + \frac{\partial y}{\partial x'}\frac{\partial y}{\partial y'}\right)\frac{dx'}{dl'}\frac{dy'}{dl'} + \left[\left(\frac{\partial x}{\partial y'}\right)^2 + \left(\frac{\partial y}{\partial y'}\right)^2\right]\left(\frac{dy'}{dl'}\right)^2}$$

$$\times\, dl' \,. \tag{9.2.10}$$

If we put $t'_1 = dx'/dl'$ and $t'_2 = dy'/dl'$, the vector $\boldsymbol{t} = (t'_1, t'_2)$ is of unit length and tangent to the line segment at point (x', y') in the $x'y'$-plane. Hence, we have

$$\int_W \rho(x, y) w(x, y)\,dx\,dy = \sum_{L_i \subset W} \int_{L_i} w(x(l), y(l))\,dl = \sum_{L'_i \subset W'} \int_{L'_i} w'(x'(l'), y'(l'))$$

$$\times \sqrt{\left[\left(\frac{\partial x}{\partial x'}\right)^2 + \left(\frac{\partial x}{\partial x'}\right)^2\right] t_1'^2 + 2\left(\frac{\partial x}{\partial x'}\frac{\partial x}{\partial y'} + \frac{\partial y}{\partial x'}\frac{\partial y}{\partial y'}\right) t'_1 t'_2 + \left[\left(\frac{\partial x}{\partial y'}\right)^2 + \left(\frac{\partial y}{\partial y'}\right)^2\right] t_2'^2}\; dl'$$

$$= \sum_{L'_i \subset W'} \int_{L'_i} w'(x'(l'), y'(l'))\Gamma(x'(l'), y'(l'), \boldsymbol{t}'(l'))\,dl'. \tag{9.2.11}$$

Since this defines the action, as a functional, of the density $\rho'(x', y')$ on the test function $w'(x', y')$ in the $x'y'$-plane, we obtain (9.2.7) for line segment textures.

Remark 9.4. For the usual integration of a continuous density $\rho(x, y)$, we would have

$$\rho'(x', y') = \rho(x(x', y'), y(x', y'))|\Delta(x', y')| \,, \tag{9.2.12}$$

where $\Delta(x', y')$ is the *Jacobian*:

$$\Delta(x', y') \equiv \begin{vmatrix} \partial x/\partial x' & \partial x/\partial y' \\ \partial y/\partial x' & \partial y/\partial y' \end{vmatrix} \,. \tag{9.2.13}$$

Namely, a continuous density is multiplied by the Jacobian Δ, which is the *magnification ratio of area*, while a line segment density is multiplied by Γ, which is the *elongation ratio of length*, and a dot density is multiplied by 1, which is the *increase ratio of number*. The number of dots is preserved by a continuous mapping. The elongation ratio Γ depends on the orientations of individual line segments as well as their positions.

9.2.1 Distributions

Let us recall the definition of the (*Dirac*) *delta function* $\delta(x)$. Mathematically, it is not a function; if a function takes the value 0 except at one point, its integral must be 0, since one point is not counted for integration.[1] The correct interpretation is to regard the delta function as a *linear functional* δ that maps every function $w(x)$ belonging to some class $\mathscr{W}$ to the value $w(0)$: $\delta[w(x)] = w(0)$.

However, it usually appeals more to our intuition if we *denote* it by $\int \delta(x)(\,.\,)dx$ rather than $\delta[\,.\,]$. As a result, the above definition is written as $\int \delta(x)w(x)dx = w(0)$. Thus, $\delta(x)$ is merely a symbolic notation. In fact, usually we do not use the delta function $\delta(x)$ itself; it is useful only when it is multiplied by some function and integrated. Hence, it suffices to define only the *rule of integration*, and we need not worry about its singularity. This is the standpoint introduced by the French mathematician Laurent Schwartz (1945–) in his *theory of distributions*.

Distributions of one variable x over $-\infty < x < \infty$ are defined as follows. Let $\mathscr{W}$ be the class of smooth functions (i.e., members of C^∞, the class of infinitely continuously differentiable functions) that have *finite supports* (i.e., each function takes nonzero values only in a finite interval). Members of $\mathscr{W}$ are called *test functions*. Since addition and scalar multiplication are done within $\mathscr{W}$, the class $\mathscr{W}$ is regarded as a linear space.

A mapping $T\colon \mathscr{W} \to \mathbb{R}$ is called a *functional* over the class $\mathscr{W}$ of test functions. In other words, a functional T maps test functions $w(x)$ of $\mathscr{W}$ to real numbers $T[w(x)]$. A functional is *continuous* if the sequence $T[w_i(x)]$, $i = 1, 2, 3, \ldots,$ converges when the derivatives of any order (including the zeroth) of the test functions $w_i(x) \in \mathscr{W}$, $i = 1, 2, 3, \ldots,$ having supports in a finite interval, converge uniformly. A functional T is *linear* if $T[cw(x)] = cT[w(x)]$ and $T[w_1(x) + w_2(x)] = T[w_1(x)] + T[w_2(x)]$ for any $w(x)$, $w_1(x)$, $w_2(x) \in \mathscr{W}$ and $c \in \mathbb{R}$. (Alternatively, we can say that $T[c_1 w_1(x) + c_2 w_2(x)] = c_1 T[w_1(x)] + c_2 T[w_2(x)]$ for all $w_1(x)$, $w_2(x) \in \mathscr{W}$ and $c_1, c_2 \in \mathbb{R}$.) A *continuous linear functional* $T\colon \mathscr{W} \to R$ is called a *distribution* over the class $\mathscr{W}$ of test functions.

The *delta function* δ is a distribution defined by $\delta[w(x)] = w(0)$. A locally integrable (i.e., integrable over any finite interval) function $f(x)$ is identified with a distribution by defining its action to be $f[w(x)] = \int f(x)w(x)dx$. The function $f(x)$ need not be integrable over $-\infty < x < \infty$ (e.g., $f(x) = 1$ is locally integrable).

Two distributions are *equal* if their actions as functionals are the same. Hence, two locally integrable functions that are almost the same everywhere are also the

[1] Here, integration is *in Lebesgue's sense*. Each point has *Lebesgue measure* 0. If two functions take different values only at a set whose Lebesgue measure is 0, the two functions are said to be the same *almost everywhere*. If integrated, two functions that are the same almost everywhere give the same value.

same as distributions. Scalar multiplication and addition of distributions are respectively defined by $(cT)[w(x)] = T[cw(x)]$ and $(T_1 + T_2)[w(x)] = T_1[w(x)] + T_2[w(x)]$. If $c(x)$ is in C^∞, we can also define distribution $c(x)T$ by $(c(x)T)[w(x)] = T[c(x)w(x)]$.

The *derivative* dT/dx of a distribution T is a distribution defined by $(dT/dx)[w(x)] = -T[dw/dx]$. This can be extended to the derivative of order r by $(d^rT/dx^r)[w(x)] = (-1)^r T[d^rw/dx^r]$. If $f(x)$ is in C^∞, its ordinary rth derivative d^rf/dx^r is, when viewed as a distribution, equal to the rth derivative of distribution $f(x)$.

A sequence $T_1, T_2, \ldots$ of distributions is said to *converge* if the sequence $T_1[w(x)], T_2[w(x)], \ldots$ converges for any test function $w(x) \in \mathscr{W}$. The limit value defines a distribution T called the *limit distribution*. If a sequence $f_1(x), f_2(x), \ldots$ of functions converges *as a distribution*, the convergence is said to be *weak*. (If $\lim_{n\to\infty} \| f_n(x) - f(x) \| = 0$ for the limit function $f(x)$, where $\|.\|$ is some norm for functions, we say that $f_1(x), f_2(x), \ldots$ converges *in norm*, or the convergence is *strong*.)

Let $\bar{\mathscr{W}}$ ($\supset \mathscr{W}$) be the class of functions that are in C^∞ and decrease "rapidly" as $x \to \pm\infty$, i.e., $o(|x|^k)$ for any natural number k. Then, all $w(x) \in \bar{\mathscr{W}}$ have their Fourier transforms $\mathfrak{F}w(x)$ that belong to $\bar{\mathscr{W}}$. The *Fourier transform* $\mathfrak{F}T$ of a distribution T over $\bar{\mathscr{W}}$ is defined as the distribution $(\mathfrak{F}T)[w(x)] = T[\mathfrak{F}w(x)]$.

From this definition, it is easy to see that the Fourier transform of the delta function $\delta(x)$ is 1: $\mathfrak{F}\delta[w(x)] = \delta[\mathfrak{F}w(x)] = \int \exp(-2i\pi\omega x)w(x)dx|_{\omega=0} = \int w(x)dx = 1[w(x)]$.

Extending the above definition, we can also define distributions of multiple variables. By imposing restrictions on the class of test functions, we can define many types of distributions. Most of the singular functions we encounter in physics and engineering problems can be treated as distributions over appropriate classes of test functions.

9.3 Perspective Projection and the First Fundamental Form

Take the viewer-centered Cartesian XYZ coordinate system in the scene (Fig. 9.1). The relationship between the scene coordinates (X, Y, Z) and the image coordinates (x, y) is given by the projection equations

$$x = \frac{fX}{Z}, \qquad y = \frac{fY}{Z}. \tag{9.3.1}$$

Consider a smooth surface whose equation is $Z = Z(X, Y)$. If this surface equation and the projection equations (9.3.1) are combined, a one-to-one correspondence is established between the image plane and the surface (to be strict, the surface part "visible" from the viewpoint). Let us first study how the scene coordinates (X, Y, Z) change on the surface. If we take *differentials* dX, dY,

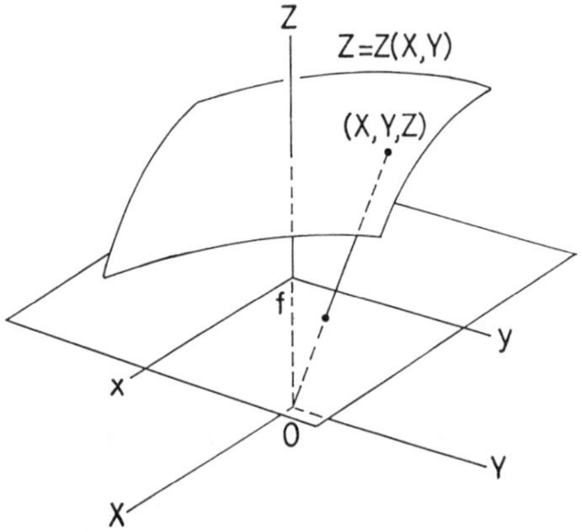

Fig. 9.1. Point (X, Y, Z) on the surface $Z = Z(X, Y)$ is projected onto point (x, y) on the image plane by perspective projection from the viewpoint $(0, 0, -f)$

dZ along the surface, the projection equations (9.3.1) yield

$$fdX - xdZ = Zdx \ , \qquad fdY - ydZ = Zdy \ . \tag{9.3.2}$$

Taking the differential of the surface equation $Z = Z(X, Y)$, we obtain

$$dZ = PdX + QdY \ , \qquad P \equiv \frac{\partial Z}{\partial X} \ , \qquad Q \equiv \frac{\partial Z}{\partial Y} \ . \tag{9.3.3}$$

Equations (9.3.2, 3) can be viewed as a set of simultaneous linear equations in dX, dY, dZ. The solution is given by

$$dX = \frac{Z}{f(f - Px - Qy)} [(f - Qy)dx + Qx\,dy] \ ,$$

$$dY = \frac{Z}{f(f - Px - Qy)} [Py\,dx + (f - Px)dy] \ ,$$

$$dZ = \frac{Z}{f - Px - Qy} [Pdx + Qdy] \ . \tag{9.3.4}$$

Consider two points (x, y), $(x + dx, y + dy)$ infinitesimally far apart on the image plane. Let ds be the distance between the corresponding points on the surface (Fig. 9.2). If (9.3.4a–c) are substituted into $ds^2 = dX^2 + dY^2 + dZ^2$, we obtain the following first *fundamental form* (Sect. 5.8):

Proposition 9.2 (*First fundamental form*).

$$ds^2 = \sum_{i,j=1}^{2} g_{ij} dx_i dx_j \ , \tag{9.3.5}$$

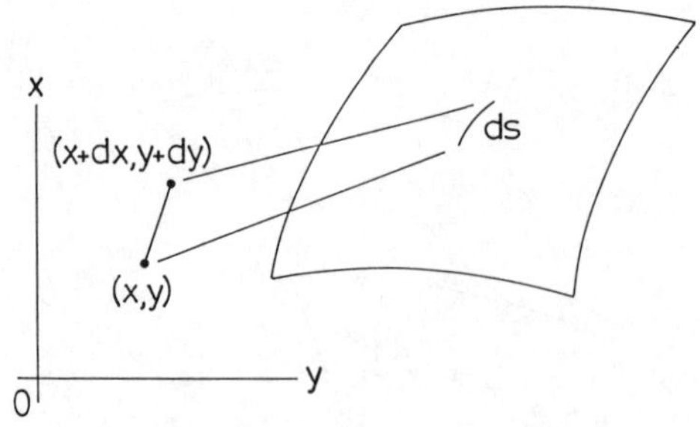

Fig. 9.2. The infinitesimal line segment connecting (x, y) and $(x + dx, y + dy)$ on the image plane corresponds to an infinitesimal line segment on the surface whose length is ds

where $x_1 = x$, $x_2 = y$, *and*

$$g_{11} = \frac{(Z/f)^2}{[1 - (Px + Qy)/f]^2}\left((1 + P^2) - 2Q\frac{y}{f} + (P^2 + Q^2)\frac{y^2}{f^2}\right),$$

$$g_{12} = \frac{(Z/f)^2}{[1 - (Px + Qy)/f]^2}\left(PQ + Q\frac{x}{y} + P\frac{y}{f} - (P^2 + Q^2)\frac{xy}{f^2}\right) = g_{21},$$

$$g_{22} = \frac{(Z/f)^2}{[1 - (Px + Qy)/f]^2}\left((1 + Q^2) - 2P\frac{x}{f} + (P^2 + Q^2)\frac{x^2}{f^2}\right).$$

(9.3.6)

The quantity $g = (g_{ij})$, $i, j = 1, 2$, is called the *first fundamental metric tensor.* The first fundamental form plays a fundamental role when we compute 3D quantities *in terms of the image coordinates.* For example, consider a smooth curve L on the image plane. The true arc length of the corresponding curve on the surface is given by integration $\int_L ds = \int_L (\sum_{i,j=1}^{2} g_{ij} dx_i dx_j)^{1/2}$ on the image plane.

Consider an infinitesimally small square on the image plane defined by four points (x, y), $(x + dx, y)$, $(x, y + dy)$, $(x + dx, y + dy)$ (Fig. 9.3). The area of this

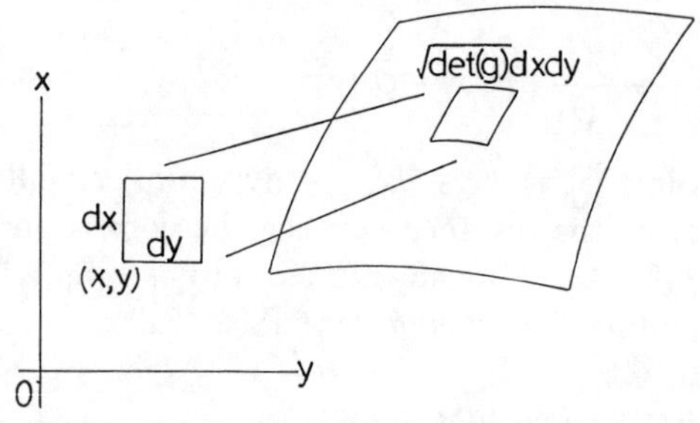

Fig. 9.3. The ratio of the area of an infinitesimal region on the surface to the are a of the corresponding region on the image plane is given by $\sqrt{\det(g)}$

square on the image plane is $dx\,dy$, but the true area of the corresponding region on the surface is equal to $\sqrt{\det(g)}dx\,dy$ (Sect. 5.8). From (9.3.6), we obtain

$$\sqrt{\det(g)} = \frac{\sqrt{1 + P^2 + Q^2}(Z/f)^2}{1 - (Px + Qy)/f} . \tag{9.3.7}$$

Hence, the true area of the region on the surface corresponding to a region S on the image plane is given by integration $\int_S \sqrt{\det(g)}\,dx\,dy$ on the image plane.

If the surface is a plane whose equation is $Z = pX + qY + r$, where p, q, r are constants, (9.3.1) can be solved for X, Y in the form

$$X = \frac{rx}{f - px - qy} , \quad Y = \frac{ry}{f - px - qy} , \quad Z = \frac{fr}{f - px - qy} . \tag{9.3.8}$$

Then, (9.3.6, 7) become

$$g_{11} = \frac{(r/f)^2}{[1 - (px + qy)/f]^4}\left(1 + p^2 - 2q\frac{y}{f} + (p^2 + q^2)\frac{y^2}{f^2}\right) ,$$

$$g_{12} = \frac{(r/f)^2}{[1 - (px + qy)/f]^4}\left(pq + q\frac{x}{f} + P\frac{y}{f} - (p^2 + q^2)\frac{xy}{f^2}\right) = g_{21} ,$$

$$g_{22} = \frac{(r/f)^2}{[1 - (px + qy)/f]^4}\left(1 + q^2 - 2P\frac{x}{f} + (p^2 + q^2)\frac{x^2}{f^2}\right) . \tag{9.3.9}$$

$$\sqrt{\det(g)} = \frac{\sqrt{1 + p^2 + q^2}(/f)^2}{[1 - (px + qy)/f]^3} . \tag{9.3.10}$$

9.4 Surface Shape Recovery from Texture

Now, we show a mathematical principle for reconstructing the surface shape from observation of inhomogeneous texture density $\rho(x, y)$ on the assumption that the true surface is homogeneously textured. For this purpose, we first study how the texture density $\rho(x, y)$ is distorted by perspective projection. Since the texture density is defined abstractly as a functional, what we need to know is how the observed density $\rho(x, y)$ acts on test functions $w(x, y)$.

Suppose the surface equation $Z = Z(X, Y)$ is known. Consider temporarily a curvilinear uv coordinate system on the surface. (We will soon do away with any surface coordinate systems.) Since the surface equation is known, there is a one-to-one correspondence between the image plane and (the visible part of) the surface. Hence, we can express the correspondence in the form $u = u(x, y)$, $v = v(x, y)$, or $x = x(u, v)$, $y = y(u, v)$.

Let W_0 be the region of the surface corresponding to the window W on the image plane. Let $\rho_0(u, v)$ be the homogeneous texture density on the surface. The homogeneity implies that

$$\int_{W_0} \rho_0(u, v) w_0(u, v) dS_0 \approx c \int_{W_0} w_0(u, v) dS_0 \ , \tag{9.4.1}$$

where $w_0(u, v)$ is a test function, and dS_0 is the area element of the surface.

The right-hand side is an ordinary integration. Hence, it can be rewritten in terms of the image coordinates (x, y) by using $dS_0 = \sqrt{\det(\boldsymbol{g})} dx\, dy$; it becomes

$$c \int_W w(x, y) \sqrt{\det(\boldsymbol{g})} dx\, dy \ , \tag{9.4.2}$$

where $w(x, y) \equiv w_0(u(x, y), v(x, y))$.

In order to rewrite the left-hand side of (9.4.1) in terms of the image coordinates (x, y), we must make a distinction of whether the texture consists of dots or line segments. For a dot texture, the left-hand side of (9.4.1) is rewritten, from (9.2.6) and Proposition 9.1, as

$$\sum_{P_i \in W} w(x_i, y_i) \ , \tag{9.4.3}$$

which is a quantity computable on the image plane (i.e., an *observable*).

For a line segment texture, the left-hand side (9.4.1) is rewritten, from Proposition 9.1, as

$$\sum_{L_i \subset W} \int_{L_i} w(x(l), y(l)) \Gamma(x(l), y(l), \boldsymbol{t}(l)) dl \ , \tag{9.4.4}$$

where l is the arc length, $\boldsymbol{t}$ is the unit tangent vector along the line segments on the image plane, and

$$\Gamma(x, y, \boldsymbol{t}) \equiv \sqrt{g_{11}(x, y) t_1^2 + 2 g_{12}(x, y) t_1 t_2 + g_{22}(x, y) t_2^2} \ . \tag{9.4.5}$$

Function $\Gamma(x, y, \boldsymbol{t})$ describes the elongation ratio between the line segments on the surface and their projections on the image plane. Since $\Gamma(x, y, \boldsymbol{t})$ depends on the orientations of the distributed line segments, the subsequent analysis is very difficult. Here, we adopt the approximation

$$\Gamma(x, y, \boldsymbol{t}) \approx [\sqrt{\det(\boldsymbol{g})}]^{1/2} \ . \tag{9.4.6}$$

The interpretation of approximation (9.4.6) is that the line segments are distributed nearly isotropically: if the area is enlarged $\sqrt{\det(\boldsymbol{g})}$ times, the individual line segments become roughly $[\sqrt{\det(\boldsymbol{g})}]^{1/2}$ times as long. As a result, if we regard $w(x, y)\Gamma(x, y, \boldsymbol{t}) \approx w(x, y)[\sqrt{\det(\boldsymbol{g})}]^{1/2}$ as a new test function

$w(x, y)$ (ignoring the dependence on orientation $\boldsymbol{t}$), we can express (9.4.4) in the form

$$\int_W \rho(x, y) w(x, y) dx\, dy \ , \tag{9.4.7}$$

which is a quantity computable on the image plane (i.e., an *observable*). The right-hand side of (9.4.1) then becomes

$$c \int_W w(x, y) [\sqrt{\det(\boldsymbol{g})}]^{1/2} dx\, dy \ . \tag{9.4.8}$$

Remark 9.5. Equations (9.4.2, 8) are interpreted as follows. Consider a small region S on the image plane, and let S_0 be the corresponding region on the surface. For a dot texture, the number of dots in S is equal to the number of dots in S_0, while the area of S is $1/\sqrt{\det(\boldsymbol{g})}$ times that of S_0. Hence, the texture density in S is $\sqrt{\det(\boldsymbol{g})}$ times that in S_0. For a line segment texture, if the texture is nearly isotropic, the total length of the line segments in S is approximately $1/[\sqrt{\det(\boldsymbol{g})}]^{1/2}$ times that of S_0, while the area of S is $1/\sqrt{\det(\boldsymbol{g})}$ times that of S_0. Hence, the texture density in S is $[\sqrt{\det(\boldsymbol{g})}]^{1/2}$ times that in S_0.

Our principle of surface recovery is as follows.

1. Let the object surface be modeled as a parametrized surface in the form

$$Z = Z(X, Y; \alpha, \beta, \gamma, \ldots) \ , \tag{9.4.9}$$

   where $\alpha, \beta, \gamma, \ldots$ are the surface parameters to be determined.
2. From the above surface equation and the geometry of perspective projection, compute the first fundamental metric tensor

$$g(x, y; \alpha, \beta, \gamma, \ldots) = (g_{ij}(x, y; \alpha, \beta, \gamma, \ldots)) \ , \qquad i, j = 1, 2, \tag{9.4.10}$$

   which is a function of the surface parameters $\alpha, \beta, \gamma, \ldots$.
3. Provide appropriate test functions $w_0(x, y), w_1(x, y), w_2(x, y), \ldots$, and compute the corresponding *observables*[2]

$$J_i = \int_W \rho(x, y) w_i(x, y) dx\, dy \ , \qquad i = 0, 1, 2, \ldots \ . \tag{9.4.11}$$

   These observables are computed by evaluating the test functions $w_i(x, y)$ at the texture points or integrating them along the line segments on the image plane.

---

[2] Equation (9.4.11) defines what we called an "image feature" in Chap. 5. Here, as in Chap. 7, we use the term "observable" to emphasize the fact that this quantity is what is actually observed on the image plane.

4. Replacing the approximation symbols in (9.4.2, 8) by equalities, and taking the ratio J_i/J_0 to eliminate the unknown true texture density c, we obtain the 3D *recovery equations* to determine the surface parameters:

$$\frac{J_i}{J_0} = \frac{\int_W w_i(x, y)[\sqrt{\det(\boldsymbol{g}(x, y; \alpha, \beta, \gamma, \ldots))}]^\kappa dx\,dy}{\int_W w_0(x, y)[\sqrt{\det(\boldsymbol{g}(x, y; \alpha, \beta, \gamma, \ldots))}]^\kappa dx\,dy},$$

$$i = 1,2, \ldots . \tag{9.4.12}$$

Here, $\kappa = 1$ for a dot texture, and $\kappa = 1/2$ for a line segment texture. Equations (4.12) provide a set of nonlinear equations in unknowns $\alpha, \beta, \gamma, \ldots$. The values of these surface parameters are determined by solving these equations.

Remark 9.6. It is intuitively easy to understand that this is the most natural formulation derivable from the assumption of texture homogeneity. Since the change of the texture density is observed through the change of the surface area under perspective projection, the equation must be expressed in terms of $\sqrt{\det(\boldsymbol{g})}$. Since the texture elements are discretely distributed, we cannot observe $\sqrt{\det(\boldsymbol{g})}$ point-wise; we need some smoothing over a finite domain. The test functions $w_i(x, y)$ do the required smoothing.

9.5 Recovery of Planar Surfaces

The simplest case is when the surface is a plane. If we model the planar surface by equation $Z = pX + qY + r$, the form of $\sqrt{\det(\boldsymbol{g})}$ is given by (9.3.10). Hence, we obtain

$$c\int_W w(x, y)\sqrt{\det(\boldsymbol{g})}dx\,dy$$

$$= c(\sqrt{1 + p^2 + q^2})^\kappa \left(\frac{r}{f}\right)^{2\kappa} \int_W \frac{w(x, y)dx\,dy}{[1 - (px + qy)/f]^3} . \tag{9.5.1}$$

From this and (9.4.12) result the following 3D recovery equations:

Propositions 9.3 (3D *recovery equations for a planar surface*).

$$\int_W \frac{w_i(x, y) - (J_i/J_0)w_0(x, y)}{[1 - (px + qy)/f]^{3\kappa}} dx\,dy = 0 , \qquad i = 1, 2. \tag{9.5.2}$$

Remark 9.7. Equations (9.5.2) are equations in unknowns p, q, and can be solved in principle, say, by iterative search in the pq-plane (often called the *gradient space*). Evidently, three test functions $w_0(x, y)$, $w_1(x, y)$, $w_2(x, y)$ are sufficient for determination of the two unknowns p, q. However, we can also use many more test functions and determine p, q by the least squares method (Sect. 7.2).

Remark 9.8. Note that the denominator of the integrand of (9.5.2) becomes zero along the line $px + qy = f$ on the image plane. Also, $\sqrt{\det(\boldsymbol{g})} = \infty$ along this line, as we can see from (9.3.10). This line is the *vanishing line* of the surface (Sect. 8.5). Evidently, we must avoid the vanishing line by appropriately confining the size and location of the window W. In fact, if the vanishing line is observed, there is no need to use the texture to estimate the surface gradient; the equation $px + qy = f$ of the vanishing line immediately tells us the gradient (p, q) (Corollary 8.7).

Suppose the surface gradient (p, q) is close to zero compared with the focal length f, i.e., $px + qy \ll f$. Consider the integrand of the 3D recovery equations (9.5.2). Taylor expansion around the image origin O yields

$$\frac{1}{[1-(px+qy)/f]^{3\kappa}} = 1 + \frac{3\kappa}{f}(px+qy) + \dots \tag{9.5.3}$$

If we put

$$L_i = \int_W w_i(x, y)dx\,dy \ ,$$

$$M_i = \frac{3\kappa}{f}\int_W xw_i(x, y)dx\,dy \ , \quad N_i = \frac{3\kappa}{f}\int_W yw_i(x, y)dx\,dy \tag{9.5.4}$$

for $i = 0, 1, 2$, and neglect higher-order terms, the 3D recovery equations (9.5.2) reduce to the following linear equations in p, q:

$$\begin{pmatrix} M_1 - (J_1/J_0)M_0 & N_1 - (J_1/J_0)N_0 \\ M_2 - (J_2/J_0)M_0 & N_2 - (J_2/J_0)N_0 \end{pmatrix}\begin{pmatrix} p \\ q \end{pmatrix} = -\begin{pmatrix} L_1 - (J_1/J_0)L_0 \\ L_2 - (J_2/J_0)L_0 \end{pmatrix} . \tag{9.5.5}$$

A simple choice of the test functions $w_i(x, y)$, $i = 0, 1, 2$, is

$$w_0(x, y) = 1 \ , \quad w_1(x, y) = x \ , \quad w_2(x, y) = y \ . \tag{9.5.6}$$

This means that we compute the following observables.

$$J_0 = \int_W \rho(x, y)dx\,dy \ ,$$

$$J_1 = \int_W x\rho(x, y)dx\,dy \ , \quad J_2 = \int_W y\rho(x, y)dx\,dy \ . \tag{9.5.7}$$

Note that $(J_1/J_0, J_2/J_0)$ is the *centroid* of the texture inside the window W. In other words, if each dot has unit mass, or each line segment has unit mass per length, the centroid $(\bar{x}, \bar{y})$ is given by $(J_1/J_0, J_2/J_0)$:

$$\bar{x} = \frac{J_1}{J_0} = \frac{\int_W x\rho(x, y)dx\,dy}{\int_W \rho(x, y)dx\,dy}\,, \qquad \bar{y} = \frac{J_2}{J_0} = \frac{\int_W y\rho(x, y)dx\,dy}{\int_W \rho(x, y)dx\,dy}\,. \tag{9.5.8}$$

In particular, if the window W is a rectangle defined by $-a \leqq x \leqq a$, $-b \leqq y \leqq b$, then (9.5.4a–c) reduce to $L_0 = 4ab$, $M_1 = 4\kappa a^3 b/f$, $N_2 = 4\kappa ab^3/f$, $L_1 = L_2 = M_0 = M_2 = N_0 = N_1 = 0$. Hence, the solution of (9.5.5) is given by

$$p = \frac{f}{\kappa a^2}\bar{x}\,, \qquad q = \frac{f}{\kappa b^2}\bar{y}\,. \tag{9.5.9}$$

This result can be interpreted as follows. Suppose $(p, q) = (0, 0)$, i.e., the surface is viewed orthogonally. Then, the centroid should coincide with the image origin if the texture is truly homogeneous. Otherwise, the orientation and the amount of its shift tell the surface gradient (p, q) in the form of (9.5.9).

Remark 9.9. The accuracy of computation depends on both the number or length of the texture elements in the window W and their distribution patterns. Let N be the number or length of the texture elements in the window W. The rule of thumb is that the error is approximately proportional to $1/\sqrt{N}$ when the texture is completely random, while it is approximately proportional to $1/N$ when the texture is very regular and periodic (Sect. 9.5.1). Textures we encounter in natural scenes and man-made objects are usually regular periodic tessellations. In such cases, high accuracy is expected. As we discussed in Sect. 9.1, *the computer need not recognize the regularity or periodicity, if they exist.*

9.5.1 Error Due to Randomness of the Texture

Let us try a rough estimation of the error which enters in the computation of the centroid. For simplicity, we consider only a dot texture. Let $(x_1, y_1), \ldots, (x_N, y_N)$ be the coordinates of the texture points in the window $-a < x \leqq a$, $-b < y \leqq b$. The centroid $(\bar{x}, \bar{y})$ is given by

$$\bar{x} = \frac{1}{N}\sum_{i=1}^{N} x_i\,, \qquad \bar{y} = \frac{1}{N}\sum_{i=1}^{N} y_i\,. \tag{9.5.10}$$

First, suppose the distribution is completely random: x_i, $i = 1, \ldots, N$ are regarded as random variables from the uniform distribution over $-a < x \leqq a$ independently of each other, and likewise y_i, $i = 1, \ldots, N$ are also random numbers from the uniform distribution over $-b < y \leqq b$. This means that

$$E[x_i] = 0 \ , \quad V[x_i] = \frac{1}{3} a^2 \ , \quad i = 1, \ldots, N \ ,$$

$$E[y_i] = 0 \ , \quad V[y_i] = \frac{1}{3} b^2 \ , \quad i = 1, \ldots, N \ , \tag{9.5.11}$$

where $E[.]$ and $V[.]$ respectively designate the expectation value and the variance. It follows then that the expectation values and variances of $\bar{x}$, $\bar{y}$ are given by

$$E[\bar{x}] = 0 \ , \quad V[\bar{x}] = \frac{1}{3N} a^2 \ ,$$

$$E[\bar{y}] = 0 \ , \quad V[\bar{y}] = \frac{1}{3N} b^2 \ . \tag{9.5.12}$$

Hence, the centroid is at the image origin on the average. Since the magnitude of error is approximated by the standard deviation, errors of magnitude of $a/\sqrt{3N}$ and $b/\sqrt{3N}$ are expected for $\bar{x}$ and $\bar{y}$, respectively.

Consider the other extreme in which x_i, $i = 1, \ldots, N$, are distributed at equal intervals of distance $2a/N$, and y_i, $i = 1, \ldots, N$, at equal intervals of distance $2b/N$. Then, the centroid must be located within the range

$$-\frac{a}{N} < \bar{x} \leqq \frac{a}{N} \ , \qquad -\frac{b}{N} < \bar{y} \leqq \frac{b}{N} \ . \tag{9.5.13}$$

From (9.5.12, 13), it can be said that the errors $\delta\bar{x}$, $\delta\bar{y}$ of $\bar{x}$, $\bar{y}$, respectively, are

$$\delta\bar{x} = \mathrm{O}\left(\frac{1}{N^{\varepsilon}}\right) \ , \qquad \delta\bar{y} = \mathrm{O}\left(\frac{1}{N^{\varepsilon}}\right) \ , \tag{9.5.14}$$

where $1/2 < \varepsilon < 1$. The parameter ε approaches $1/2$ as the distribution becomes very random, while approaches 1 as the distribution becomes very regular.

9.6 Numerical Scheme of Planar Surface Recovery

As we showed in the preceding section, the solution of the 3D recovery equations is immediately obtained if the gradient (p, q) is close to zero. If it is not close to zero, we need iterations.

The first idea is to use the camera rotation transformation. Suppose the surface gradient is not small. We first apply the method of the preceding section. Let (p_0, q_0) be the computed gradient. Suppose the camera is rotated in such a way that the estimated surface becomes parallel to the image plane. If this new image is regarded as the input, the surface gradient should be small. Hence, the method of the preceding section can be applied. Let (p_0', q_0') be the computed surface gradient. If this newly estimated surface is rotated back into the original camera orientation, we obtain a better estimation (p_1, q_1). This process can be applied repeatedly, producing a sequence of estimations (p_i, q_i), $i = 0, 1, 2, \ldots,$ until no further improvement is made.

We should recall the fact that *the camera need not actually be rotated*, because the image transformation is given in an analytical expression. However, we should also note that *the transformed image need not be generated*, because *all we need is the values of the observables* J_i, $i = 0, 1, 2$. As we will now show, we can derive the transformation rule of observables. Hence, we only need to compute the "transformed observables" J_i', $i = 0, 1, 2$.

To be specific, if we want to compute observable J_i' for a test function $w'(x', y')$ over the transformed image $\rho'(x', y')$ inside the transformed window W', we want to compute it by integrating a modified test function $\tilde{w}(x, y)$ applied to the original image $\rho(x, y)$ inside the original window W:

$$\int_{W'} w'(x', y')\rho'(x', y')dx'\,dy' = \int_{W} \tilde{w}(x, y)\rho(x, y)dx\,dy \;. \tag{9.6.1}$$

Let (p_0, q_0) be the initial estimate of the gradient. Note the following fact:

Lemma 9.2. *The unit surface normal* $\boldsymbol{n} = (n_1, n_2, n_3)$ *to a planar surface whose gradient is* (p_0, q_0) *is given by*

$$n_1 = \frac{p_0}{\sqrt{1 + p_0^2 + q_0^2}}\,, \quad n_2 = \frac{q_0}{\sqrt{1 + p_0^2 + q_0^2}}\,,$$
$$n_3 = -\frac{1}{\sqrt{1 + p_0^2 + q_0^2}}\,. \tag{9.6.2}$$

Proof. If we take differentials of both sides of the surface equation $Z = p_0 X + q_0 Y + r$, we obtain $dZ = p_0 dX + q_0 dY$, or

$$(p_0 \quad q_0 \quad -1)\begin{pmatrix} dX \\ dY \\ dZ \end{pmatrix} = 0 \;. \tag{9.6.3}$$

This means that vector $(p_0, q_0, -1)$ is orthogonal to the line segment connecting (X, Y, Z) and $(X + dX, Y + dY, Z + dZ)$. Since this is true for arbitrary differentials dX, dY, dZ along the surface, vector $(p_0, q_0, -1)$ is perpendicular to the surface. Equation (9.6.2) is obtained by normalizing this vector.

Proposition 9.4. *The camera rotation that transforms a planar surface of gradient* (p, q) *into a planar surface parallel to the image plane is given by*

$$\boldsymbol{R} = \begin{pmatrix} r_{11} & r_{12} & n_1 \\ r_{12} & r_{22} & n_2 \\ -n_1 & -n_2 & n_3 \end{pmatrix}, \tag{9.6.4}$$

where

$$r_{11} = \frac{p_0^2 n_3 + q_0^2}{p_0^2 + q_0^2}, \quad r_{12} = \frac{p_0 q_0 (n_3 - 1)}{p_0^2 + q_0^2}, \quad r_{22} = \frac{q_0^2 n_3 + p_0^2}{p_0^2 + q_0^2}, \tag{9.6.5}$$

and n_1, n_2, n_3 *are given by* (9.6.2).

Proof. The unit surface normal $\boldsymbol{n}$ given by (9.6.2) makes angle

$$\Omega = \cos^{-1} n_3 \tag{9.6.6}$$

with the unit vector $\boldsymbol{k} = (0, 0, 1)$ along the Z-axis. The unit vector perpendicular to both $\boldsymbol{n}$ and $\boldsymbol{k}$ is given by

$$\boldsymbol{l} = \frac{\boldsymbol{k} \times \boldsymbol{n}}{\|\boldsymbol{k} \times \boldsymbol{n}\|} = \frac{(q_0, -p_0, 0)}{\sqrt{p_0^2 + q_0^2}}. \tag{9.6.7}$$

If the camera is rotated around this vector $\boldsymbol{l}$ by angle Ω screw-wise, the estimated surface becomes parallel to the new image plane. The corresponding rotation matrix is given by (9.6.4) (Proposition 6.5).

Hence, from Theorem 1.1, we have:

Corollary 9.1. *The image transformation induced by the camera rotation* (9.6.4) *is given by*

$$x'(x, y) = f\frac{r_{11}x + r_{12}y + n_1 f}{n_1 x + n_2 y + n_3 f}, \quad y'(x, y) = f\frac{r_{12}x + r_{22}y - n_2 f}{n_1 x + n_2 y + n_3 f}. \tag{9.6.8}$$

It follows from (9.6.8) that if the original window W is placed near the image origin, the transformed window W' is located near point (fp_0, fq_0). Let (p', q') be the true surface gradient relative to the new $x'y'$-plane. Our next step is to estimate this gradient (p', q'). Taylor expansion around point (fp_0, fq_0) yields

$$\frac{1}{[1 - (p'x' + q'y')/f]^{3\kappa}} = \frac{1}{(1 - p_0 p' - q_0 q')^{3\kappa}}[1 + A(x' - fp_0) + B(y' - fq_0) + \ldots], \tag{9.6.9}$$

where

$$A = \frac{3\kappa p'}{f(1 - p_0 p' - q_0 q')}, \qquad B = \frac{3\kappa q'}{f(1 - p_0 p' - q_0 q')}. \tag{9.6.10}$$

Let $w_i'(x', y')$, $i = 0, 1, 2$, be the test functions. If (9.6.9) is substituted into the 3D recovery equations (9.5.2) in the $x'y'$-plane, we obtain a set of linear equations that has the form of (9.5.5), namely

$$\begin{pmatrix} M_1' - (J_1'/J_0')M_0' & N_1' - (J_1'/J_0')N_0' \\ M_2' - (J_2'/J_0')M_0' & N_2' - (J_2'/J_0')N_0' \end{pmatrix} \begin{pmatrix} A \\ B \end{pmatrix} = - \begin{pmatrix} L_1' - (J_1'/J_0')L_0' \\ L_2' - (J_2'/J_0')L_0' \end{pmatrix}, \tag{9.6.11}$$

where

$$J_i' = \int_{W'} w_i'(x', y')\rho'(x', y')dx'\,dy', \tag{9.6.12}$$

$$L_i' = \int_{W'} w_i'(x', y')dx'\,dy',$$

$$M_i' = \int_{W'} (x' - fp_0)w_i'(x', y')dx'\,dy',$$

$$N_i' = \int_{W'} (y' - fq_0)w_i'(x', y')dx'\,dy' \tag{9.6.13}$$

for $i = 0, 1, 2$, and $\rho'(x', y')$ is the transformed texture density.

If A and B are determined from (9.6.11), the gradient (p', q') is determined by solving (9.6.10a, b), which are rewritten as

$$\begin{pmatrix} Ap_0 + 3\kappa/f & Aq_0 \\ Bp_0 & Bq_0 + 3\kappa/f \end{pmatrix} \begin{pmatrix} p' \\ q' \end{pmatrix} = \begin{pmatrix} A \\ B \end{pmatrix}. \tag{9.6.14}$$

If the computed gradient (p', q') is sufficiently close to zero, the estimate is correct. Otherwise, the camera is rotated back into the original orientation. The surface gradient is transformed as follows:

Lemma 9.3. *The surface gradient* (p', q') *is transformed into* (p_1, q_1) *by the inverse of the camera rotation* (9.6.4) *as follows*:

$$p_1 = \frac{r_{11}p' + r_{12}q' - n_1}{n_1 p' + n_2 q' + n_3}, \qquad q_1 = \frac{r_{12}p' + r_{22}q' - n_2}{n_1 p' + n_2 q' + n_3}. \tag{9.6.15}$$

Proof. Note that the camera was rotated by $\boldsymbol{R}$. This means that the surface was rotated by $\boldsymbol{R}^{-1}(= \boldsymbol{R}^{\mathrm{T}})$ relative to the camera. Hence, the surface is rotated back into its original position by rotation $(\boldsymbol{R}^{-1})^{-1} = \boldsymbol{R}$. Since the unit surface normal to a planar surface whose gradient is (p', q') is $(p', q', -1)/\sqrt{1 + p'^2 + q'^2}$ (Lemma 9.2), the unit surface normal $\boldsymbol{n}' = (n'_1, n'_2, n'_3)$ to the rotated surface is given by

$$\begin{pmatrix} n'_1 \\ n'_2 \\ n'_3 \end{pmatrix} = \frac{1}{\sqrt{1 + p'^2 + q'^2}} \begin{pmatrix} r_{11} & r_{12} & n_1 \\ r_{12} & r_{22} & n_2 \\ -n_1 & -n_2 & n_3 \end{pmatrix} \begin{pmatrix} p' \\ q' \\ -1 \end{pmatrix} \tag{9.6.16}$$

The new gradient is given by $(p_1, q_1) = (-n'_1/n'_3, -n'_2/n'_3)$.

These values are a better approximation to the surface gradient. This process can be iterated to compute (p_2, q_2), (p_3, q_3), ... until convergence. In this procedure, the only quantities that must be computed for the transformed image are (9.6.12, 13). We now show that *they can be computed from the original image* $\rho(x, y)$ *over the original window* W *by changing variables.* First, consider a dot texture. From Proposition 9.1, we see that if we put

$$\tilde{w}_i(x, y) \equiv w'_i(x'(x, y), y'(x, y)) , \qquad i = 0, 1, 2, \tag{9.6.17}$$

(9.6.12) is written as

$$J'_i = \sum_{P_i \in W} \tilde{w}_i(x_i, y_i) , \qquad i = 0, 1, 2, \tag{9.6.18}$$

which can be computed over the original image.

Next, consider a line segment texture. From Proposition 9.1, we obtain the following rule:[3]

$$J'_i = \sum_{L_i \subset W} \int_{L_i} \tilde{w}_i(x(l), y(l)) \sqrt{Et_1^2 + 2Ft_1t_2 + Gt_2^2}\, dl , \qquad i = 0, 1, 2. \tag{9.6.19}$$

Here, l is the arc length, and $\boldsymbol{t} = (t_1, t_2)$ is the unit tangent vector along the individual line segments of the original image, while E, F, G are functions of x, y defined by

$$E(x, y) \equiv \left(\frac{\partial x'}{\partial x}\right)^2 + \left(\frac{\partial y'}{\partial x}\right)^2$$

$$= f^2 \frac{1 + n_1^2(x'^2 + y'^2 - 1) - 2n_1(r_{11}x' + r_{12}y')}{(n_1x + n_2y + n_3f)^2} ,$$

[3] Variables x, y in Proposition 9.1 correspond to x', y' here, while variables x', y' there correspond to x, y here.

$$F(x, y) \equiv \frac{\partial x'}{\partial x}\frac{\partial x'}{\partial y} + \frac{\partial y'}{\partial x}\frac{\partial y'}{\partial y}$$

$$= f^2 \frac{n_1 n_2(x'^2 + y'^2 - 1) - n_1(r_{12}x' + r_{22}y') - n_2(r_{11}x' + r_{12}y')}{(n_1 x + n_2 y + n_3 f)^2} ,$$

$$G(x, y) \equiv \left(\frac{\partial x'}{\partial y}\right)^2 + \left(\frac{\partial y'}{\partial y}\right)^2$$

$$= f^2 \frac{1 + n_2^2(x'^2 + y'^2 - 1) - 2n_2(r_{12}x' + r_{22}y')}{(n_1 x + n_2 y + n_3 f)^2} . \tag{9.6.20}$$

If we use the coordinate transformation (9.6.8), these are all expressed in terms of the original coordinates x, y. Hence, (9.6.19) can be computed over the original image.

On the other hand, (9.6.13a–c) are integration of continuous functions. Hence, we immediately rewrite them in terms of the original coordinates x, y as

$$L_i' = \int_W \tilde{w}_i(x, y)\Delta(x, y)dx\,dy ,$$

$$M_i' = \int_W [x'(x, y) - fp_0]\tilde{w}_i(x, y)\Delta(x, y)dx\,dy ,$$

$$N_i' = \int_W [y'(x, y) - fq_0]\tilde{w}_i(x, y)\Delta(x, y)dx\,dy \tag{9.6.21}$$

for $i = 0, 1, 2$, where $\Delta(x, y)$ is the Jacobian of the coordinate transformation of (9.6.8) given by

$$\Delta(x, y) \equiv \begin{vmatrix} \partial x'/\partial x & \partial x'/\partial y \\ \partial y'/\partial x & \partial y'/\partial y \end{vmatrix} = -f\frac{n_1 x'(x, y) + n_2 y'(x, y) - n_3 f}{(n_1 x + n_2 + n_3 f)^2} . \tag{9.6.22}$$

Thus, (9.6.21) are easily computed by a numerical integration scheme.

Remark 9.10. Equation (9.6.19) is a rigorous relation. Since the Jacobian $\Delta(x, y)$ plays the role of $\sqrt{\det(\boldsymbol{g})}$ of Sect. 9.4, a simple approximation corresponding to (9.4.6) is

$$\sqrt{Et_1^2 + 2Ft_1t_2 + Gt_2^2}\,dl \approx \sqrt{\Delta(x, y)} . \tag{9.6.23}$$

Hence, we could use the approximation

$$J_i' \approx \sum_{L_i \subset W} \int_{L_i} \tilde{w}_i(x(l), y(l))\sqrt{\Delta(x(l), y(l))}\,dl . \tag{9.6.24}$$

However, this approximation is not used here. See Remark 9.11.

The method described above is somewhat complicated. There exists an alternative scheme. Suppose (p_k, q_k) is the estimate at the kth step. If we expand the left-hand side of (9.5.3) at (p_k, q_k) rather than at (0, 0), we obtain

$$\frac{1}{[1-(px+qy)/f]^{3\kappa}} = L(x, y) + M(x, y)\delta p_k + N(x, y)\delta q_k + \ \ldots\ , \tag{9.6.25}$$

where $\delta p_k = p - p_k$, $\delta q_k = q - q_k$, and

$$L(x, y) = \frac{1}{[1-(p_k x + q_k y)/f]^{3\kappa}} ,$$
$$M(x, y) = \frac{3\kappa x}{f[1-(p_k x + q_k y)/f]^{3\kappa+1}} ,$$
$$N(x, y) = \frac{3\kappa y}{f[1-(p_k x + q_k y)/f]^{3\kappa+1}} . \tag{9.6.26}$$

Then, the 3D recovery equations (9.5.2) reduce to a set of linear equations that has the form of (9.5.5), namely

$$\begin{pmatrix} M_1 - (J_1/J_0)M_0 & N_1 - (J_1/J_0)N_0 \\ M_2 - (J_2/J_0)M_0 & N_2 - (J_2/J_0)N_0 \end{pmatrix} \begin{pmatrix} \delta p_k \\ \delta q_k \end{pmatrix} = - \begin{pmatrix} L_1 - (J_1/J_0)L_0 \\ L_2 - (J_2/J_0)L_0 \end{pmatrix} , \tag{9.6.27}$$

where

$$L_i = \int_W w_i(x, y) L(x, y) dx\, dy ,$$
$$M_i = \int_W w_i(x, y) M(x, y) dx\, dy , \quad N_i = \int_W w_i(x, y) N(x, y) dx\, dy , \tag{9.6.28}$$

for $i = 0, 1, 2$. If δp_k, δq_k are sufficiently close to zero, the kth estimate is accurate. Otherwise, $p_{k+1} = p_k + \delta p_k$, $q_{k+1} = q_k + \delta q_k$, are better approximations, and the process can be iterated until convergence.

Remark 9.11. This method is exactly the *Newton iterations* of the 3D recovery equations over two variables p, q. It is known that if we start from an initial guess that is sufficiently close to the true solution, the iterations converge *quadratically* (the error is roughly proportional to the square of the error at the preceding step). However, it is difficult to guess a good initial estimate, and if we start from an inaccurate guess, the behavior of the iterations is unpredictable. Also, the geometrical meaning of functions $M(x, y)$, $N(x, y)$ is not clear.

In contrast, the geometrical interpretation of the camera rotation simulation method is very clear: essentially, we are incrementally "undoing" perspective distortion so that the centroid of the texture elements coincides with the center of the window, though it seems difficult to give a rigorous proof of convergence. While the Taylor expansion method still involves the approximation (9.4.6), the camera rotation simulation method is not so much affected by this approximation; the texture image is *exactly* transformed repeatedly by (9.6.19, 20), so that the true surface becomes more and more parallel to the image plane. In the limit of $p \to 0$, $q \to 0$, the approximation (9.4.6) reduces to a trivial identity.

Example 9.1. Figures 9.4–7 are synthetic images of textured planar surfaces. The focal length f is taken as the unit of length, and the window size is $a = b = f \tan 10° = 0.176$. (i.e., the actual window is a rectangle of size $2a \times 2b$). The Z-axis is assumed to pass through the center of the window. The true surface gradient is $(p, q) = (1.500, 0.866)$ for all the figures.

Figure 9.4 shows a projected image of a regularly aligned dot texture on a planar surface. If we use test functions $w_0(x, y) = 1$, $w_1(x, y) = x$, $w_2(x, y) = y$, an initial estimate is obtained by computing the centroid of the texture (9.5.5). If the camera rotation simulation method and the Taylor expansion method with $w_0(x, y) = L(x, y)$, $w_1(x, y) = M(x, y)$, $w_2(x, y) = N(x, y)$ are applied, the following values of (p, q) are computed successively. The left column is for the camera rotation simulation method, while the right column is for the Taylor expansion method.

1	(1.610, 1.965)	(1.610, 0.965)
2	(1.552, 0.875)	(1.549, 0.897)
3	(1.548, 0.871)	(1.542, 0.887)
4	(1.548, 0.871)	(1.541, 0.887)

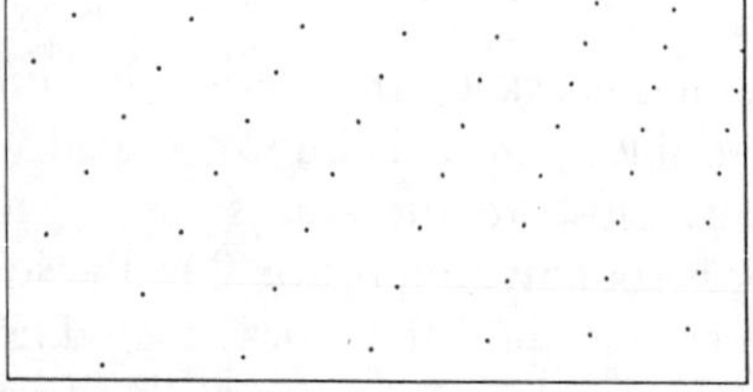

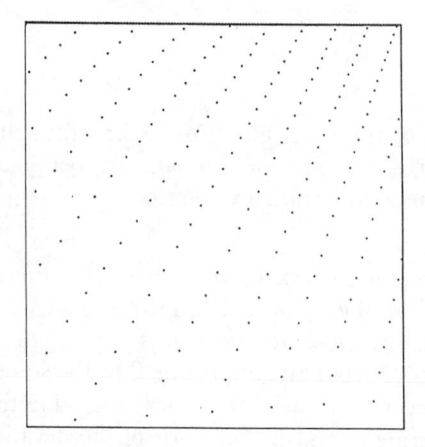

Fig. 9.4. A regularly aligned dot texture on a planar surface

Figure 9.5 shows a random dot texture on a planar surface. Applying the same procedures, we obtain

1	(1.662, 1.945)	(1.662, 0.945)
2	(1.617, 0.843)	(1.674, 0.832)
3	(1.613, 0.840)	(1.667, 0.834)
4	(1.613, 0.840)	(1.667, 0.834)

Figure 9.6 shows a regularly aligned line segment texture on a planar surface. We obtain

1	(1.762, 1.048)	(1.762, 1.048)
2	(1.534, 0.899)	(1.672, 0.960)
3	(1.505, 0.875)	(1.669, 0.956)
4	(1.504, 0.874)	(1.669, 0.956)

Figure 9.7 shows a random line segment texture on a planar surface. We obtain

1	(2.196, 0.768)	(2.196, 0.768)
2	(1.906, 0.590)	(2.033, 0.561)
3	(1.856, 0.559)	(2.022, 0.561)
4	(1.856, 0.559)	(2.022, 0.567)

Both methods yield almost the same results for dot textures, but for line segment textures the camera rotation simulation method predicts more accurate values (Remark 9.11). The convergence is very rapid for both methods; only two or three iterations yield the surface gradient up to two or three decimal places.

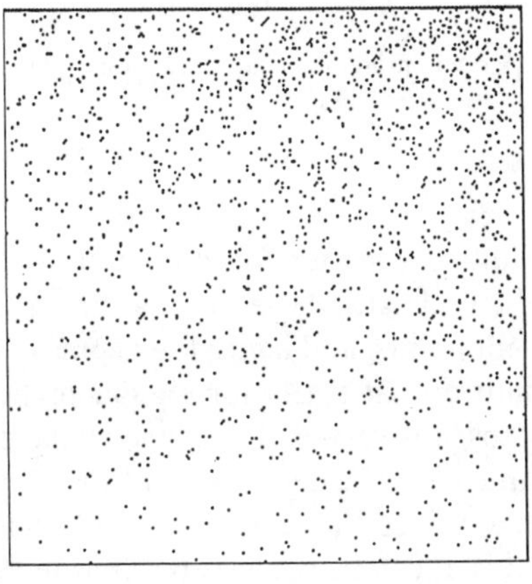

Fig. 9.5. A random dot texture on a planar surface

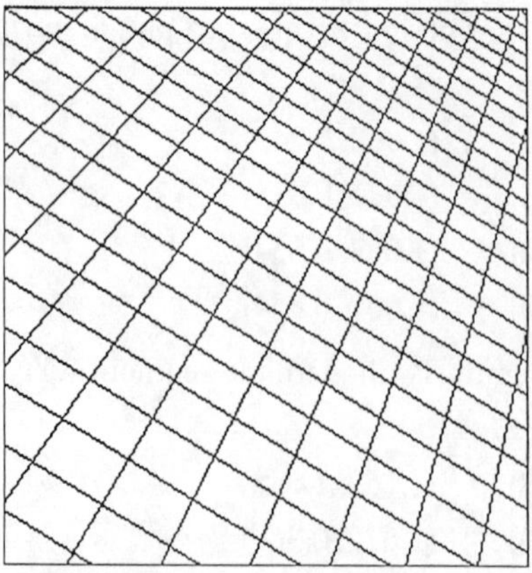

Fig. 9.6. A regularly aligned line segment texture on a planar surface

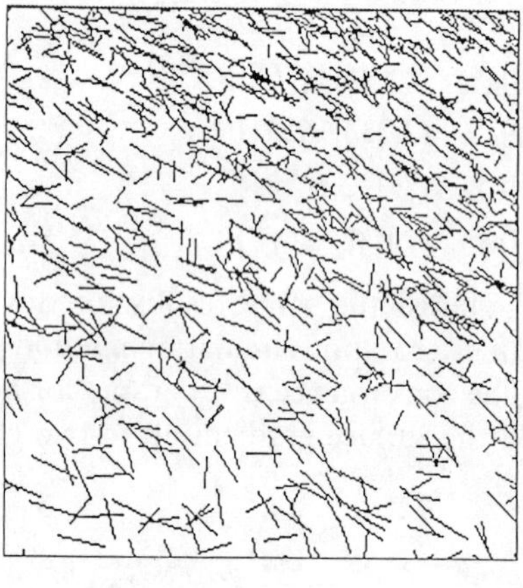

Fig. 9.7. A random line segment texture on a planar surface

On the whole, the results are better for dot textures than for line segment textures. This is easily understood because a line segment is a restricted coalescence of constituent points, and hence the degree of homogeneity is lower.

As expected, the results are far better for regular textures than random ones (Remark 9.9). For humans, it is very easy to perceive the depth by seeing regular patterns like Figs. 9.4 and 9.6, while depth perception is very difficult for random textures like Figs. 9.5 and 9.7. In this sense, our method is "superior" to human perception. In view of this, it can be conjectured that human perception may be based on the "structure" of the texture (Sect. 9.1).

9.6.1 Technical Aspects of Implementation

We should not forget the fact that appropriate preprocessing is necessary if our method is to be used. In our method, a texture is assumed to be composed of dots without area and line segments without width. If the dots have area, we can consider their centroids, or regard their boundaries as texture elements. If the line segments have width, we can regard their center lines (*skeletons* or *medial axes*) or boundaries as texture elements. For natural texture images having continuous gray levels, a simple scheme may be to apply some edge detection algorithm (e.g., of *zero-crossing*). Then, the detected edges serve as line segment texture elements.

As we have shown, dot textures and line segment textures cannot be treated in the same manner. In particular, *pixels constituting line segments cannot be regarded as pixels of a dot texture*, because pixels constituting a line segment are necessarily *uniformly sampled*. If pixels on a line segment could be sampled with converging intervals in such a way that the effect of perspective projection is reflected, no distinction would be needed between dots and line segments. That is why we must introduce the exponent κ in order to distinguish these two types of texture.

One of the difficulties of shape-from-texture analysis is the effect of the *resolution threshold*: the part of the texture far apart from the viewer becomes dense on the image plane, and hence its spatial frequency becomes high. As a result, true texture elements are likely to coalesce into larger apparent texture elements, or *supertexture*. In contrast, the part near the viewer is likely to be processed with excessive resolution, resulting in *subtexture* (variations of gray levels within individual texture elements). Hence, the surface gradient tends to be underestimated if computation is based on such false clues.

There are many ways to avoid this effect, although none may be complete—for example, to confine the window so that only a small depth range is observed, or to focus on only clearly visible clues by lowering the resolution. In view of the fact that the appearance of subtexture is greatly affected by the level of resolution, the true texture can be picked out by changing the resolution level and choosing only stable patterns.

9.6.2 Newton Iterations

The Newton iterations for a set of nonlinear equations

$$f_1(x_1, \ldots, x_n) = 0, \ldots, f_n(x_1, \ldots, x_n) = 0 \ , \tag{9.6.29}$$

are described as follows. Let $x_i^{(k)}$, $i = 1, \ldots, n$, be the kth approximate solutions, and let $\delta x_i^{(k)} = x_i - x_i^{(k)}$, $i = 1, \ldots, n$, be the discrepancies from the true values. Then, (9.6.29) can be written as

$$f_j(x_1^{(k)} + \delta x_1^{(k)}, \ldots, x_n^{(k)} + \delta x_n^{(k)}) = 0 \ , \qquad j = 1, \ldots, n \ . \tag{9.6.30}$$

By Taylor expansion, we have

$$f_j + \frac{\partial f_j}{\partial x_1}\delta x_1^{(k)} + \dots \frac{\partial f_j}{\partial x_n}\delta x_n^{(k)} + \dots = 0 \;, \qquad j = 1, \dots, n \;, \tag{9.6.31}$$

where f_j and $\partial f_j/\partial x_1, \dots \partial f_j/\partial x_n$ are evaluated at the kth estimates $x_1 = x_1^{(k)}, \dots, x_n = x_n^{(k)}$, and the remaining terms have orders equal to or higher than 2 in $\delta x_1^{(k)}, \dots, \delta x_n^{(k)}$.

If the higher order terms are neglected, $\delta x_i^{(k)}$, $i = 1, \dots, n$, are obtained by solving the following set of simultaneous linear equations:

$$\begin{pmatrix} \partial f_1/\partial x_1 \dots\dots \partial f_1/\partial x_n \\ \dots\dots\dots \\ \partial f_n/\partial x_1 \dots\dots \partial f_n/\partial x_n \end{pmatrix} \begin{pmatrix} \delta x_1^{(k)} \\ \vdots \\ \delta x_n^{(k)} \end{pmatrix} = - \begin{pmatrix} f_1 \\ \vdots \\ f_n \end{pmatrix} . \tag{9.6.32}$$

Using the solution of this, we obtain the $(k+1)$th estimates in the form

$$x_i^{(k+1)} = x_i^{(k)} + \delta x_i^{(k)} \;, \qquad i = 1, \dots, n \;. \tag{9.6.33}$$

It can be shown that if we begin with an initial guess sufficiently close to the true solution, the iterations converge *quadratically*. Namely, if $\varepsilon^{(k)}$ is the error of the kth estimate, then $\varepsilon^{(k+1)} \approx c(\varepsilon^{(k)})^2$, where c is a constant. There are many ways to define the "error" $\varepsilon^{(k)}$. Typical ones are $\sum_{i=1}^{n} |x_i - x_i^{(k)}|/n$ (L_1-norm), $(\sum_{i=1}^{n} (x_i - x_i^{(k)})^2)^{1/2}$ (L_2-norm), and $\max_i |x_i - x_i^{(k)}|$ (L_∞-norm). Whichever is used, the convergence behavior is qualitatively the same.

Exercises

9.1* Prove (9.2.12, 13). Namely, show that the Jacobian enters if the variables are changed in the ordinary double integral.

9.2 Show that the derivative $h'(x)$ of the (*Heaviside*) *step function*

$$h(x) = \begin{cases} 1 & x \geqq 0 \\ 0 & x < 0 \end{cases}$$

is the (Dirac) delta function $\delta(x)$ in the sense of a distribution.

9.3 Prove the rule of differentiation of the product

$$\frac{d}{dx}(c(x)T) = \frac{dc}{dx}T + c(x)\frac{dT}{dx}$$

in the sense of a distribution, where $c(x)$ is a C^∞ function.

9.4* Prove that the Fourier transform of $f(x) = 1$ is the delta function $\delta(x)$, i.e., $\mathfrak{F}1 = \delta(x)$, in the sense of a distribution.

9.5 Derive (9.3.4) from (9.3.2, 3), and show that the first fundamental form is given by (9.3.5, 6).

9.6 Derive (9.3.7).

9.7 Derive (9.3.8a–c), and show from them that the coefficients of the first fundamental form are given by (9.3.9) for planar surfaces. Also show (9.3.10) by using (9.3.9).

9.7 Show that if (9.5.1) is used, the 3D recovery equations (9.4.12) take the form of (9.5.2) for planar surfaces.

9.8 Derive the Taylor expansion (9.5.3) and show that the 3D recovery equations become (9.5.5) if the gradient is small.

9.9 Confirm (9.5.9) when the test functions (9.5.6) are used for a rectangular window.

9.10 (a) If X is a random variable, show that the expectation value and variance of the scalar multiple cX are given respectively by $E[cX] = cE[X]$ and $V[cX] = c^2 V[X]$.

(b) If X and Y are random variables, show that the expectation value and variance of the sum $S = X + Y$ are given respectively by $E[X + Y] = E[X] + E[Y]$ and, $V[X + Y] = V[X] + C[X, Y] + V[Y]$, where $C[.,.]$ is the *covariance* defined by $C[X, Y] \equiv E[(X - E[X]) \times (Y - E[Y])]$.

(c) Let $X_1, \ldots, X_N$ be mutually independent random variables obeying an identical distribution with expectation value 0 and variance σ^2. Show that the variance of the average $\sum_{i=1}^{N} X_i/N$ is σ^2/N.

9.11 Show that the Jacobian of (9.6.8) is given by (9.6.22).

9.12 Derive the Taylor expansion (9.6.9) and show that the 3D recovery equations (9.5.2) take the form of (9.6.11).

9.13 Prove (9.6.14).

9.14 Prove that (9.6.19, 20) are obtained by the change of variables.

9.15 Derive the Taylor expansion (9.6.25) and show that the 3D recovery equations (9.5.2) take the form of (9.6.27).

10. Shape from Surface

In this chapter, we show that although the 3D orientations of edges and surfaces are theoretically sufficient for reconstructing the 3D object shape, this does not mean that the 3D object shape can actually be reconstructed. Specifying the edge and surface orientations is often "over-specification", and inconsistency may result if image data contain errors. We propose a scheme of optimization to construct a consistent object shape from inconsistent data. Our optimization is achived by solving a set of linear equations; no searches and iterations are necessary. This technique is first applied to the problem of shape from motion and then to the 3D recovery based on the rectangularity hypothesis and the parallelism hypothesis.

10.1 What Does 3D Shape Recovery Mean?

So far, we have studied various 3D shape recovery techniques called *shape from* . . . .[1] Now, we must ask the following question: Do these techniques really enable us to recover the 3D object shape? They certainly provide us with sufficient information for 3D shape recovery. However, "to provide sufficient information" does not necessarily mean "to recover the 3D shape".

If we look closely into the "shape from . . ." paradigms, we find that the information we obtain is the *3D orientations* of edges and surfaces constituting the object. The surface gradient can be recovered from optical flow (*shape from motion*, Chap. 7); the 3D orientations of edges can be recovered from the 2D orientations of their projections (*shape from angle*, Chap. 8); the surface gradient can be recovered from an observed texture density (*shape from texture*, Chap. 9). All these "shape from . . . " paradigms can be called *passive vision* in the sense that the analysis is based solely on a single image or a sequence of images presented to us arbitrarily; we have no conrol over the image-taking process. Under such circumstances, we are in general unable to recover the absolute depth: all we can estimate are the 3D orientations of edges and surfaces.

---

[1] We have not discussed one well-known 3D recovery paradigm called *shape from shading*. We can recover the surface gradient (not necessarily uniquely) from the intensity of light reflectance from the object surface. We do not discuss this here because there already exist sufficient accounts in many books. See the Bibliography.

If we can control the image-taking process, the image analysis is called *active vision*. For example, we can directly obtain a depth map by analyzing the phase of the wave emitted from a radar or sonar and reflected by the object surface. The depth map can also be obtained from *stereo* by using two or more cameras.[2] Projecting a slit pattern over the object surface from one direction and taking its image from another is essentially stereo, and we can also obtain a depth map.

However, many active methods only enable us to estimate the surface gradient. Consider *photometric stereo*, for example. Providing two light sources, we take an image of an object illuminated by one light source and then take another image by switching to the other light source. The surface gradient can be estimated from the difference in light reflectance combined with the knowledge of the positions of the light sources ȧnd the reflectance property of the object surface. Another active method is to project a light pattern of a known shape over the object surface. For instance, if a light pattern consisting of circles or squares is projected from a known direction, the surface gradient can be estimated by observing the deformation of the observed pattern.

Mathematically speaking, the object shape is uniquely determined if the 3D orientations of its constituent edges and surfaces are specified. (Here and in the following, we disregard the absolute size when we talk about the 3D shape. In other words, the "shape" is the same if it is multiplied by an arbitrary scale factor.) Thus, the 3D orientations of edges and surfaces are sufficient for 3D shape recovery. In spite of this trivial fact, there arises a serious difficulty in real situations: specifying the 3D orientations of edges and surfaces is often *over-specification*.

For example, suppose the image is segmented into "planar patches" (regions resulting from projection of a planar part of the object surface).[3] Assume that the surface gradient is known for each planar patch. Can we reconstruct the 3D object shape? Theoretically, yes. We can reconstruct the shape by "patching together" appropriate planar surfaces one by one in the scene according to the prescribed orientations, in such a way that each of them is projected exactly onto the corresponding region of the image plane.

However, the image analysis is based on the data supplied by low-level image processing stages, and these data necessarily contain errors. If the surface gradient estimates are not accurate, the reconstructed object shape depends on the order of the patching, and there may arise *incompatibility* of face adjacency: two faces may not meet at a common boundary (Fig. 10.1a).

---

[2] However, since no "action" is taken on the object other than placing two cameras, many researchers classify stereo as passive vision.

[3] If the object is a polyhedron, its faces themselves constitute planar patches. If the object has a smooth surface, it is approximated by a collection of planar patches (i.e., by an appropriate polyhedron).

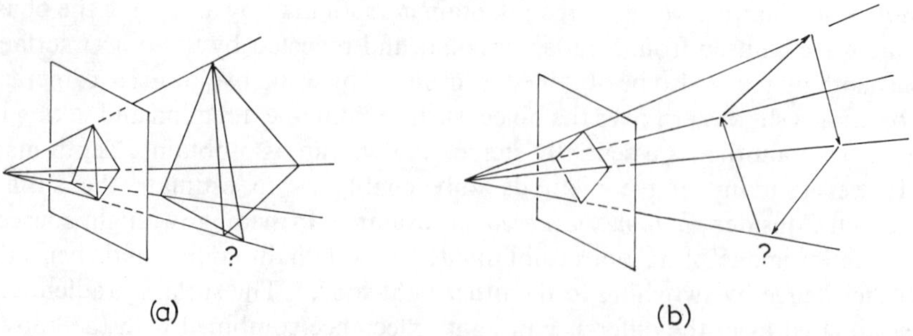

Fig. 10.1a. Incompatibility of face adjacency. If faces are placed in the scene according to their surface gradient estimates in such a way that they are projected onto the image plane as observed, adjacent faces may not meet with common boundaries. **b** Incompatibility of edge adjacency. If edges are placed in the scene according to their 3D orientation estimates in such a way that they are projected onto the plane as observed, edges constituting the boundary of a face may not form a closed loop

A Similar difficulty arises when 3D edge orientations are estimated. Theoretically, we can reconstruct the 3D object shape by placing its edges in the scene according to their prescribed 3D orientations, in such a way that each edge is projected exactly onto the corresponding edge on the image plane. After this process, however, edges constituting the boundary of a face may not form a closed loop; the endpoint may not coincide with the starting point (Fig. 10.1b). Even if they do coincide, the resulting face may not be planar.

It follows that there is a big gap between obtaining "sufficient" information for 3D recovery and actually reconstructing the 3D shape. This consideration leads to the following observation: as long as the object is approximated by a polyhedron, *the surface gradient values cannot be assigned arbitrarily*. In the next section, we study this strong constraint on polyhedron images.

Then, we present a scheme of *optimization* for computing a consistent polyhedron shape from inconsistent image data. We show that the optimization reduces to solving a set of simultaneous *linear* equations. We first apply our technique to the shape-from-motion problem and demonstrate how a consistent object is reconstructed from inaccurate data.

Then, our optimization technique is applied to the shape-from-angle problem. First, we adopt the rectangularity hypothesis, regarding corners as rectangular as long as no inconsistency results. Next, we adopt the parallelism hypothesis, and show that the 3D shape is reconstructed if parallel edges are identified. We also present a heuristic algorithm for finding parallel edges.

10.2 Constraints on a $2\frac{1}{2}$D Sketch

As we pointed out in the preceding section, the "shape from . . ." paradigms usually present us with object images equipped with the following types of 3D shape cues:

(i) The surface gradient (p, q), or equivalently the unit surface normal $\boldsymbol{n}$, is "densely" estimated (i.e., as a function of location) over the region corresponding to the object surface (Fig. 10.2a).

(ii) The region corresponding to the object surface is segmented into planar patches, and the surface gradient (p, q), or equivalently the unit surface normal $\boldsymbol{n}$, is estimated for each patch (Fig. 10.2b). In other words, the object surface is approximated by a polyhedron.[4]

(iii) The region corresponding to the object surface is segmented into planar patches and approximated by a polyhedron with estimated 3D edge orientations (Fig. 10.2c).

We call an image equipped with such 3D information a $2\frac{1}{2}$D *sketch*.[5]

Remark 10.1. Among these three, case (ii) may be the one we most often encounter, because what we can estimate, whether by active or passive vision, is the *local surface gradient*—the surface gradient obtained on the assumption that

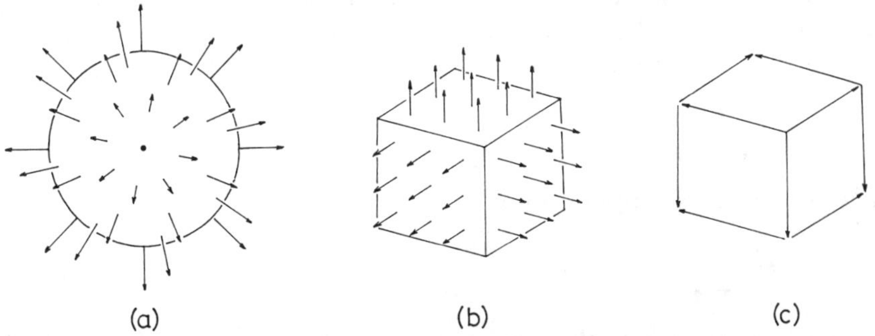

Fig. 10.2a–c. $2\frac{1}{2}$D sketch. **a** The surface gradient, or equivalently the surface normal, is estimated densely over the region corresponding to an object surface. **b** The image region corresponding to an object surface is segmented into planar patches, and the surface gradient, or equivalently the surface normal, is estimated for each patch. **c** The image region corresponding to an object surface is segmented into planar patches, and the 3D orientation is estimated for each edge

---

[4] We mean the *visible part* of the object surface. In the following, we ignore the invisible part occluded by the object itself.

[5] This term was first coined out by David Marr, although he meant only case (i). According to him, obtaining a $2\frac{1}{2}$D sketch is the goal of the low-level image processing stage, or *early vision*, which is supposed to be done bottom-up over primal images without assuming particular knowledge of the domain.

the object surface is locally flat. If the object is itself a polyhedron, its faces themselves constitute planar patches. If the object has a smooth surface, it is approximated by a collection of planar patches.

First, consider case (ii), and assume that the object surface is approximated by a polyhedron. As pointed out in the preceding section, "being a polyhedron" is a very strong contraint, and the surface gradient cannot be assigned arbitrarily. Then, there arises a natural question: How much freedom do we have in assigning values? This question leads to the definition of the *degree of freedom* of a polyhedron image.

In order to formulate this problem, we first define the *incidence structure* of a polyhedron. We say that a vertex is *incident* to a face if the vertex is on the boundary of the face. Consider a polyhedron image. Let $\mathscr{V} = \{V_1, \ldots, V_n\}$ be the set of its vertices, and $\mathscr{F} = \{F_1, \ldots, F_m\}$ be the set of its faces. The incidence structure is specified by a set $\mathscr{R}$ of *incidence pairs* (F_α, V_i), meaning that vertex V_i is incident to face F_α[6]. Let l be the number of such incident pairs.

Let (X_i, Y_i, Z_i) be the scene coordinates of vertex V_i, $i = 1, \ldots, n$, and let $Z = p_\alpha X + q_\alpha Y + r_\alpha$ be the equation of face F_α, $\alpha = 1, \ldots, m$. Let us call p_α, q_α, r_α the *surface parameters* of face F_α. The incidence pair (F_α, V_i) states that vertex V_i is incident to face F_α:

$$Z_i = p_\alpha X_i + q_\alpha Y_i + r_\alpha \,. \tag{10.2.1}$$

In this chapter, we take the image-centered coordinate system, $(0, 0, -f)$ being the viewpoint. This means that the image coordinates (x_i, y_i) of vertex V_i are related to its scene coordinates (X_i, Y_i, Z_i) by the projection equations

$$x_i = \frac{fX_i}{f + Z_i}\,, \qquad y_i = \frac{fY_i}{f + Z_i}\,. \tag{10.2.2}$$

Now, we introduce a new quantity

$$z \equiv \frac{fZ}{f + Z}\,, \tag{10.2.3}$$

and call it the *reduced depth*. There exists a one-to-one correspondence between the scene coordinates (X_i, Y_i, Z_i) and (x_i, y_i, z_i):

$$x_i = \frac{fX_i}{f + Z_i}\,, \qquad y_i = \frac{fY_i}{f + Z_i}\,, \qquad z_i = \frac{fZ_i}{f + Z_i}\,, \tag{10.2.4}$$

$$X_i = \frac{fx_i}{f - z_i}\,, \qquad Y_i = \frac{fy_i}{f - z_i}\,, \qquad Z_i = \frac{fz_i}{f - z_i}\,. \tag{10.2.5}$$

---

[6] To put it in algebraic terms, the incidence structure $\mathscr{R}$ is a *relation* over $\mathscr{F}$ and $\mathscr{V}$, i.e., a subset of the Cartesian product $\mathscr{F} \times \mathscr{V}$, (see the Appendix, Sect. A.1).

Remark 10.2. Equation (10.2.4, 5) can be regarded as defining a one-to-one mapping between the XYZ-space and xyz-space:

$$x = \frac{fX}{f+Z}, \quad y = \frac{fY}{f+Z}, \quad z = \frac{fZ}{f+Z}, \tag{10.2.6}$$

$$X = \frac{fx}{f-z}, \quad Y = \frac{fy}{f-z}, \quad Z = \frac{fz}{f-z}. \tag{10.2.7}$$

This is a projective transformation of the XYZ-space, and hence preserves collinearity and coplanarity; a line is mapped onto a line, and a plane is mapped onto a plane. From (10.2.6, 7), we find that a plane $Z =$ const. is mapped onto a plane $z =$ const. In particular, the XY-plane ($Z = 0$) is mapped onto the xy-plane ($z = 0$), and the "plane" $Z = \infty$ at infinity is mapped onto the plane $z = f$, while the plane $Z = -f$ is mapped onto the "plane" $z = \infty$ at infinity. We can also prove easily that lines passing through the viewpoint $(0, 0, -f)$ in the XYZ-space are mapped onto lines parallel to the z-axis in the xyz-space. Thus, the transformation is like the one shown in Fig. 10.3. This transformation is frequently used in computer graphics, because it makes the removal of hidden lines very easy: rays starting from the viewpoint become parallel to the z-axis, and visibility can be checked as if the projection were orthographic.

Substituting (10.2.5) into (10.2.1), we can express the reduced depth z_i in terms of its image coordinates (x_i, y_i):

$$z_i = \frac{fp_\alpha}{f+r_\alpha} x_i + \frac{fq_\alpha}{f+r_\alpha} y_i + \frac{fr_\alpha}{f+r_\alpha}. \tag{10.2.8}$$

Now, we define new parameters

$$P_\alpha \equiv \frac{fp_\alpha}{f+r_\alpha}, \quad Q_\alpha \equiv \frac{fq_\alpha}{f+r_\alpha}, \quad R_\alpha \equiv \frac{fr_\alpha}{f+r_\alpha}, \tag{10.2.9}$$

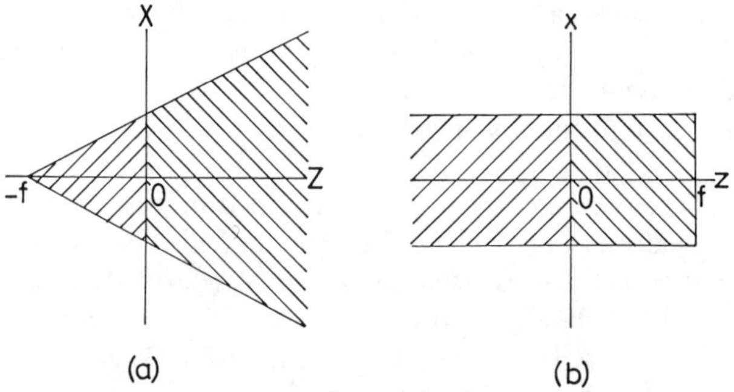

Fig. 10.3a,b. Reduced depth z. The XYZ-space **a** is mapped onto the xyz-space **b** by a projective transformation that preserves collinearity and coplanarity. The infinite depth $Z = \infty$ is reduced to $z = f$. In the xyz-space, the projection can be regarded as orthographic

and call these the *reduced surface parameters*. The inverse relationship to the original surface parameters p_α, q_α, r_α is given by

$$p_\alpha = \frac{P_\alpha}{f - R_\alpha}, \quad q_\alpha = \frac{Q_\alpha}{f - R_\alpha}, \quad r_\alpha = \frac{R_\alpha}{f - R_\alpha}. \tag{10.2.10}$$

In terms of the reduced surface parameters P_α, Q_α, R_α, the reduced depth z_i is written as

$$z_i = P_\alpha x_i + Q_\alpha y_i + R_\alpha . \tag{10.2.11}$$

Since the image coordinates (x_i, y_i) are known, the 3D vertex positions are determined (10.2.5) if their reduced depths z_i are known. Hence, the reduced depths z_i, $i = 1, \ldots, n$, can be taken as unknowns in the place of the original depths Z_i, $i = 1, \ldots$ n. Similarly, the reduced surface parameters P_α, Q_α, R_α, $\alpha = 1, \ldots, m$, can serve as unknowns in the place of the original surface parameters p_α, q_α, r_α, $\alpha = 1, \ldots, m$. Thus, we obtain l (the number of incidence pairs) equations

$$x_i P_\alpha + y_i Q_\alpha + R_\alpha - z_i = 0 , \qquad (F_\alpha, V_i) \in \mathcal{R} . \tag{10.2.12}$$

These equations are linear in the reduced depths z_i, $i = 1, \ldots, n$, and the reduced surface parameters P_α, Q_α, R_α, $\alpha = 1, \ldots, m$. Let us call these l equations the *constraints* of the polyhedron image.

Suppose the l constraints are linearly independent. The number of unknowns is $N = n + 3m$ (the number of vertices plus three times the number of faces). In other words, the solution is a point in an N-dimensional space. Since (10.2.12) imposes l linear constraints, the solution is constrained to an $(N - l)$-dimensional linear subspace. The dimensionality of the solution space is called the *degree of freedom*. If the degree of freedom is r, the solution contains r indeterminate parameters. This means that we can assign *at most* r values to the polyhedron image. (We cannot assign surface gradients to arbitrary r faces. The interdependence of faces depends on the incidence structure.)

If the l constraints are linearly independent, the degree of freedom is $r = n + 3m - l\,(= N - l)$. If the l constraints are not independent, say only l' $(< l)$ are independent, the degree of freedom is $r = n + 3m - l'\,(> n + 3m - l)$. Thus, the degree of freedom is *at least* $n + 3m - 1$.

Remark 10.3. The degree of freedom of a polyhedron image should be *at least four* if the polyhedron *is not coplanar* (i.e., not "flattened out"). This can be seen as follows. Suppose there exists a solution polyhedron that satisfies all the constraints for given data (x_i, y_i), $i = 1, \ldots, n$. We can freely translate and deform the solution polyhedron into another while keeping the projection equations and the incidence structure preserved (Fig. 10.4). Such a "deformation" involves four parameters. Hence, there exist at least four degrees of freedom.

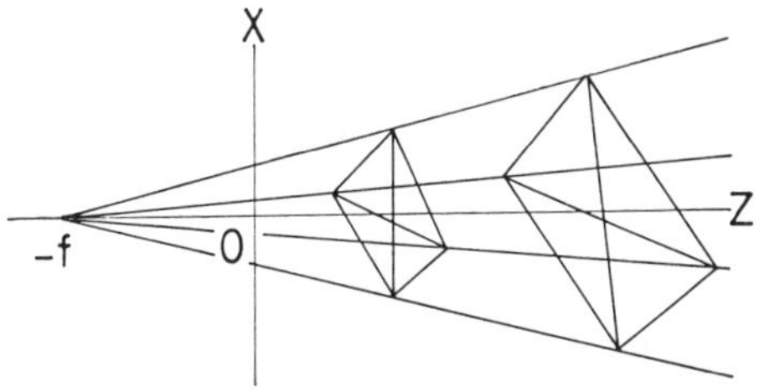

Fig. 10.4. The deformation of a non-flat polyhedron that preserves the projection relationship and the incidence structure involves at least four independent parameters. Hence, the degree of freedom of any non-flat polyhedron is at least four

The formal proof is as follows. We can assume that $n \geqq 4$. (If $n \leqq 3$, we can only define a point, a line segment, or a triangular plane.) Let z_i, $i = 1, \ldots, n$, P_α, Q_α, R_α, $\alpha = 1, \ldots, m$, be a solution satisfying the constraints (10.2.12). It is easy to confirm that the following are also a solution (Exercise 10.5):

$$z_i' = Ax_i + By_i + Cz_i + D\ , \qquad i = 1, \ldots, n\ , \tag{10.2.13}$$

$$P_\alpha' = CP_\alpha + A\ , \qquad Q_\alpha' = CQ_\alpha + B\ ,$$
$$R_\alpha' = CR_\alpha + D\ , \qquad \alpha = 1, \ldots, m\ . \tag{10.2.14}$$

Here, A, B, C, D are arbitrary constants such that $C > 0$.

Our proof is complete if we can show that the four constants A, B, C D are independent, i.e., there is no redundancy among them. Assume the contrary, and suppose A, B, C, D and A', B', C', D' define the same deformation. From (10.2.13), this means that

$$Ax_i + By_i + Cz_i + D = A'x_i + B'y_i + C'z_i + D'\ ,$$
$$i = 1, \ldots, n\ , \tag{10.2.15}$$

or

$$(A - A')x_i + (B - B')y_i + (C - C')z_i + (D - D') = 0\ ,$$
$$i = 1, \ldots, n\ . \tag{10.2.16}$$

Thus, we obtain n ($\geqq 4$) homogeneous linear equations in $A - A'$, $B - B'$, $C - C'$, $D - D'$. Among them, there should be at least four independent equations, and hence $A - A' = 0$, $B - B' = 0$. $C - C' = 0$, $D - D' = 0$, or $A = A'$, $B = B'$, $C = C'$, $D = D'$. In other words, any changes of these four parameters will produce different deformations. This means that the degree of freedom is at least four. The fact that at least four of (10.2.16) are independent can be seen easily: if not, any four of them are linearly dependent. Consequently, all minors of degree four are zero:

$$\begin{vmatrix} x_{i_1} & y_{i_1} & z_{i_1} & 1 \\ x_{i_2} & y_{i_2} & z_{i_2} & 1 \\ x_{i_3} & y_{i_3} & z_{i_3} & 1 \\ x_{i_4} & y_{i_4} & z_{i_4} & 1 \end{vmatrix} = 0 \ , \qquad \{i_1, \ldots, i_4\} \subset \{1, \ldots, n\} \ . \qquad (10.2.17)$$

This means that all the vertices are coplanar in the xyz-space (hence in the XYZ-space as well, cf. Remark 10.2) (Exercise 10.6). This contradicts our assumption that the polyhedron is not coplanar.

Remark 10.4. The use of the image coordinates (x_i, y_i) and the reduced depths z_i as well as the reduced surface parameters P_α, Q_α, R_α enables us to forget the difference between perspective and orthographic projections. In fact, orthographic projection is attained in the limit of $f \to \infty$, and in this limit we can see from (10.2.4, 5, 9, 10) that all the reduced parameters coincide with the original parameters: $x_i \to X_i$, $y_i \to Y_i$, $z_i \to Z_i$, $P_i \to p_i$, $Q_i \to q_i$, $R_i \to r_i$. Hence, *all the discussions in this chapter also hold for orthographic projection if this limit is taken.* This is the main reason why we adopt the image-centered coordinate system instead of the viewer-centered coordinate system.

It should be clear by now why we switched the coordinate system between viewer-centered and image-centered from chapter to chapter. The viewer-centered coordinate system is useful if the properties we are studying are *unique to perspective projection*, such as the camera rotation invariance, while the image-centered coordinate system is essential if we are discussing projection properties *common to both perspective and orthographic projections.* (It would certainly be possible to stick to either throughout, but then we would miss many critical observations buried in notational complexity.)

10.2.1 Singularity of the Incidence Structure

Consider the polyhedron image of Fig. 10.5a. The numer of vertices is $n = 6$, the number of faces is $m = 4$, and the incidence structure consists of the following 15 incidence pairs:

$$\begin{aligned} \mathscr{R} = \{&(F_1, V_1), (F_1, V_2), (F_1, V_3), (F_2, V_1), (F_2, V_2), (F_2, V_4), \\ &(F_2, V_5), (F_3, V_2), (F_3, V_3), (F_3, V_5), (F_3, V_6), (F_4, V_3), \\ &(F_4, V_1), (F_4, V_6), (F_4, V_4)\}. \end{aligned}$$

We omit the numerical data, but it can be shown that the 15 constraints are linearly independent. It follows that the degree of freedom is $r = 6 + 3 \times 4 - 15 = 3$. According to Remark 10.3, however, it must be at least

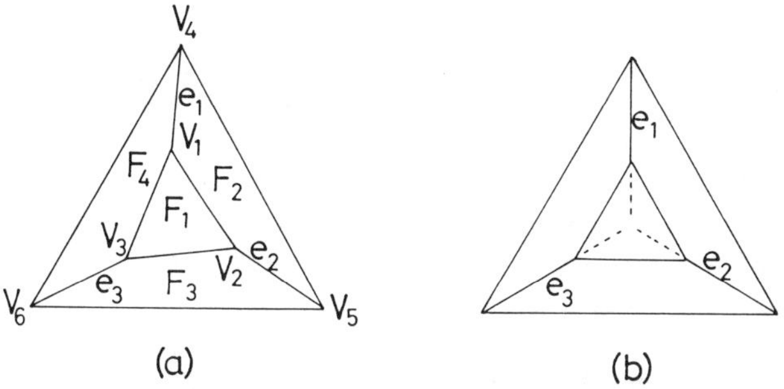

Fig. 10.5a. A singular incidence structure. The only possible interpretation is a "flat polyhedron", all its faces lying on a plane. **b** A singular incidence structure admits a non-flat interpretation if and only if the vertex coordinates take special values, which is highly unlikely in real situations

4 if the polyhedron is not coplanar. This means that the polyhedron must necessarily be "flat". This can easily be understood: if any of the vertices V_1, V_2, V_3 is out of the plane defined by the vertices V_4, V_5, V_6, the faces F_2, F_3, F_4 cannot all be planar. This is because edges e_1, e_2, e_3 are not concurrent, while they should be if faces F_2, F_3, F_4 are all to be planar.

Consider Fig. 10.5b. This polyhedron has exactly the same incidence structure as that of Fig. 10.5a. This time, the image coordinates of vertices V_1, V_2, V_3 are such that edges e_1, e_2, e_3 are concurrent, and hence a "non-flat" interpretation is possible. This is because the 15 constraints become linearly dependent; only 14 of them are independent, and the degree of freedom is $6 + 3 \times 4 - 14 = 4$.
only 14 of them are independent, and the degree of freedom is $6 + 3 \times 4 - 14 = 4$.

In a real problem, however, *this situation is highly unlikely to occur*, because only a small disturbance to the image coordinates will make the 15 constraints linearly independent, reducing the degree of freedom to 3. In other words, the situation of Fig. 10.5b is *practically impossible.* (We are able to say that the occurrence of this event is a *measure* 0 or of *probability* 0 by introducing an appropriate measure or probability.) Hence, we must regard this polyhedron as having an "inherently bad structure". To put it differently, we must assume in general that the l constraints are always linearly independent; we cannot expect the very small possibility of degeneracy to occur.

Earlier, we showed that the degree of freedom of a polyhedron that has n vertices, m faces, and l incidence pairs is at least $n + 3m - l$. Hence, if $n + 3m - 1 > 4$, it is expected that a non-flat interpretation can exist. However, a part of it may have a bad structure, allowing only the flat interpretation for that part. We say that a polyhedron (or, to be precise, its incidence structure) is *singular* if some part of it has this bad structure. Requiring the above inequality for all "subpolyhedra", we can expect that a polyhedron is nonsingular if and only if

$$|\mathscr{V}(\mathscr{F}')| + 3|\mathscr{F}'| - |\mathscr{R}(\mathscr{F}')| \geqq 4 \qquad (10.2.18)$$

for any subset $\mathscr{F}' \subset \mathscr{F} = \{F_1, \ldots, F_m\}$ such that $|\mathscr{F}'| \geqq 2$. Here, $\mathscr{V}(\mathscr{F}')$ is the set of the vertices incident to at least one of the faces in subset $\mathscr{F}'$, and $\mathscr{R}(\mathscr{F}')$ is the set of the incidence pairs involving the faces in subset $\mathscr{F}'$. The vertical bars $|.|$ denote the number of elements.

Indeed, we can prove that the above condition is indeed the necessary and sufficient condition for an incidence structure to be nonsingular. This finding, attributed to Kokichi Sugihara, is important because the criterion of (10.2.18) is easy to check: the process is *combinatorial*, and no numerical computation is involved.[7]

If the incidence structure is singular, we can make it nonsingular by partitioning some of its faces. For example, the incidence structure of Fig. 10.5a becomes nonsingular if face F_2 is partitioned into two faces F_2' and F_2'' (Fig. 10.6a). In general, the incidence structure always becomes nonsingular if all the faces are partitioned into triangular subfaces. If the $2\frac{1}{2}$D sketch is obtained as a polyhedral approximation of a smooth surface, we can always do this because the partitioning is completely arbitrary.

On the other hand, if we want to preserve the original partitioning, we can make its incidence structure nonsingular by removing some of the incidence pairs. As a result, some faces will meet in different places, so the vertex positions must also be displaced in the 3D reconstruction. For example, the incidence structure of Fig. 10.5a becomes nonsingular if the incidence pair (F_2, V_1) is removed, but vertex V_1 must be displaced to a new position V_1' (Fig. 10.6b).

As we noted in Remark 10.4, none of the above argument depends on whether the projection is perspective or orthographic.

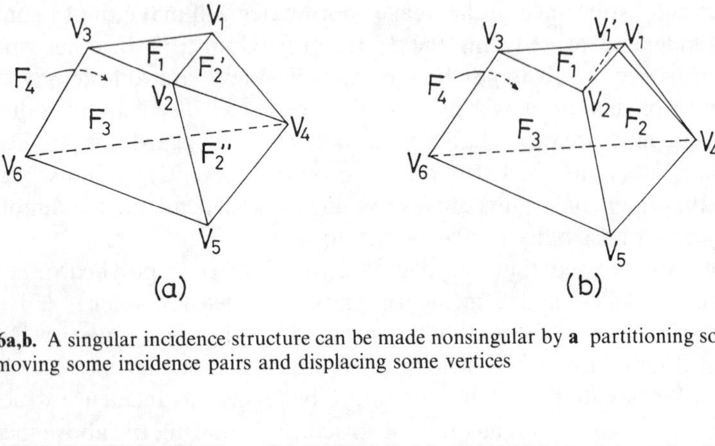

Fig. 10.6a,b. A singular incidence structure can be made nonsingular by **a** partitioning some faces, or **b** removing some incidence pairs and displacing some vertices

[7] If we are to exhaust all subpolyhedra, $O(2^m)$ steps of computation are required, where m is the number of faces. However, there exists an efficient algorithm, due to Sugihara, that requires only $O(l^2)$ steps, where l is the number of incidence pairs.

10.2.2 Integrability Condition

Consider case (i)—the $2\frac{1}{2}$D sketch with densely estimated surface gradient. Suppose the surface gradient (p, q) is estimated as a smooth function. Then, we are questioning the *integrability condition*: Does a smooth surface that has the specified surface gradient $(p(x, y), q(x, y))$ exist? Suppose the region corresponding to the surface is simply connected (i.e., with no "holes").

First, let us consider orthographic projection.[8] We seek a solution $z = z(x, y)$ such that

$$p(x, y) = \frac{\partial z}{\partial x}, \qquad q(x, y) = \frac{\partial z}{\partial y}. \tag{10.2.19}$$

Such a surface exists if and only if

$$\frac{\partial p}{\partial y} = \frac{\partial q}{\partial x}. \tag{10.2.20}$$

From (10.2.19), the necessity of (10.2.20) is obvious. The sufficiency is shown as follows. Choose an arbitrary point (x_0, y_0) on the image plane and prescribe its reduced depth z_0 arbitrarily. Take a point (x, y) on the image plane, and draw an arbitrary path C starting from (x_0, y_0) and ending at (x, y). Put

$$z(x, y; C) = \int_{(x_0, y_0)}^{(x, y)} p(x, y)dx + q(x, y)dy, \tag{10.2.21}$$

where the curvilinear integration is performed along the path C. If (10.2.21) defines a single-valued function, i.e., if its value at (x, y) does not depend on the path C, we can easily confirm that (10.2.21) satisfies (10.2.19). Since the domain in question is assumed to be simply connected, the requirement that the integration (10.2.21) be independent of the path is equivalent to requiring that (10.2.21) be zero when integrated along an arbitrary closed loop. If C is a closed loop, *Green's theorem* states that

$$\int_C pdx + qdy = \int_S \left(\frac{\partial q}{\partial x} - \frac{\partial p}{\partial y}\right) dx\, dy, \tag{10.2.22}$$

where the right-hand side is a double integral over the region S encircled by the loop C. If (10.2.20) is satisfied, the right-hand side of (10.2.22) is identically zero for an arbitrary closed loop C. Hence, (10.2.21) defines the desired surface.

Next, consider perspective projection. We seek a solution $Z = Z(X, Y)$ such that $p(x, y)$ and $q(x, y)$ are respectively equal to the values of $\partial Z/\partial X$ and $\partial Z/\partial Y$

[8] If the projection is orthographic, there no longer exists a distinction between the uppercase variables X, Y, Z, P, Q, R and the lowercase variables x, y, z, p, q, r (Remark 10.4). Here, we use the lowercase letters.

evaluated at the point (X, Y, Z) corresponding to point (x, y). This means that we are looking for a surface $Z = Z(X, Y)$ such that

$$dZ = p\left(\frac{fX}{f+Z}, \frac{fY}{f+Z}\right)dX + q\left(\frac{fX}{f+X}, \frac{fY}{f+Z}\right)dY . \tag{10.2.23}$$

We do not give the proof here, but the following integrability condition is well known in the theory of differential equations: The necessary and sufficient condition that there exists a surface along which differentials dX, dY, dZ satisfy

$$A(X, Y, Z)dX + B(X, Y, Z)dY + C(X, Y, Z)dZ = 0 \tag{10.2.24}$$

is

$$\left(\frac{\partial C}{\partial Y} - \frac{\partial B}{\partial Z}\right)A + \left(\frac{\partial A}{\partial Z} - \frac{\partial C}{\partial X}\right)B + \left(\frac{\partial B}{\partial X} - \frac{\partial A}{\partial Y}\right)C = 0 \tag{10.2.25}$$

(*Frobenius' theorem*). From this theorem, we conclude that the desired surface exists if and only if the following condition is satisfied (Exercise 10.10):

$$\frac{\partial q}{\partial x} - \frac{\partial p}{\partial y} = \frac{1}{f}\left[\left(p\frac{\partial q}{\partial x} - q\frac{\partial p}{\partial x}\right)x + \left(p\frac{\partial q}{\partial y} - q\frac{\partial p}{\partial y}\right)y\right] . \tag{10.2.26}$$

If we take the orthographic limit $f \to \infty$, (10.2.26) reduces to (10.2.20) as expected. Thus, whether the projection is orthographic or perspective, the surface gradient (p, q) cannot be assigned arbitrarily to the $2\frac{1}{2}$D sketch.

10.3 Optimization of a $2\frac{1}{2}$D Sketch

Suppose we are given a $2\frac{1}{2}$D sketch. Let $\mathscr{V} = \{V_1, \ldots, V_n\}$ and $\mathscr{F} = \{F_1, \ldots, F_m\}$ be the sets of its vertices and faces, respectively. Let $\mathscr{R} = \{(F_\alpha, V_i)\}$ be its incidence structure. We assume that this incidence structure is nonsingular. (If it is singular, we can always make it nonsingular by a small modification as discussed in Sect. 10.2.1.) Let $(\hat{p}_\alpha, \hat{q}_\alpha)$ be the estimate of the surface gradient of face F_α. As we argued in Sect. 10.2, there does not in general exist a polyhedron that is compatible with the given projection image and has the prescribed surface gradients exactly.

Here, we assume that the surface gradient estimates $(\hat{p}_\alpha, \hat{q}_\alpha)$, $\alpha = 1, \ldots, m$, are not accurate, while the observed image coordinates (x_i, y_i), $i = 1, \ldots, n$ are accurate.[9] Let (p_α, q_α) be the true surface gradient of face F_α. Here, we seek, from among the infinitely many consistent polyhedron solutions that are exactly projected onto the observed image, the one whose surface gradients are the

[9] Of course, the observed image coordinates (x_i, y_i), $i = 1, \ldots, n$, may also contain errors. However, most surface gradient estimation processes involve these image coordinates. Consequently, the computed surface gradient estimates are "more indirect" than these coordinates.

closest to the estimates on the average (Fig. 10.7). Specifically, let us consider the least squares method to minimize

$$J = \frac{1}{2}\sum_{\alpha=1}^{m} W_\alpha[(p_\alpha - \hat{p}_\alpha)^2 + (q_\alpha - \hat{q}_\alpha)^2] , \tag{10.3.1}$$

where W_α is the weight for face F_α. If (10.2.12) is substituted, (10.3.1) is rewritten in terms of the reduced surface parameters as

$$J = \frac{1}{2}\sum_{\alpha=1}^{m} W_\alpha\left(\frac{f + r_\alpha}{f}\right)^2\left[\left(P_\alpha + \frac{\hat{p}_\alpha}{f}R_\alpha - \hat{p}_\alpha\right)^2 + \left(Q_\alpha + \frac{\hat{q}_\alpha}{f}R_\alpha - \hat{q}_\alpha\right)^2\right] . \tag{10.3.2}$$

In order to make our analysis easy, we replace r_α in the above equation by its "estimate" $\hat{r}_\alpha$, assuming that it is somehow available. Then, put

$$W_\alpha\left(\frac{f + \hat{r}_\alpha}{f}\right)^2 = \frac{1}{\zeta_\alpha} . \tag{10.3.3}$$

Now, ζ_α is a constant assigned to face F_α whose value is yet to be determined.

Remark 10.5. One heuristic for assigning the value of ζ_α is as follows. If face F_α has a large depth $\hat{r}_\alpha$, estimation of the surface gradient may not be very accurate. It follows that a small weight W_α should be assigned. At the same time, if either $\hat{p}_\alpha$ or $\hat{q}_\alpha$ has a very large magnitude, the masurement may also be inaccurate, and a small weight W_α should be assigned, too. In view of these considerations, it may be appropriate to choose the following weight:[10]

$$W_\alpha = \frac{1}{\hat{p}_\alpha^2 + \hat{q}_\alpha^2 + 1}\left(\frac{f}{f + \hat{r}_\alpha}\right)^2 . \tag{10.3.4}$$

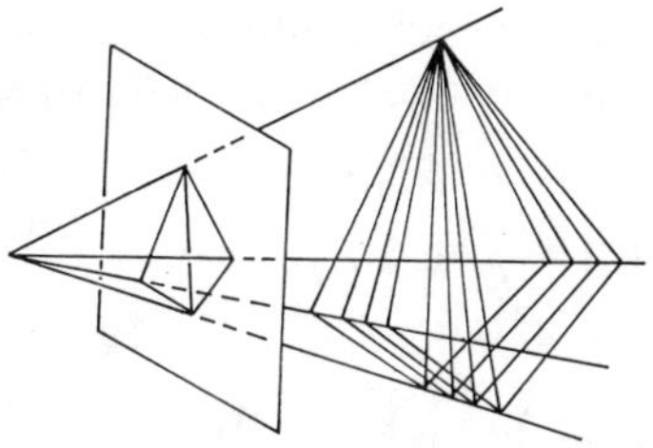

Fig. 10.7. From among infinitely many consistent polyhedron solutions that are exactly projected onto the image plane as observed, we seek the one whose surface gradients are the closest to the given estimates on the average

[10] In the denominator, + 1 is added to avoid zero division in the case $\hat{p}_\alpha = \hat{q}_\alpha = 0$.

For this weight, the constant ζ_α becomes simply

$$\zeta_\alpha = \hat{p}_\alpha^2 + \hat{q}_\alpha^2 + 1 \,. \tag{10.3.5}$$

Of course, the choice is not unique. Different choices may result in different solutions. The validity of the choice must be checked a posteriori.

The problem now reduces to the minimization of J under the constraints (10.2.12). However, surface gradient cues alone cannot determine the absolute depth; the projection image and the surface gradient of each face is kept identical if the object is translated in the scene away from the viewer and at the same time its size is proportionally increased. Hence, the depth Z, or equivalently the reduced depth z, must be given to one vertex. Let that vertex be V_n. Since J is quadratic and the constraints (10.2.12) are linear, the minimization is achieved by solving a set of *linear* equations. The final result is given as follows:

Proposition 10.1. *The reduced depths z_i of vertices V_i, $i = 1, \ldots, n-1$, are given by solving the following $n + 3m + l - 1$ linear equations in $n + 3m + l - 1$ unknowns z_i, $i = 1, \ldots, n-1$, P_α, Q_α, R_α, $\alpha = 1, \ldots, m$, and $\Lambda_{\alpha i}$ for $(F_\alpha, V_i) \in \mathscr{R}$:*

$$x_i P_\alpha + y_i Q_\alpha + R_\alpha - z_i = 0 \,, \qquad (F_\alpha, V_i) \in \mathscr{R} \,, \tag{10.3.6}$$

$$P_\alpha + \frac{\hat{p}_\alpha}{f} R_\alpha + \zeta_\alpha \sum_{i:(F_\alpha, V_i)\in\mathscr{R}} x_i \Lambda_{\alpha i} = \hat{p}_\alpha, \qquad \alpha = 1, \ldots, m \,, \tag{10.3.7}$$

$$Q_\alpha + \frac{\hat{q}_\alpha}{f} R_\alpha + \zeta_\alpha \sum_{i:(F_\alpha, V_i)\in\mathscr{R}} y_i \Lambda_{\alpha i} = \hat{q}_\alpha \,, \qquad \alpha = 1, \ldots, m \,, \tag{10.3.8}$$

$$\sum_{i:(F_\alpha, V_i)\in\mathscr{R}} \left(\frac{\hat{p}_\alpha x_i + \hat{q}_\alpha y_i}{f} - 1 \right) \Lambda_{\alpha i} = 0 \,, \qquad \alpha = 1, \ldots, m \,, \tag{10.3.9}$$

$$\sum_{\alpha:(F_\alpha, V_i)\in\mathscr{R}} \Lambda_{\alpha i} = 0 \,, \qquad i = 1, \ldots, n-1 \,. \tag{10.3.10}$$

Proof. If we introduce Lagrangian multipliers $\Lambda_{\alpha i}$ to all the incidence pairs $(F_\alpha, V_i) \in \mathscr{R}$, the problem is converted into unconstrained optimization of

$$\tilde{J} = \frac{1}{2} \sum_{\alpha=1}^{m} \frac{1}{\zeta_\alpha} \left[\left(P_\alpha + \frac{\hat{p}_\alpha}{f} R_\alpha - \hat{p}_\alpha \right)^2 + \left(Q_\alpha + \frac{\hat{q}_\alpha}{f} R_\alpha - \hat{q}_\alpha \right)^2 \right]$$

$$+ \sum_{\alpha:i:(F_\alpha, V_i)\in\mathscr{R}} \Lambda_{\alpha i} (P_\alpha x_i + Q_\alpha y_i + R_2 - z_i) \,. \tag{10.3.11}$$

Equation (10.3.6) expresses the required constraints themselves. Equations (10.3.7, 8, 10) are obtained by taking derivatives of $\tilde{J}$ with respect to P_α, Q_α, $\alpha = 1, \ldots, m$, and z_i, $i = 1, \ldots, n-1$, and setting the respective results to zero. Equation (10.3.9) is obtained by differentiating $\tilde{J}$ with respect to R_α, $\alpha = 1, \ldots, m$, setting the result to zero, and substituting (10.3.7, 8) into it.

Remark 10.6. In Proposition 10.1, we assumed that surface gradient estimates $(\hat{p}_\alpha, \hat{q}_\alpha)$ are given for all faces, but this is not always necessary; if some faces are given no estimates, the corresponding terms are simply dropped from (10.3.1). Hence, if face F_α is given no surface gradient estimate, (10.3.7–9) are respectively replaced by the following equations (Exercise 10.12):

$$\sum_{i:(F_\alpha, V_i)\in \mathscr{R}} x_i\Lambda_{\alpha i} = 0 \ , \qquad \sum_{i:(F_\alpha, V_i)\in \mathscr{R}} y_i\Lambda_{\alpha i} = 0 \ , \qquad \sum_{i:(F_\alpha, V_i)\in \mathscr{R}} \Lambda_{\alpha i} = 0 \ . \tag{10.3.12}$$

Remark 10.7. In the orthographic limit $f \to \infty$, (3.7–9) reduce to

$$P_\alpha + \zeta_\alpha \sum_{i:(F_\alpha, V_i)\in \mathscr{R}} x_i\Lambda_{\alpha i} = \hat{p}_\alpha \ , \qquad \alpha = 1, \ldots, m \ , \tag{10.3.13}$$

$$Q_\alpha + \zeta_\alpha \sum_{i:(F_\alpha, V_i)\in \mathscr{R}} y_i\Lambda_{\alpha i} = \hat{q}_\alpha \ , \qquad \alpha = 1, \ldots, m \ , \tag{10.3.14}$$

$$\sum_{i:(F_\alpha, V_i)\in \mathscr{R}} \Lambda_{\alpha i} = 0 \ , \qquad \alpha = 1, \ldots, m \ , \tag{10.3.15}$$

Next, consider case (iii)—the 3D edge orientations are estimated. Again, let $\mathscr{V} = \{V_1, \ldots, V_n\}$ and $\mathscr{F} = \{F_1, \ldots, F_m\}$ be the sets of vertices and faces, respectively. Let $\mathscr{E}_\alpha = \{e_1, \ldots, eN_\alpha\}$ be the set of edges constituting the boundary of face F_α, and let $\hat{\boldsymbol{e}}_k = (\hat{e}_{k(1)}, \hat{e}_{k(2)}, \hat{e}_{k(3)})$ be the unit vector indicating the estimated 3D orientation of edge e_k. If (p_α, q_α) is the surface gradient of face F_α, the unit surface normal $\boldsymbol{n}_\alpha = (n_{\alpha(1)}, n_{\alpha(2)}, n_{\alpha(3)})$ to face F_α is given by

$$\boldsymbol{n}_\alpha = \left(\frac{p_\alpha}{\sqrt{p_\alpha^2 + q_\alpha^2 + 1}} \ , \ \frac{q_\alpha}{\sqrt{p_\alpha^2 + q_\alpha^2 + 1}} \ , \ \frac{-1}{\sqrt{p_\alpha^2 + q_\alpha^2 + 1}} \right) \tag{10.3.16}$$

(Lemma 9.2). The vectors $\hat{\boldsymbol{e}}_k$ for $e_k \in \mathscr{E}_\alpha$ are supposed to be all perpendicular to $\boldsymbol{n}_\alpha$, but this is not necessarily guaranteed in the presence of noise. Hence, it is reasonable to estimate the surface normal $\boldsymbol{n}_\alpha$ by the least squares method to minimize

$$\frac{1}{2} \sum_{e_k \in \mathscr{E}_\alpha} (\hat{\boldsymbol{e}}_k, \boldsymbol{n}_\alpha) = \frac{1}{2} \sum_{e_k \in \mathscr{E}_\alpha} (\hat{e}_{k(1)} n_{\alpha(1)} + \hat{e}_{k(2)} n_{\alpha(2)} + \hat{e}_{k(3)} n_{\alpha(3)})^2 \ , \tag{10.3.17}$$

However, the minimization need not be done for each face separately. The surface normals of all the faces are estimated by minimizing

$$\begin{aligned} J &= \frac{1}{2} \sum_{\alpha=1}^{m} W_\alpha \sum_{e_k \in \mathscr{E}_\alpha} (\hat{e}_{k(1)} n_{\alpha(1)} + \hat{e}_{k(2)} n_{\alpha(2)} + \hat{e}_{k(3)} n_{\alpha(3)})^2 \\ &= \frac{1}{2} \sum_{\alpha=1}^{m} \frac{W_\alpha}{p_\alpha^2 + q_\alpha^2 + 1} \sum_{e_k \in \mathscr{E}_\alpha} (\hat{e}_{k(1)} p_\alpha + \hat{e}_{k(2)} q_\alpha - \hat{e}_{k(3)})^2 \ , \end{aligned} \tag{10.3.18}$$

where W_α is the weight for face F_α. If (10.2.10) is substituted, this expression is rewritten in terms of the reduced surface parameters $P_\alpha, Q_\alpha, R_\alpha, \alpha = 1, \ldots, m$, as

$$J = \frac{1}{2}\sum_{\alpha=1}^{m} \frac{W_\alpha}{p_\alpha^2 + q_\alpha^2 + 1}\left(\frac{f + \hat{r}_\alpha}{f}\right)^2 \sum_{e_k \in \mathscr{E}_\alpha} \left(\hat{e}_{k(1)} P_\alpha + \hat{e}_{k(2)} Q_\alpha + \frac{1}{f}\hat{e}_{k(3)} R_\alpha - \hat{e}_{k(3)}\right)^2 . \tag{10.3.19}$$

It follows that optimization can be achieved as a single step.

Again, in order resort to make the subsequent analysis easy, we replace r_α by its estimate $\hat{r}_\alpha$ as before. We also replace p_α and q_α by their estimates $\hat{p}_\alpha, \hat{q}_\alpha$, assuming that they are somehow available, and put

$$\frac{W_\alpha}{\hat{p}_\alpha^2 + \hat{q}_\alpha^2 + 1}\left(\frac{f + \hat{r}_\alpha}{f}\right)^2 = \frac{1}{\zeta_\alpha} . \tag{10.3.20}$$

Then, ζ_α is a constant for face F_α whose value is yet to be determined.

Remark 10.8. For example, we can reason that if face F_α has a large depth value $\hat{r}_\alpha$, the orientation estimation of its boundary edges may not be very accurate, and hence a small weight W_α should be assigned. On the other hand, if $\hat{p}_\alpha$ and $\hat{q}_\alpha$ are close to zero, the variation of the 3D edge orientation in the Z-direction will cause only a small effect on its projection image, so a small weight W_α should be assigned. In view of these considerations, one candidate is

$$W_\alpha = (\hat{p}_\alpha^2 + \hat{q}_\alpha^2 + 1)\left(\frac{f}{f + \hat{r}_\alpha}\right)^2 . \tag{10.3.21}$$

For this weight, the constant ζ_α becomes simply

$$\zeta_\alpha = 1 . \tag{10.3.22}$$

Of course, other factors—the projected areas of the faces, the projected lengths of the edges, etc.—can also be taken into account if considered to be relevant.

Thus, the problem again reduces to quadratic minimization under linear constraints. Since the absolute depth cannot be determined from 3D edge orientation cues alone, the depth Z, or equivalently the reduced depth z, must be assigned to one vertex. Let that vertex be V_n. As before, the minimization is achieved by solving a set of linear equations. The final result is given as follows:

Proposition 10.2. *The reduced depths z_i of vertices V_i, $i = 1, \ldots, n-1$, are given by solving the following $n + 3m + l - 1$ linear equations in $n + 3m + l - 1$ unknowns z_i, $i = 1, \ldots, n-1$, $P_\alpha, Q_\alpha, R_\alpha$, $\alpha = 1, \ldots, m$, and $\Lambda_{\alpha i}$ for $(F_\alpha, V_i) \in \mathscr{R}$:*

$$x_i P_\alpha + y_i Q_\alpha + R_\alpha - z_i = 0 , \qquad (F_\alpha, V_i) \in \mathscr{R} , \tag{10.3.23}$$

$$\Big(\sum_{e_k \in \mathscr{E}_\alpha} \hat{e}_{k(1)^2}\Big) P_\alpha + \Big(\sum_{e_k \in \mathscr{E}_\alpha} \hat{e}_{k(1)}\hat{e}_{k(2)}\Big) Q_\alpha + \frac{1}{f}\Big(\sum_{e_k \in \mathscr{E}_\alpha} \hat{e}_{k(1)}\hat{e}_{k(3)}\Big) R_\alpha$$

$$+ \zeta_\alpha \sum_{i:(F_\alpha, V_i) \in \mathscr{R}} x_i \Lambda_{\alpha i} = \sum_{e_k \in \mathscr{E}_\alpha} \hat{e}_{k(1)}\hat{e}_{k(3)} \,, \quad \alpha = 1, \ldots, m \,, \tag{10.3.24}$$

$$\Big(\sum_{e_k \in \mathscr{E}_\alpha} \hat{e}_{k(1)}\hat{e}_{k(2)}\Big) P_\alpha + \Big(\sum_{e_k \in \mathscr{E}_\alpha} \hat{e}_{k(2)^2}\Big) Q_\alpha + \frac{1}{f}\Big(\sum_{e_k \in \mathscr{E}_\alpha} \hat{e}_{k(2)}\hat{e}_{k(3)}\Big) R_\alpha$$

$$+ \zeta_\alpha \sum_{i:(F_\alpha, V_i) \in \mathscr{R}} y_i \Lambda_{\alpha i} = \sum_{e_k \in \mathscr{E}_\alpha} \hat{e}_{k(2)}\hat{e}_{k(3)} \,, \quad \alpha = 1, \ldots, m \,, \tag{10.3.25}$$

$$\Big(\sum_{e_k \in \mathscr{E}_\alpha} \hat{e}_{k(1)}\hat{e}_{k(3)}\Big) P_\alpha + \Big(\sum_{e_k \in \mathscr{E}_\alpha} \hat{e}_{k(2)}\hat{e}_{k(3)}\Big) Q_\alpha + \frac{1}{f}\Big(\sum_{e_k \in \mathscr{E}_\alpha} \hat{e}_{k(3)^2}\Big) R_\alpha$$

$$+ \zeta_\alpha \sum_{i:(F_\alpha, V_i) \in \mathscr{R}} f \Lambda_{\alpha i} = \sum_{e_k \in \mathscr{E}_\alpha} \hat{e}_{k(3)^2} \,, \quad \alpha = 1, \ldots, m \,, \tag{10.3.26}$$

$$\sum_{\alpha:(F_\alpha, V_i) \in \mathscr{R}} \Lambda_{\alpha i} = 0 \,, \qquad i = 1, \ldots, n-1 \,. \tag{10.3.27}$$

Proof. If we introduce Lagrangian multipliers $\Lambda_{\alpha i}$ to all the incidence pairs for $(F_\alpha, V_i) \in \mathscr{R}$, the problem is converted into unconstrained optimization of

$$\tilde{J} = \frac{1}{2}\sum_{\alpha=1}^{m} \frac{1}{\zeta_\alpha} \sum_{e_k \in \mathscr{E}_\alpha} \Big(\hat{e}_{k(1)} P_\alpha + \hat{e}_{k(2)} Q_\alpha + \frac{1}{f}\hat{e}_{k(3)} R_\alpha - \hat{e}_{k(3)}\Big)^2$$

$$+ \sum_{\alpha, i:(F_\alpha, V_i) \in \mathscr{R}} \Lambda_{\alpha i}(P_\alpha x_i + Q_\alpha y_i + R_\alpha - z_i) \,. \tag{10.3.28}$$

Equation (10.3.23) expresses the required constraints themselves. Equations (10.3.23–28) are obtained by taking derivatives with respect to P_α, Q_α, R_α, $\alpha = 1, \ldots, m$, and z_i, $i = 1, \ldots, n-1$, and setting the respect results to 0.

Remark 10.9. Again, 3D orientation estimates need not always be given for all edges. If some edges are given no orientation estimates, the set $\mathscr{E}_\alpha$ in the above equations is interpreted as the set of edges for which the 3D orientation estimates are given. If no boundary edges of face F_α are given 3D orientation estimates, the corresponding terms are simply dropped from (10.3.19). This means that (10.3.24–26) are replaced by (10.3.12).

Remark 10.10. In the orthographic limit $f \to \infty$, (10.3.24–26) are simply replaced by

$$\Big(\sum_{e_k \in \mathscr{E}_\alpha} \hat{e}_{k(1)^2}\Big) P_\alpha + \Big(\sum_{e_k \in \mathscr{E}_\alpha} \hat{e}_{k(1)}\hat{e}_{k(2)}\Big) Q_\alpha + \zeta_\alpha \sum_{i:(F_\alpha, V_i) \in \mathscr{R}} x_i \Lambda_{\alpha i}$$

$$= \sum_{e_k \in \mathscr{E}_\alpha} \hat{e}_{k(1)}\hat{e}_{k(3)} \,, \quad \alpha = 1, \ldots, m \,, \tag{10.3.29}$$

$$\left(\sum_{e_k \in \mathscr{E}_\alpha} \hat{e}_{k(1)}\hat{e}_{k(2)}\right)P_\alpha + \left(\sum_{e_k \in \mathscr{E}_\alpha} \hat{e}_{k(2)^2}\right)Q_\alpha + \zeta_\alpha \sum_{i:(F_\alpha, V_i) \in \mathscr{R}} y_i \Lambda_{\alpha i}$$

$$= \sum_{e_k \in \mathscr{E}_\alpha} \hat{e}_{k(2)}\hat{e}_{k(3)} \ , \quad \alpha = 1, \ldots, m \ , \tag{10.3.30}$$

$$\sum_{i:(F_\alpha, V_i) \in \mathscr{R}} \Lambda_{\alpha i} = 0 \ , \qquad \alpha = 1, \ldots, m \ . \tag{10.3.31}$$

10.3.1 Regularization

The term *regularization* has often been used in the context of computer vision. As discussed in Chap. 1, the 3D recovery problem is viewed as an estimation of object parameters $\alpha_1, \ldots, \alpha_n$ from observed image characteristics $c_1, \ldots, c_m$. A set of object parameters $\alpha_1, \ldots, \alpha_n$ is identified with a point $\boldsymbol{\alpha}$ in an n-dimensional *object parameter space* $\mathscr{O}$, and a set of image characteristics $c_1, \ldots, c_m$ is identified with a point $\boldsymbol{c}$ in an m-dimensional *image characteristic space* $\mathscr{I}$. If an object is placed in the scene according to a specified object parameter vector $\boldsymbol{\alpha}$, its projection image determines its image characteristic vector $\boldsymbol{c}$. Hence, a mapping $\Pi: \mathscr{O} \to \mathscr{I}$ is defined from the object parameter space $\mathscr{O}$ to the image characteristic space $\mathscr{I}$. Hence, the 3D recovery problem can be viewed as the *inverse problem* to find the inverse mapping Π^{-1} that should give $\boldsymbol{\alpha} \in \mathscr{O}$ for observed $\boldsymbol{c} \in \mathscr{I}$ such that $\boldsymbol{c} = \Pi\boldsymbol{\alpha}$.

However, this inverse problem is often *ill-posed*. A problem is *well-posed* if the solution exists, is unique and depends continuously on input data. Otherwise, it is called ill-posed.[11] The "regularization" refers to techniques for making an ill-posed problem well-posed by adding extra constraints and employing variational principles. There are many factors that make a problem ill-posed. The following two are the most typical.

First, if observed information is insufficient, the inverse problem is *underspecified*, and the solution is indeterminate. In mathematical terms, the mapping Π is not one-to-one. This happens typically when $n > m$. In such cases, the usual procedure is to restrict the solution to a subset $\mathscr{M}$, say a manifold or subspace, of the object parameter space $\mathscr{O}$ by requiring reasonable constraints or assuming (often heuristically) a physically feasible *model* in such a way that the mapping Π becomes one-to-one from $\mathscr{M}$ to $\mathscr{I}$.

[11] Do not confuse this with other concepts that sound similar. A problem is *well-defined* if the problem has a definite meaning (i.e., it is defined) for all possible input values. Otherwise, it is *ill-defined*. A well-posed problem is *well-conditioned* if the solution is stable in the sense that the magnitude of output disturbance caused by input disturbance is bounded in a required form (the definition depends on the problem). Otherwise, it is *ill-conditioned*.

However, the resulting mapping Π: $\mathcal{M} \to \mathcal{I}$ may not be onto, which means that the set $\Pi\mathcal{M}$ is a proper subset of $\mathcal{I}$. Since noise is inevitable in real problems, observed image characteristics may be found outside the subset $\Pi\mathcal{M} \subset \mathcal{I}$. Consequently, we may not be able to find a solution in the assumed model $\mathcal{M}$.

This difficulty can be overcome by adopting an appropriate optimization. The observed data $\boldsymbol{c} \in \mathcal{I}$ is modified into $\boldsymbol{c}' \in \Pi\mathcal{M}$ "minimally" so that a solution $\boldsymbol{\alpha}$ can exist. In mathematical terms, we introduce an appropriate *norm* $\|\,.\,\|$ in the image characteristic space $\mathcal{I}$, and seek for a solution $\boldsymbol{\alpha} \in \mathcal{M}$ that minimizes $\|\boldsymbol{c} - \boldsymbol{c}'\| = \|\boldsymbol{c} - \Pi\boldsymbol{\alpha}\|$ (Fig. 10.8). In other words, we compute

$$\min_{\alpha \in \mathcal{M}} \|\boldsymbol{c} - \Pi\boldsymbol{\alpha}\|^2 \,. \tag{10.3.32}$$

The other major source of ill-posedness is excessive information. As a result, the problem is *overspecified*. This occurs typically when $n < m$. For example, if we do redundant measurements in an attempt to enhance reliability, different measurements may yield mutually incompatible results. In mathematical terms, the mapping Π: $\mathcal{O} \to \mathcal{I}$ is not onto, and $\Pi\mathcal{O}$ is a proper subset of $\mathcal{I}$. Ideally, the observed image characteristics should be in $\Pi\mathcal{O} \subset \mathcal{I}$, but no solution exists if the observed data are outside it. The usual remedy is the same as in the case of underspecification: the observed data $\boldsymbol{c}$ is modified into $\boldsymbol{c}' \in \Pi\mathcal{O}$ "minimally". In other words, we compute

$$\min_{\alpha \in \mathcal{O}} \|\boldsymbol{c} - \Pi\boldsymbol{\alpha}\|^2 \,. \tag{10.3.33}$$

In this case, the entire object parameter space $\mathcal{O}$ plays the role of the model.

In any case, we can regularize the problem by computing (10.3.32) ((10.3.33) is a special case of it). Let M be a mapping from $\mathcal{O}$ to some space equipped with a norm $\|\,.\,\|$, and assume that the model $\mathcal{M}$ is given in the form of $\|M\boldsymbol{\alpha}\| = 0$. Then, minimization (10.3.32) can be achieved by computing

$$\min_{\alpha \in \mathcal{O}} \left(\|\boldsymbol{c} - \Pi\boldsymbol{\alpha}\|^2 + \Lambda \|M\boldsymbol{\alpha}\|^2\right) , \tag{10.3.34}$$

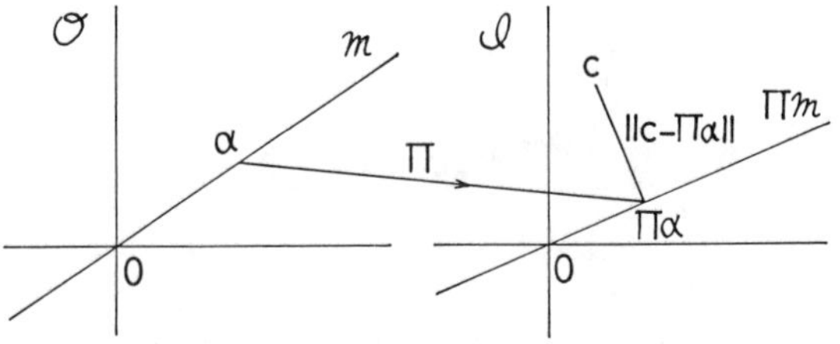

Fig. 10.8. By assigning a model $\mathcal{M}$ to the solution space $\mathcal{O}$, we seek a solution $\alpha \in \mathcal{M}$ that minimizes $\|\boldsymbol{c} - \Pi\boldsymbol{\alpha}\|$ for an observed piece of data $\boldsymbol{c}$

where Λ is the Lagrangian multiplier with respect to the model; its value is determined so that the model $\|M\boldsymbol{\alpha}\| = 0$ is satisfied. Let us call this type of regularization *regularization without compromise*. If Π and M are both linear mappings, the solution $\boldsymbol{\alpha}$ is given as a linear form $\Pi^- \boldsymbol{c}$ in input $\boldsymbol{c}$. The linear mapping $\Pi^-: \mathscr{I} \to \mathscr{M}$ is called the *pseudo-inverse* (or *generalized inverse*) of mapping Π: $\mathscr{M} \to \mathscr{I}$ with respect to this regularization (see Sect. 7.2).

Still another difficulty may arise: we may not know exactly what the model should be. Suppose we are to reconstruct a very smooth surface, but do not know anything about it except for smoothness. If $Z = Z(X, Y)$ is the surface to be reconstructed, we want to require $\partial^2 Z/\partial X^2 \approx 0$, $\partial^2 Z/\partial X\,\partial Y \approx 0$ and $\partial^2 Z/\partial Y^2 \approx 0$, but we cannot demand strict equality; if we did so, only a planar surface could result. In general terms, we want to require $M\boldsymbol{\alpha} \approx \mathbf{0}$ for some operator M, but the requirement $M\boldsymbol{\alpha} = \mathbf{0}$ would be incompatible with the projection relation $\Pi\boldsymbol{\alpha} = \boldsymbol{c}$.

In such a case, a possible solution would be to compute

$$\min_{\alpha \in \mathscr{O}} \left(\|\boldsymbol{c} - \Pi\boldsymbol{\alpha}\|^2 + \Lambda \|M\boldsymbol{\alpha}\|^2 \right) . \tag{10.3.35}$$

This looks exactly like (10.3.34). However, Λ is no longer a Lagrangian multiplier; it is a fixed positive constant that controls the "compromise" between the projection relation $\Pi\boldsymbol{\alpha} = \boldsymbol{c}$ and the model requirement $M\boldsymbol{\alpha} = \mathbf{0}$ (Fig. 10.9). Let us call this type of regularization *regularization with compromise*.

If Π and M are both linear mappings, the problem is called the *standard regularization*; the mapping M and the constant Λ are respectively called the *stabilizer* and its *regularization parameter*. The solution $\boldsymbol{\alpha}$ is given as a linear form $\Pi_\Lambda^- \boldsymbol{c}$ in input $\boldsymbol{c}$, where $\Pi_\Lambda^-: \mathscr{I} \to \mathscr{M}$ is the corresponding pseudo-inverse, but it now depends on the regularization parameter Λ.

This type of regularization is very flexible and can be used to stabilize many kinds of ill-posed problem. Once a regularized solution $\boldsymbol{\alpha}$ is obtained, we can compute $\|M\boldsymbol{\alpha}\|$ to see how well the model $\mathscr{M}$ fits. If the fitting is not satisfactory, we can replace the model by another. For instance, we can detect surface

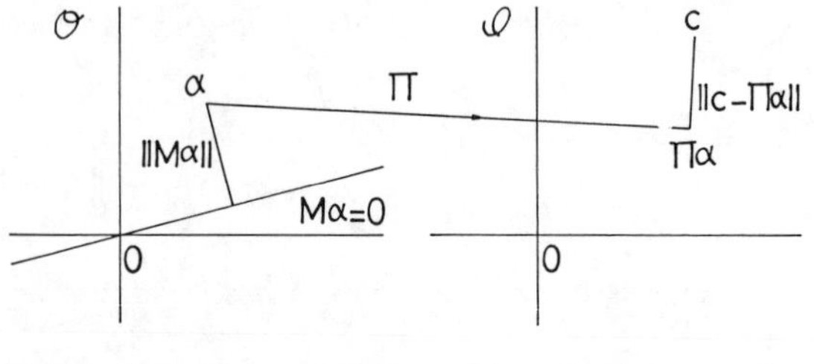

Fig. 10.9 If the projection relation $\Pi\boldsymbol{\alpha} = \boldsymbol{c}$ and the model requirement $M\boldsymbol{\alpha} = 0$ are incompatible, we introduce a regularization parameter that controls the compromise between the two

discontinuities by this technique. However, the cost of this flexibility is the arbitrariness of the formulation: the stabilizer is chosen for convenience, and the value of the regularization parameter may not have a particular meaning. In contrast, the optimization technique in Sect. 10.3 is a regularization without compromise, and its geometrical meaning is clear. However, it is often very difficult to specify a model $\mathcal{M}$ into which the solution is to be confined, so in many cases we must be content with regularization with compromise.

10.4 Optimization for Shape from Motion

In this section, we apply our optimization technique to the shape-from-motion problem. Suppose we are given a sequence of images of a polyhedron moving in a scene. Let us assume that the point-to-point correspondence has already been detected between consecutive frames. Although we do not attempt it in this book, we could write down the equations that determine the 3D shape and motion from displacements of the vertices on the image plane. In the presence of noise, however, the computed vertices of one face will not necessarily be coplanar.

We can avoid this inconsistency by choosing, as unknowns, the surface gradients of the object faces. Then, computed solutions necessarily have planar faces, resulting in a $2\frac{1}{2}$D sketch. If noise is present, the computed surface gradients are in general inaccurate, and the obtained $2\frac{1}{2}$D sketch does not necessarily define a consistent polyhedron. This difficulty is overcome by the optimization technique discussed in Sect. 10.3.

Consider a face that has four or more corners. If the image velocities are observed at (at least four) vertices, and if no three of these vertices are collinear, the hypothetical optical flow is uniquely determined, and the flow parameters can be computed (Sect. 7.7). Rearranging them into invariants, we can compute the rotation velocity $(\omega_1, \omega_2, \omega_3)$ and the surface gradient (p, q) of this face in an analytical form (Theorem 7.1). There exist two sets of solutions, but the spurious solutions can be discarded if two or more faces of the polyhedron are observed, since the true rotation velocity $(\omega_1, \omega_2, \omega_3)$ must be common to all faces (Sect. 7.6). Thus, we can estimate the surface gradient (p, q) for the faces that have four or more corners, obtaining a $2\frac{1}{2}$D sketch.

Remark 10.11. The theory of Chap. 7 is based on the assumption that *instantaneous velocities* are observed on the image plane. For a sequence of images, the instantaneous velocity must be approximated by the displacement between consecutive frames (by taking the time lapse between frames as unit time). This approximation introduces considerable error even if measurements on individual images are very accurate.

Remark 10.12. The rotation velocity $(\omega_1, \omega_2, \omega_3)$ should be common to all the faces. In the presence of noise, however, we may not be able to find a strictly common rotation velocity. It follows that we need some clustering technique in the three-dimensional $\omega_1\omega_2\omega_3$-space to find the one that is most likely to be the common rotation velocity.

Remark 10.13. If the surface gradients are computed for individual faces, we can obtain a $2\frac{1}{2}$D sketch by assigning them to either the first image or the second, or we can assign them to the "middle image" obtained by connecting the midpoints of the vertex displacements.

Remark 10.14. If the object has a triangular face, its surface gradient is not assigned. However, our optimization works even if some faces do not have assigned surface gradient estimates (Remarks 10.6 and 10.9).

Example 10.1. Consider the two images of Fig. 10.10a,b. Let us label the vertices and the faces as shown in Fig. 10.11a. The incidence structure is given by

$$\begin{aligned}\mathscr{R} = \{&(F_1, V_1)\ , (F_1, V_4)\ , (F_1, V_5), (F_1, V_7)\ , (F_1, V_8)\ ,\\ &(F_2, V_3)\ , (F_2, V_4)\ , (F_2, V_5)\ , (F_2, V_6)\ , (F_3, V_1),\\ &(F_3, V_2)\ , (F_3, V_3)\ , (F_3, V_4)\ , (F_4, V_1)\ ,\\ &(F_4, V_2)\ , (F_4, V_8)\ , (F_4, V_9)\}\ .\end{aligned}$$

Since we can easily check that condition (10.2.19) is satisfied, the incidence structure is nonsingular. The displacements of the vertices are shown in Fig. 10.11b. Regarding the displacements as instantaneous velocities, and applying

(a) (b)

Fig. 10.10a, b. Two object images

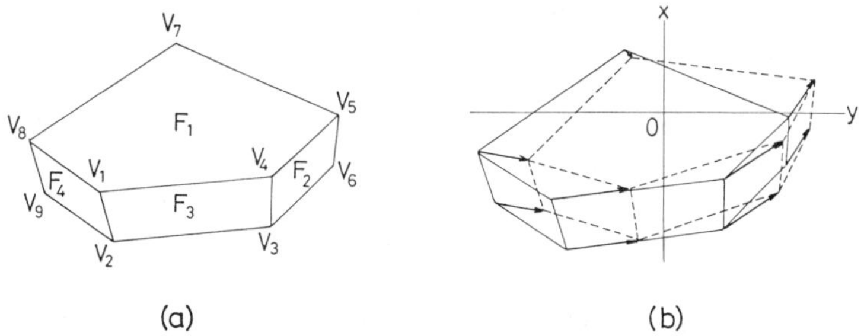

Fig. 10.11a. Labeled line drawing for the object image of Fig. 10.10a. **b** Displacements of vertices obtained from the two images of Fig. 10.10

the procedure described above, we can reconstruct the 3D shape uniquely up to a single scale factor. If surface gradient estimates are assigned to the first image, and (10.3.4) (i.e., ζ_α of (10.3.5)) is used for the weight W_α of optimization, the 3D shape shown in Fig. 10.12 is obtained. Figure 10.12a shows the top view (orthographic projection onto the YZ-plane), and Fig. 10.12b shows the side view (orthographic projection onto the ZX-plane). In spite of the presence of noise and the inaccuracy of the estimated surface gradient values, the final result is fairly correct.

10.5 Optimization of Rectangularity Heuristics

In this section, we consider the 3D reconstruction of a polyhedron from a single image by invoking the *rectangularity hypothesis*. If one corner is known to be rectangular and has three visible edges, we can compute the 3D orientations of the three edges up to the mirror image (Sect. 8.4). The computation is based on

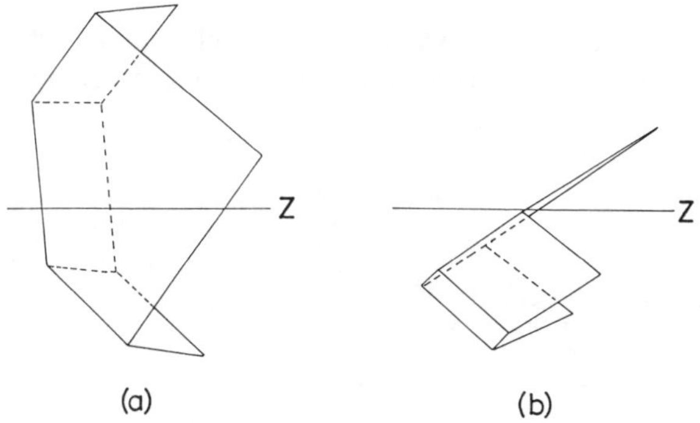

Fig. 10.12a,b. The 3D shape reconstructed from Fig. 10.11. **a** Top view (orthographic projection onto the YZ-plane). **b** Side view (orthographic projection onto the ZX-plane)

the *canonical angle* defined in terms of the camera rotation transformation. Each edge orientation indicates the surface normal to the face defined by the other two edges. Hence, we can determine the surface gradient of the three faces.

For each rectangular corner, we obtain two mirror image solutions. However, it is usually easy to choose the true solution. For example, we can apply the line drawing interpretation of Sect. 8.3 or Huffman–Clowes edge labeling (Sect. 8.3.1). We can also use other sources of information if available—range sensing, shading information, etc. If neither of the two mirror image solutions can be eliminated, we simply retain both and produce multiple solutions. In the following, let us assume for simplicity that the true solution can be distinguished from its mirror image.

If a face has two or more rectangular corners with three visible edges, multiple surface gradients are obtained. Then, their average is assigned, and the optimization technique is applied to the resulting a $2\frac{1}{2}$D sketch. Alternatively, the optimization technique can be applied to the $2\frac{1}{2}$D sketch with 3D edge orientation estimates. All faces and edges need not be given estimates; our optimization works if some faces or edges lack their orientation estimates (Remarks 10.6 and 10.9).

Thus, the remaining question is how to find rectangular corners. The first criterion is the *rectangularity test* based on Lemma 8.1. Namely, we compute the canonical angles of the three edges and reject those whose configuration in the canonical position is neither a fork nor an arrow as defined in Sect. 8.2. Specifically, a corner cannot be rectangular if the three angles between edges satisfy, in the canonical position, one of the following conditions:

(i) One is an acute angle, and the remaining two are obtuse angles (Fig. 10.13a). (This is not a fork.)
(ii) One is an acute angle, another is an obtuse angle, and the other is larger than π (Fig. 10.13b). (This is not an arrow.)
(iii) Two are acute angles, and the other is larger than $3\pi/2$ (Fig. 10.13c). (This is not an arrow, either.)

Still, the remaining corners are not necessarily all rectangular. The next step is the *compatibility test* for two corners. We choose two corners that share at least

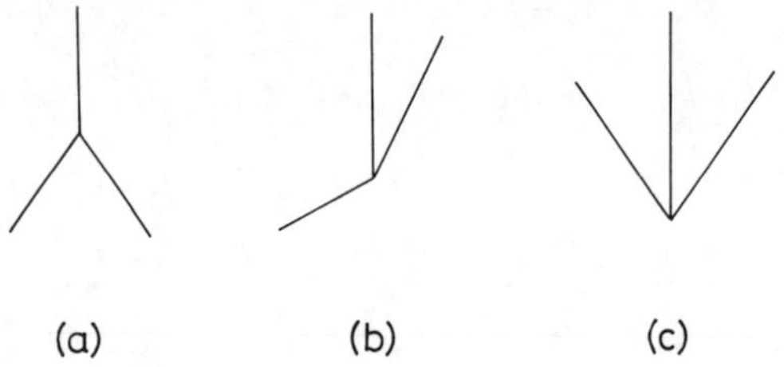

Fig. 10.13a–c. Rectangularity test. **a** This is not a fork. **b** This is not an arrow. **c** This is not an arrow. If these configurations arise in the canonical position, none of these can be the projection of a rectangular corner

one face, and compute the 3D edge orientations and the surface gradients at these corners, assuming that both are rectangular corners. If this assumption is correct, an identical 3D orientation should be obtained for the connecting edge (if the two corners are connected), and identical surface gradients should be obtained for the common faces (within some fixed tolerance). If not, we say that they are *incompatible* as rectangular corners.

After this test, many different strategies are possible. One heuristic is to assume that the object is very likely to be rectangular and form maximal compatible sets of corners in such a way that as many corners are included as possible unless incompatible pairs arise among them. Then, assuming that the corners of each set are all rectangular, we end up with as many $2\frac{1}{2}$D sketches as the maximal compatible sets.

Example 10.2. Figure 10.14a is a real image of a polyhedron. Suppose the line drawing of Fig. 10.14b is obtained, and its vertices and faces are labeled as indicated in the figure. Its incidence structure is

$$\begin{aligned}\mathscr{R} = \{&(F_1, V_1)\ , (F_1, V_2)\ , (F_1, V_3)\ , (F_1, V_4)\ , (F_2, V_5)\ ,\\ &(F_2, V_6)\ , (F_2, V_7)\ , (F_2, V_8)\ , (F_3, V_3)\ , (F_3, V_4)\ ,\\ &(F_3, V_6)\ , (F_3, V_7)\ , (F_3, V_9)\ , (F_3, V_{10})\ , (F_4, V_1)\ ,\\ &(F_4, V_4)\ , (F_4, V_{10})\ , (F_4, V_{11})\}\ .\end{aligned}$$

Since the condition (10.2.19) is satisfied, this incidence structure is nonsingular. Vertices V_1, V_3, V_4, V_6, V_{10} have three visible edges,[12] but the rectangularity test rejects vertex V_3 as nonrectangular. The rest are combined into pairs sharing at least one common face, resulting in

$$\{V_1, V_4\}\ , \{V_1, V_{10}\}\ , \{V_4, V_6\}\ , \{V_4, V_{10}\}\ , \{V_6, V_{10}\}\ .$$

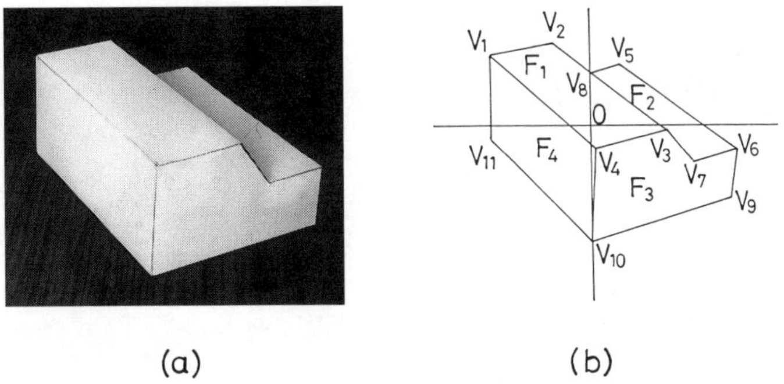

Fig. 10.14. a A polyhedron image. **b** The labeling of its line drawing

[12] Vertex V_8 is a T-junction, so it is incident to F_2 alone.

Then, we apply the compatibility test by computing the canonical angles and estimating the surface gradients as described in Sect. 8.4. Then, all the pairs pass this test. So, all these corners are assumed to be rectangular. Then, surface gradient estimates are assigned to the faces to which these corners are incident (the average is taken if necessary), and the optimization technique is applied to the resulting $2\frac{1}{2}$D sketch. If (10.3.4), i.e., ζ_α of (10.3.5), is used for the weight W_α, the 3D shape indicated in Fig. 10.15 is obtained. Figure 10.15a shows the top view (orthographic projection onto the YZ-plane), and Fig. 10.15b shows the side view (orthographic projection onto the ZX-plane).

Example 10.3. Figure 10.16a is another real image of a polyhedron. Suppose the line drawing of Fig. 10.16b is obtained, and its vertices and faces are labeled as indicated in the figure. Its incidence structure is

$$\begin{aligned}\mathscr{R} = \{&(F_1, V_1)\ , (F_1, V_2)\ , (F_1, V_3)\ , (F_1, V_4)\ , (F_2, V_3)\ ,\\ &(F_2, V_4)\ , (F_2, V_5)\ , (F_2, V_6)\ , (F_3, V_5), (F_3, V_6)\ ,\\ &(F_3, V_7)\ , (F_3, V_8)\ , (F_4, V_7)\ , (F_4, V_8)\ , (F_4, V_9)\ ,\\ &(F_4, V_{10})\ , (F_5, V_2)\ , (F_5, V_3)\ , (F_5, V_6)\ , (F_5, V_7)\ ,\\ &(F_5, V_{10})\ , (F_5, V_{11})\}\ .\end{aligned}$$

Vertices $V_2, V_3, V_4, V_5, V_6, V_7, V_8, V_{10}$, have three visible edges. This time, the rectangularity test cannot reject any vertices as definitely nonrectangular. So, these vertices are combined into pairs sharing at least one common face, resulting in

$$\begin{aligned}&\{V_2, V_3\}\ , \{V_2, V_4\}\ , \{V_2, V_6\}\ , \{V_2, V_7\}\ , \{V_2, V_{10}\}\ ,\\ &\{V_2, V_4\}\ , \{V_3, V_5\}\ , \{V_3, V_6\}\ , \{V_3, V_7\}\ , \{V_3, V_{10}\}\ ,\\ &\{V_4, V_5\}\ , \{V_4, V_6\}\ , \{V_5, V_6\}\ , \{V_5, V_7\}\ ,\\ &\{V_5, V_8\}\ , \{V_6, V_7\}\ , \{V_6, V_8\}\ , \{V_6, V_{10}\}\ , \{V_7, V_8\}\ ,\\ &\{V_7, V_{10}\}\ , \{V_8, V_{10}\}\ .\end{aligned}$$

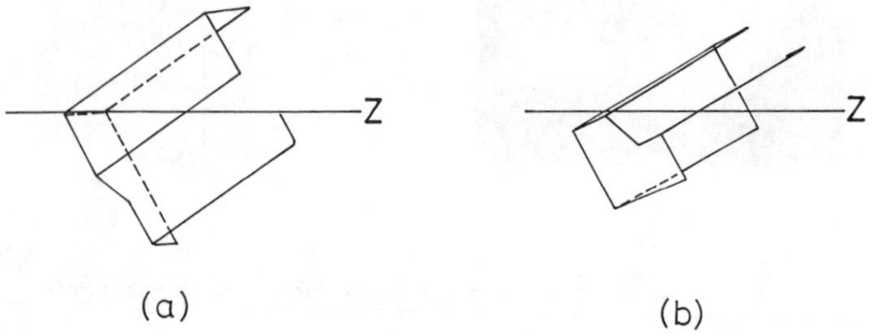

Fig. 10.15a,b. The 3D shape reconstructed from Fig. 10.14. **a** Top view (orthographic projection onto the YZ-plane). **b** Side view (orthographic projection onto the ZX-plane)

(a)

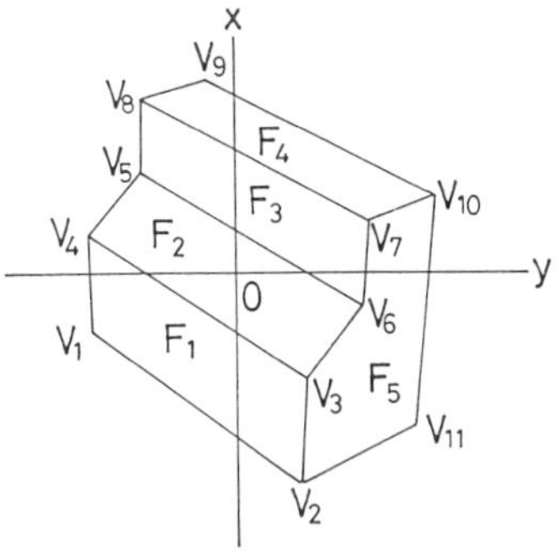

(b)

Fig. 10.16a. A polyhedron image. **b** The labeling of its line drawing

The compatibility test tells us that these vertices are split into two compatible groups

$$\{V_2, V_7, V_8, V_{10}\} \ , \{V_3, V_4, V_5, V_6\} \ .$$

Hence, we obtain two solutions by assuming that the corners of each group are rectangular. Applying the optimization technique, we obtain the two 3D shapes shown in Figs. 10.17 and 10.18. In both of them, (a) shows the top view (orthographic projection onto the YZ-plane), and (b) shows the side view (orthographic projection onto the ZX-plane).

Remark 10.15. The objects shown in Figs. 10.14a and 10.16a are actually the same but placed differently, and the reconstructions of Figs. 10.15 and 10.17 are correct. From this observation, we notice that false solutions can be removed if two images of the same object are taken from different angles, since the true solutions must have an identical shape. Alternatively, we can eliminate false

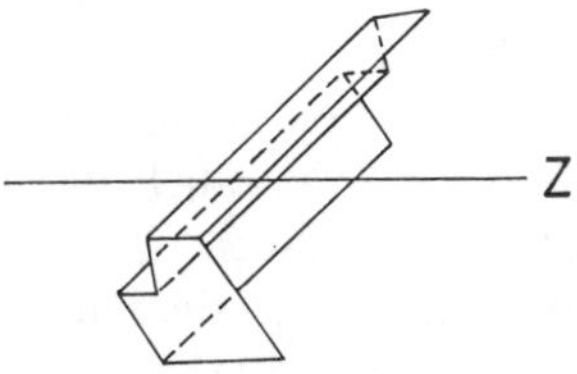

(a)

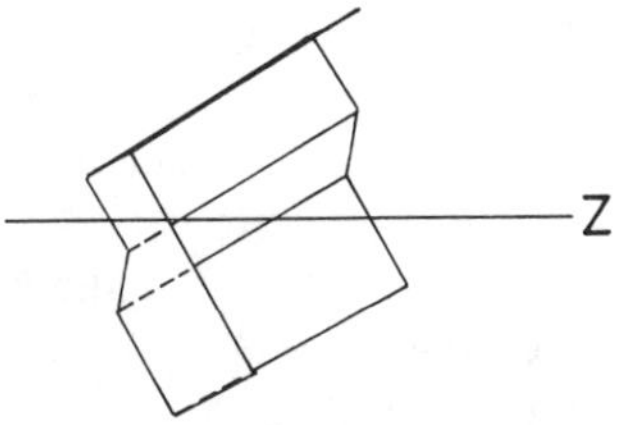

(b)

Fig. 10.17a,b. One 3D shape reconstructed from Fig. 10.16. **a** Top view (orthographic projection onto the YZ-plane). **b** Side view (orthographic projection onto the ZX-plane)

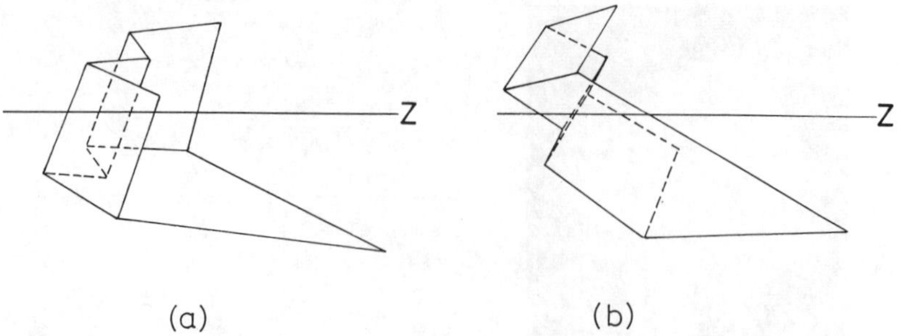

Fig. 10.18a,b. Another 3D shape reconstructed from Fig. 10.16. **a** Top view (orthographic projection onto the YZ-plane). **b** Side view (orthographic projection onto the ZX-plane)

solutions if some global 3D characteristic, say the aspect ratio (the maximum of the ratio of the height measured in one direction to the width measured perpendicularly to it), is given or estimated beforehand.

10.6 Optimization of Parallelism Heuristics

Now, let us consider 3D recovery based on the *parallelism hypothesis* coupled with the optimization technique. Suppose we observe a polyhedron image. If we can find a set of edges that are parallel in the scene, their 3D orientation is estimated from their vanishing points (Sect. 8.5). Hence, if a parallel edge finding algorithm is available, we can obtain a $2\frac{1}{2}$D sketch with estimated 3D edge orientations. The 3D edge orientations computed from vanishing points may not be consistent with each other in the presence of noise, but this inconsistency can be overcome by the optimization technique of Sect. 10.3.

Thus, it remains to construct an algorithm for finding parallel edges. A naive heuristic is to group together those edges that are "nearly parallel" on the image plane. However, if two edges are far apart on the image plane, they can be parallel in the scene even if they make a large angle on the image plane. On the other hand, if two edges meet at a corner of the object, they cannot be parallel, however small the angle between them is. In general, two parallel edges in the scene can be projected to make any angle on the image plane.

However, there exists a powerful clue to parallelism. As we noted in Sect. 8.5, projections of parallel lines in the scene must be concurrent on the image plane, defining a common vanishing point. It follows that we can detect parallelism by the *concurrency test*: *If three or more edges are concurrent on the image plane, they are judged to be parallel in the scene.* This heuristic is reasonable because it is very unlikely (though not entirely impossible) that three or more nonparallel edges happen to be concurrent when projected.

However, a serious problem arises. As we have already pointed out many times, parallel edges, when extended, may not necessarily intersect at a single point in the presence of noise (Fig. 10.19). What criterion should we use to accept them as concurrent or reject them as non-concurrent? One solution is to decide that lines are concurrent if the maximam separation among their mutual intersections is smaller than a threshold value ε.

The trouble is that the threshold value cannot be fixed. For instance, if two lines intersect at a point far apart from the image origin, slight disturbance to the orientations of these lines will cause very large displacements of their intersection. Hence, the threshold value ε must be very large, but then we cannot distinguish intersections near the image origin. Thus, the threshold value must depend on the distance of the intersection from the mage origin.

In order to confine the computation to the finite domain, let us write all relationships of points and lines in terms of the unit vectors associated with them (Sect. 8.5). This is equivalent to using, instead of the usual image plane, the *image sphere* of radius f centered at the viewpoint (Sects. 4.6, 8.5). The following argument involves many trial-and-error type heuristics and therefore has much room for improvement; we do not intend the result here to be taken as definitive.

In order to find an appropriate threshold value for deviation of the intersection of two lines, consider a right *spherical triangle* ΔOPQ drawn on the image sphere (Fig. 10.20). Here, O is a right-angle corner and located at the intersection of the image sphere with the Z-axis. Let $s = \rho(O, P)$, $L = \rho(O, Q)$, and $\varphi = \theta(PO, PQ)$, where $\rho(.,.)$ and $\theta(.,.)$ are respectively the arc length and the angle measured on the image sphere (i.e., the *invariant distance* and the *invariant angle*, see Sect. 4.6).

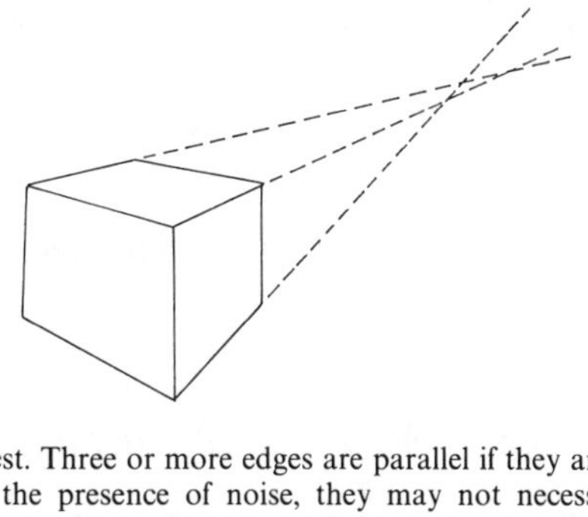

Fig. 10.19. Concurrency test. Three or more edges are parallel if they are concurrent on the image plane when extended. In the presence of noise, they may not necessarily intersect at a single vanishing point. How can we tolerate the error to judge concurrency?

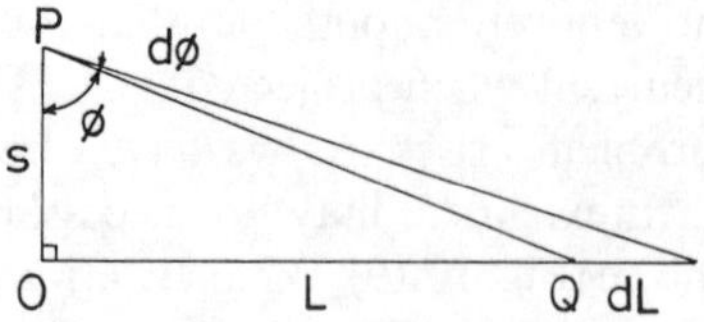

Fig. 10.20. A right spherical triangle drawn on an image sphere of radius f centered at the viewpoint

Lemma 10.1.

$$\frac{dL}{d\varphi} = f\left[\frac{1}{\sin(s/f)} - \left(\frac{1}{\sin(s/f)} - \sin\frac{s}{f}\right)\cos^2\frac{L}{f}\right]. \tag{10.6.1}$$

Proof. Invoking spherical trigonometry, we obtain

$$\tan\varphi = \frac{\tan(L/f)}{\sin(s/f)}. \tag{10.6.2}$$

Keeping s fixed, and differentiating both sides, we obtain

$$\frac{d\varphi}{\cos^2\varphi} = \frac{dL/f}{\sin(s/f)\,\cos^2(L/f)}. \tag{10.6.3}$$

Equation (10.6.1) is obtained if φ is eliminated from this by using (10.6.2) and rearranging the result.

If $s/f \ll 1$, we observe

$$\frac{dL}{d\varphi} \approx \frac{f^2}{s}\left(1 - \cos^2\frac{L}{f}\right). \tag{10.6.4}$$

We can interpret this expression as the "sensitivity" of the location of the intersection of two lines, separated by distance s, to possible inaccuracy of edge orientation φ. Motivated by this interpretation, we identify L with the invariant distance (the arc length measured on the image sphere) of the intersection of two lines from the image origin O. First, we observe:

Lemma 10.2. *The unit vector* $\boldsymbol{m}$ *associated with the intersection of two lines* l, l' *is*

$$\boldsymbol{m} = \pm\frac{\boldsymbol{n} \times \boldsymbol{n}'}{\|\boldsymbol{n} \times \boldsymbol{n}'\|}, \tag{10.6.5}$$

where $\boldsymbol{n}$ *and* $\boldsymbol{n}'$ *are the unit vectors associated with* l *and* l'.

Proof. The unit vector $\boldsymbol{m}$ associated with the intersection must be perpendicular to both of the unit vectors $\boldsymbol{n}, \boldsymbol{n}'$ associated with the two lines (Fig. 10.21).

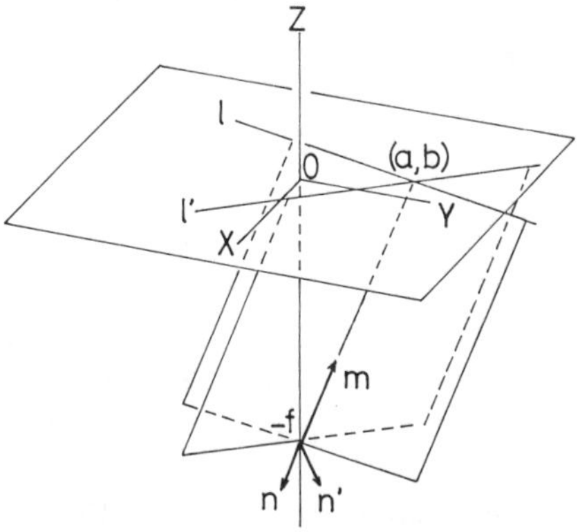

Fig. 10.21. The unit vector $\boldsymbol{m}$ associated with the intersection (a, b) must be perpendicular to both of the unit vectors $\boldsymbol{n}$, $\boldsymbol{n}'$ associated with the two lines

Proposition 10.3. *The invariant distance L of the intersection of two edges e, e' from the image origin O is given by*

$$L = f \cos^{-1} \eta_{ee'} , \tag{10.6.6}$$

where

$$\eta_{ee'} \equiv \left| \frac{|\boldsymbol{n}\boldsymbol{n}'\boldsymbol{k}|}{\|\boldsymbol{n} \times \boldsymbol{n}'\|} \right| . \tag{10.6.7}$$

Here, $\boldsymbol{n}$ and $\boldsymbol{n}'$ are the unit vectors associated with the two edges, and $\boldsymbol{k} = (0, 0, 1)$.

Proof. The unit vector $\boldsymbol{m}$ associated with the intersection is given by (10.6.5). Since the angle of $\boldsymbol{m}$ from the Z-axis is L/f or $\pi - L/f$, we see that

$$\cos\frac{L}{f} = |(\boldsymbol{m}, \boldsymbol{k})| = \left|\frac{(\boldsymbol{n} \times \boldsymbol{n}', \boldsymbol{k})}{\|\boldsymbol{n} \times \boldsymbol{n}'\|}\right| = \left|\frac{|\boldsymbol{n}\boldsymbol{n}'\boldsymbol{k}|}{\|\boldsymbol{n} \times \boldsymbol{n}'\|}\right| . \tag{10.6.8}$$

($|\boldsymbol{n}\boldsymbol{n}'\boldsymbol{k}|$ is the scalar triple product, i.e., the determinant of the matrix whose columns are $\boldsymbol{n}$, $\boldsymbol{n}'$, $\boldsymbol{k}$ in this order.)

Remark 10.16. For measuring the distance of the intersection from the image origin O, the quantity $\eta_{ee'}$ of (10.6.7) is more convenient than the invariant distance L of (10.6.6), since it is a nondimensional quantity whose value is always in the range $0 \leqq \eta(l_1, l_2) \leqq 1$. The value 0 corresponds to an infinitely distant intersection, while the value 1 corresponds to the image origin O. The computation involves only the unit vectors $\boldsymbol{n}$, $\boldsymbol{n}'$ associated with the two edges alone; the coordinates of the intersection need not be computed.

It follows that we can adopt

$$\varepsilon_{ee'} = \frac{f^2 \Delta\varphi_{ee'}}{S_{ee'}}(1 - \eta_{ee'}^2) \tag{10.6.9}$$

as the threshold value $\varepsilon_{ee'}$ for two edges e, e', where $s_{ee'}$ is an appropriately defined "average separation" between edges e and e', and $\Delta\varphi_{ee'}$ is the admissible error of edge orientation.

First, consider $s_{ee'}$. Let (x_0, y_0) and (x_1, y_1) be the two endpoints of edge e, and let (x'_0, y'_0) and (x'_1, y'_1) be those of edge e'. Put

$$\boldsymbol{e} = (x_1 - x_0, y_1 - y_0) , \qquad \boldsymbol{e}' = (x'_1 - x'_0, y'_1 - y'_0) . \tag{10.6.10}$$

We define their *average orientation* $\boldsymbol{l}$ (weighted by their lengths) as

$$\boldsymbol{l} = \begin{cases} \dfrac{\boldsymbol{e} + \boldsymbol{e}'}{\|\boldsymbol{e} + \boldsymbol{e}'\|} & (\boldsymbol{e}, \boldsymbol{e}') \geqq 0 , \\ \dfrac{\boldsymbol{e} - \boldsymbol{e}'}{\|\boldsymbol{e} - \boldsymbol{e}'\|} & (\boldsymbol{e}, \boldsymbol{e}') < 0 . \end{cases} \tag{10.6.11}$$

The *average separation* $s_{ee'}$ between edges e and e' is defined to be the projection of the distance between their midpoints onto the axis perpendicular to the average orientation $\boldsymbol{l} = (l_1, l_2)$ (Fig. 10.22):

$$s_{ee'} \equiv \left(\frac{x_0 + x_1}{2} - \frac{x'_0 + x'_1}{2}\right) l_2 - \left(\frac{y_0 + y_1}{2} - \frac{y'_0 + y'_1}{2}\right) l_1 . \tag{10.6.12}$$

Finally, the admissible error $\Delta\varphi_{ee'}$ must be determined. Let $|e|$ and $|e'|$ be the respective lengths of edges e and e'. It makes sense to assume that the orientation of an edge becomes less accurate as its length becomes shorter, so we put

$$\Delta\varphi_{ee'} = \frac{\text{const.}}{\min(|e|, |e'|)} . \tag{10.6.13}$$

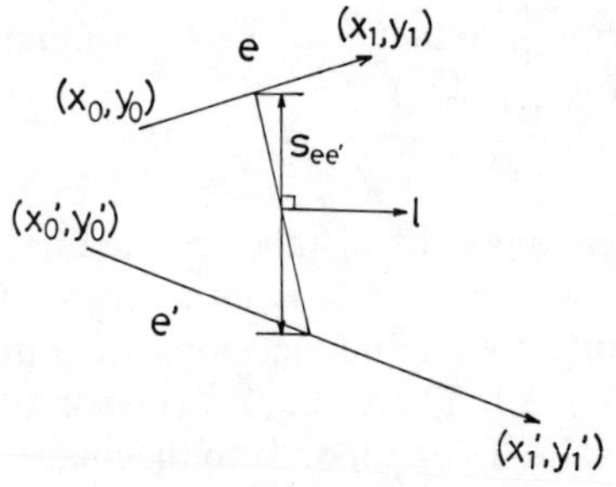

Fig. 10.22. The average separation $s_{ee'}$ of edges e, e' is defined to be the projection of the distance between their midpoints onto the axis perpendicular to the average orientation $\boldsymbol{l}$

Now, we obtain the following procedure for the concurrency test. First, we choose two edges e, e' that do not meet at a corner of the object, and regard them as candidate parallel edges. Let $\boldsymbol{m}''$ be the unit vector associated with their intersection P'' (when they are extended) computed by (10.6.5). Let us adopt the convention that the one with $m_3 \geqq 0$ is chosen whenever we compute such a vector.

Consider the third edge e'', and let $\boldsymbol{m}$ and $\boldsymbol{m}'$ be the unit vectors, respectively, associated with the intersections P, P' of edge e'' with edges e, e' (when these edges are extended). We judge the three edges to be concurrent if

$$\rho(P, P') < \varepsilon_{ee''} + \varepsilon_{e'e''} , \qquad \rho(P, P'') < \varepsilon_{ee'} + \varepsilon_{e''e'} ,$$
$$\rho(P', P'') < \varepsilon_{e'e} + \varepsilon_{e''e} \tag{10.6.14}$$

(Fig. 10.23). In terms of the vectors $\boldsymbol{m}$, $\boldsymbol{m}'$, $\boldsymbol{m}''$, these are rewritten as

$$|(\boldsymbol{m}, \boldsymbol{m}')| > \cos\frac{\varepsilon_{ee''} + \varepsilon_{e'e''}}{f} , \qquad |(\boldsymbol{m}, \boldsymbol{m}'')| > \cos\frac{\varepsilon_{ee'} + \varepsilon_{e''e'}}{f} ,$$
$$|(\boldsymbol{m}', \boldsymbol{m}'')| > \cos\frac{\varepsilon_{e'e} + \varepsilon_{e''e}}{f} . \tag{10.6.15}$$

Those edges that satisfy this condition are judged to be parallel to the candidate pair e, e'. If two edges that share a common endpoint are judged to be parallel to e and e', the one that passes this test with a lower threshold is chosen. If no edge is judged to be parallel to the candidate pair e, e', edges e, e' are regarded as nonparallel.

Remark 10.17. In the above process, the order of choosing candidate edges can be arbitrary, but it is desirable to start from edges that are "most likely" to be parallel. One heuristic is to start from the pair e, e' that has the smallest value of $\eta_{ee'}$. This is equivalent to assuming that *two edges are more likely to be parallel in the scene if their intersection is farther apart from the image origin O*. Let us call this heuristic the *vanishing point heuristic*. At the same time, it makes sense to

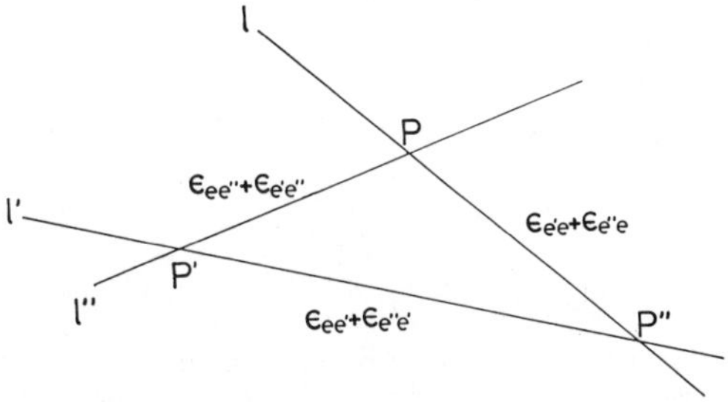

Fig. 10.23. Concurrency of three lines is judged by the pairwise arc lengths of the three intersections measured on the image sphere

reject as nonparallel those edges intersecting within an appropriately set distance L_0 from the image origin O, or $\eta_{ee'} > \cos(L_0/f)$. For example, if (x_i, y_i), $i = 1, \ldots, n$, are the image coordinates of the vertices, we can set $L_0 = \max_{1 \leq i \leq n} \sqrt{x_i^2 + y_i^2}$, assuming that the polyhedron is drawn near the center of the image.

Since this concurrency test is based on the occurrence of three or more parallel edges, it is incapable of detecting a pair of parallel edges if no other edges are parallel to them. Let us consider how to detect such pairs of parallel edges. It is reasonable to check only those pairs that share a common face but no common vertex, because it is highly unlikely that two edges belonging to different faces are parallel, yet no other edges are parallel to them.

After finding such candidate pairs, the following two tests must be applied. The first is the *parallelogram test*: *If two pairs of parallel lines lying on the same plane are projected onto half-lines starting from their respective vanishing points, they must intersect with each other at exactly four points on the image plane.* This constraint comes from the fact that two pairs of parallel lines on a plane must define a parallelogram. For example, suppose edges e_1, e_2 are already judged to be parallel. In Fig. 10.24a, edges e_3, e_4 cannot be parallel in the scene because they do not define a parallelogram with edges e_1, e_2, while edges e_3, e_4 of Fig. 10.24b pass this test.

The second test is the *collinearity test*: *If three or more sets of parallel lines belong to the same face, their vanishing points must be collinear.* Hence, a candidate pair of edges cannot be parallel if their intersection is not on the vanishing line already established from other vanishing points (within some tolerance) (Fig. 10.25a). Let $\boldsymbol{m}$ be the unit vector associated with the intersection of two candidate edges e, e' computed by (10.6.5), and let $\boldsymbol{n}$ be the unit vector associated with the already established vanishing line (estimated by the method in Sect.

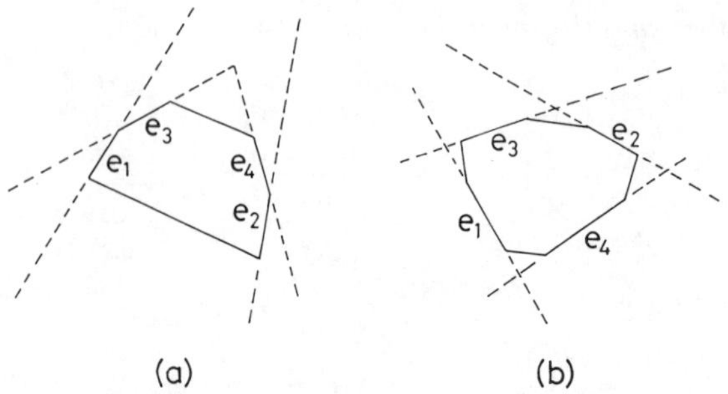

Fig. 10.24a,b. Parallelogram test. Two pairs of parallel edges regarded as half-lines starting from their respective vanishing points must intersect each other at exactly four intersecting points. If edges e_1, e_2 are parallel, edges e_3, e_4 cannot be parallel in **a** but can be parallel in **b**

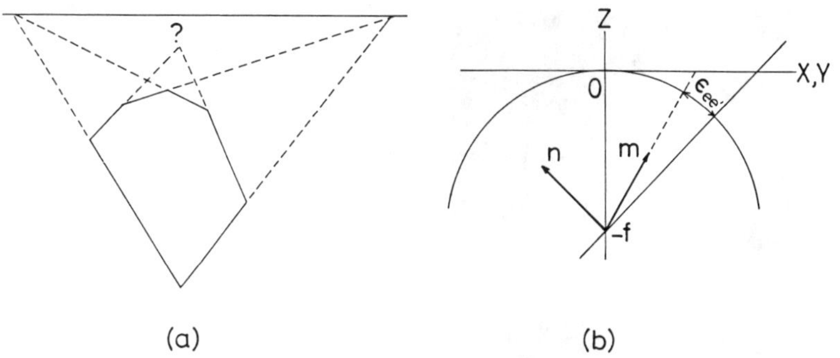

Fig. 10.25a. Collinearity test. The vanishing points of parallel edges belonging to the same face must be collinear (within some tolerance). **b** The criterion for the collinearity test

8.5). We use the following criterion (Fig. 10.25b):

$$|(\boldsymbol{m}, \boldsymbol{n})| < \sin \frac{\varepsilon_{ee'}}{f} \ . \tag{10.6.16}$$

Let us assume that those edge pairs that have passed these two tests are parallel in the scene. If one edge appears in multiple pairs, we choose, by invoking the "vanishing point heuristic", the pair whose intersection is located farthest apart from the image origin O (i.e., the one for which $\eta_{ee'}$ is the smallest) (Fig. 10.26).

Example 10.4. Figure 10.27a is a real image of a polyhedron. Suppose the line drawing of Fig. 10.27b is obtained. Its vertices, edges, and faces are labeled as indicated in this figure. Its incidence structure is

$$\begin{aligned} \mathscr{R} = \{ & (F_1, V_1) \, , \quad (F_1, V_2) \, , \quad (F_1, V_3) \, , \quad (F_1, V_4) \, , \\ & (F_2, V_3) \, , \quad (F_2, V_4) \, , \quad (F_2, V_5) \, , \quad (F_2, V_6) \, , \\ & (F_3, V_5) \, , \quad (F_3, V_6) \, , \quad (F_3, V_7) \, , \quad (F_3, V_8) \, , \\ & (F_4, V_1) \, , \quad (F_4, V_4) \, , \quad (F_4, V_5) \, , \quad (F_4, V_8) \, , \quad (F_4, V_9)\} \ . \end{aligned}$$

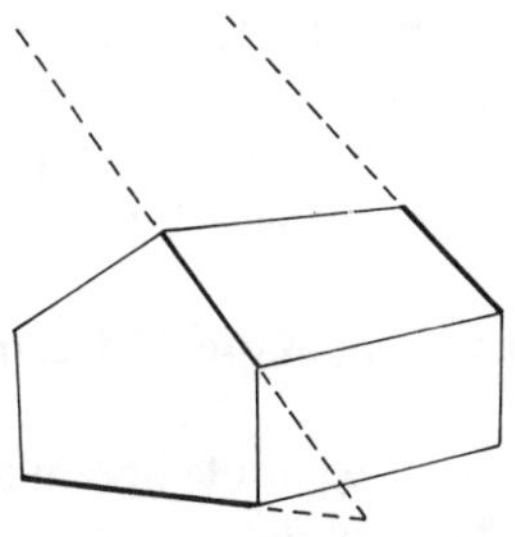

Fig. 10.26. Vanishing point heuristic. Two edges are more likely to be parallel in the scene if their intersection is farther apart from the image origin O

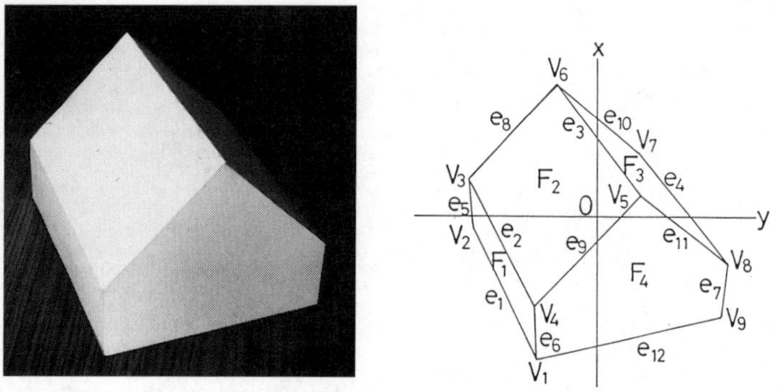

(a) (b)

Fig. 10.27a. A polyhedron image. **b** The labeling of its line drawing

Since condition (10.2.19) is satisfied, this incidence structure is nonsingular. Applying the concurrency test, we detect the following set of parallel edges:

$$\{e_1, e_2, e_3, e_4\} , \quad \{e_5, e_6, e_7\} .$$

From among the remaining edges, those pairs that share common faces but no common vertices are the next candidates for parallel pairs:

$$\{e_8, e_9\} , \quad \{e_{10}, e_{11}\} , \quad \{e_9, e_{12}\} , \quad \{e_{11}, e_{12}\} .$$

All of these pairs pass both the parallelogram test and the collinearity test. Edge e_9 belongs to two pairs $\{e_8, e_9\}$ and $\{e_9, e_{12}\}$. Invoking the vanishing point heuristic, we choose the former, because its intersection is farther apart than that of the latter. Similarly, edge e_{11} belongs to two pairs $\{e_{10}, e_{11}\}$ and $\{e_{11}, e_{12}\}$, but the former is chosen. Thus, we conclude that the sets of parallel edges are

$$\{e_1, e_2, e_3, e_4\} , \quad \{e_5, e_6, e_7\} , \quad \{e_8, e_9\} , \quad \{e_{10}, e_{11}\} .$$

Applying the optimization technique, we can recover the 3D shape up to a single scale factor. If (10.3.21), i.e., ζ_α of (10.3.22), is used for the weight W_α, the 3D shape shown in Fig. 10.28 is obtained. Figure 10.28a shows the top view orthographic projection onto the YZ-plane), while Fig. 10.28b shows the side view (orthographic projection on to the ZX-plane).

10.6.1 Parallelogram Test and Computational Geometry

Let us construct an algorithm for the parallelogram test. This test is decomposed into repeated tests for the existence of the intersection of two half-lines represented by their endpoints and points on them. Let l and l' be two half-lines starting from points (a, b) and (a', b'), respectively, and let (x, y) and (x', y') be points on

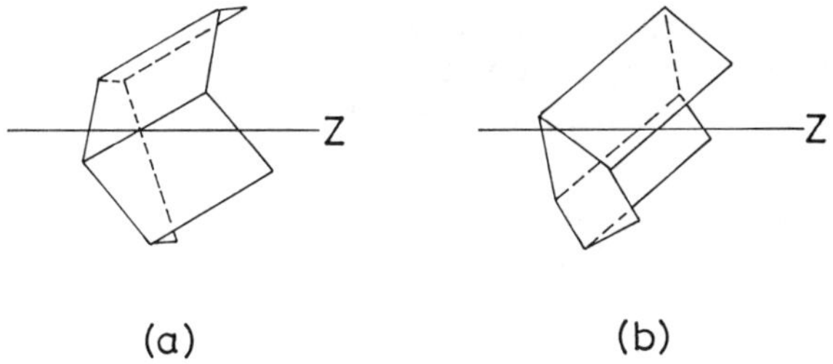

Fig. 10.28a,b. The 3D shape reconstructed from Fig. 10.17. **a** Top view (orthographic projection onto the YZ-plane). **b** Side view (orthographic projection onto the ZX-plane)

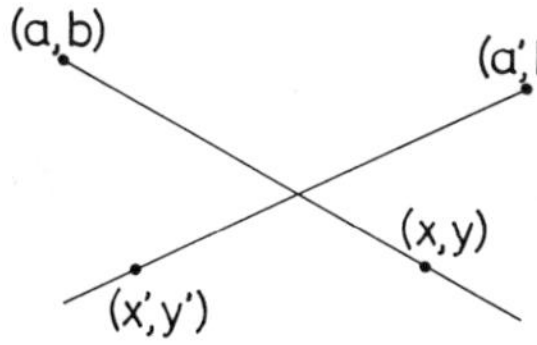

Fig. 10.29. The parallelogram test reduces to repeated tests for intersection of two half-lines starting from the passing through given points

lines l and l', respectively (Fig. 10.29). We want to know whether or not lines l and l' intersect when (a, b), (x, y), (a', b'), (x', y') are given as input data.

The intersection exists and is given by

$$\begin{pmatrix} a + t(x - a) \\ b + t(y - b) \end{pmatrix} = \begin{pmatrix} a' + t'(x' - a') \\ b' + t'(y' - b') \end{pmatrix} , \tag{10.6.17}$$

if and only if there exist such positive t and t'. From this relation, t and t' are obtained in the form

$$t = \frac{1}{\Delta}[(x' - a)(b' - b) - (y' - b)(a' - a)] ,$$

$$t' = \frac{1}{\Delta}[(x - a)(b' - b) - (y - b)(a' - a)] ,$$

$$\Delta = (x - a)(y' - b) - (y - b)(x' - a) . \tag{10.6.18}$$

Hence, the test is performed by checking whether the three expressions

$$(x' - a)(b' - b) - (y' - b)(a' - a) ,$$

$$(x - a)(b' - b) - (y - b)(a' - a) ,$$

$$(x - a)(y' - b') - (y - b)(x' - a') \tag{10.6.19}$$

have the same sign.

However, if (a, b) and (a', b') are the vanishing points of parallel lines, they may be located on the image plane very far away from the image origin. As we have often pointed out, the computation can be done in the finite domain if we introduce (not necessarily unit) vectors $\boldsymbol{n} = (n_1, n_2, n_3)$ and $\boldsymbol{n}' = (n'_1, n'_2, n'_3)$ associated with points (a, b) and (a', b') (Sect. 8.5). Since $\boldsymbol{n}$ and $-\boldsymbol{n}$ represent the same point, let us choose the one with $n_3 \geqq 0$. Similarly, we choose $\boldsymbol{n}'$ with $n'_3 \geqq 0$. If we substitute

$$a = f\frac{n_1}{n_3}\ , \quad b = f\frac{n_2}{n_3}\ , \quad a' = f\frac{n'_1}{n'_3}\ , \quad b' = f\frac{n'_2}{n'_3} \tag{10.6.20}$$

into expressions (10.6.19) and multiply them by n_3 and n'_3 appropriately, we see that the test is reduced to checking whether the following three expressions have the same sign:

$$\begin{aligned}
&(n_3x - fn_1)(n'_2n_3 - n_2n'_3) - (n_3y - fn_2)(n'_1n_3 - n_1n'_3)\ ,\\
&(n'_3x' - fn'_1)(n'_2n_3 - n_2n'_3) - (n'_3y' - fn'_2)(n'_1n_3 - n_1n'_3)\ ,\\
&(n_2x - fn_1)(n'_3y' - fn'_2) - (n_3y - fn_2)(n'_3x' - fn'_1)\ .
\end{aligned} \tag{10.6.21}$$

Thus, all computations are done in the finite domain.

The study of algorithms for testing geometrical properties of figures consisting of points and lines on a plane, measuring their geometrical quantities, and generating figures that possess prescribed geometrical properties is called *computational geometry*. The most fundamental requirement on such algorithms is the speed of computation; algorithms should involve as few arithmetic operations as possible. At the same time, all computations must be done in a reasonably limited domain in order to prevent overflow or underflow; computation should not involve values which are either too large or too small. Furthermore, we must take into account the fact that all input data have only limited accuracy and computation is done with fixed finite accuracy. Hence, care must be taken to prevent numerical instability and inconsistency in the result. This is a challenging area of research and is also vital to the study of image understanding.

Exercises

10.1 Show that the inverse relationship of (10.2.4) is given by (10.2.5).

10.2 (a) Show that the equation of a line in the XYZ-space passing through point (a, b, c) and extending in the direction $\boldsymbol{m} = (m_1, m_2, m_3)$ is given by

$$\frac{X-a}{m_1} = \frac{Y-b}{m_2} = \frac{Z-c}{m_3}\ .$$

(b) Show that line

$$\frac{X}{m_1} = \frac{Y}{m_2} = \frac{Z+f}{m_3}\ ,$$

which passes through the viewpoint (0, 0, $-f$) and extends in the direction $\boldsymbol{m} = (m_1, m_2, m_3)$, is transformed by the mapping of (10.2.6) into the line

$$x = f\frac{m_1}{m_3}, \qquad y = f\frac{m_2}{m_3},$$

which is parallel to the z-axis.

10.3 Show that the reduced depth z_i of vertex V_i is given by (10.2.8).

10.4 Show that the inverse relationship of (10.2.9) is given by (10.2.10).

10.5 Show that if the reduced depth z_i, $i = 1, \ldots, n$, and the reduced surface parameters P_α, Q_α, R_α, $\alpha = 1, \ldots, m$, satisfy the constraints (10.2.12), so do z'_i, $i = 1, \ldots, n$, P'_α, Q'_α, R'_α, $\alpha = 1, \ldots, m$, given by (10.2.13, 14).

10.6 (a) Show that the equation of a plane in space that passes through three points (X_1, Y_1, Z_1), (X_2, Y_2, Z_2), (X_3, Y_3, Z_3) is given by

$$\begin{vmatrix} X & Y & Z & 1 \\ X_1 & Y_1 & Z_1 & 1 \\ X_2 & Y_2 & Z_2 & 1 \\ X_3 & Y_3 & Z_3 & 1 \end{vmatrix} = 0 .$$

Hint: Show that this is a *linear* equation in X, Y, Z and three points (X_1, Y_1, Z_1) (X_2, Y_2, Z_2), (X_3, Y_3, Z_3) satisfy this equation.

(b) Show that four points (X_1, Y_1, Z_1), (X_2, Y_2, Z_2), (X_3, Y_3, Z_3), (X_4, Y_4, Z_4) in space are coplanar if and only if

$$\begin{vmatrix} X_1 & Y_1 & Z_1 & 1 \\ X_2 & Y_2 & Z_2 & 1 \\ X_3 & Y_3 & Z_3 & 1 \\ X_4 & Y_4 & Z_4 & 1 \end{vmatrix} = 0 .$$

Hint: Use the result of (a), or show that the vectors indicating points (X_1, Y_1, Z_1), (X_2, Y_2, Z_2), (X_3, Y_3, Z_3) relative to point (X_4, Y_4, Z_4) are linearly dependent.

10.7 Give an example of a singular polyhedron other than the one shown in Fig. 10.5.

10.8* Prove Sugihara's theorem that the incidence structure of a polyhedron is nonsingular if and only if inequality (10.2.18) holds for any subset $\mathscr{F}' \subset \mathscr{F} = \{F_1, \ldots, F_m\}$ such that $|\mathscr{F}'| \geqq 2$.

10.9 (a) Show that (10.2.20) is necessary for (10.2.19) to hold.

(b) Show that (10.2.21) satisfies (10.2.19) if the integration does not depend on the path C connecting points (x_0, y_0) and (x, y) on the image plane.

(c) Prove Green's theorem (10.2.22).

10.10 (a) Prove that (10.2.25) holds if the differential equation (10.2.24) is integrable. *Hint*: Let $F(X, Y, Z) = \text{const.}$ be the solution. Then, we must have $A = \lambda\partial F/\partial X$, $B = \lambda\partial F/\partial Y$, $C = \lambda\partial F/\partial Z$, where λ is a common factor that depends on X, Y, Z.

(b)* Prove that the differential equation (10.2.24) is integrable if (10.2.25) holds. *Hint*: Recall that the two-dimensional differential equation $a(x, y)dx + b(x, y)dy = 0$ is always integrable. Hence, we can always find a function $G(X, Y, Z)$ such that $A = \lambda\partial G/\partial X$, $B = \lambda\partial G/\partial Y$, where λ is a common factor that depends on X, Y, Z. From (10.2.25), we can find a function $H(G, Z)$ such that $C = \lambda(\partial G/\partial Z + H)$. Then, the solution of (10.2.24) is given by $F(G(X, Y, Z), Z) = \text{const.}$, where $F(G, Z) = \text{const.}$ is the solution of the two-dimensional differential equation $dG + H(G, Z)dZ = 0$.

(c) Show that the integrability condition of (10.2.23) is given by (10.2.26).

10.11 Show that (10.3.1) is rewritten as (10.3.2) in terms of the reduced surface parameters.

10.12 (a) Derive (10.3.6–10), assuming that all faces are given their surface gradient estimates.

(b) Show that (3.7–9) must be replaced by (10.3.12) if face F_α is not given its surface gradient estimate.

10.13 Show that (10.3.18) is rewritten as (10.3.19) in terms of the reduced surface parameters $P_\alpha, Q_\alpha, R_\alpha$.

10.14 (a) Derive (10.3.23–27), assuming that all the edges are given their 3D orientation estimates.

(b) Show that (10.3.24–26) must be replaced by (10.3.12) if no boundary edges of face F_α are given their 3D orientation estimates.

10.15* Construct a clustering technique in the three-dimensional space. How can it be modified if all data are unit vectors (i.e., points on a unit sphere)?

10.16 Show that a corner cannot be rectangular if its projection image satisfies, in the canonical position, one of the three conditions (i–iii) of Sect. 10.5.

10.17 (a)* Prove the following equations of spherical trigonometry. Here, ABC is a right spherical triangle ($C = \pi/2$) drawn on a sphere of unit radius, and a, b, c are respectively the arc lengths of the sides opposite to corners A, B, C.

$$\sin A = \frac{\sin a}{\sin c}, \qquad \cos A = \frac{\tan b}{\tan c}, \qquad \tan A = \frac{\tan a}{\tan b}.$$

(b) Derive (10.6.3) by differentiating (10.6.2).

(c) Prove (10.6.1).

10.18 Show that inequalities (10.6.14) are equivalent to inequalities (10.6.15).

10.19* Construct an algorithm for the concurrency test described in Sect. 10.6 in such a way that all computation is done in a finite domain.

10.20* Construct an algorithm for the collinearity test described in Sect. 10.6 in such a way that all computation is done in a finite domain.

10.21 (a) Show that t and t' that satisfy (10.6.17) are given by (10.6.18).
(b) Show that testing the signs of expressions (10.6.19) reduces to testing the signs of expressions (10.6.21).
(c)* Construct an algorithm for the parallelogram test in such a way that all computation is done in the finite domain.

Appendix. Fundamentals of Group Theory

In the following, we explain some fundamental facts about group theory, which form a theoretical background for all the chapters of this book. As the first goal, we attempt to derive the theorem known as *Schur's lemma*, which plays a central role in group representation theory. The treatment is almost self-contained, and no specific knowledge is required except for basic matrix algebra—determinants and eigenvalues, in particular. Then, we give precise definitions of basic concepts related to topology, manifolds, Lie groups, Lie algebras, and spherical harmonics.

A.1 Sets, Mappings, and Transformations

A collection of objects is called a *set* if the condition of membership is clearly defined. (This definition is somewhat ambiguous, and causes some subtleties in set theory.) We write $S=\{x|\ldots\}$ to specify a set, where ... states the conditions imposed on x. A member x of set S is also called an *element*[1] of S, and we write $x\in S$. If x is not a member of set S, we write $x\notin S$. Fundamental sets in mathematics include the set of natural numbers $\mathbb{N}$, the set of integers $\mathbb{Z}$, the set of rational numbers $\mathbb{Q}$, the set of real numbers $\mathbb{R}$, and the set of complex numbers $\mathbb{C}$.

We say that two sets S, S' are *equal* (and write $S = S'$) if $x\in S$ implies $x\in S'$ and if $x\in S'$ also implies $x\in S$. Otherwise, $S \neq S'$. We say that a set S is a *subset* of set S' (or the set S' is a *superset* of S) and write $S\subset S'$ (or $S'\supset S$), if $x\in S$ implies $x\in S'$.[2] If $S\subset S'$ and $S\neq S'$, the set S is said to be a *proper subset* of S'. The set consisting of no elements is called the *empty* (or *null*) *set* and denoted by $\varnothing$. The empty set $\varnothing$ is a subset of any set. A familiar example of set inclusion is $\varnothing\subset\mathbb{N}\subset\mathbb{Q}\subset\mathbb{R}\subset\mathbb{C}$.

The *union* $S\cup S'$, the *intersection* $S\cap S'$, and the *difference* $S-S'$ of two sets S, S' are defined as follows:

$$S\cup S'=\{x|\quad x\in S\quad \text{or}\quad x\in S'\}\ , \tag{A.1.1}$$

$$S\cap S'=\{x|\quad x\in S\quad \text{and}\quad x\in S'\}\ , \tag{A.1.2}$$

$$S-S'=\{x|\quad x\in S\quad \text{but}\quad x\notin S'\}\ . \tag{A.1.3}$$

---

[1] We also say that element x *belongs* to set S or is *contained* in S.

[2] We also say that set S is *included* in S'. In some books, $S\subset S'$ and $S'\supset S$ are respectively denoted by $S\subseteq S'$ and $S'\supseteq S$. In such books, $S\subset S'$ or $S'\supset S$ implies that S is a *proper* subset of S'.

If $S \cap S' = \emptyset$, two sets S, S' are said to be *disjoint*. The *Cartesian product* $S \times S'$ of two sets S, S' is the set of all pairs consisting of elements of S and S':[3]

$$S \times S' = \{(x, y) | \quad x \in S, \quad y \in S'\} \ . \tag{A.1.4}$$

The set of all subsets of set S is called the *power set* of S.

Consider a fixed set S. We call its members *points*. They may be infinite in number. A *mapping* T of the set S to itself is a rule by which each $x \in S$ is associated with another $x' \in S$. (Mappings can be defined more generally from one set S, called the *domain*, to another set S', called the *range*. In the following, we consider only mappings for which the domain S and the range S' are the same.) The point x' is called the *image* of x under the mapping T, and we write

$$x' = Tx \ . \tag{A.1.5}$$

The set of all images of elements of set S under mapping T is called the *image* of set S under the mapping T, and we write

$$TS = \{x' | \quad x' = Tx, \quad x \in S\} \ . \tag{A.1.6}$$

Two mappings T, T' of set S are identified (written as $T = T'$) if and only if $Tx = T'x$ for all $x \in S$.

We denote by I the mapping that maps each $x \in S$ onto itself:

$$x = Ix \ . \tag{A.1.7}$$

This mapping is called the *identity*.

Let T, T' be mappings. Successive application

$$x' = T'(Tx) \tag{A.1.8}$$

of T and T' is also a mapping. This mapping is called the *composition* (or *product*) of the two mappings and denoted by $T' \circ T$:

$$x' = (T' \circ T)x \ . \tag{A.1.9}$$

Let T, T', T'' be mappings. Apply these successively:

$$x' = Tx \ , \qquad x'' = T'x' \ , \qquad x''' = T''x'' \ . \tag{A.1.10}$$

Since point x'' is the image of point x under the mapping $T' \circ T$, point x''' is the image of point x under the mapping $T'' \circ (T' \circ T)$. At the same time, point x''' is the image of point x' under the mapping $T'' \circ T'$ and hence the image of point x under the mapping $(T'' \circ T') \circ T$. Thus,

$$T'' \circ (T' \circ T) = (T'' \circ T') \circ T \ . \tag{A.1.11}$$

This rule is called the *associative law* of composition. The mapping of (A.1.11) is written simply as $T'' \circ T' \circ T$.

---

[3] A subset of $S \times S'$ is called a *(binary) relation* over the sets S, S'. The Cartesian product of multiple sets $S_1, \ldots, S_n$ is similarly defined: $S_1 \times \ldots \times S_n = \{(x_1, \ldots, x_n) | x_k \in S_k, k = 1, \ldots, n\}$. A subset of $S_1 \times \ldots \times S_n$ is also called an *(n-ary) relation* over $S_1, \ldots, S_n$.

Let T and T' be mappings. If application of these two does not depend on the order of application, i.e., if

$$T(T'x) = T'(Tx) \tag{A.1.12}$$

for all $x \in S$, or

$$T \circ T' = T' \circ T \ , \tag{A.1.13}$$

we say that mappings T and T' *commute* with each other.

The set of elements that are mapped onto x' by mapping T is called the *inverse image* of x' under the mapping T and denoted by $T^{-1}(x')$:

$$T^{-1}(x') = \{x | \quad Tx = x'\} \ . \tag{A.1.14}$$

A mapping T is *one-to-one* if the inverse image $T^{-1}(x')$ consists of a single element for all $x' \in S$, i.e., no two points have the same image:

$$x \neq x' \rightarrow Tx \neq Tx' \ . \tag{A.1.15}$$

A mapping T is *onto* if every $x' \in S$ is the image of some $x \in S$, i.e., $TS = S$. A mapping T is called a *transformation* of set S if it is both one-to-one and onto. (A one-to-one mapping is called an *injection*, while an onto mapping is called a *surjection*. A one-to-one and onto mapping is also called a *bijection*.)

Let T be a transformation of set S. Since for each $x' \in S$ there exists one (and only one) $x \in S$ such that $x' = Tx$, we can define a transformation that maps x' onto x. This transformation is called the *inverse* of T and denoted by T^{-1}:

$$x' = Tx \rightarrow x = T^{-1}x' \ . \tag{A.1.16}$$

By definition, $x = T^{-1}(Tx)$ and $x' = T(T^{-1}x')$, i.e.,

$$T^{-1} \circ T = T \circ T^{-1} = I \ . \tag{A.1.17}$$

Let T, T' be transformations of set S. Their composition $T' \circ T$ is also a transformation of set S: if $x' = Tx$ and $x'' = T'x'$, then $x'' = (T' \circ T)x$. On the other hand, from $x = T^{-1}x'$ and $x' = T'^{-1}x''$, we find that $x = (T^{-1} \circ T'^{-1})x''$. Hence, we observe that

$$(T' \circ T)^{-1} = T^{-1} \circ T'^{-1}. \tag{A.1.18}$$

Consider n-dimensional space $\mathbb{R}^n$. A *linear mapping* of $\mathbb{R}^n$ is a mapping that maps a point with coordinates $(x_1, \ldots, x_n)$ onto a point with coordinates $(x'_1, \ldots, x'_n)$ in the form

$$x'_i = \sum_{j=1}^{n} a_{ij} x_j \ , \qquad i = 1, \ldots, n \ . \tag{A.1.19}$$

Here, a_{ij}, $i, j = 1, \ldots, n$, are real coefficients. If these coefficients are arranged as an n-dimensional *matrix* with n rows and n columns, (A.1.19) is written as

$$\begin{pmatrix} x'_1 \\ \vdots \\ x'_n \end{pmatrix} = \begin{pmatrix} a_{11} & \dots & a_{1n} \\ \vdots & & \vdots \\ a_{n1} & \dots & a_{nn} \end{pmatrix} \begin{pmatrix} x_1 \\ \vdots \\ x_n \end{pmatrix} , \tag{A.1.20}$$

or simply

$$\boldsymbol{x}' = \boldsymbol{A}\boldsymbol{x} , \tag{A.1.21}$$

where

$$\boldsymbol{x} = \begin{pmatrix} x_1 \\ \vdots \\ x_n \end{pmatrix} , \quad \boldsymbol{x}' = \begin{pmatrix} x'_1 \\ \vdots \\ x'_n \end{pmatrix} , \quad \boldsymbol{A} = \begin{pmatrix} a_{11} & \dots & a_{1n} \\ \vdots & & \vdots \\ a_{n1} & \dots & a_{nn} \end{pmatrix} . \tag{A.1.22}$$

The linear mapping (A.1.21) is a transformation if and only if the *determinant* $|\boldsymbol{A}|$ of matrix $\boldsymbol{A}$ is not zero. A matrix whose determinant is not zero is said to be *nonsingular* (or *regular*). Linear transformations of n-dimensional space are called *n-dimensional linear transformations.* If $\boldsymbol{A}$ is a nonsingular matrix, the inverse of the linear transformation (A.1.21) is also a linear transformation. Its coefficient matrix is called the *inverse* of $\boldsymbol{A}$ and denoted by $\boldsymbol{A}^{-1}$:

$$\boldsymbol{x} = \boldsymbol{A}^{-1}\boldsymbol{x}' . \tag{A.1.23}$$

Let $\boldsymbol{B} = (b_{ij})$, $i, j = 1, \dots, n$, be another n-dimensional matrix. Suppose the corresponding linear mapping is applied to $\boldsymbol{x}'$ following the linear mapping (A.1.19):

$$x''_i = \sum_{j=1}^{n} b_{ij}x'_j , \quad i = 1, \dots, n , \tag{A.1.24}$$

or

$$\boldsymbol{x}'' = \boldsymbol{B}\boldsymbol{x}' . \tag{A.1.25}$$

The product of these two mappings is given by

$$x''_i = \sum_{j=1}^{n} b_{ij}\left(\sum_{k=1}^{n} a_{jk}x_k \right) = \sum_{k=1}^{n} \left(\sum_{j=1}^{n} b_{ij}a_{jk} \right) x_k . \tag{A.1.26}$$

Hence, the product is again a linear mapping defined by

$$x''_i = \sum_{k=1}^{n} c_{ik}x_k , \quad i = 1, \dots, n , \tag{A.1.27}$$

or

$$\boldsymbol{x}'' = \boldsymbol{C}\boldsymbol{x} , \tag{A.1.28}$$

where

$$c_{ik} = \sum_{j=1}^{n} b_{ij}a_{jk} , \quad i = 1, \dots, n . \tag{A.1.29}$$

This equation defines the *product*

$$C = BA \tag{A.1.30}$$

or matrices A and B.

A.2 Groups

A *group* G is a set for which a rule, called *multiplication* (or *composition*), is defined in such a way that any two elements $a, b \in G$ are associated with an element of G denoted by ab, which is called the *product* of a and b. Also, the following conditions are required:

(1) Multiplication is *associative*, i.e.,

$$a(bc) = (ab)c \tag{A.2.1}$$

for all $a, b, c \in G$.
(2) The set G contains a unique element e called the *unit* (or *identity*) *element* such that[4]

$$ae = ea = a \tag{A.2.2}$$

for every element $a \in G$.
(3) For every element $a \in G$, there exists a unique element $b \in G$ such that[5]

$$ab = ba = e \ . \tag{A.2.3}$$

This element b is called the *inverse* of a and denoted by a^{-1}.

Two elements $a, b \in G$ are said to *commute* with each other if their product is independent of their order:

$$ab = ba \ . \tag{A.2.4}$$

By definition, the unit element e commutes with all the elements of G. A group is said to be *Abelian* (or *commutative*) if all of its elements commute with each other. For Abelian groups, "multiplication" is often called *addition* and denoted by $+$; the "product" $a + b$ is called the *sum* of a and b. Furthermore, the "unit element" is denoted by 0 and called *zero*, while the "inverse" of a is denoted by $-a$ and called the *negative* of a.

Let g_0 be an element of a group G. If we write $g_0G = \{g_0g \mid g \in G\}$, it is easy to confirm that $g_0G = G$, i.e., g_0G and G are exactly the same set, and element g_0

---

[4] Actually, either $ae = a$ or $ea = a$ is sufficient. If $ae = a$ for every $a \in G$, element e is called the *left unit element*, while it is called the *right unit element* if $ea = a$ for every $a \in G$. It is easy to prove that, for a group, the existence of one implies the existence of the other and they are identical.

[5] Actually, either $ab = e$ or $ba = e$ is sufficient. If $ba = e$, element b is called the *left inverse* of a, while it is called the *right inverse* of a if $ab = e$. It is easy to prove that, for a group, the existence of one implies the existence of the other and they are identical.

acts on G as a transformation. A group G is said to be *finite* if it consists of a finite number of elements—the number of elements is called the *order* of G, and denoted by $|G|$. For a finite group G, an element $g_0 \in G$ acts on G as a permutation.

Let G be a group. A subset $H \subset G$ is called a *subgroup* of G if H is itself a group under the multiplication rule of G. (It is easy to prove that a subset H of a group is a subgroup if and only if $ab^{-1} \in H$ for all $a, b \in H$.) The set $\{e\}$ consisting of the unit element e alone is a trivial subgroup. Since G itself is a subset of G, the group G is also a subgroup of G. Subgroups other than $\{e\}$ and G are called *proper subgroups.*

Let $H_1, \ldots, H_r$ be subgroups of G. The group G is said to be the *direct product* of subgroups $H_1, \ldots, H_r$ if the following two conditions are satisfied:

(1) The elements of different subgroups commute with each other:

$$h_i h_j = h_j h_i \ , \quad h_i \in H_i \ , \quad h_j \in H_j \ , \quad i \neq j \ . \tag{A.2.5}$$

(2) Every element $g \in G$ is expressed in one and only one way as

$$g = h_1 \ldots h_r \ , \quad h_i \in H_i \ , \quad i = 1, \ldots, r \ . \tag{A.2.6}$$

If so, we write symbolically[6]

$$G = H_1 \otimes \ldots \otimes H_r \ . \tag{A.2.7}$$

The subgroups $H_1, \ldots, H_r$ are called the *direct factors*. It follows from this definition that direct factors $H_1, \ldots, H_r$ have no elements in common except the unit element. (They are all *invariant* (or *normal*) *subgroups of* G, but we skip the details.)

Let $H_1, \ldots, H_r$ be groups. We regard them as distinct groups, although some of them may actually be the same. Define a group G consisting of r-tuples of the form

$$(h_1, \ldots, h_r) \ , \quad h_i \in H_i \ , \quad i = 1, \ldots, r \ . \tag{A.2.8}$$

The multiplication of two r-tuples $(h_1, \ldots, h_r)$, $(h'_1, \ldots, h'_r)$ is defined by

$$(h'_1, \ldots, h'_r)(h_1, \ldots, h_r) = (h'_1 h_1, \ldots, h'_r h_r) \ . \tag{A.2.9}$$

The unit element of G is $(e_1, \ldots, e_r)$, where e_k is the unit element of H_k, $k = 1, \ldots, r$. The inverse of $(h_1, \ldots, h_r)$ is $(h_1^{-1}, \ldots, h_r^{-1})$, where each h_i^{-1} is the inverse of h_i under the multiplication of group H_i.

For each H_i, let us define a group $\tilde{H}_i$ consisting of r-tuples of the form

$$(e_1, \ldots, e_{i-1}, h_i, e_{i+1}, \ldots, e_r) \ , \qquad h_i \in H_i \ . \tag{A.2.10}$$

The unit element of $\tilde{H}_i$ is $(e_1, \ldots, e_r)$, and the multiplication rule is the same as in G. All $\tilde{H}_1, \ldots, \tilde{H}_r$ are subgroups of G. It is easy to confirm that conditions

[6] Some authors write $G = H_1 \times \ldots \times H_r$.

(1), (2) are satisfied. Namely, two elements $(e_1, \ldots, h_i, \ldots, e_r) \in H_i$ and $(e_1, \ldots, h_j, \ldots, e_r) \in H_j$ commute with each other if $i \neq j$, and every element $(h_1, \ldots, h_r)$ is expressed in one and only one way as $(h_1, e_2, \ldots, e_r) \ldots (e_1, \ldots, e_{r-1}, h_r)$. Hence, G is the direct product of $\tilde{H}_1, \ldots, \tilde{H}_r$. However, we often identify $\tilde{H}_i$ with H_i and write[6]

$$G = H_1 \otimes \ldots \otimes H_r \ . \tag{A.2.11}$$

The group G is also called the *direct product* of groups $H_1, \ldots, H_r$.

If $H_1, \ldots, H_r$ are all Abelian groups, the direct product G is also an Abelian group and called the *direct sum*. Instead of (A.2.11), we write[7]

$$G = H_1 \oplus \ldots \oplus H_r \ . \tag{A.2.12}$$

Let G and G' be groups. A mapping T from group G to group G' is called a *homomorphism* if

$$T(ab) = (Ta)(Tb) \tag{A.2.13}$$

for all $a, b \in G$. The right-hand side is the product under the multiplication of group G'. From this definition, it is easy to prove that the unit element $e \in G$ is mapped onto the unit element $e' \in G'$,

$$e' = Te \ , \tag{A.2.14}$$

and the inverse of $a \in G$ is mapped onto the inverse of $Ta \in G'$,

$$(Ta)^{-1} = T(a^{-1}) \ . \tag{A.2.15}$$

A homomorphism is called an *isomorphism* if it is onto and one-to-one. If there exists an isomorphism T from group G to group G', its inverse T^{-1} is also an isomorphism from group G' to group G. Then, we say that the two groups are *isomorphic*, and write

$$G \cong G' \ . \tag{A.2.16}$$

If two groups are isomorphic, they can be viewed as essentially the same group with different notational appearances. For example, the group $\tilde{H}_i$ consisting of elements of the form of (A.2.10) is isomorphic to group H_i. In general, properties of a group that are preserved by isomorphisms are regarded as describing the "structure" of the group. (We omit the details, but we can define such invariant notions as *cosets*, *conjugate classes*, *invariant* (or *normal*) *subgroups*, and *quotient* (or *factor*) *groups* to describe the structures of groups.)

A set G is said to be a *group of transformations* over a set S if it is a set of transformations of S and is a group under composition. In other words, a set G of transformations is a group if the following three conditions are satisfied:[8]

---

[7] Some authors write $G = H_1 \dotplus \ldots \dotplus H_r$ or simply $G = H_1 + \ldots + H_r$.

[8] The associative law need not be included because it always holds for the composition of mappings, see (A.1.10).

(1) If T, $T' \in G$, then $T \circ T' \in G$.
(2) The set G contains the identity I.
(3) For every $T \in G$, the set G contains its inverse transformation T^{-1}.

Let $\boldsymbol{A}$, $\boldsymbol{A}'$ be n-dimensional nonsingular matrices. Since

$$|\boldsymbol{A}'\boldsymbol{A}| = |\boldsymbol{A}'| \cdot |\boldsymbol{A}| \; , \tag{A.2.17}$$

the product $\boldsymbol{A}'\boldsymbol{A}$ is also an n-dimensional nonsingular matrix. It is easy to confirm that the set of linear transformations defined by all n-dimensional nonsingular matrices is a group of transformations. This is not an Abelian group because $\boldsymbol{A}\boldsymbol{A}' = \boldsymbol{A}'\boldsymbol{A}$ does not necessarily hold. This group is called the *n-dimensional general linear group* and denoted by $GL(n, \mathbb{R})$. [Similarly, if we identify complex matrices with linear transformations of a complex linear space, we can also define the general linear group $GL(n, \mathbb{C})$.]

An n-dimensional matrix A is said to be *orthogonal* if

$$\boldsymbol{A}^{\mathrm{T}} = \boldsymbol{A}^{-1} \quad \text{or} \quad \boldsymbol{A}^{\mathrm{T}}\boldsymbol{A} = \boldsymbol{A}\boldsymbol{A}^{\mathrm{T}} = \boldsymbol{I} \; , \tag{A.2.18}$$

where ^{T} designates the transpose of the matrix. If $\boldsymbol{A}$ and $\boldsymbol{A}'$ are n-dimensional orthogonal matrices, we observe that

$$(\boldsymbol{A}'\boldsymbol{A})^{\mathrm{T}}(\boldsymbol{A}'\boldsymbol{A}) = (\boldsymbol{A}'\boldsymbol{A})(\boldsymbol{A}'\boldsymbol{A})^{\mathrm{T}} = \boldsymbol{I} \; . \tag{A.2.19}$$

Hence, $\boldsymbol{A}'\boldsymbol{A}$ is also an n-dimensional orthogonal matrix. It is easy to confirm that the set of linear transformations defined by all n-dimensional orthogonal matrices is also a group of transformations. This is a proper subgroup of $GL(n, \mathbb{R})$. This group is called the *orthogonal group* and denoted by $O(n)$.

If $\boldsymbol{A}$ is an orthogonal matrix, we see from (A.2.17, 18) that

$$|\boldsymbol{A}| \cdot |\boldsymbol{A}^{\mathrm{T}}| = |\boldsymbol{A}\boldsymbol{A}^{\mathrm{T}}| = |\boldsymbol{I}| = 1 \; . \tag{A.2.20}$$

Since $|\boldsymbol{A}^{\mathrm{T}}| = |\boldsymbol{A}|$, we can see that

$$|\boldsymbol{A}| = \pm 1 \; . \tag{A.2.21}$$

We can also see from (A.2.17) that the set of linear transformations defined by all n-dimensional orthogonal matrices that are *unimodular* (i.e. of determinant 1) is a proper subgroup of $O(n)$. This group is called the *special orthogonal group* and denoted by $SO(n)$ (the term "special" means that the determinant is 1).

A.3 Linear Spaces

As defined earlier, a set V is an *Abelian group* if *addition* of any two elements $\boldsymbol{x}, \boldsymbol{y} \in V$ is defined, i.e., their *sum* $\boldsymbol{x} + \boldsymbol{y} \in V$ exists,

$$\boldsymbol{x}, \boldsymbol{y} \in V \rightarrow \boldsymbol{x} + \boldsymbol{y} \in V \; , \tag{A.3.1}$$

and if the following four conditions are satisfied:

(1) The addition is *associative*, i.e., for all $\boldsymbol{x}, \boldsymbol{y} \in V$

$$\boldsymbol{x} + (\boldsymbol{y} + \boldsymbol{z}) = (\boldsymbol{x} + \boldsymbol{y}) + \boldsymbol{z} . \tag{A.3.2}$$

(2) The addition is *commutative*, i.e., for all $\boldsymbol{x}, \boldsymbol{y} \in V$

$$\boldsymbol{x} + \boldsymbol{y} = \boldsymbol{y} + \boldsymbol{x} . \tag{A.3.3}$$

(3) There exists an element $\mathbf{0} \in V$, called *zero*, such that for all $\boldsymbol{x} \in V$

$$\boldsymbol{x} + \mathbf{0} = \mathbf{0} + \boldsymbol{x} = \boldsymbol{x} . \tag{A.3.4}$$

(4) Each element $\boldsymbol{x} \in V$ has its *negative* $-\boldsymbol{x} \in V$ such that

$$\boldsymbol{x} + (-\boldsymbol{x}) = (-\boldsymbol{x}) + \boldsymbol{x} = \mathbf{0} . \tag{A.3.5}$$

We write the sum $\boldsymbol{x} + (-\boldsymbol{y})$ simply as $\boldsymbol{x} - \boldsymbol{y}$ and call it the *difference* of $\boldsymbol{x}$ and $\boldsymbol{y}$.

An Abelian group V is a *real linear* (or *vector*) *space* if each element $\boldsymbol{x} \in V$ can be "multiplied" by a real number $c \in \mathbb{R}$, i.e.,

$$\boldsymbol{x} \in V , \quad c \in \mathbb{R} \rightarrow c\boldsymbol{x} \in V , \tag{A.3.6}$$

and if the following conditions are satisfied for all $\boldsymbol{x}, \boldsymbol{y} \in V$ and all $c, c' \in \mathbb{R}$:

$$(c + c')\boldsymbol{x} = c\boldsymbol{x} + c'\boldsymbol{x} , \tag{A.3.7}$$

$$(cc')\boldsymbol{x} = c(c'\boldsymbol{x}) , \tag{A.3.8}$$

$$c(\boldsymbol{x} + \boldsymbol{y}) = c\boldsymbol{x} + c\boldsymbol{y} , \tag{A.3.9}$$

$$1\boldsymbol{x} = \boldsymbol{x} . \tag{A.3.10}$$

Elements of a linear space are called *vectors*, and the numbers to be multiplied are called *scalars*. If the scalars are complex numbers, V is called a *complex linear* (or *vector*) *space*.

Let $\boldsymbol{x}_1, \ldots, \boldsymbol{x}_r \in V$ be vectors, and $c_1, \ldots, c_r$ be scalars. The vector

$$\boldsymbol{x} = c_1\boldsymbol{x}_1 + \ldots + c_r\boldsymbol{x}_r \tag{A.3.11}$$

is called the *linear combination* of $\boldsymbol{x}_1, \ldots, \boldsymbol{x}_r$ with coefficients $c_1, \ldots, c_r$. Vectors are said to be *linearly independent* if we cannot construct $\mathbf{0}$ from any linear combination of them unless all the coefficients are zero. In other words, vectors $\boldsymbol{x}_1, \ldots, \boldsymbol{x}_r$ are linearly independent if

$$c_1\boldsymbol{x}_1 + \ldots + c_r\boldsymbol{x}_r = \mathbf{0} \tag{A.3.12}$$

implies $c_1 = \ldots = c_r = 0$. Otherwise, vectors $\boldsymbol{x}_1, \ldots, \boldsymbol{x}_r$ are said to be *linearly dependent*.

The maximum number of linearly independent vectors is called the *dimension* of the linear space V and denoted by $\dim V$ (it can be infinity). Let $\boldsymbol{e}_1, \ldots, \boldsymbol{e}_n$ be linearly independent vectors of an n-dimensional linear space V. We can see that *any vector* $\boldsymbol{x} \in V$ *can be expressed as a linear combination of these vectors.* Indeed, since $n+1$ vectors $\boldsymbol{x}, \boldsymbol{e}_1, \ldots, \boldsymbol{e}_n$ cannot be linearly independent by definition, there must exist a linear combination

$$c_0\boldsymbol{x} + c_1\boldsymbol{e}_1 + \ldots + c_n\boldsymbol{e}_n = \boldsymbol{0} \tag{A.3.13}$$

such that not all $c_0, c_1, \ldots, c_n$ are zero. But c_0 cannot be zero, because if so, vectors $\boldsymbol{e}_1, \ldots, \boldsymbol{e}_n$ would become linearly dependent. Thus, we obtain

$$\boldsymbol{x} = x_1\boldsymbol{e}_1 + \ldots + x_n\boldsymbol{e}_n , \qquad x_i = -\frac{c_i}{c_0} , \qquad i = 1, \ldots, n . \tag{A.3.14}$$

Moreover, these coefficients $x_1, \ldots, x_n$ are *unique*. Indeed, if we have another expression

$$\boldsymbol{x} = x'_1\boldsymbol{e}_1 + \ldots + x'_n\boldsymbol{e}_n , \tag{A.3.15}$$

subtraction of (A.3.15) from (A.3.14) yields

$$(x_1 - x'_1)\boldsymbol{e}_1 + \ldots + (x_n - x'_n)\boldsymbol{e}_n = \boldsymbol{0} . \tag{A.3.16}$$

Since vectors $\boldsymbol{e}_1, \ldots, \boldsymbol{e}_n$ are linearly independent, all the coefficients must be zero, and hence $x_i = x'_i$ for $i = 1, \ldots, n$.

A set $\{\boldsymbol{e}_1, \ldots, \boldsymbol{e}_n\}$ of vectors is a *basis* of V if any vector $\boldsymbol{x} \in V$ can be expressed as a linear combination of these vectors uniquely. As shown above, any n linearly independent vectors of an n-dimensional linear space V can serve as its basis.[9]

Thus, if we fix a basis $\{\boldsymbol{e}_1, \ldots, \boldsymbol{e}_n\}$, every vector $\boldsymbol{x} \in V$ is uniquely represented by the coefficients $x_1, \ldots, x_n$ of the linear combination

$$\boldsymbol{x} = x_1\boldsymbol{e}_1 + \ldots + x_n\boldsymbol{e}_n . \tag{A.3.17}$$

Hence, we can identify the vector $\boldsymbol{x} \in V$ with the column

$$\begin{pmatrix} x_1 \\ \vdots \\ x_n \end{pmatrix} . \tag{A.3.18}$$

The (real or complex) numbers $x_1, \ldots, x_n$ are called the *coordinates* (or *components*) of $\boldsymbol{x} \in V$ with respect to basis $\{\boldsymbol{e}_1, \ldots, \boldsymbol{e}_n\}$. It follows that addition and scalar multiplication are performed by addition and scalar multiplication of each coordinate. Under component-wise addition and scalar multiplication, the

---

[9] But m vectors cannot be a basis if $m < n$. This follows if we show that for any linearly independent vectors $\boldsymbol{x}_1, \ldots, \boldsymbol{x}_m \in V$ for $m < n$, we can always choose vectors $\boldsymbol{x}_{m+1}, \ldots, \boldsymbol{x}_n \in V$ such that $\boldsymbol{x}_1, \ldots, \boldsymbol{x}_n$ are linearly independent. This fact is proved, in turn, by showing that (1) if $\{\boldsymbol{x}_1, \ldots, \boldsymbol{x}_r\}$ is a set of linearly independent vectors, any subset of it is also a linearly independent set, and (2) if $\{\boldsymbol{x}_1, \ldots, \boldsymbol{x}_r\}$ and $\{\boldsymbol{y}_1, \ldots, \boldsymbol{y}_{r+1}\}$ are both linearly independent sets, we can choose some $\boldsymbol{y}_i$ such that $\{\boldsymbol{x}_1, \ldots, \boldsymbol{x}_r, \boldsymbol{y}_i\}$ is also a linearly independent set. These two are the essence of the notion of "independence". An algebraic structure called a *matroid* is defined by required these as axioms.

set of all n-dimensional columns becomes an n-dimensional linear space (denoted by $\mathbb{R}^n$ or $\mathbb{C}^n$, depending on whether V is a real or a complex linear space).

Two linear spaces V and V' are *isomorphic* if there exists between them an isomorphism as Abelian groups such that if $\boldsymbol{x} \in V$ is mapped onto $\boldsymbol{x}' \in V'$, then $c\boldsymbol{x}$ is also mapped onto $c\boldsymbol{x}'$ for all scalars c. Real linear spaces of the same finite dimension n are isomorphic to each other, and they are all isomorphic to $\mathbb{R}^n$. Similarly, all n-dimensional complex linear spaces are isomorphic to $\mathbb{C}^n$.

Let $\{\boldsymbol{e}'_1, \ldots, \boldsymbol{e}'_n\}$ be another basis of V. Since this is also a basis, each original basis vector $\boldsymbol{e}_j$, $j = 1, \ldots, n$, must be expressed as a linear combination of the new basis vectors:

$$\boldsymbol{e}_j = \sum_{i=1}^{n} p_{ij}\boldsymbol{e}'_i \,, \qquad j = 1, \ldots, n \,. \tag{A.3.19}$$

Similarly, each new basis vector $\boldsymbol{e}'_i$, $i = 1, \ldots, n$, must be expressed as a linear combination of the original basis vectors $\boldsymbol{e}_1, \ldots, \boldsymbol{e}_n$. This means that the n equations (A.3.19) can be inverted, and $\boldsymbol{e}'_1$, $i = 1, \ldots, n$, can be solved in terms of $\boldsymbol{e}_1, \ldots, \boldsymbol{e}_n$. This is possible if and only if the matrix $\boldsymbol{P} = (p_{ij})$, $i, j = 1, \ldots, n$, is nonsingular. Thus, a basis can be replaced by another if and only if the two are related by a nonsingular matrix $\boldsymbol{P} = (p_{ij})$.

Let $(x_1, \ldots, x_n)$ and $(x'_1, \ldots, x'_n)$ be the coordinates of the same vector $\boldsymbol{x} \in V$ with respect to different bases $\{\boldsymbol{e}_1, \ldots, \boldsymbol{e}_n\}$ and $\{\boldsymbol{e}'_1, \ldots, \boldsymbol{e}'_n\}$, i.e.,

$$\boldsymbol{x} = x_1\boldsymbol{e}_1 + \ldots + x_n\boldsymbol{e}_n = x'_1\boldsymbol{e}'_1 + \ldots + x'_n\boldsymbol{e}'_n \,. \tag{A.3.20}$$

If we substitute (A.3.19), we obtain

$$\boldsymbol{x} = \sum_{j=1}^{n} x_j\boldsymbol{e}_j = \sum_{j=1}^{n} x_j\left(\sum_{i=1}^{n} p_{ij}\boldsymbol{e}'_i\right) = \sum_{i=1}^{n}\left(\sum_{j=1}^{n} p_{ij}x_j\right)\boldsymbol{e}'_i \,. \tag{A.3.21}$$

Comparing this with (A.3.20), we see that

$$x'_i = \sum_{j=1}^{n} p_{ij}x_j \,, \qquad i = 1, \ldots, n \,. \tag{A.3.22}$$

This means that the corresponding column is multiplied by matrix $\boldsymbol{P} = (p_{ij})$:

$$\begin{pmatrix} x'_1 \\ \vdots \\ x'_n \end{pmatrix} = \boldsymbol{P}\begin{pmatrix} x_1 \\ \vdots \\ x_n \end{pmatrix} \,. \tag{A.3.23}$$

A.4 Metric Spaces

In this section, we consider complex linear spaces, but all the results also hold for real linear spaces if the asterisk $*$ denoting the complex conjugate is ignored.

An *inner* (or *scalar*) *product* is a scalar $(\boldsymbol{x}, \boldsymbol{y})$ defined for every pair of vectors $\boldsymbol{x}, \boldsymbol{y} \in V$ in such a way that the following conditions are satisfied for all $\boldsymbol{x}, \boldsymbol{y}, \boldsymbol{x}_1, \boldsymbol{x}_2 \in V$ and all scalars c:[10]

$$(\boldsymbol{x}, \boldsymbol{y}) = (\boldsymbol{y}, \boldsymbol{x})^* \ , \tag{A.4.1}$$

$$(c\boldsymbol{x}, \boldsymbol{y}) = c^*(\boldsymbol{x}, \boldsymbol{y}) \ , \qquad (\boldsymbol{x}, c\boldsymbol{y}) = c(\boldsymbol{x}, \boldsymbol{y}) \ , \tag{A.4.2}$$

$$(\boldsymbol{x}_1 + \boldsymbol{x}_2, \boldsymbol{y}) = (\boldsymbol{x}_1, \boldsymbol{y}) + (\boldsymbol{x}_2, \boldsymbol{y}) \ , \qquad (\boldsymbol{x}, \boldsymbol{y}_1 + \boldsymbol{y}_2) = (\boldsymbol{x}, \boldsymbol{y}_1) + (\boldsymbol{x}, \boldsymbol{y}_2) \ , \tag{A.4.3}$$

$$(\boldsymbol{x}, \boldsymbol{x}) \geqq 0 \ . \tag{A.4.4}$$

In addition, the equality in (A.4.4) is required to hold if and only if $\boldsymbol{x} = \boldsymbol{0}$. Two vectors $\boldsymbol{x}, \boldsymbol{y}$ are *orthogonal* if $(\boldsymbol{x}, \boldsymbol{y}) = 0$. A linear space equipped with an inner product as called a *metric (linear) space.*

The *norm* is a scalar $\|x\|$ defined for every $\boldsymbol{x} \in V$ in such a way that the following conditions are satisfied for all $\boldsymbol{x}, \boldsymbol{y} \in V$ and all scalars c:

$$\|\boldsymbol{x}\| \geqq 0 \ , \tag{A.4.5}$$

$$\|c\boldsymbol{x}\| = |c| \, \|\boldsymbol{x}\| \ , \tag{A.4.6}$$

$$\|\boldsymbol{x} + \boldsymbol{y}\| \leqq \|\boldsymbol{x}\| + \|\boldsymbol{y}\| \ . \tag{A.4.7}$$

In addition, the equality in (A.4.5) is required to hold if and only if $\boldsymbol{x} = \boldsymbol{0}$. The inequality (A.4.7) is called the *triangular inequality.* A linear space equipped with a norm is called a *normed (linear) space.* Vectors whose norm is 1 are called *unit vectors.*

If V is a metric space, it becomes a normed space by introducing the following norm:[11]

$$\|\boldsymbol{x}\| \equiv \sqrt{(\boldsymbol{x}, \boldsymbol{x})} \ . \tag{A.4.8}$$

It is easy to confirm that conditions (A.4.5–7) are satisfied. The triangular inequality (A.4.7) is proved from *Schwarz's inequality*[12]

$$|(\boldsymbol{x}, \boldsymbol{y})| \leqq \|\boldsymbol{x}\| \cdot \|\boldsymbol{y}\| \ . \tag{A.4.9}$$

The equality of (A.4.7) holds if and only if x and y are *parallel*, i.e., $\boldsymbol{x} = t\boldsymbol{y}$ for some $t \geqq 0$, or $\boldsymbol{y} = \boldsymbol{0}$.

---

[10] Actually, (A.4.2b, 3b) are redundant; they can be derived from (A.4.2a, 3a, 1). Conditions (A.4.2, 3) can be combined into the single condition $(\boldsymbol{x}, c_1\boldsymbol{y}_1 + c_2\boldsymbol{y}_2) = c_1(\boldsymbol{x}, \boldsymbol{y}_1) + c_2(\boldsymbol{x}, \boldsymbol{y}_2)$ for all $\boldsymbol{x}, \boldsymbol{y}_1, \boldsymbol{y}_2 \in V$ and all scalars c_1, c_2. Equations (A.4.2a, b) are preferred by physicists, while mathematicians prefer $(c\boldsymbol{x}, \boldsymbol{y}) = c(\boldsymbol{x}, \boldsymbol{y})$ and $(\boldsymbol{x}, c\boldsymbol{y}) = c^*(\boldsymbol{x}, \boldsymbol{y})$.

[11] Conversely, if the *parallelogram rule* $\|\boldsymbol{x} + \boldsymbol{y}\|^2 + \|\boldsymbol{x} - \boldsymbol{y}\|^2 = 2\|\boldsymbol{x}\|^2 + 2\|\boldsymbol{y}\|^2$ is added to (A.4.5–7), a normed space V becomes a metric space with inner product $(\boldsymbol{x}, \boldsymbol{y}) = (\|\boldsymbol{x} + \boldsymbol{y}\|^2 - \|\boldsymbol{x} - \boldsymbol{y}\|^2)/4$.

[12] Schwarz's inequality (A.4.9) is proved from $\|\boldsymbol{x}\|^2\|\boldsymbol{y}\|^2 - |(\boldsymbol{x}, \boldsymbol{y})|^2 = \|[\|\boldsymbol{y}\|\boldsymbol{x} - (\boldsymbol{x}, \boldsymbol{y})\boldsymbol{y}/\|\boldsymbol{y}\|]\|^2 \geqq 0$. The equality holds if and only if $\boldsymbol{x} = c\boldsymbol{y}$ for some scalar c, or $\boldsymbol{y} = \boldsymbol{0}$. The triangular inequality is proved from $\|\boldsymbol{x} + \boldsymbol{y}\|^2 = (\boldsymbol{x} + \boldsymbol{y}, \boldsymbol{x} + \boldsymbol{y}) = \|\boldsymbol{x}\|^2 + 2\mathrm{Re}\{(\boldsymbol{x}, \boldsymbol{y})\} + \|\boldsymbol{y}\|^2 \leqq \|\boldsymbol{x}\|^2 + 2|(\boldsymbol{x}, \boldsymbol{y})| + \|\boldsymbol{y}\|^2 \leqq \|\boldsymbol{x}\|^2 + 2\|\boldsymbol{x}\| \cdot \|\boldsymbol{y}\| + \|\boldsymbol{y}\|^2 = (\|\boldsymbol{x}\| + \|\boldsymbol{y}\|)^2$.

Let $\{e_1, \ldots, e_n\}$ be a basis of a metric space V, and let $(x_1, \ldots, x_n)$ and $(y_1, \ldots, y_n)$ be the coordinates of vectors $x, y \in V$ with respect to this basis: $x = \sum_{i=1}^n x_i e_i$, $y = \sum_{j=1}^n y_j e_j$. From the linearity (A.4.2, 3) of the inner product, we see that

$$(x, y) = \left(\sum_{i=1}^{n} x_i e_i, \sum_{j=1}^{n} y_j e_j \right) = \sum_{i=1}^{n} \sum_{j=1}^{n} x_i^* y_j (e_i, e_j)$$

$$= \sum_{i=1}^{n} \sum_{j=1}^{n} g_{ij} x_i^* y_j \ , \tag{A.4.10}$$

where we put

$$g_{ij} \equiv (e_i, e_j) \ , \qquad i, j = 1, \ldots, n \ . \tag{A.4.11}$$

From (A.4.1–3), we can easily prove that $G = (g_{ij})$, $i, j = 1, \ldots, n$, is a *positive-definite Hermitian matrix*. Conversely, an inner product is uniquely defined through (A.4.10, 11) if a positive-definite Hermitian matrix $G = (g_{ij})$, i, $j = 1, \ldots, n$, is given. (See Sect. A.5 for the definition of a Hermitian matrix. For real linear spaces, G is a *positive-definite symmetric matrix*. Matrix G is often called the *metric* (or *Gram*) *tensor* because it is transformed as a tensor under coordinate changes.) In terms of coordinates, the norm $\|x\|$ is expressed as

$$\|x\| = \sqrt{\sum_{i=1}^{n} \sum_{j=1}^{n} g_{ij} x_i^* x_j} \ . \tag{A.4.12}$$

A basis $\{e_1, \ldots, e_n\}$ is said to be *orthonormal* if

$$(e_i, e_j) = \delta_{ij} \ , \qquad i, j = 1, \ldots, n \ , \tag{A.4.13}$$

where δ_{ij} is the *Kronecker delta* taking value 1 if $i = j$ and value 0 otherwise. In other words, a basis is orthonormal if it consists of unit vectors that are orthogonal to each other. For an orthonormal basis, the matrix $G = (g_{ij})$ becomes the unit matrix $I = (\delta_{ij})$.

If a nonorthonormal basis $\{e_1, \ldots, e_n\}$ is given, an orthonormal basis is always constructed by the following (*Gram-*)*Schmidt orthogonalization*. Let $e_1' = e_1/\|e_1\|$. Suppose we have already constructed $e_1', \ldots, e_{k-1}'$ that are orthonormal. Then, e_k' is constructed by

$$e_k' = c_k \left(e_k - \sum_{i=1}^{k-1} c_i e_i' \right) . \tag{A.4.14}$$

The constants $c_1, \ldots, c_k$ are determined so that $e_1', \ldots, e_k'$ become orthonormal. The requirement that e_k' be orthogonal to all e_j', $j = 1, \ldots, k-1$, is given by

$$(e_k', e_j') = c_k \left((e_k, e_j') - \sum_{i=1}^{k-1} c_i (e_i', e_j') \right) = 0 \tag{A.4.15}$$

for $j = 1, \ldots, k-1$. Since $(e'_i, e'_j) = \delta_{ij}$ by the inductive hypothesis, we obtain

$$c_j = (e_k, e'_j) , \qquad j = 1, \ldots, k-1 . \tag{A.4.16}$$

The remaining constant c_k is chosen so that e'_k becomes a unit vector.

For an orthonormal basis, we see from (A.4.10, 12) that

$$(x, y) = \sum_{i=1}^{n} x_i^* y_i , \qquad \|x\| = \sqrt{\sum_{i=1}^{n} |x_i|^2} . \tag{A.4.17}$$

These are respectively called the *Euclidean inner product* and the *Euclidean norm*, or both are simply referred to as the *Euclidean metric*. Linear space $\mathbb{R}^n$ equipped with the Euclidean metric is called the *Euclidean (metric) space* and denoted by E^n.

A subset V' of a linear space V is called a *subspace* if it is itself a linear space under the addition and scalar multiplication of V. The set consisting of 0 alone and the set of entire V are trivial subspaces. Other subspaces are called *proper subspaces*.

Let V' be an l-dimensional subspace of an n-dimensional linear space V, and let $\{e_1, \ldots, e_l\}$ be a basis of V'. A basis of V can be constructed by augmenting this basis. Namely, we can choose $e_{l+1}, \ldots, e_n \in V$ such that $\{e_1, \ldots, e_n\}$ is a basis of V. Let V'' be the subspace generated by $e_{l+1}, \ldots, e_n$.[13] Then, any vector x is expressed uniquely as a sum $x = x' + x''$, $x' \in V'$, $x'' \in V''$, and subspaces V' and V'' share $\mathbf{0}$ alone. We say that the linear space V is resolved into the *direct sum* of subspaces V', V'', and write

$$V = V' \oplus V'' . \tag{A.4.18}$$

If every vector of subspace V' is orthogonal to every vector of subspace V'', the two subspaces V', V'' are said to be *orthogonal*, and we write

$$V' \perp V'' . \tag{A.4.19}$$

Orthogonal subspaces share $\mathbf{0}$ alone: $V' \cap V'' = \{\mathbf{0}\}$. In fact, if $x \in V' \cap V''$, vector x is orthogonal to itself, i.e., $(x, x) = 0$, which implies $x = \mathbf{0}$ according to the requirement on the inner product.

If the entire space V is resolved into the direct sum of subspaces V', V'' that are orthogonal to each other, they are called *orthogonal complements* of each other, and we write

$$V' = V''^{\perp} , \qquad V'' = V'^{\perp} . \tag{A.4.20}$$

For any l-dimensional subspace $V' \subset V$, we can choose an orthonormal basis $\{e_1, \ldots, e_l\}$ for V', say by Schmidt orthogonalization, and also choose

---

[13] A subspace V' *generated* (or *spanned*) by vectors $x_1, \ldots, x_r$ is the set of all linear combinations of the form $c_1 x_1 + \ldots + c_r x_r$. This is the *minimal* subspace in the sense that it is included in every subspace containing vectors $x_1, \ldots, x_r$.

$e_{l+1}, \ldots, e_n \in V$ such that $\{e_1, \ldots, e_n\}$ is an orthonormal basis of V. In other words, we can always resolve V into the direct sum of V' and its orthogonal complement $V'^{\perp}$:

$$V = V' \oplus V'^{\perp} . \tag{A.4.21}$$

A.5 Linear Operators

A mapping $\boldsymbol{y} = T\boldsymbol{x}$ of a linear space V is a *linear mapping* if

$$T(\boldsymbol{x}_1 + \boldsymbol{x}_2) = T\boldsymbol{x}_1 + T\boldsymbol{x}_2 , \tag{A.5.1}$$

$$T(c\boldsymbol{x}) = cT\boldsymbol{x} \tag{A.5.2}$$

for all $\boldsymbol{x}_1$, $\boldsymbol{x}_2 \in V$ and all scalars c. (This can be combined into the single condition $T(c_1\boldsymbol{x}_1 + c_2\boldsymbol{x}_2) = c_1 T\boldsymbol{x}_1 + c_2 T\boldsymbol{x}_2$ for all $\boldsymbol{x}_1$, $\boldsymbol{x}_2 \in V$ and all scalars c_1, c_2.) We also say that such a $\boldsymbol{T}$ is a *linear operator*.

Let $\{\boldsymbol{e}_1, \ldots, \boldsymbol{e}_n\}$ be a basis of V. Since $T\boldsymbol{e}_j$, $j = 1, \ldots, n$, are all vectors of V, they are expressed as linear combinations of $\boldsymbol{e}_1, \ldots, \boldsymbol{e}_n$:

$$T\boldsymbol{e}_j = \sum_{i=1}^{n} t_{ij}\boldsymbol{e}_i , \qquad j = 1, \ldots, n . \tag{A.5.3}$$

The coefficients t_{ij}, i, $j = 1, \ldots, n$, are called the *matrix elements* of linear operator T with respect to this basis. If the basis is orthonormal, we obtain

$$t_{ij} = (\boldsymbol{e}_i, T\boldsymbol{e}_j) , \qquad i, j = 1, \ldots, n , \tag{A.5.4}$$

by taking the inner product of $\boldsymbol{e}_i$ and the left-hand side (A.5.3) and noting that $(\boldsymbol{e}_i, \boldsymbol{e}_j) = \delta_{ij}$. (In quantum mechanics, the notation $t_{ij} = \langle i | T | j \rangle$ is often used. For a nonorthonormal basis, (A.5.4) is replaced by $t_{ij} = (\tilde{\boldsymbol{e}}_i, T\boldsymbol{e}_j)$, where $\{\tilde{\boldsymbol{e}}_1, \ldots, \tilde{\boldsymbol{e}}_n\}$ is the *reciprocal* (or *dual*) basis defined by $(\tilde{\boldsymbol{e}}_i, \boldsymbol{e}_j) = \delta_{ij}$.)

Let $(x_1, \ldots, x_n)$ be the coordinates of $\boldsymbol{x}$, and $(y_1, \ldots, y_n)$ be the coordinates of $\boldsymbol{y} = T\boldsymbol{x}$. Then,

$$\begin{aligned} \boldsymbol{y} &= T\left(\sum_{j=1}^{n} x_j\boldsymbol{e}_j\right) = \sum_{j=1}^{n} x_j(T\boldsymbol{e}_j) = \sum_{j=1}^{n} x_j\left(\sum_{i=1}^{n} t_{ij}\boldsymbol{e}_i\right) \\ &= \sum_{i=1}^{n}\left(\sum_{j=1}^{n} t_{ij}x_j\right)\boldsymbol{e}_i . \end{aligned} \tag{A.5.5}$$

Hence, we find that

$$y_i = \sum_{j=1}^{n} t_{ij}x_j , \qquad i = 1, \ldots, n . \tag{A.5.6}$$

In other words, any linear operator T acts on coordinates as a multiplication of matrix $\boldsymbol{T} = (t_{ij})$:

$$\begin{pmatrix} y_1 \\ \vdots \\ y_n \end{pmatrix} = \boldsymbol{T} \begin{pmatrix} x_1 \\ \vdots \\ x_n \end{pmatrix} . \tag{A.5.7}$$

Since the correspondence between a linear operator T and its matrix elements t_{ij}, i, $j = 1, \ldots, n$, is defined for some basis $\{\boldsymbol{e}_1, \ldots, \boldsymbol{e}_n\}$, different matrix elements result from the same operator if different bases are used. Let $\{\boldsymbol{e}'_1, \ldots, \boldsymbol{e}'_n\}$ be another basis. Let the relationship between two bases be

$$\boldsymbol{e}_j = \sum_{i=1}^{n} p_{ij}\boldsymbol{e}'_i, \qquad j = 1, \ldots, n , \tag{A.5.8}$$

where $\boldsymbol{P} = (p_{ij})$, $i, j = 1, \ldots, n$, is a nonsingular matrix, cf. (A.3.19). Let $(x'_1, \ldots, x'_n)$ and $(y'_1, \ldots, y'_n)$ be the coordinates of $\boldsymbol{x}, \boldsymbol{y}$ with respect to basis $\{\boldsymbol{e}'_1, \ldots, \boldsymbol{e}'_n\}$. Since matrix $\boldsymbol{P} = (p_{ij})$ is nonsingular, we have from (A.3.23)

$$\begin{pmatrix} x_1 \\ \vdots \\ x_n \end{pmatrix} = \boldsymbol{P}^{-1} \begin{pmatrix} x'_1 \\ \vdots \\ x'_n \end{pmatrix}, \quad \begin{pmatrix} y_1 \\ \vdots \\ y_n \end{pmatrix} = \boldsymbol{P}^{-1} \begin{pmatrix} y'_1 \\ \vdots \\ y'_n \end{pmatrix} . \tag{A.5.9}$$

Substituting these into (A.5.7), we see that

$$\boldsymbol{P}^{-1} \begin{pmatrix} y'_1 \\ \vdots \\ y'_n \end{pmatrix} = \boldsymbol{T}\boldsymbol{P}^{-1} \begin{pmatrix} x'_1 \\ \vdots \\ x'_n \end{pmatrix} , \tag{A.5.10}$$

or

$$\begin{pmatrix} y'_1 \\ \vdots \\ y'_n \end{pmatrix} = \boldsymbol{P}\boldsymbol{T}\boldsymbol{P}^{-1} \begin{pmatrix} x'_1 \\ \vdots \\ x'_n \end{pmatrix} \tag{A.5.11}$$

This means that the matrix $\boldsymbol{T}' = (t'_{ij})$, i, $j = 1, \ldots, n$, defined for the new basis is given by

$$\boldsymbol{T}' = \boldsymbol{P}\boldsymbol{T}\boldsymbol{P}^{-1} . \tag{A.5.12}$$

This is called the *similarity transformation* of matrix $\boldsymbol{T}$ by matrix $\boldsymbol{P}$. (We omit the details, but matrix properties invariant under this transformation can be regarded as characteristics of the corresponding linear mapping itself—*determinant, rank, nullity, eigenvalues*, etc. By choosing an appropriate basis, a matrix is transformed into its *canonical form* by the similarity transformation.)

Now, consider a metric space V. Linear operator $T^\dagger$ is called the *adjoint* of linear operator T if

$$(Tx, y) = (x, T^\dagger y) \tag{A.5.13}$$

for all $x, y \in V$. It follows easily from this definition that

$$(T^\dagger)^\dagger = T \ , \qquad (T_1 T_2)^\dagger = T_2^\dagger T_1^\dagger \ . \tag{A.5.14}$$

A linear operator T is *Hermitian* (or *self-adjoint*) if it is identical to its adjoint: $T^\dagger = T$. Namely,

$$(Tx, y) = (x, Ty) \tag{A.5.15}$$

for all $x, y \in V$. A matrix is said to be *Hermitian* (or *self-adjoint*) if it represents a Hermitian operator *with respect to an orthonormal basis.* It is easily confirmed that a matrix $\boldsymbol{T}$ is Hermitian if and only if

$$\boldsymbol{T} = (\boldsymbol{T}^{\mathrm{T}})^* \ . \tag{A.5.16}$$

The right-hand side is the element-wise complex conjugate of matrix $\boldsymbol{T}$ transposed, which is called the *Hermitian conjugate* of matrix $\boldsymbol{T}$ and is also denoted by $\boldsymbol{T}^\dagger$. (For a real linear space, a Hermitian matrix reduces to a symmetric matrix, and Hermitian conjugate simply means matrix transpose.)

A linear operator U is said to be *unitary* if it preserves the metric, i.e.,

$$(Ux, Uy) = (x, y) \tag{A.5.17}$$

for all $x, y \in V$. Equation (A.5.17) also implies that a unitary operator U preserves the norm, i.e.,

$$\| Ux \| = \| x \| \tag{A.5.18}$$

for all $x \in V$. From (A.5.13), we can say that a linear operator U is unitary if and only if

$$U^\dagger U = U U^\dagger = I \ . \tag{A.5.19}$$

Note that if U is unitary so is U^{-1}. A matrix $\boldsymbol{U}$ is said to be *unitary* if it satisfies (A.5.19) when identified with a matrix equation. (For a real linear space, a unitary matrix reduces to an orthogonal matrix.) Hence, a matrix $\boldsymbol{U}$ is unitary if it represents a unitary operator *with respect to an orthonormal basis.*

Let V' be an l-dimensional subspace of V. Subspace V' is said to be an *invariant subspace* of linear operator T if the image Tx of every vector $x \in V'$ is also contained in V', namely,

$$x \in V' \rightarrow Tx \in V' \ , \tag{A.5.20}$$

or $TV' \subset V'$. Let $\{e_1, \ldots, e_l\}$ be a basis of V'. If we construct a basis $\{e_1, \ldots, e_l, e_{l+1}, \ldots, e_n\}$ of V by augmenting it, condition (A.5.20) implies

$$Te_r = \sum_{i=1}^{l} t_{ir} e_i \ , \qquad r = 1, \ldots, l \ . \tag{A.5.21}$$

This means that the corresponding matrix $\boldsymbol{T}$ has the form

$$\boldsymbol{T} = \left(\begin{array}{c|c} \boldsymbol{T}' & \tilde{\boldsymbol{T}} \\ \hline \boldsymbol{0} & \boldsymbol{T}'' \end{array}\right) . \tag{A.5.22}$$

Let V'' be the subspace generated by $\boldsymbol{e}_{l+1}, \ldots, \boldsymbol{e}_n$. As shown earlier, the linear space V is resolved into the direct sum of V' and V'': $V = V' \oplus V''$. If V'' is also an invariant subspace of T, we have

$$\boldsymbol{T}\boldsymbol{e}_s = \sum_{i=l+1}^{n} t_{is}\boldsymbol{e}_i \ , \qquad s = l+1, \ldots, n \ , \tag{A.5.23}$$

and the matrix representation of $\boldsymbol{T}$ has the form

$$\boldsymbol{T} = \left(\begin{array}{c|c} \boldsymbol{T}' & \boldsymbol{0} \\ \hline \boldsymbol{0} & \boldsymbol{T}'' \end{array}\right) . \tag{A.5.24}$$

Thus, the submatrices $\boldsymbol{T}'$ and $\boldsymbol{T}''$ respectively act as linear operators on subspaces V' and V''. If this is the case, we write

$$T = T' \oplus T'' \ , \tag{A.5.25}$$

and say that linear operator T is resolved into the *direct sum* of T' and T''.

A.6 Group Representation

A group of linear transformations of linear space V is called a *linear representation* (or simply *representation*) of group G if there exists a homomorphism from G onto it. The linear space V is called its *representation space*. The dimension n of V is called the *degree* of the representation, and the representation is said to be *n-dimensional*. Let $\{T_g\}$, $g \in G$ be a representation of group G. By definition, we have

$$T_{g'g} = T_{g'}T_g \ , \qquad g, g' \in G \ , \tag{A.6.1}$$

$$T_e = I \ , \qquad (T_g)^{-1} = T_{g^{-1}} \ , \qquad g \in G \ , \tag{A.6.2}$$

where e is the unit element of group G. Hence, all linear operators T_g, $g \in G$, are nonsingular. If the homomorphism is also an isomorphism, the representation is said to be *faithful*. Otherwise, multiple elements of G correspond to the identity I. (Such elements form a subgroup H of G. The represented is faithful to the *quotient* (or *factor*) *group* G/H.)

If we fix a basis of the representation space V, all linear operators T_g, $g \in G$, can be identified with n-dimensional matrices. However, we must note that *the matrices that represent the same linear operator take different forms if we use different bases.* Namely, although the representation itself is the same, each operator is represented by different matrices depending on the bases.

Let $\boldsymbol{T}_g$ be the matrix representing $g \in G$ with respect to a basis $\{\boldsymbol{e}_1, \ldots, \boldsymbol{e}_n\}$, and let $\boldsymbol{T}'_g$ be the matrix representing the same element g but with respect to another basis $\{\boldsymbol{e}'_1, \ldots, \boldsymbol{e}'_n\}$. As shown in (A.5.11), there exists an n-dimensional nonsingular matrix $\boldsymbol{P}$ such that

$$\boldsymbol{T}'_g = \boldsymbol{P}\boldsymbol{T}_g\boldsymbol{P}^{-1} , \qquad g \in G . \tag{A.6.3}$$

The matrix $\boldsymbol{P}$ *does not depend on the group element* g.

Let $\{T_g\}$ and $\{T'_g\}$ be two representations of group G in the same representation space V. The two representations are said to be *equivalent* if we can take two bases $\{\boldsymbol{e}_i\}$ and $\{\boldsymbol{e}'_i\}$ such that, for every $g \in G$, the matrix $\boldsymbol{T}_g$ with respect to basis $\{\boldsymbol{e}_i\}$ is identical to matrix $\boldsymbol{T}'_g$ with respect to basis $\{\boldsymbol{e}'_i\}$. We then write

$$T'_g \cong T_g . \tag{A.6.4}$$

Two equivalent representations can be regarded as essentially the same because one is obtained from the other by merely changing the basis of the representation space. From the above definition, we see that two representations $\{T_g\}$ and $\{T'_g\}$ are equivalent if and only if there exists a linear transformation P of the representation space such that

$$T'_g = PT_gP^{-1} , \qquad g \in G . \tag{A.6.5}$$

A subspace V' of the representation space V is called an *invariant subspace* of representation $\{T_g\}$ if it is an invariant subspace of linear operator T_g *for every* $g \in G$. If V' is an invariant subspace of representation $\{T_g\}$, we see that $T_g V' = V'$ for all $g \in G$, since every linear operator T_g, $g \in G$, is nonsingular.

We say that a representation $\{T_g\}$ is *reducible* if there exists a "proper" invariant subspace V' for it. If there exists no invariant subspaces other than $\{\boldsymbol{0}\}$ and V itself, the representation is said to be *irreducible*.

Suppose $\{T_g\}$ is a reducible representation of G, and let V' be an invariant subspace. Let $\{e_1, \ldots, e_l\}$ be its basis, and let $\{e_1, \ldots, e_l, e_{l+1}, \ldots, e_n\}$ be the augmented basis of the entire representation space V. Then, as shown in the preceding section, every linear operator T_g is represented by a matrix of the form

$$\boldsymbol{T}_g = \left(\begin{array}{c|c} \boldsymbol{T}'_g & \tilde{\boldsymbol{T}}_g \\ \hline 0 & \boldsymbol{T}''_g \end{array}\right) . \tag{A.6.6}$$

Hence, we have

$$T_{g'g} = T_{g'} T_g = \left(\begin{array}{c|c} T'_{g'} T'_g & T'_{g'} \tilde{T}_g + \tilde{T}_{g'} T''_g \\ \hline 0 & T''_{g'} T''_g \end{array}\right) . \tag{A.6.7}$$

Let V'' be the subspace generated by $(e_{l+1}, \ldots, e_n\}$ so that $V = V' \oplus V''$. From (A.6.7), we see that T'_g and T''_g respectively act as linear operators on subspaces V' and V''. This means that they separately define representations of G in V' and V''. If subspace V'' is also an invariant subspace, i.e., if $T_g V'' = V''$ for every $g \in G$, then (A.6.7) becomes

$$T_{g'g} = T_{g'} T_g = \left(\begin{array}{c|c} T'_{g'} T'_g & 0 \\ \hline 0 & T''_{g'} T''_g \end{array}\right) . \tag{A.6.8}$$

If this is the case, we say that the representation $\{T_g\}$ is *fully reducible* (or *decomposable*) and write

$$T_g \cong T'_g \oplus T''_g . \tag{A.6.9}$$

We can repeat this process, called *reduction*, if representation $\{T'_g\}$ in V' and/or representation $\{T''_g\}$ in V'' are not irreducible.

Let the representation space V be a metric space. A representation $\{T_g\}$ is said to be *unitary* if T_g is a unitary operator for every $g \in G$. *All unitary representations are fully reducible.* Indeed, if a unitary representation $\{T_g\}$ is reducible, there exists a proper invariant subspace $V' \subset V$. Let $V'^{\perp}$ be the orthogonal complement of V' with respect to the metric (x, y) of V. As shown earlier, the representation space V is resolved into the direct sum $V = V' \oplus V'^{\perp}$. Now,

$$\begin{aligned} T_g V'^{\perp} &= \{T_g x \,|\, (x, y) = 0 \quad \text{for all } y \in V'\} \\ &= \{x' \,|\, (T_g^{-1} x', y) = 0 \quad \text{for all } y \in V'\} \\ &= \{x' \,|\, (x', T_g y) = 0 \quad \text{for all } y \in V'\} \\ &= \{x' \,|\, (x', y') = 0 \quad \text{for all } y' \in V'\} = V'^{\perp} . \end{aligned} \tag{A.6.10}$$

Here, we used the fact that T_g is a unitary operator. Note that $T_g V' = V'$, since V' is an invariant subspace of $\{T_g\}$. Thus, the orthogonal complement $T'^{\perp}$ is also an invariant subspace, and the representation $\{T_g\}$ is fully reducible.

Any representation of a finite group is fully reducible into irreducible representations. To show this, let (x, y) be an (arbitrarily introduced) metric of the representation space V. Let $\{T_g\}$ be a representation of a finite group G consisting of $|G|$ members. Let us introduce a new metric by the new inner product

$$\langle x, y \rangle \equiv \frac{1}{|G|} \sum_{g \in G} (T_g x, T_g y) . \tag{A.6.11}$$

It is easy to confirm that this inner product satisfies all of (A.4.1–4). Moreover, we see that

$$\langle T_g x, T_g y\rangle = \frac{1}{|G|}\sum_{g'\in G}(T_{g'}T_g x, T_{g'}T_g y) = \frac{1}{|G|}\sum_{g'\in G}(T_{g'g}x, T_{g'g}y)$$
$$= \frac{1}{|G|}\sum_{g''\in G}(T_{g''}x, T_{g''}y) = \langle x, y\rangle \ . \qquad \text{(A.6.12)}$$

Here, we put $g'' = g'g$ and used the fact that $g'G$ and G are exactly the same set. Equation (A.6.12) states that all linear operators T_g are unitary with respect to the metric (A.6.11). Hence, the representation is fully reducible. The metric (A.6.11) is said to be *invariant* with respect to G.

Any representation of a compact group is also fully reducible into irreducible representation.[14] The proof is the same as for a finite group except that, instead of (A.6.11), we use

$$\langle x, y\rangle = \frac{1}{|G|}\int_G (T_g x, T_g y)d\mu(g) \ , \qquad |G| \equiv \int_G d\mu(g) \ , \qquad \text{(A.6.13)}$$

where $\int_G(\ldots)d\mu(g)$ is the *invariant integration*, and $d\mu(g)$ is the *invariant measure*, which is guaranteed to exist for a compact group. It is easy to confirm that this metric also satisfies all of (A.4.1–4) and the representation is unitary with respect to this metric.

A.7 Schur's Lemma

Now, we are in a position to prove a well-known theorem called *Schur's lemma.* This theorem plays a central role in group representation theory. We state this lemma in three separate forms.

Theorem A.1 (*Schur's lemma* 1). *Let* $\{T_g\}$ *and* $\{T'_g\}$ *be irreducible representations of different degrees. If there exists, for some matrix representations of* T_g *and* T'_g, *a matrix* $\boldsymbol{P}$ *such that*

$$\boldsymbol{T}_g\boldsymbol{P} = \boldsymbol{P}\boldsymbol{T}'_g \qquad \text{(A.7.1)}$$

for all $g\in G$, *then* $\boldsymbol{P} = \boldsymbol{O}$.

Proof. Let l and l' be the respective degrees of irreducible representations $\{T_g\}$ and $\{T'_g\}$. Let $(\boldsymbol{T}_g)_{ij}$ be the matrix elements of $\boldsymbol{T}_g$, i.e.,

[14] The definition of a compact group is given in Sect. A.8. However, we omit the discussion about the invariant measure altogether. Just note that the 3D rotation group $SO(3)$ is a compact group and hence all its representations are fully reducible.

$$T_g e_j = \sum_{i=1}^{l} (\boldsymbol{T}_g)_{ij} \boldsymbol{e}_i \,, \qquad j = 1, \ldots, l \,, \tag{A.7.2}$$

for some basis $\{\boldsymbol{e}_1, \ldots, \boldsymbol{e}_l\}$ of the representation space V. Similarly, let $(\boldsymbol{T}_g')_{ij}$ be the matrix elements of T_g' for some basis of its representation space. Suppose matrix $\boldsymbol{P} = (p_{ij})$, $i = 1, \ldots, l$, $j = 1, \ldots, l'$, satisfies (A.7.1), i.e.,

$$\sum_{k=1}^{l} (\boldsymbol{T}_g)_{ik} p_{kj} = \sum_{k=1}^{l'} p_{ik} (\boldsymbol{T}_g')_{kj} \,, \tag{A.7.3}$$

for all $g \in G$.

Case (i). Suppose $l > l'$. Define l' vectors $\boldsymbol{x}_1, \ldots, \boldsymbol{x}_{l'}$ by

$$\boldsymbol{x}_j = \sum_{k=1}^{l} p_{kj} \boldsymbol{e}_k, \qquad j = 1, \ldots, l' \,. \tag{A.7.4}$$

Then, we see that

$$\begin{aligned} T_g \boldsymbol{x}_j &= \sum_{k=1}^{l} p_{kj} T_g \boldsymbol{e}_k = \sum_{k=1}^{l} p_{kj} \left(\sum_{i=1}^{l} (\boldsymbol{T}_g)_{ik} \boldsymbol{e}_i \right) = \sum_{i=1}^{l} \left(\sum_{k=1}^{l} (\boldsymbol{T}_g)_{ik} p_{kj} \right) \boldsymbol{e}_i \\ &= \sum_{i=1}^{l} \left(\sum_{k=1}^{l'} p_{ik} (\boldsymbol{T}_g')_{kj} \right) \boldsymbol{e}_i = \sum_{k=1}^{l'} (\boldsymbol{T}_g')_{kj} \left(\sum_{i=1}^{l} p_{ik} \boldsymbol{e}_i \right) = \sum_{k=1}^{l'} (\boldsymbol{T}_g')_{kj} \boldsymbol{x}_k \,. \end{aligned} \tag{A.7.5}$$

This means that the subspace V' generated by $l'(< l)$ vectors $x_1, \ldots, x_{l'}$ is an invariant subspace of T_g. In other words, the representation space V of irreducible representation T_g contains an invariant subspace V' of a lower dimensionality, which is impossible unless $\boldsymbol{x}_i = \boldsymbol{0}$ for all $i = 1, \ldots, l$. From (A.7.4), this means that $p_{kj} = 0$ for all $k = 1, \ldots, l$, $j = 1, \ldots, l'$, or $\boldsymbol{P} = \boldsymbol{O}$.

Case (ii). Suppose $l < l'$. We can easily confirm that linear operator $\tilde{T}_g \equiv (T_{g^{-1}})^\dagger$, also defines a representation of G.[15] As shown in the previous section, if we take an orthonormal basis, the matrix representing $\tilde{T}_g$ is the Hermitian conjugate of the matrix representing $T_{g^{-1}}$. Taking the Hermitian conjugate of both sides of (A.7.1), and replacing g by g^{-1}, we obtain

$$\boldsymbol{P}^\dagger (\boldsymbol{T}_{g^{-1}})^\dagger = (\boldsymbol{T}'_{g^{-1}})^\dagger \boldsymbol{P}^\dagger \,, \tag{A.7.6}$$

see (A.5.14). Hence, if we put $\tilde{\boldsymbol{P}} = \boldsymbol{P}^\dagger$, then (A.7.6) becomes

$$\tilde{\boldsymbol{T}}_g' \tilde{\boldsymbol{P}} = \tilde{\boldsymbol{P}} \, \tilde{\boldsymbol{T}}_g \,, \tag{A.7.7}$$

and the problem is reduced to case (i).

---

[15] The representation $\{(T_g)^\dagger\}$ is called the *conjugate representation* of $\{T_g\}$. It is easy to show that $\{(T_g)^\dagger\}$ is irreducible if and only if $\{T_g\}$ is irreducible, since if $V' \subset V$ is an invariant subspace of $\{T_g\}$, its orthogonal complement $V'^\perp \subset V$ is an invariant subspace of $\{(T_g)^\dagger\}$.

Theorem A.2 (*Schur's lemma* 2). *Let* $\{T_g\}$ and $\{T'_g\}$ *be irreducible representations of the same degree. If there exists, for some matrix representations of* T_g *and* T'_g, *a matrix* $\boldsymbol{P}$ *such that*

$$\boldsymbol{T}_g \boldsymbol{P} = \boldsymbol{P} \boldsymbol{T}'_g \tag{A.7.8}$$

for all $g \in G$, *then either the two representations are equivalent or* $\boldsymbol{P} = \boldsymbol{O}$.

Proof. If we set $l = l'$ in the proof of Theorem 1, we find that the subspace V' generated by the l vectors $\boldsymbol{x}_1, \ldots, \boldsymbol{x}_l$ must be either $\{\boldsymbol{0}\}$ or V itself. In the former case, we have $\boldsymbol{P} = \boldsymbol{O}$ from (A.7.4). In the latter case, $\boldsymbol{P}$ is a nonsingular matrix, and (A.7.8) implies

$$\boldsymbol{T}_g = \boldsymbol{P} \boldsymbol{T}'_g \boldsymbol{P}^{-1} \ . \tag{A.7.9}$$

Hence, the two representations are equivalent.

Theorem A.3 (*Schur's lemma* 3). *Let* $\{T_g\}$ *be an irreducible representation. If there exists, for some matrix representation of* T_g, *a matrix* $\boldsymbol{P}$ *such that*

$$\boldsymbol{T}_g \boldsymbol{P} = \boldsymbol{P} \boldsymbol{T}_g \tag{A.7.10}$$

for all $g \in G$, *then* $\boldsymbol{P} =$ const. $\times$ $\boldsymbol{I}$.

Proof. If V is the representation space of $\{T_g\}$, the matrix $\boldsymbol{P}$ defines a linear mapping of V. Let $\boldsymbol{x}(\neq \boldsymbol{0})$ be an eigenvector of $\boldsymbol{P}$ for eigenvalue λ: $\boldsymbol{P}\boldsymbol{x} = \lambda \boldsymbol{x}$. Then, $T_g \boldsymbol{x}$ is also its eigenvector for the same eigenvalue: $\boldsymbol{P}(T_g \boldsymbol{x}) = T_g \boldsymbol{P}\boldsymbol{x} = \lambda(T_g \boldsymbol{x})$. This means that the eigenspace of $\boldsymbol{P}$ for eigenvalue λ is an invariant subspace. Since representation $\{T_g\}$ is irreducible, either such an eigenvector does not exist or the eigenspace is V itself. The former case implies $\boldsymbol{P} = \boldsymbol{O}$, while the latter case is possible only when $\boldsymbol{P} = \lambda \boldsymbol{I}$.

Note that Theorems A.1 and A.2 hold whether the representation space is real or complex. However, Theorem A.3 holds only when the representation space is a *complex* linear space, because eigenvalues and eigenvectors do not always exist in the real domain.

An important corollary is obtained from Theorem A.3 when the group G is an Abelian group. If G is Abelian, all T_g, $g \in G$, commute with each other. Hence, for any irreducible representation $\{T_g\}$, each operator must be represented by a matrix of the form of const. $\times$ $\boldsymbol{I}$. But such a representation is reducible unless the degree is 1. Thus, *all irreducible representations of an Abelian group are one-dimensional if the representation spaces are complex linear spaces.*

There are many systematic ways to reduce a given representation into the direct sum of irreducible representations—for example, the use of the *character* and the *orthogonality* of irreducible representations. The irreducible reduction of a (*Kronecker*) *product representation* is known as the *Clebsch–Gordon series.* These techniques are found in the literature on group representation theory—often in relation to quantum mechanics (e.g., angular momentum), quantum chemistry (e.g., atomic orbitals), and solid-state physics (e.g., crystal vibration and scattering).

A.8 Topology, Manifolds, and Lie Groups

In this section, we briefly summarize basic notions, definitions and terminologies of topology and Lie groups.

A set X is a *topological space* if a set $\mathcal{O}$ of subsets of X is given such that

(1) the set $\mathcal{O}$ contains the empty set $\varnothing$ and the set X itself: $\varnothing, X \in \mathcal{O}$;
(2) the set $\mathcal{O}$ contains the union of any of its members: $U_i \in \mathcal{O}, i \in I \rightarrow \cup_{i \in I} U_i \in \mathcal{O}$, where I is an arbitrary index set;
(3) the set $\mathcal{O}$ contains the intersection of any "finite" number of its members: $U_i \in \mathcal{O}, i \in J \rightarrow \cap_{i \in J} U_i \in \mathcal{O}$, where J is an arbitrary "finite" index set.

The set $\mathcal{O}$ is called the *topology* of set X. The members of $\mathcal{O}$ are called *open sets*. The complement $X - U$ of an open set U is said to be *closed*. A subset X' of a topological space X itself becomes a topological space by taking $\{X' \cap U | U \in \mathcal{O}\}$ as its topology, where $\mathcal{O}$ is the topology of X. This topology is called the *relative topology*. The topology can be defined in many different ways.[16]

For n-dimensional space $\mathbb{R}^n$, a topology $\mathcal{O}$ can be defined as all unions of *open boxes* $\{(x_1, \ldots, x_n) | a_i < x_i < b_i, i = 1, \ldots, n\}$ for all $a_i, b_i \in \mathbb{R}$. Alternatively, we can define a topology $\mathcal{O}$ as all unions of *open balls* $\{(x_1, \ldots, x_n) | [\sum_{i=1}^{n} (x_i - c_i)^2]^{1/2} < r\}$ for all $c_i, r \in \mathbb{R}, r > 0$. However, these two topologies are *equivalent*: the set $\mathcal{O}$ of open sets is actually the same. This topology is called the *Euclidean topology*. A subset of $\mathbb{R}^n$ is said to be *bounded* if it is contained in some open ball (or equivalently in some open box).

An (*open*) *neighborhood* of a point x in X is an open set containing the point x. A point $x \in A \subset X$ is an *interior* (or *inner*) *point* of A if there exists a neighborhood of x that is included in A. The *interior* A° of set A is the set of all the interior points of A. A points $x \in X$ is a *limit* (or *accumulation*) *point* of set $A \subset X$ if every neighborhood of point x intersects with set A. The *closure* $\bar{A}$ of set A is the set of all the limit points of A. The set $\bar{A} - A^\circ$ is called the *boundary* of set A, and its members are called *boundary points* of A.

[16] We obtain the *discrete topology* if $\mathcal{O}$ includes all subsets of X, and the *trivial topology* if $\mathcal{O}$ consists of X and $\varnothing$ only. Topology $\mathcal{O}$ is said to be *stronger* than topology $\mathcal{O}'$ (or $\mathcal{O}'$ is *weaker* than $\mathcal{O}$) if $\mathcal{O} \supset \mathcal{O}'$. The discrete topology is the strongest topology, while the trivial topology is the weakest topology. We can define the *metric topology* for a *metric space*, the *order topology* for a *linearly* (or *totally*) *ordered set*, the *direct product topology* for the *direct product* of topological spaces, and the *induced topology* by a continuous mapping.

A sequence $\{x_i\}$, $i = 1, 2, \ldots$, of points in X is said to *converge* to a *limit point* $x \in X$ if every neighborhood U of x contains a subsequence $\{x_k\}$, $k = N, N+1, \ldots$, starting from some element x_N.

A topological space X is *Hausdorff* (or *separable*) if any two distinct points x_1, $x_2 \in X$ have disjoint neighborhoods U_1, U_2: $U_1 \cap U_2 = \emptyset$. In a Hausdorff space, every point is a closed set. The Euclidean topology on R^n is Hausdorff.

A collection $\{U_i\}$, $i \in I$, of open sets of X is said to be a *covering* if each point in X belongs to at least one U_i, i.e., $\cup_{i \in I} U_i = X$. A Hausdorff space X is *compact* if, for an arbitrary covering $\{U_i\}$, $i \in I$, we can always choose from it a "finite" set of members that covers X, namely, if we can choose a "finite" subset J of the index set I such that $\cup_{j \in J} U_j = X$. Any sequence x_i, $i = 1, 2, \ldots$, of points in a compact space X has a converging subsequence (whose limit point is called an *accumulation point*). Bounded closed subsets of $\mathbb{R}^n$ are compact (*Heine–Borel theorem*). Hence, any infinite set of points of a bounded closed subset of $\mathbb{R}^n$ has an accumulation point (*Balzano–Weierstrass theorem*).

A topological space X is *disconnected* if it has two disjoint non-empty open subsets X_1, X_2 such that $X_1 \cup X_2 = X$. Otherwise, it is said to be *connected*. Let X be a disconnected space, having disjoint non-empty open subsets X_1, X_2 such that $X_1 \cup X_2 = X$ and $X_1 \cap X_2 = \emptyset$. Since the open subsets X_1, X_2 are complements of each other, they are at the same closed sets. They are called the *connected components* of X if they are both connected. If not, they are further decomposed into connected components, each being open as well as closed.

A set of a topological space X is said to be *connected* or *disconnected* depending on whether it is connected or disconnected when regarded as a topological space by the relative topology. A topological space X is *locally connected* if any neighborhood of any point $x \in X$ contains a connected neighborhood.

A mapping f from a topological space X to a topological space Y is *continuous* at point $x \in X$ if, for any neighborhood $V_{f(x)}$ of image point $f(x) \in Y$, there exists a neighborhood U_x of x such that the image of U_x by f is contained in $V_{f(x)}$, i.e., $f(U_x) \subset V_{f(x)}$. A mapping $f: X \to Y$ is continuous if and only if for every open set $V \subset Y$, the inverse image $f^-(V)$ is open in X. The image of a compact set under a continuous map is also compact.

A mapping $f: X \to Y$ is a *homeomorphism*, if it is one-to-one and onto, and if f and f^{-1} are both continuous. Two topological spaces X and Y are *homeomorphic* to each other if there exists a homeomorphism between them. Since the image and the inverse image of an open set under a homeomorphism are both open, we also say that homeomorphic spaces X, Y *have the same topology* in the sense that all the open sets of X are in one-to-one correspondence with all the open sets of Y. Characteristics of a topological space are *topological invariants* if they are preserved by all homeomorphisms, i.e., if they take the same values for spaces homeomorphic to each other. For instance, the number of connected components, called the *connectivity*, is a topological invariant. Other topological invariants include the *Euler (–Poincaré) characteristic* (or *Euler number*) and the *Betti number*.

A Hausdorff space is a *manifold*, if every point x has a neighborhood U from which a homeomorphism φ exists onto open set of $\mathbb{R}^n$ for some fixed finite n, and if the following condition is satisfied: If point $x \in M$ belongs to two such neighborhoods U_1, U_2 equipped, respectively, with homeomorphisms φ_1, φ_2, the mapping $\varphi_2\varphi_1^{-1}$ is a homeomorphism from $\varphi_1(U_1 \cap U_2) \subset \mathbb{R}^n$ onto $\varphi_2(U_1 \cap U_2) \subset \mathbb{R}^n$. The integer n is called the *dimension* of manifold M. The image $\varphi(x) = (x_1, \ldots, x_n) \in \mathbb{R}^n$ defines *coordinates* of point $x \in U$. The pair (U, φ) is called the *local coordinates system* at $x \in M$. If point $x \in M$ has two local coordinate systems (U_1, φ_1) and (U_2, φ_2), the homeomorphic mapping $\varphi_2\varphi_1^{-1}$ defines the *coordinate change* of $x \in M$ from local coordinate system (U_1, φ_1) to local coordinate system (U_2, φ_2).

A manifold M is a *differentiable manifold* if it is equipped with a local coordinate system for every point such that the coordinate change $\varphi_2\varphi_1^{-1}$ is continuously differentiable up to some degree k (i.e., having continuous partial derivatives up to degree k). A differentiable manifold M is said to be C^k*-differentiable* if the coordinate change is continuously differentiable up to degree k. A manifold M is *smooth* if it is C^∞-differentiable.

A continuous mapping f from a smooth manifold M to a smooth manifold N is C^k*-differentiable* if it is C^k-differentiable as a function that expresses local coordinates of N in terms of local coordinates of M. (Since M, N are both C^∞-differentiable, this definition does not depend on the choice of the local coordinate systems of M and N.) If a C^∞-differentiable mapping $f: M \to N$ is a homeomorphism, and if f^{-1} is also C^∞-differentiable, f is called a *diffeomorphism*, and M and N are said to be *diffeomorphic* to each other.

A *Lie group* G is a group that is also a differentiable manifold such that the multiplication $(a, b) \in G \times G \to ab \in G$ and the inverse $a \in G \to a^{-1} \in G$ are both *differentiable mappings*, which means that the mapping f: $\varphi_a(U_a) \times \varphi_b(U_b) \to \varphi_{ab}(U_{ab})$ defined by $f(\boldsymbol{x}, \boldsymbol{y}) = \varphi_{ab}(\varphi_a^{-1}(\boldsymbol{x})\varphi_b^{-1}(\boldsymbol{y}))$ for $\boldsymbol{x}, \boldsymbol{y} \in \mathbb{R}^n$ and the mapping g: $\varphi_a(U_a) \to \varphi_{a^{-1}}(U_{a^{-1}})$ defined by $g(\boldsymbol{x}) = \varphi_{a^{-1}}(\varphi_a(\boldsymbol{x})^{-1})$ for $\boldsymbol{x} \in \mathbb{R}^n$ are both differentiable functions with respect to the coordinate system of $\mathbb{R}^n$.[17] Here, (φ_a, U_a), (φ_b, U_b), (φ_{ab}, U_{ab}), $(\varphi_{a^{-1}}, U_{a^{-1}})$ are local coordinate systems of elements $a, b, ab, a^{-1} \in G$, respectively.

The general linear group $GL(n, \mathbb{R})$ can be identified with a subset of $\mathbb{R}^{n^2}$ if the n^2 elements of each matrix are regarded as its coordinates. It consists of points of $\mathbb{R}^{n^2}$ at which the determinant is not zero. The hypersurface $A \subset \mathbb{R}^{n^2}$ defined as the set of points at which the determinant vanishes is a closed subset of $\mathbb{R}^{n^2}$ (with respect to the Euclidean topology of $\mathbb{R}^{n^2}$). Since $GL(n, \mathbb{R})$ is the complement $\mathbb{R}^{n^2} - A$, it is an open subset of $\mathbb{R}^{n^2}$.

---

[17] Actually, the requirement of differentiability of both $(a, b) \to ab$ and $a \to a^{-1}$ is redundant: it can be replaced by the differentiability of $(a, b) \to ab^{-1}$ alone.

The general linear group $GL(n, \mathbb{R}) \subset \mathbb{R}^{n^2}$ becomes a topological space by the relative topology. It also becomes a manifold if a local coordinate system is defined by the *inclusion mapping*—the identity regarded as a mapping from $GL(n, \mathbb{R})$ into $\mathbb{R}^{n^2}$. Since the identity is infinitely differentiable, this manifold is C^∞-differentiable. Since each element of the product of two matrices is a polynomial in the elements of the two matrices, and each element of the inverse of a matrix is a rational function of the elements of the original matrix, the matrix multiplication is a differentiable mapping from $\mathbb{R}^{n^2} \times \mathbb{R}^{n^2}$ into $\mathbb{R}^{n^2}$, and the matrix inversion is a differentiable mapping from $\mathbb{R}^{n^2}$ into itself. Thus, $GL(n, \mathbb{R})$ is a Lie group.

The orthogonal group $O(n)$ and the special orthogonal group $SO(n)$ are proper subsets of $GL(n, \mathbb{R})$ and hence topological spaces under the relative topology. They are also C^∞-differentiable manifolds. Since the matrix multiplication and matrix inversion are differentiable mappings, they are Lie groups. The definition of an orthogonal matrix implies that the absolute value of each element is equal to or less than 1. This means that the orthogonal group $O(n)$ is a bounded subset of $\mathbb{R}^{n^2}$. Hence, it is a compact Lie group. Since the determinant of each element is either 1 or -1, the an orthogonal group $O(n)$ is disconnected. It consists of two connected components—one with determinant 1 and the other with determinant -1. The special orthogonal group $SO(n)$ is a connected compact Lie group; it is the connected component of $O(n)$ of determinant 1. Typical Lie groups represented as subgroups of $GL(n, \mathbb{R})$ or $GL(n, \mathbb{C})$ are called *classical groups*—besides $GL(n, \mathbb{R})$, $O(n)$ and $SO(n)$, they include the *special linear* (or *unimodular*) *group* $SL(n)$, the *unitary group* $U(n)$, the *special unitary group* $SU(n)$, *the Lorentz group*, and the *symplectic group* $Sp(n)$.

A *path* p in a topological space X is a continuous mapping from interval $[0, 1]$ into X, and $p(0), p(1) \in X$ are respectively called the *initial* and the *terminal points* of path p. A path p is called a *loop* (or *closed path*) if its initial and terminal points coincide; the point $p(0) = p(1) \in X$ is called the *base* of loop p. Each point $x \in X$ can be identified with a path $x(t) = x$, $0 \leqq t \leqq 1$, its base being itself.

We say that two points x, y in a topological space X can be *joined by an arc* if there exists a path p in X such that $p(0) = x$, $p(1) = y$. A topological space X is *arcwise* (or *path*) *connected* if any two points $x, y \in X$ can be joined by an arc.[18] The set of all the points that can be joined to point $x \in X$ is called the *arcwise* (or *path*) *connected component* of x. A topological space X is *locally arcwise* (or *path*) *connected* if, for any point $x \in X$ and any neighborhood U of x, there exists a neighborhood V of x such that any two points of V can be joined by an arc in U.

If the terminal point of path p coincides with the initial point of path q, their *product* (or *concatenation*) is defined as the path pq such that $pq(t) = p(2t)$ for

[18] An arcwise connected space is also a connected space, but the converse is not true in general. However, open connected sets are always arcwise connected.

$0 \leqq t \leqq 1/2$ and $pq(t) = q(2t - 1)$ for $1/2 \leqq t \leqq 1$. The *inverse* of path p is defined as the path p^{-1} such that $p^{-1}(t) = p(1 - t)$.

Path p_1 is said to be *homotopic* to path p_2 if there exists a continuous mapping h: $[0, 1] \times [0, 1] \to X$ such that $h(0, t) = p_1(t)$ and $h(1, t) = p_2(t)$. The function h is called the *homotopy* of the two paths. If path p_1 is homotopic to path p_2, we write $p_1 \sim p_2$ and say that path p_1 can be *continuously deformed* (or *deformable*) into path p_2. We say that loop p is *homotopic to 0* and write $p \sim 0$ if it is deformable to a single point. An arcwise connected topological space X is said to be *simply connected* if any loop of X is homotopic to 0.

Loops having the same base $x \in X$ are classified into *homotopic classes* such that members of each class are homotopic to each other.[19] We write as $[p]$ the homotopic class containing loop p, and loop p is called the *representative* of the homotopic class $[p]$. Let $\pi_1(X)$ be the set of homotopic classes of loops with base $x \in X$. The *product* of two members $[p]$, $[q] \in \pi_1(X)$ is defined by $[p][q] = [pq]$. This product does not depend on the choice of the representatives of the two homotopic classes. (Namely, if $p \sim p'$ and $q \sim q'$, then $pq \sim p'q'$.) The homotopic class $[x]$ is the unit element for this multiplication rule. The *inverse* of a homotopic class $[p] \in \pi_1(X)$ is defined by $[p]^{-1} = [p^{-1}]$. Again, this does not depend on the choice of the representative. (Namely, if $p \sim p'$, then $p^{-1} \sim p'^{-1}$.) It is easy to prove that (A.2.1–3) are satisfied for this multiplication and inverse. Hence, $\pi_1(X)$ is a group. This group is called the *fundamental group* (or *first homotopy group*) of arcwise connected space X.[20] If it consists of the unit element alone, we write $\pi_1(X) \cong 0$. (However, the fundamental group is not necessarily Abelian.) This occurs if and only if space X is simply connected. *The fundamental group is topologically invariant*: if X and X' are homeomorphic to each other, then $\pi_1(X) \cong \pi_1(X')$.

Let X, $\tilde{X}$ be arcwise and locally arcwise connected spaces. Space $\tilde{X}$ is said to be a *covering space* of space X if there exists a continuous mapping π: $\tilde{X} \to X$, called *projection*, such that (1) projection π is onto, and (2) projection π is locally homeomorphic, i.e., for every $x \in X$, there exists a neighborhood U of x such that $\pi^{-1}(U)$ is a disjoint union of open sets of $\tilde{X}$, each of which is projected by π homeomorphically onto U. A covering space $\tilde{X}$ is said to be *universal* if it is simply connected.

A covering space $\tilde{G}$ of a Lie group G is called a *covering group* if, in addition to the conditions for a covering space, the projection π: $\tilde{G} \to G$ is a homomorphism. (Covering groups are also defined for general *continuous* (or *topological*)

[19] The homotopy relation $\sim$ is an *equivalence relation*. Namely, (1) $p \sim p$ (the *reflexive law*), (2) $p \sim q \to q \sim p$ (the *symmetric law*), and (3) $p \sim q$, $q \sim r \to p \sim r$ (the *transitive law*). An equivalence relation classifies elements into *equivalence classes*. If $\mathscr{P}_x$ is the set of loops with base x, the set $\pi_1(X)$ is the *quotient set* $\mathscr{P}_x/\sim$ with respect to the homotopy relation.

[20] Although the definition of $\pi_1(X)$ involves the base $x \in X$, group $\pi_1(X)$ with base x is isomorphic to group $\pi_1(X)$ with another base y if x, $y \in X$ can be joined by an arc. Hence, the fundamental group $\pi_1(X)$ is *uniquely defined up to isomorphism* for an arcwise connected space X.

groups that are not necessarily Lie groups.) Covering group $\tilde{G}$ is said to be *universal* if it is simply connected. As discussed in Sect. 6.9, $SU(2)$ is homeomorphic to a 3-sphere S^3, for which $\pi_1(S^3) \cong 0$. Hence, $SU(2)$ is also simply connected: $\pi_1(SU(2)) \cong 0$. However, $SO(3)$ is not simply connected; its fundamental group is isomorphic to the group of integers under addition modulo 2: $\pi_1(SO(3)) \cong \mathbb{Z}/2$. As shown in Sect. 6.9, $SU(2)$ is a universal covering group of $SO(3)$.

A.9 Lie Algebras and Lie Groups

Let V be a real linear space. A mapping $[.,.]: V \times V \to V$ is called a *Lie bracket* if

$$[\boldsymbol{x}, \boldsymbol{y}] = -[\boldsymbol{y}, \boldsymbol{x}] \ , \tag{A.9.1}$$

$$[c\boldsymbol{x}, \boldsymbol{y}] = c[\boldsymbol{x}, \boldsymbol{y}], \qquad [\boldsymbol{x}, c\boldsymbol{y}] = c[\boldsymbol{x}, \boldsymbol{y}] \ , \tag{A.9.2}$$

$$[\boldsymbol{x} + \boldsymbol{y}, \boldsymbol{z}] = [\boldsymbol{x}, \boldsymbol{z}] + [\boldsymbol{y}, \boldsymbol{z}], \qquad [\boldsymbol{x}, \boldsymbol{y} + \boldsymbol{z}] = [\boldsymbol{x}, \boldsymbol{y}] + [\boldsymbol{x}, \boldsymbol{z}] \ , \tag{A.9.3}$$

$$[\boldsymbol{x}, [\boldsymbol{y}, \boldsymbol{z}]] + [\boldsymbol{y}, [\boldsymbol{z}, \boldsymbol{x}]] + [\boldsymbol{z}, [\boldsymbol{x}, \boldsymbol{y}]] = 0 \tag{A.9.4}$$

for all $\boldsymbol{x}, \boldsymbol{y}, \boldsymbol{z} \in V$ and all $c \in \mathbb{R}$.[21] Equation (A.9.1) states that the Lie bracket is *anticommutative*, and (A.9.2, 3) state that it is *bilinear*. Equation (A.9.4) is called the *Jacobi identity*. A real linear space V equipped with a Lie bracket is called a *Lie algebra*. If V is n-dimensional as a linear space, it is called an n-dimensional Lie algebra.

It is easy to see that the *vector* (or *outer*) *product* $\boldsymbol{a} \times \boldsymbol{b}$ for $\boldsymbol{a}, \boldsymbol{b} \in \mathbb{R}^3$ satisfies (A.9.1–4). Hence, $\mathbb{R}^3$ is a three-dimensional Lie algebra under the Lie bracket $[\boldsymbol{a}, \boldsymbol{b}] = \boldsymbol{a} \times \boldsymbol{b}$.

A group G of linear transformations of a linear space V is itself regarded as a linear space in a natural way. We define

$$(cT)(\boldsymbol{x}) = T(c\boldsymbol{x}) \ , \qquad (T_1 + T_2)(\boldsymbol{x}) = T_1(\boldsymbol{x}) + T_2(\boldsymbol{x}) \tag{A.9.5}$$

for all T, T_1, $T_2 \in G$, all $\boldsymbol{x} \in V$, and all $c \in \mathbb{R}$. Then, it is easy to see that the *commutator* defined by

$$[T_1, T_2] \equiv T_1 T_2 - T_2 T_1 \tag{A.9.6}$$

satisfies (A.9.1–4). Hence, a group of linear transformations is a Lie algebra under the commutator.

Let V be an n-dimensional Lie algebra, and let $\{\boldsymbol{e}_1, \ldots, \boldsymbol{e}_n\}$ be its basis. Since all elements of V are expressed as linear combinations of the basis vectors, and

---

[21] The term *Poisson bracket* is sometimes used, depending on the context. Equations (A.9.2b, 3b) are actually redundant; they are derived from (A.9.2a, 3a, 1).

since the Lie bracket is bilinear, the Lie bracket of any two vectors is completely determined once the Lie bracket among these basis vectors are prescribed. Let us put

$$[e_i, e_j] = \sum_{k=1}^{n} c_{ij}^k e_k \ , \qquad i, j = 1, \ldots, n \ . \tag{A.9.7}$$

Then n^3 constants c_{ij}^k, $i, j, k = 1, \ldots, n$, are called the *structure constants*. Equation (A.9.7) is called the *commutation relations* of the basis vectors.

In terms of the structure constants, the Lie bracket of vectors whose components are (x_i) and (y_i), $i = 1, \ldots, n$, is a vector whose components are $(\sum_{i,j=1}^{n} c_{ij}^k x_i y_j)$, $k = 1, \ldots, n$:

$$\left(\sum_{i=1}^{n} x_i e_i, \sum_{j=1}^{n} y_j e_j\right) = \sum_{i=1}^{n} \sum_{j=1}^{n} x_i y_j [e_i, e_j] = \sum_{k=1}^{n} \left(\sum_{i=1}^{n} \sum_{j=1}^{n} c_{ij}^k x_i y_j\right) e_k \ . \tag{A.9.8}$$

Although the Lie algebra is uniquely determined by prescribing its structure constants, the n^3 structure constants c_{ij}^k cannot be independently chosen; they must be so chosen that (A.9.1–4) are satisfied. Equations (A.9.2, 3) are automatically satisfied. Equations (A.9.1, 4) are respectively equivalent to

$$c_{ij}^k = -c_{ji}^k \ , \qquad i, j, k = 1, \ldots, n \ , \tag{A.9.9}$$

$$\sum_{l=1}^{n} c_{ij}^l c_{lk}^m + \sum_{l=1}^{n} c_{jk}^l c_{li}^m + \sum_{l=1}^{n} c_{ki}^l c_{lj}^m = 0 \ , \quad i, j, k, m = 1, \ldots, n \ . \tag{A.9.10}$$

If all the structure constants are zero (hence $[x, y] = 0$ for all $x, y \in V$), the Lie algebra is said to be *Abelian*.

Let V, V' be Lie algebras. A mapping f: $V \to V$ is a *homomorphism* if it is a homomorphism from V to V' viewed as linear spaces and if

$$f([x, y]) = [f(x), f(y)] \tag{A.9.11}$$

for all $x, y \in V$. A homomorphism is an *isomorphism* if it is also one-to-one and onto. If an isomorphism exists between two Lie algebras V, V', we say that they are *isomorphic* to each other and write $V \cong V'$. It is easy to see that two Lie algebras are isomorphic to each other if and only if we can take bases in such a way that the structure constants become identical.

Let c_{ij}^k, $i, j, k = 1, \ldots, n$, be the structure constants of Lie algebra V with respect to basis $\{e_i\}$, $i = 1, \ldots, n$, and let $c_{ij}^{k\prime}$, $i, j, k = 1, \ldots, n$, be the structure constants of the same Lie algebra with respect to another basis $\{e_i'\}$, $i = 1, \ldots, n$. Since there exists a nonsingular matrix $P = (p_{ij})$ such that $e_j = \sum_{i=1}^{n} p_{ij} e_i'$, the two sets of the structure constants must be related by

$$c_{ij}^{k\prime} = \sum_{l=1}^{n} \sum_{m=1}^{n} \sum_{s=1}^{n} p_{li}^{-1} p_{mj}^{-1} p_{ks} c_{lm}^s \ , \qquad i, j, k = 1, \ldots, n \ , \tag{A.9.12}$$

where $\boldsymbol{P}^{-1} = (p_{ij}^{-1})$ is the inverse of matrix $\boldsymbol{P}$.[22] Hence, two Lie algebras are isomorphic to each other if and only if there exists a nonsingular matrix $\boldsymbol{P}$ such that (A.9.12) is satisfied between their structure constants. Mutually isomorphic Lie algebras are regarded as the "same" Lie algebra.

Now, we define Lie algebras associated with Lie groups. By definition, a Lie group is a differential manifold on which the group operation acts as a differential mapping. Its Lie algebra is defined in terms of its differentiability, but to this end we first define *tangent vectors* as differentiation operators, and then *tangent spaces* at each point—or more generally the *fiber bundle*. The *Lie algebra* of a Lie group is defined as a set of *invariant vector fields*, for which the Lie bracket is defined as the commutator. Then, the relationship between Lie groups and their Lie algebras is described in terms of *differential* (or *Pfaffian*) *forms* and their *integrability conditions* (*Frobenius' theorem*). Here, however, we take another approach. In the following, we only consider Lie groups that can be represented as subgroups of $GL(n, \mathbb{C})$, and define their Lie algebras in terms of their matrix representations.

Let $G\,[\subset GL(n, \mathbb{C})]$ be a Lie group. Let L be the set of n-dimensional matrices $\boldsymbol{X}$ such that

$$\exp(t\boldsymbol{X}) \equiv \sum_{k=0}^{\infty} \frac{t^k}{k!} \boldsymbol{X}^k \in G \tag{A.9.13}$$

for any real number t. (Exponentiation of any matrix is always (absolutely) convergent.) It is easy to prove that L is a (real) linear space:

$$\boldsymbol{X} \in L \rightarrow c\boldsymbol{X} \in L\ , \qquad c \in \mathbb{R}\ , \tag{A.9.14}$$

$$\boldsymbol{X}, \boldsymbol{Y} \in L \rightarrow \boldsymbol{X} + \boldsymbol{Y} \in L\ . \tag{A.9.15}$$

It is also easy to prove that L is a Lie algebra under the commutator $[\boldsymbol{X}, \boldsymbol{Y}] \equiv \boldsymbol{X}\boldsymbol{Y} - \boldsymbol{Y}\boldsymbol{X}$:[23]

$$\boldsymbol{X}, \boldsymbol{Y} \in L \rightarrow [\boldsymbol{X}, \boldsymbol{Y}] \in L\ . \tag{A.9.16}$$

The set L is called the *Lie algebra* of Lie group G. The members of its basis are called (by physicists) *infinitesimal generators* of G. Since (A.9.13) can express only those elements that belong to the *connected component* containing the unit element, the Lie algebras of two Lie groups are the same if the two Lie groups share the same connected component that contains the unit element e.

---

[22] In tensor notation with the Einstein summation convention (Sect. 5.3.1), the relationship between the two bases is written as $\boldsymbol{e}_{i'} = p^i_{i'}\boldsymbol{e}_i$ or $\boldsymbol{e}_i = p^{i'}_i \boldsymbol{e}_{i'}$, and accordingly (A.9.12) is written as $c^{k'}_{i'j'} = p^i_{i'} p^j_{j'} p^{k'}_k c^k_{ij}$.

[23] This can be proved from the fundamental relations $\exp(t\boldsymbol{X})\exp(t\boldsymbol{Y}) = \exp[t(\boldsymbol{X}+\boldsymbol{Y}) + t^2[\boldsymbol{X}, \boldsymbol{Y}]/2 + O(t^3)]$ and $\{\exp(t\boldsymbol{X}), \exp(t\boldsymbol{Y})\} = \exp\{t^2[\boldsymbol{X}, \boldsymbol{Y}] + O(t^3)\}$, where $\{\boldsymbol{A}, \boldsymbol{B}\} \equiv \boldsymbol{A}\boldsymbol{B}\boldsymbol{A}^{-1}\boldsymbol{B}^{-1}$ (also called the *commutator*).

The Lie algebra $gl(n, \mathbb{R})$ (or $gl(n, \mathbb{C})$) of the general linear group $GL(n, \mathbb{R})$ (or $GL(n, \mathbb{C})$) consists of all n-dimensional nonsingular real (or complex) matrices. The Lie algebra $sl(n)$ of the special linear group $SL(n)$ consists of all n-dimensional nonsingular real matrices X of trace 0 ($\mathrm{Tr}\{X\} = 0$). The Lie algebra $u(n)$ of the orthogonal group $O(n)$ consists of all n-dimensional real antisymmetric matrices X ($X^{\mathrm{T}} = -X$). The Lie algebra $so(n)$ of the special orthogonal group $SO(n)$ is the same as $o(n)$, because $SO(n)$ is the connected component of $O(n)$ that contains the unit element. The Lie algebra $u(n)$ of the unitary group $U(n)$ consists of all n-dimensional anti-Hermitian (complex) matrices $X(X^{\dagger} = -X)$. The Lie algebra $su(n)$ of the special unitary group $SU(n)$ consists of all n-dimensional anti-Hermitian matrices X of trace $0(X^{\dagger} = -X, \mathrm{Tr}\{X\} = 0)$.

The importance of the Lie algebra L of Lie group G lies in the fact that the Lie algebra L uniquely determines the Lie group G *locally*. To be specific, there exists a neighborhood U_e of the unit element e of G such that any $g \in U_e$ is uniquely expressed as $g = \exp(\sum_{i=1}^{n} t_i A_i)$ for $|t_i| < 1, i = 1, \ldots, n$, if an appropriate basis $\{A_1, \ldots, A_n\}$ *of* L is chosen. If the Lie group G is *connected*, any neighborhood $U_e \subset G$ of the unit element e can generate the entire group G (i.e., any $g \in G$ is expressed as the product $g = u_1 \ldots u_N$ of a finite number of elements $u_i \in U_e$, $i = 1, \ldots, N$), and hence any element $g \in G$ is expressed (not necessarily uniquely) as $g = \exp(X_1) \ldots \exp(X_N)$ in terms of $X_i \in L$, $i = 1, \ldots, N$. *The Lie group is Abelian if and only if its Lie algebra is Abelian.*

Let G, G' be Lie groups. A continuous mapping $f: G \to G'$ is a *local homomorphism* if there exists a neighborhood $U_e \subset G$ of the unit element $e \in G$ such that if $g, g', gg' \in G$, then

$$f(g)f(g') = f(gg') \ . \tag{A.9.17}$$

If local homomorphism $f: G \to G'$ is one-to-one mapping, and if $f^{-1}: G' \to G$ is also a local homomorphism, the mapping f is said to be a *local isomorphism*, and the two Lie groups G, G' are said to be *locally isomorphic*. It follows that two Lie groups G and G' have the same Lie algebra (up to isomorphism) *if and only if G and G' are locally isomorphic.*[24] If two connected groups G and G' have the same Lie algebra, and if G is *simply connected*, it is easy to prove that G is the *universal covering group* of G'. For example, $SO(3)$ and $SU(2)$ have the same Lie algebra. Hence, they are locally isomorphic. Since $SU(2)$ is simply connected, it is the universal covering group of $SO(3)$.

Let G be a Lie group, and L be its Lie algebra. An element $g \in G$ defines a transformation (called the *adjoint transformation*) $\mathrm{Ad}_g: L \to L$, $g \in G$, by

$$\mathrm{Ad}_g(X) \equiv gXg^{-1} \ . \tag{A.9.18}$$

---

[24] This is proved by showing that if G and G' are Lie groups, and L, L' are their respective Lie algebras and if $f: G \to G'$ is a local homomorphism, then the mapping $df: L \to L'$ defined by $(df)(X) = df(\exp(tX))/dt|_{t=0}$ (called the *derivation* of f) is a homomorphism from L to L'. Thus, a local isomorphism between Lie groups defines an isomorphism between their Lie algebras.

It is easy to prove that this is a transformation of L.[25] Also, it is easy to see that this is a linear transformation of L:

$$\mathrm{Ad}_g(cX) = c\mathrm{Ad}_g(X)\ , \qquad c \in \mathbb{R}\ , \tag{A.9.19}$$

$$\mathrm{Ad}_g(X+Y) = \mathrm{Ad}_g(X) + \mathrm{Ad}_g(Y)\ . \tag{A.9.20}$$

Furthermore, the correspondence from g to Ad_g is a homomorphism:

$$\mathrm{Ad}_{g'}\mathrm{Ad}_g = \mathrm{Ad}_{g'g}\ , \tag{A.9.21}$$

$$\mathrm{Ad}_e = I\ , \qquad (\mathrm{Ad}_g)^{-1} = \mathrm{Ad}_{g^{-1}}\ . \tag{A.9.22}$$

Hence, $\{\mathrm{Ad}_g\}$, $g \in G$, defines a representation of G over its Lie algebra L. This representation is called the *adjoint representation.*

Since $\mathrm{Ad}_g: L \to L$, $g \in G$, defines a group of (linear) transformations, it is itself regarded as a Lie group. Hence, it has its Lie algebra. The Lie algebra consists of $\mathrm{d}_X: L \to L$, $X \in L$, called the *derivation*[26] and defined by

$$\mathrm{d}_X(Y) = [X, Y]\ , \tag{A.9.23}$$

The Jacobi identity (A.9.4) is rewritten as

$$\mathrm{d}_X([Y, Z]) = [\mathrm{d}_X(Y), Z] + [Y, \mathrm{d}_X(Z)]\ . \tag{A.9.24}$$

The derivation d_X is a linear transformation of L:

$$\mathrm{d}_X(cY) = c\mathrm{d}_X(Y)\ , \qquad c \in \mathbb{R}\ , \tag{A.9.25}$$

$$\mathrm{d}_X(Y+Z) = \mathrm{d}_X(Y) + \mathrm{d}_X(Z)\ . \tag{A.9.26}$$

Furthermore, the correspondence from X to d_X is a (Lie algebra) homomorphism:

$$\mathrm{d}_{cX} = c\mathrm{d}_X\ , \qquad c \in R\ , \tag{A.9.27}$$

$$\mathrm{d}_{X+Y} = \mathrm{d}_X + \mathrm{d}_Y\ , \tag{A.9.28}$$

$$\mathrm{d}_{[X+Y]} = [\mathrm{d}_X, \mathrm{d}_Y]\ . \tag{A.9.29}$$

A set of linear transformations is a *representation* of Lie algebra L if there exists a (Lie algebra) homomorphism from L to its action. Thus, the set of derivations $\{\mathrm{d}_X\}$, $X \in L$, defines a representation of L over itself. This representation is called the *differential representation.*

The above consideration can be extended to an arbitrary representation of Lie group G. Let $\{T_g\}$, $g \in G$, be a representation of G over a representation space

[25] The product on the right-hand side means matrix multiplication. The mapping $\mathrm{Ad}_g: L \to L$ is in fact the *derivation* of the *inner automorphism* $\mathrm{Aut}_g: G \to G$ defined by $\mathrm{Aut}_g(h) = ghg^{-1}$. Namely, $\mathrm{Ad}_g(X) = d\mathrm{Aut}_g(\exp(tX))/dt|_{t=0}$.

[26] In fact, $\mathrm{d}_X(Y) = d\mathrm{Ad}_{\exp(tX)}(Y)/dt|_{t=0}$.

V. Let L be the Lie algebra of G. Then we can define a linear transformation dT_X: $V \to V$, $X \in L$, called the *derivation*, by

$$dT_X(\boldsymbol{x}) = \frac{d}{dt} T_{\exp(tX)}(\boldsymbol{x})\Big|_{t=0} \ . \tag{A.9.30}$$

It can be proved easily that $\{dT_X\}$, $X \in L$, is a representation of Lie algebra L, which is called the *differential representation*. It can also be proved that if G is connected, a subspace V' of the representation space V is an invariant subspace for $\{T_g\}$ if and only if it is an invariant subspace for its differential representation $\{dT_X\}$. Hence, *a representation $\{T_g\}$ of a connected Lie group G is irreducible, reducible, or fully reducible, if and only if its differential representation $\{dT_X\}$ is irreducible, reducible, or fully reducible, respectively.*

A.10 Spherical Harmonics

Let $f(X, Y, Z)$, or briefly $f(\boldsymbol{r})$, be a function defined in three-dimensional space. Let $\tilde{f}(\boldsymbol{r})$ be the function obtained by "rotating" function $f(\boldsymbol{r})$ around the coordinate origin O. In other words, if $\boldsymbol{R}$ is a rotation matrix, the value $\tilde{f}(\boldsymbol{r}')$ at $\boldsymbol{r}' = \boldsymbol{R}\boldsymbol{r}$ is given by $f(\boldsymbol{r})$. Hence, $\tilde{f}(\boldsymbol{r}) = f(\boldsymbol{R}^{-1}\boldsymbol{r}) = f(\boldsymbol{R}^{\mathrm{T}}\boldsymbol{r})$ ($\boldsymbol{R}$ is an orthogonal matrix: $\boldsymbol{R}^{-1} = \boldsymbol{R}^{\mathrm{T}}$). Thus, we define the *rotation operator* T_R by

$$T_R f(\boldsymbol{r}) = f(\boldsymbol{R}^{\mathrm{T}}\boldsymbol{r}) \ . \tag{A.10.1}$$

A rotation around an axis $\boldsymbol{n} = (n_1, n_2, n_3)$ (unit vector) by a small angle Ω screw-wise is given by

$$\boldsymbol{R} = \boldsymbol{I} + \Omega \begin{pmatrix} 0 & -n_3 & n_2 \\ n_3 & 0 & -n_1 \\ -n_2 & n_1 & 0 \end{pmatrix} + \mathrm{O}(\Omega^2) \ . \tag{A.10.2}$$

Hence, the rotation operation T_R for a small rotation $\boldsymbol{R}$ becomes

$$\begin{aligned} T_R f(X, Y, Z) &= f(X + \Omega[n_3 Y - n_2 Z] + \mathrm{O}(\Omega^2), Y + \Omega[n_1 Z - n_3 X] \\ &\quad + \mathrm{O}(\Omega^2), Z + \Omega[n_2 X - n_1 Y] + \mathrm{O}(\Omega^2)) \\ &= f(X, Y, Z) + \Omega(n_1 D_1 + n_2 D_2 + n_3 D_3) f(X, Y, Z) \\ &\quad + \mathrm{O}(\Omega^2) \ , \end{aligned} \tag{A.10.3}$$

where D_1, D_2, D_3 are *infinitesimal generators* defined as differential operators

$$D_1 \equiv Z\frac{\partial}{\partial Y} - Y\frac{\partial}{\partial Z} \ , \quad D_2 \equiv X\frac{\partial}{\partial Z} - Z\frac{\partial}{\partial X} \ , \quad D_3 \equiv Y\frac{\partial}{\partial X} - X\frac{\partial}{\partial Y} \ . \tag{A.10.4}$$

They satisfy the *commutation relations*

$$[D_2, D_3] = D_1 , \qquad [D_3, D_1] = D_2 , \qquad [D_1, D_2] = D_3 , \tag{A.10.5}$$

where $[.,.]$ denotes the commutator: $[D, D'] \equiv DD' - D'D$.

The *spherical coordinates* (r, θ, φ) are related to the Cartesian coordinates (X, Y, Z) by

$$X = r \sin\theta \cos\varphi , \qquad Y = r \sin\theta \sin\varphi , \qquad Z = r\cos\theta . \tag{A.10.6}$$

In terms of these spherical coordinates, (A.10.4, 5) are rewritten as

$$T_R f(\boldsymbol{r}) = f(\boldsymbol{r}) + \Omega(n_1 D_1 + n_2 D_2 + n_3 D_3) f(\boldsymbol{r}) + \mathrm{O}(\Omega^2) , \tag{A.10.7}$$

$$D_1 = \sin\varphi \frac{\partial}{\partial\theta} + \frac{\cos\varphi}{\sin\theta}\frac{\partial}{\partial\theta} , \qquad D_2 = -\cos\varphi\frac{\partial}{\partial\theta} + \frac{\sin\varphi}{\sin\theta}\frac{\partial}{\partial\theta} ,$$

$$D_3 = -\frac{\partial}{\partial\varphi} . \tag{A.10.8}$$

The commutation relations (A.10.5) are also satisfied.

The *Casimir operator* $H \equiv -(D_1^2 + D_2^2 + D_3^2)$ becomes

$$H = -\frac{1}{\sin\theta}\frac{\partial}{\partial\theta}\sin\theta\frac{\partial}{\partial\theta} - \frac{1}{\sin^2\theta}\frac{\partial^2}{\partial\varphi^2} . \tag{A.10.9}$$

The differential equation

$$HF(\theta, \varphi) = l(l+1)F(\theta, \varphi) \tag{A.10.10}$$

for an integer or half-integer l has $2l+1$ linearly independent solutions $F_l^m(\theta, \varphi)$, $m = -l, \ldots, l$, which generate a $(2l+1)$-dimensional eigenspace of H for eigenvalue $l(l+1)$. The rotation operator T_R defines an irreducible representation of $SO(3)$ in this space. This representation is equivalent to $\mathscr{D}_l$. The differential equation (A.10.10) has the form

$$\frac{1}{\sin\theta}\frac{\partial}{\partial\theta}\sin\theta\frac{\partial F}{\partial\theta} + \frac{1}{\sin^2\theta}\frac{\partial^2 F}{\partial\varphi^2} + l(l+1)F = 0 . \tag{A.10.11}$$

Solutions of this differential equation are called *spherical harmonics* (or *spherical functions*) or degree l. Equation (A.10.11) has single-valued solutions only when l is an integer.

The *Laplace equation* is

$$\nabla^2 f(X, Y, Z) = 0 , \tag{A.10.12}$$

where ∇^2 is the *Laplacian* defined by

$$\nabla^2 \equiv \frac{\partial^2}{\partial X^2} + \frac{\partial^2}{\partial Y^2} + \frac{\partial^2}{\partial Z^2} . \tag{A.10.13}$$

Solutions of the Laplace equation are called *harmonics* (or *harmonic functions*).

In terms of the spherical coordinates (r, θ, φ), the Laplacian ∇^2 of (A.10.13) becomes

$$\nabla^2 = \frac{1}{r^2}\frac{\partial}{\partial r}r^2\frac{\partial}{\partial r} + \frac{1}{r^2 \sin\theta}\frac{\partial}{\partial\theta}\sin\theta\frac{\partial}{\partial\theta} + \frac{1}{r^2\sin^2\theta}\frac{\partial^2}{\partial\varphi^2}, \tag{A.10.14}$$

and the Laplace equation (A.10.12) becomes

$$\frac{1}{r^2}\frac{\partial}{\partial r}r^2\frac{\partial f}{\partial r} + \frac{1}{r^2 \sin\theta}\frac{\partial}{\partial\theta}\sin\theta\frac{\partial f}{\partial\theta} + \frac{1}{r^2\sin^2\theta}\frac{\partial^2 f}{\partial\varphi^2} = 0 . \tag{A.10.15}$$

If we seek a solution in the form $f(X, Y, Z) = G(r)F(\theta, \varphi)$ (the *method of separation of variables*), (A.10.15) is rewritten as

$$\frac{1}{G}\frac{\partial}{\partial r}r^2\frac{\partial G}{\partial r} = -\frac{1}{F}\left(\frac{1}{\sin\theta}\frac{\partial}{\partial\theta}\sin\theta\frac{\partial F}{\partial\theta} + \frac{1}{\sin^2\theta}\frac{\partial^2 F}{\partial\varphi^2}\right) . \tag{A.10.16}$$

The left-hand side is a function of r alone, while the right-hand side is a function of θ, φ alone. Hence, both sides must be some constant λ. Thus, (A.10.16) splits into two equations

$$r^2\frac{d^2 G}{\partial r^2} + 2r\frac{\partial G}{\partial r} - \lambda G = 0 , \tag{A.10.17}$$

$$\frac{1}{\sin\theta}\frac{\partial}{\partial\theta}\sin\theta\frac{\partial F}{\partial\theta} + \frac{1}{\sin^2\theta}\frac{\partial^2 F}{\partial\varphi^2} + \lambda F = 0 . \tag{A.10.18}$$

It can be proved that (A.10.18) has single-valued solutions if and only if $\lambda = l(l+1)$, where l is an integer. Thus, these two equations become, respectively,

$$r^2\frac{d^2 G}{\partial r^2} + 2r\frac{\partial G}{\partial r} - l(l+1)G = 0 , \tag{A.10.19}$$

$$\frac{1}{\sin\theta}\frac{\partial}{\partial\theta}\sin\theta\frac{\partial F}{\partial\theta} + \frac{1}{\sin^2\theta}\frac{\partial^2 F}{\partial\varphi^2} + l(l+1)F = 0 . \tag{A.10.20}$$

Equation (A.10.19) has two solutions r^l, $1/r^{l+1}$. Equation (A.10.20) is identical with (A.10.11). Hence, its solutions are spherical harmonics of degree l. If we find $2l+1$ linearly independent solution $F_l^m(\theta, \varphi)$, $m = -l, \ldots, l$, of (A.10.20), any harmonic can be expressed as a (finite or infinite) linear combination of

$$f^l F_l^2(\theta, \varphi) , \quad \frac{1}{r^{l+1}}F_l^m(\theta, \varphi) , \quad l = 0, 1, \ldots , \quad m = -l, \ldots, l . \tag{A.10.21}$$

If we put $F(\theta, \varphi) = P(\theta)Q(\varphi)$ in (A.10.20) (the method of separation of variables), we have

$$\frac{\sin\theta}{P}\frac{d}{d\theta}\sin\theta\frac{dP}{d\theta} + l(l+1)\sin^2\theta = -\frac{1}{Q}\frac{d^2 Q}{d\varphi^2} . \tag{A.10.22}$$

The left-hand side is a function of θ alone, while the right-hand side is a function of φ alone. Hence, both sides must be some constant, which we put as m^2. Then, (A.10.22) splits into two equations

$$\frac{1}{\sin\theta}\frac{d}{d\theta}\sin\theta\frac{dP}{d\theta}+\left(l(l+1)-\frac{m^2}{\sin^2\theta}\right)P=0\ , \tag{A.10.23}$$

$$\frac{d^2Q}{d\varphi^2}+m^2Q=0\ . \tag{A.10.24}$$

The solution of (A.10.24) is given by $\exp(\pm\, im\varphi)$. Since the solution must be periodic in φ with period 2π, the number m must be an integer. Since m and $-m$ play the same role, we take m to be nonnegative.

If we put $z = \cos\theta$ as a new independent variable, (A.10.23) is rewritten as

$$\frac{d}{dz}(1-z^2)\frac{dP}{dz}+\left(l(l+1)-\frac{m^2}{1-z^2}\right)P=0\ . \tag{A.10.25}$$

This equation is called the *associated Legendre equation*. Its solution is called the *associated Legendre function* and denoted by $P_l^m(z)$.

Thus, the $2l+1$ spherical harmonics of degree l are given by

$$Y_l^{\pm m}(\theta,\varphi)=\frac{1}{\sqrt{2\pi}}\exp(\pm\, im\varphi)P_l^m(\cos\theta)\ ,\quad m=0,\ldots,l\ . \tag{A.10.26}$$

This expression is called the lth *Laplace spherical harmonic*. They are *complete*: any function defined over a unit sphere can be expressed as a (finite or infinite) linear combination of $Y_l^m(\theta,\varphi)$ for $l=0,1,2,\ldots$, $m=-l,\ldots,l$. (To be precise, we must exclude pathological functions, and consider the *convergence in norm* (or *strong convergence*) in $L_2(S^2)$.)

If we put $m=0$ in (A.10.25), we have

$$\frac{d}{dz}(1-z^2)\frac{dP}{dz}+l(l+1)P=0\ . \tag{A.10.27}$$

This equation is called the lth *Legendre equation*. It has a solution expressed in the form

$$P_l(z)=\frac{1}{2^l l!}\frac{d^l(z^2-1)^l}{dz^l}\ . \tag{A.10.28}$$

This is a polynomial in z of degree l and is called the lth *Legendre polynomial*.[27] This expression of the lth Legendre polynomial is called the *Rodrigues formula*. The Legendre polynomials $P_l(z)$ satisfy the following *orthogonality relation*:

$$\int_{-1}^{1}P_l(z)P_{l'}(z)dz=\frac{2}{2l+1}\delta_{ll'}\ . \tag{A.10.29}$$

[27] The constant $1/2^l l!$ is so chosen that $P_l(0)=1$ (and consequently $P_l(-1)=(-1)^l$).

Hence, the Legendre polynomials $P_0(z)$, $P_1(z)$, $P_2(z)$, . . . are constructed from 1, z, z^2, . . . by Schmidt orthogonalization with respect to the inner product $(p(z), q(z)) = \int_{-1}^{1} p(z)\, q(z) dz$.

In terms of the Legendre polynomial, the associated Legendre function is written as

$$P_l^m(z) = (z^2 - 1)^{m/2} \frac{d^m P_l(z)}{dz^m} , \qquad m = 0, \ldots, l . \tag{A.10.30}$$

The associated Legendre functions $P_l^m(z)$ satisfy the following *orthogonality relations*:

$$\int_{-1}^{1} P_l^m(z) P_{l'}^m(z)\, dz = \frac{2(l+m)!}{(l-m)!(2n+1)} \delta_{ll'} , \tag{A.10.31}$$

$$\int_{-1}^{1} \frac{P_l^m(z) P_l^{m'}(z)}{1 - z^2} dz = \frac{(l+m)!}{(l-m)!\, m} \delta_{mm'} . \tag{A.10.32}$$

From these follows the *orthogonality relation* of the Laplace spherical harmonics $Y_l^m(\theta, \varphi)$:

$$\int_0^{2\pi} \int_0^{\pi} Y_l^m(\theta, \varphi)^* \, Y_{l'}^{m'}(\theta, \varphi) \sin\theta \, d\theta \, d\varphi = \frac{2\pi(l+m)!}{(l-m)!(2l+1)} \delta_{ll'} \delta_{mm'} , \tag{A.10.33}$$

where * denotes the complex conjugate.

Bibliography

Chapter 1

As stated in Sect. 1.1, *pattern recognition* has a long history, and there exists a vast body of literature. Since it is almost impossible to choose the most typical or appropriate books, we exclude mention of them altogether. For image understanding, numerous papers and books have already been published. Books that cover a wide range of topics include:

D. H. Ballard, C. M. Brown. *Computer Vision* (Prentice-Hall, Englewood Cliffs, NJ 1982)
R. Nevatia: *Machine Perception* (Prentice-Hall, Englewood Cliffs, NJ 1982)
M. D. Levine: *Vision in Man and Machine* (McGraw-Hill, New York 1985)
B. K. P. Horn: *Robot Vision* (MIT Press, Cambridge, MA 1986)
Y. Shirai: *Three-Dimensional Computer Vision* (Springer, Berlin, Heidelberg 1987)

Image understanding is often regarded as a branch of *artificial intelligence*, and is introduced in books under that title, for example:

P. R. Cohen, E. A. Feigenbaum (eds.): *The Handbook of Artificial Intelligence*, Vol. 3 (Pitman, London 1982)
P. H. Winston: *Artificial Intelligence*, 2nd ed. (Addison-Wesley, Reading, MA 1984)
E. Charniak, D. McDermott: *Introduction to Artificial Intelligence* (Addison-Wesley, Reading, MA 1985)

For discussions on specific topics, many books of collected papers and contributed articles have been published. For example:

P. H. Winston (ed.): *The Psychology of Computer Vision* (McGraw-Hill, New York 1975)
A. R. Hanson, E. M. Riseman (eds.): *Computer Vision Systems* (Academic, New York 1978)
S. Tanimoto, A. Klinger (eds.): *Structured Computer Vision: Machine Perception through Hierarchical Computation Structures* (Academic, New York 1980)
J. M. Brady (ed.): *Computer Vision* (North-Holland, Amsterdam 1981)
J. Beck, B. Hope, A. Rosenfeld (eds.): *Human and Machine Vision* (Academic, New York 1983)
T. S. Huang (ed.): *Image Sequence Processing and Dynamic Scene Analysis*, NATO ASI Ser., Ser. F, Vol. 2 (Springer, New York 1983)
S. Ullman, W. Richards (eds.): *Image Understanding* 1984 (Ablex, Norwood, NJ 1984)
A. Rosenfeld (ed.): *Techniques for* 3-D *Machine Perception* (North-Holland, Amsterdam 1986)
T. Kanade (ed.): *Three-Dimensional Machine Vision* (Kluwer Academic, Norwell, MA 1987)

However, the material in these books varies greatly in emphasis, from image processing techniques, data structure and system organization, to knowledge base, inference strategies, and speculative discussions on human perception.

The following books discuss psychological aspects of human visual perception, and have had great influence on the study of computer vision:

J. J. Gibson: *The Ecological Approach to Visual Perception* (Houghton Mifflin, Boston, MA 1979)
S. Ullman: *The Interpretation of Visual Motion* (MIT Press, Cambridge, MA 1979)

D. Marr: *Vision: A Computational Investigation into the Human Representation and Processing of Visual Information* (W. H. Freeman, San Francisco 1982)

In this book, we do not deal with image processing techniques. Although image processing is the initial and vital state of image understanding, we concentrate on the mathematical aspects, assuming that appropriate image processing techniques are available whenever necessary. Here, we just list the following two volumes:

A. Rosenfeld, A. C. Kak: *Digital Picture Processing*, 2nd ed., Vols. 1,2 (Academic, Orlando, FL 1982)

Although the topic is very limited, the following book well deserves attention. The treatment is very mathematical.

K. Sugihara: *Machine Interpretation of Line Drawings* (MIT Press, Cambridge, MA 1986)

While the present book focuses on *geometrical* aspects, Sugihara illuminates *algebraic* aspects of 3D recovery. Some of his results are also introduced in Chap. 10.

Most of the material in the present book is based on the author's own work, and has been presented as conference papers and journal articles on various occasions. Also, papers dealing with similar topics have been presented by other researchers. Since the research in this area is currently in the development stage, we do not try to list them all. However, readers can make an almost complete survey by checking the annual bibliographical report *Image Analysis and Computer Vision* (formerly called *Picture Processing*) written by Azriel Rosenfeld, which appears in the journal

Computer Vision, Graphics, and Image Processing (Academic Press, San Diego, CA)

Other major journals which carry papers in this area are

Artificial Intelligence (North-Holland, Amsterdam)
IEEE Transactions on Pattern Analysis and Machine Intelligence (IEEE Computer Society, New York)
International Journal of Computer Vision (Kluwer Academic, Norwell, MA)

Other related journals include

Robotics Research (MIT Press, Cambridge, MA)
Machine Vision and Applications (Springer, Heidelberg, Berlin)
Spatial Vision (VNU Science Press, Utrecht, The Netherlands)
Image and Vision Computing (Butterworth, London)
International Journal of Pattern Recognition and Artificial Intellience (World Scientific Publishing, Singapore)

Periodic conferences in this research area include

ICCV: *International Conference on Computer Vision*, IEEE Computer Society
CVPR: *Conference on Computer Vision and Pattern Recognition*, IEEE Computer Society
IJCAI: *International Joint Conference on Artificial Intelligence*, American Association for Artificial Intelligence (AAAI)
NCAI: *National Conference on Artificial Intelligence*, American Association for Artificial Intelligence (AAAI)
ICPR: *International Conference on Pattern Recognition*, International Association for Pattern Recognition (IAPR)
DARPA IU: *Image Understanding Workshop*, Defense Advanced Research Projects Agency (DARPA)

Chapter 2

There are many books on group theory and group representation theory. Here, we cite

M. Hamermesh: *Group Theory and Its Application to Physical Problems* (Addison-Wesley, Reading, MA 1962)

The above book also describes the irreducible reduction of tensor representations. Strongly motivated readers are advised to read Weyl's classic books:

H. Weyl: *The Classical Groups, Their Invariants and Representations*, 2nd ed. (Princeton University Press, Princeton, NJ 1946)

H. Weyl: *The Theory of Groups and Quantum Mechanics* (Dover, New York 1950)

However, a considerable effort may be required to read these books. The philosophy which we called *Weyl's thesis* clearly underlies these books but is not explicitly stated, nor is the term "Weyl's thesis" used.

This chapter is mostly based on

K. Kanatani: Structure and motion from optical flow under orthographic projection. Comput. Vision, Graphics Image Process. **35**, 181–199 (1986)

K. Kanatani: Coordinate rotation invariance of image characteristics for 3D shape and motion recovery, in Proc. 1st Intl. Conf. Comput. Vision, June 1987, London, pp. 55–64

K. Kanatani, T.-C. Chou: Shape from texture: General principle. Artif. Intell. **38** (1), 1–48 (1989)

Chapter 3

Classic books on Lie groups are

C. Chevalley: *Theory of Lie Groups I* (Princeton University Press, Princeton, NJ 1946)

L. Pontrjagin: *Topological Groups*, 2nd ed. (Gordon and Breach, New York 1966)

There are many other textbooks. The following is probably easiest to read for gaining a general idea:

M. Hamermesh: *Group Theory and Its Application to Physical Problems* (Addison-Wesley, Reading, MA 1962)

In Sect. 3.3, we introduced many results that are valid specifically for $SO(3)$. An excellent treatment of this subject is found in the first half of

I. M. Gel'fand, R. A. Minlos, Z. Ya. Shapiro: *Representations of the Rotation and Lorenz Groups and Their Applications* (Macmillan, New York 1963)

The second half deals with the *Lorenz group*, which we need not consider in image understanding applications. It also deals with spinors, but a full account is found in

E. Cartan: *The Theory of Spinors* (Dover, New York 1981)

Applications of group theory to physics, in particular quantum mechanics, are found in many textbooks on group theory, Classic ones are

H. Weyl: *The Theory of Groups and Quantum Mechanics* (Dover, New York 1950)

E. P. Wigner: *Group Theory and Its Applications to the Quantum Mechanics of Atomic Spectra* (Academic, New York 1959)

A. R. Edmonds: *Angular Momentum in Quantum Mechanics* (Princeton University Press, Princeton, NJ 1974)

This chapter is mostly based on

K. Kanatani: Transformation of optical flow by camera rotation. *Trans. IEEE* PAMI-10 (2), 131–143 (1988)

Chapter 4

The best known classical work on *invariants* is

H. Weyl: *The Classical Groups, Their Invariants and Representations* (Princeton University Press, Princeton, NJ 1946)

which specifically deals with *polynomial invariants*, as we discussed in Sect. 4.7.2. There are many other old books on invariants, but most of them are written from the viewpoint of projective geometry; the theory of quadrics (or conics) is a typical example. In those days, the study of invariants was done mostly from mathematical interest, and few people conceived applications to engineering problems.

In the 1960s, the theory of invariants suddenly attracted the attention of researchers studying continuum mechanics from a very general viewpoint. Their study was called *rational mechanics*. They were interested in expressing mechanical responses of materials in a way independent of the coordinate system. The choice of the coordinate system in the material is dictated by the apparatus used for the measurement. If experimental data are not converted into invariants, data obtained with reference to different coordinate systems (resulting from different apparatuses) cannot be compared, even if the materials are identical. Well-known books on this subject are

C. Truesdell, R. A. Toupin: *Classical Field Theories*, Handbuch der Physik, Group 2, Vol. 3, Part (Springer, Berlin, Heidelberg 1960)
A. C. Eringen: *Nonlinear Theory of Continuous Media* (McGraw-Hill, New York 1962)
C. Truesdell, W. Noll: *Nonlinear Field Theories of Mechanics*, Handbuch der Physik, Group 2, Vol. 3, Part 3 (Springer, Berlin, Heidelberg 1965)
E. Kroner (ed.): *Mechanics of Generalized Continua* (Springer, Berlin, Heidelberg 1968)
A. C. Eringen (ed.): *Continuum Physics*, Vols. I–IV (Academic, New York 1971, 1975, 1976, 1977)

A comprehensive treatment of polynomial invariants and integral bases for vectors and tensors was given by Spencer (1971), and mechanical responses of isotropic materials were expressed in terms of functional invariants by Wang:

A. J. M. Spencer: "Theory of Invariants", in *Continuum Physics*, Vol. 1, ed. by A. C. Eringen (Academic, New York 1971) pp. 239–353
C.-C. Wang: On a general representation theorem for constitutive relations. Arch. Ration. Mech. Anal. **33**, 1–25 (1969)
C.-C. Wang: On representations for isotropic functions, Part I. Isotropic functions of symmetric tensors and vectors. Arch. Ration. Mech. Anal. **33**, 249–267 (1969)
C.-C. Wang: On representations for isotropic functions, Part II. Isotropic functions of skew-symmetric tensors, symmetric tensors, and vectors. Arch. Ration. Mech. Anal. **33**, 249–267 (1969)

Later, mistakes in Wang's papers were pointed out, and a revised version was presented:

G. F. Smith: On a fundamental error in two papers of C.-C. Wang "On representations for isotropic functions, Parts I and II", Arch. Ration. Mech. Anal. **36**, 161–165 (1970)
C.-C. Wang: A new representation theorem for isotropic functions: An answer to Professor G. F. Smith's criticism of my papers on representations for isotropic functions, Part 1, Scalar-valued isotropic functions. Arch. Ration. Mech. Anal. **36**, 166–197
C.-C. Wang: A new representation theorem for isotropic functions: An answer to Professor G. F. Smith's criticism of my papers on representations for isotropic functions, Part 2, Vector-valued isotropic functions, symmetric tensor-valued isotropic functions, and skew-symmetric tensor-valued isotropic functions. Arch. Ration. Mech. Anal. **36**, 198–223 (1970)

Again, some minor errors were pointed out, and they were corrected:

G. F. Smith: On isotropic functions of symmetric tensors, skew-symmetric tensors and vectors. Int. J. Eng. Sci. **9**, 899–916 (1971)
C.-C. Wang: Corrigendum to my recent papers on "Representations for isotropic functions". Arch. Ration. Mech. Anal. **43**, 392–395 (1971)

This chapter is mostly based on

K. Kanatani: Camera rotation invariance of image characteristics. Comput. Vision, Graphics Image Process. **39**, 328–354 (1987)

Chapter 5

Classic works on tensor calculus are

J. A. Schouten: *Tensor Analysis for Physicists* (Clarendon, Oxford 1951)
J. A. Schouten: *Ricci Calculus* (Springer, Berlin 1954)

Spherical harmonics are treated in relation to representations of $SO(3)$ in

I. M. Gel'fand, R. A. Minlos, Z. Ya. Shapiro: *Representations of the Rotation and Lorentz Groups and Their Applications* (Macmillan, New York 1966)
A. R. Edmonds: *Angular Momentum in Quantum Mechanics* (Princeton University Press, Princeton, NJ 1974)

"Image features" obtained by weighted integration of gray-level images have been studied in relation to pattern recognition. In particular, those that are constructed from *moments* (i.e., integration of $x^p y^q$) have been extensively studied. Usually, invariances are imposed with respect to such transformations as scalings, 2D translations, and 2D rotations. For this reason, the resulting quantities are often called *moment invariants*. Probably the most easily available reference book is

A. Rosenfeld, A. C. Kak: *Digital Picture Processing*, Vol. 2, 2nd ed. (Academic, Orlando, FL 1982)

in which our "image features" are called "picture properties". Most of the studies on this type of image characterization — both theories and applications — are scattered in journals. For example:

M. K. Hu, Visual pattern recognition by moment invariants. IRE Trans. Inf. Theory IT-8, 179–187 (1962)
S. Amari: Invariant structures of signal and feature spaces in pattern recognition problems. RAAG Memoirs **4**, 553–566 (1968)

S. A. Dudani, K. J. Breeding, R. B. McGhee: Aircraft identification by moment invariants. IEEE Trans. C-26, 39–46 (1977)
R. Y. Wong, E. L. Hall: Scene matching with invariant moments. Comput. Vision, Graphics Image Process. **8,** 16–24 (1978)
A. M. Shvedov, A. A. Shmidt, V. A. Yakumovich: Invariant systems of features in pattern recognition. Autom. Remote Control (USSR) **40,** 430–441 (1979)
F. A. Sadjadi, E. L. Hall: Three dimensional moment invariants. IEEE Trans. PAMI-2, 127–136 (1980)
M. Teague: Image analysis via the general theory of moments. J. Opt. Soc. Am. **70,** 920–930 (1980)
J. J. Reddi: Radial and angular moment invariants for image identification. IEEE Trans. PAMI-3, 240–242 (1981)
Y. S. Abu-Mostafa, D. Psaltis: Recognitive aspects of moment invariants. IEEE Trans. PAMI-6, 698–706 (1984)
K. Kanatani: Detecting the motion of a planar surface by line and surface integrals. Comput. Vision, Graphics Image Process. **29,** 13–22 (1985)
D. Cyganski, J. A. Orr: Applications of tensor theory to object recognition and orientation determination. IEEE Trans. PAMI-7, 662–673 (1985)
K. Kanatani: Camera rotation invariance of image characteristics. Comput. Vision, Graphics Image Process. **39,** 328–354 (1987)

This chapter is mostly based on the last paper.

Chapter 6

The following book specifically discusses representations of $SO(3)$:

I. M. Gel'fand, R. A. Minlos, Z. Ya. Shapiro: *Representations of the Rotation and Lorentz Groups and Their Applications* (Macmillan, New York 1966)

Treatments of $SO(3)$ in relation to $SU(2)$ in a more general mathematical framework — Lie groups, Lie algebras, adjoint groups, adjoint representations, derivations, etc. — are found, for example, in

C. Chevalley: *Theory of Lie Groups I* (Princeton University Press, Princeton, NJ 1946)
N. Steenrod: *The Topology of Fiber Bundles* (Princeton University Press, Princeton, NJ 1951)
L. Pontrjagin: *Topological Groups*, 2nd ed. (Gordon and Breach, New York 1966)
R. Hermann: *Lie Groups for Physicists* (Benjamin, New York 1966)
R. F. Schutz: *Geometrical Methods of Mathematical Physics* (Cambridge University Press, Cambridge 1980)
T. Bröcker, T. Tom Dieck: *Representations of Compact Lie Groups*, Graduate Texts in Mathematics, Vol. 98 (Springer, Berlin, Heidelberg 1985)

Pontrjagin (1966) discusses invariant measures of general compact Lie groups. The topological aspects of Lie groups are discussed in the framework of *fiber bundles* by Steenrod (1951). However, these books require much preliminary knowledge of mathematics.

For discussions of fundamental groups, the following article is highly recommended. This article is written in relation to the classification of singularities in *liquid crystals*, and introduces fundamental groups and homotopy groups to physicists who do not have a mathematical background. The treatment is excellent and most enlightening to all non-mathematicians, whether interested in liquid crystals or not.

N. D. Mermin: The topological theory of defects in ordered mdia. Rev. Mod. Phys. **51,** 591–648 (1979)

Chapter 7

Visual perception of the outside world from moving images (caused by motions of objects or the viewer) has long been studied in relation to psychology. For example:

J. J. Gibson: *The Perception of the Visual World* (Houghton Mifflin, Boston, MA 1950)
D. A. Gordon: Static and dynamic visual fields in human space perception. J. Opt. Soc. Am. **55,** 1296–1303 (1965)
J. J. Gibson: *The Senses Considered as Perceptual Systems* (Houghton Mifflin, Boston, MA 1966)
J. C. Hay: Optical motions and space perception: An extension of Gibson's analysis. Psychol. Rev. **73,** 550–563 (1966)
J. J. Koenderink, A. J. van Doorn: Invariant properties of the motion parallax field due to the movement of rigid bodies relative to an observer. Opt. Acta **22,** 773–791 (1975)
J. J. Koenderink, A. J. van Doorn: Visual perception of rigidity of solid shape. J. Math. Biol. **3,** 79–85 (1976)
D. N. Lee: A theory of visual control of braking based on information about time-to-collision. Perception **5,** 437–459 (1976)
J. J. Koenderink, A. J. van Doorn; Local structure of movement parallax of the plane. J. Opt. Soc. Am. **66,** 717–723 (1976)
J. J. Gibson: *The Ecological Approach to Visual Perception* (Houghton Mifflin, Boston, MA 1979)
S. Ullman: *The Interpretation of Visual Motion* (MIT Press, Cambridge, MA 1979)
D. Regan, K. Beverly, M. Cynader: The visual perception of the motion in depth. Sci. Am. **241** (1), 122–133 (1979)
W. F. Clocksin: Perception of surface slant and edge labels from optical flow: A computational approach. Perception **9,** 253–269 (1980).
I. Hadani, G. Ishai, M. Gur: Visual stability and space perception in monocular vision: Mathematical model. J. Opt. Soc. Am. **70,** 60–65 (1980)
D. N. Lee: The optic flow field: The foundation of vision. Philos. Trans. R. Soc. Lond. **B290,** 169–179 (1980)
H. C. Longuet-Higgins: The interpretation of a moving retinal image. Proc. R. Soc. Lond. **B208,** 385–397 (1980)
K. Prazdny: Egomotion and relative depth map from optical flow. Biol. Cybern. **36,** 87–102 (1980)
J. J. Koenderink, A. J. van Doorn: Exterospecific component of the motion parallax field. J. Opt. Soc. Am. **71,** 953–957 (1981)
D. D. Hoffman, B. E. Flinchbaugh: The interpretation of biological motion. Biol. Cybern **42,** 195–204 (1982)
D. D. Hoffman: Inferring local surface orientation from motion fields. J. Opt. Soc. Am. **72,** 888–892 (1982)
J. H. Rieger: Information in optical flows induced by curved paths of observation. J. Opt. Soc. Am. **73,** 339–344 (1983)
H. C. Longuet-Higgins: The visual ambiguity of a moving plane. Proc. R. Soc. Lond. **B223,** 165–175 (1984)

Mathematical schemes of 3D recovery from optical flow have been proposed by many researchers of computer vision. However, the purpose of 3D recovery differs from researcher to researcher: some aim at the reconstruction of object shape, some are interested in object motion, and others try to compute the motion of the viewer relative to a stationary scene, called *egomotion* (Sect. 2.6). Mathematical principles are also different depending on what type of information is available — a dense flow field, image velocities of sparsely distributed points, orthographic projection or perspective projection, etc. — and also what type of assumption is used — the object shape is known but its motion unknown, the motion is known but the shape unknown, the motion is translation without rotation, rotation without translation, or constrained on a horizontal plane, etc. This chapter is based on Kanatani (1985b, 1987, 1988). The analysis of Sect. 2.6 is based on Kanatani (1986).

J. W. Roach, J. K. Aggarwal: Determining the movement of objects from a sequence of images. IEEE Trans. PAMI-2, 554–562 (1980)

H.-H. Nagel: Representation of moving rigid objects based on visual observations. Computer **14**(8), 29–39 (1981)

T. S. Huang, R. Y. Tsai: "Image sequence analysis: Motion estimation", in Image sequence Analysis, ed. by T. S. Huang, Springer Ser. Info. Sci., Vol. 5 (Springer, Berlin, Heidelberg 1981)

H. C. Longuet-Higgins: A computer algorithm for reconstructing a scene from two projections. Nature **293**(10), 133–135 (1981)

K. Prazdny: Determining the instantaneous direction of motion from optical flow generated by a curvilinear moving observer. Comput. Vision, Graphics Image Process. **17**, 124–248 (1981)

A. R. Bruss, B. K. P. Horn: Passive navigation. Comput. Vision, Graphics Image Process. **21**, 3–20 (1983)

B. L. Yen, T. S. Huang: Determining 3-D motion and structure of a rigid body using the spherical projection. Comput. Vision, Graphics Image Process. **21**, 21–32 (1983)

J.-Q. Fang, T. S. Huang: Solving three-dimensional small-rotation motion equations: Uniqueness, algorithm, and numerical results. Comput. Vision, Graphics Image Process. **26**, 183–206 (1984)

R. Y. Tsai, T. S. Huang: Uniqueness and estimation of three-dimensional motion parameters of rigid objects with curved surfaces. IEEE Trans. PAMI-6, 13–27 (1984)

K. Sugihara, N. Sugie: Recovery of rigid structure from orthographically projected optical flow. Comput. Vision, Graphics Image Process. **27**, 309–320 (1984)

K. Kanatani: Tracing planar surface motion from a projection without knowing the correspondence. Comput. Vision, Graphics Image Process **29**, 1–12 (1985a)

K. Kanatani: Detecting the motion of a planar surface by line and surface integrals. Comput. Vision, Graphics Image Process. **29**, 13–22 (1985b)

A. M. Waxman, S. Ullman: Surface structure and three-dimensional motion from image flow kinematics. Int. J. Robotics Res. **4**, 72–94 (1985)

A. M. Waxman, K. Wohn: Contour evolution, neighborhood deformation, and global image flow: planar surfaces in motion. Int. J. Robotics Res. **4**, 95–108 (1985)

S. J. Maybank: The angular velocity associated with the optical flowfield arising from motion through a rigid environment. Proc. R. Soc. Lond. **A401**, 317–326 (1985)

M. Subbarao, A. M. Waxman: Closed form solution to image flow equations for planar surface in motion. Comput. Vision, Graphics Image Process. **36**, 208–228 (1986)

K. Kanatani: Structure and motion from optical flow under orthographic projection. Comput. Vision, Graphics Image Process. **35**, 181–199 (1986)

K. Kanatani: Structure and motion from optical flow under perspective projection. Comput. Vision, Graphics Image Process. **38**, 122–146 (1987)

B. K. P. Horn: Motion fields are hardly ambiguous. Int. J. Comput. Vision **1**, 259–274 (1987)

K. Kanatani, K. Yamada: Model-based determination of object position and orientation without matching. J. Infor. Process. **12**(1), 1–8 (1988)

There are many studies of optical flow detection techniques, see Sect. 2.6. There are also numerous works on the use of image motion for image segmentation and object recognition. Much experimental research has also been done to implement the theoretically proposed 3D recovery schemes by hardware and examine their accuracy and applicability for practical problems. Since these studies are now actively in progress, it is difficult to list ongoing work.

As pointed out in Remark 7.9, the classification of possible views of a single object seen from different angles become important if motion detection techniques are to be combined with model-based object recognition problems. The infinite number of possible views must be grouped into a finite number of *topological equivalence classes* (or *characteristic views*, *attitudes*, etc.). Related works include:

J. J. Koenderink, A. J. Van Doorn: The singularities of the visual mapping. Biol. Cybern. **24**, 51–59 (1979).

J. J. Koenderink, A. J. Van Doorn: Internal representation of solid shape with respect to vision. Biol. Cybern. **32**, 211–216 (1979)

K. Sugihara: Automatic construction of junction dictionaries and their exploitation for analysis for range data. Proc. 6th Intl. Joint Conf. Artificial Intelligence, August 1979, Tokyo, Japan, pp. 859–864

I. Chakravarty, H. Freeman: Characteristic views as a basis for three-dimensional object representation. Proc. SPIE Conf. Robot Vision, 336, May 1982, Arlington, VA, pp. 37–45.

C. Thorpe, S. Shafer: Correspondence in line drawings of multiple views. Proc. 8th Intl. Joint Conf. Artificial Intelligence, August 1983, Karlsruhe, FRG, pp. 959–965

G. Fekete, L. S. Davis: Property spheres: A new representation for 3-D object recognition, Proc. IEEE Workshop on Computer Vision: Representation and Control, April–May 1984, Annapolis, MD, pp. 192–201

R. Scott: Graphics and prediction from models. Proc. DARPA Image Understanding Workshop, October 1984, New Orleans, LA, pp. 98–106

P. J. Besl, R. C. Jain: Three-dimensional object recognition. Comput. Surveys **17**, 75–145 (1985)

K. Ikeuchi: Generating an interpretation tree from a CAD model for 3D-object recognition in bin-picking tasks, Int. J. Comput. Vision **1**, 145–165 (1987)

Chapter 8

The *Huffman–Clowes edge labeling* was given independently by Huffman (1971) and Clowes (1971). It had developed out of similar ideas of Guzman (1968) and others. Various generalizations of the Huffman–Clowes edge labeling were proposed: Waltz (1975) considered line drawings with shading and shadows; Kanade (1980) allowed thin paper-like objects (the *origami world*); Sugihara (1978) considered images where hidden lines were also given. The NP-completeness of Huffman–Clowes edge labeling was pointed out by Kirousis and Papadimitriou (1985).

Attempts to reject "impossible objects" were made first by the use of the *gradient space* proposed by Huffman (1977). Later, Sugihara (1982a, 1982b, 1984, 1986) presented a consistent algebraic approach to classify all impossible objects.

The use of the *rectangularity hypothesis* was suggested by Mackworth (1976). Barnard (1985) attempted 3D recovery based on the rectangularity hypothesis. Both argued on possible relationships to the human visual perception mechanism. Shakunaga and Kaneko (1989) tried to solve a wider class of related problems. Sections 8.2 and 8.3 of this chapter are based on Kanatani (1986), and Section 8.4 is based on Kanatani (1988).

The use of the *parallelism hypothesis* was applied to a vision expert system by Mulgaonkar et al. (1986). The *skew-symmetry hypothesis* was formulated by Kanade (1981) and Kanade and Kender (1983). For hypotheses about closed contours, see Brady and Yuille (1984).

A. Guzman: Computer Recognition of Three-Dimensional Objects in a Visual Scene, Technical Report 228, Artificial Intelligence Laboratory, Massachusetts Institute of Technology, December 1968
M. B. Clowes: On seeing things. Artif. Intell. **2,** 79–116 (1971)
D. A. Huffman: "Impossible Objects as Nonsense Sentences", in *Machine Intelligence*, Vol. 6, ed. by B. Meltzer, D. Michie (Edinburgh University Press, Edinburgh, U.K. 1971) pp. 295–323
A. K. Mackworth: Interpreting pictures of polyhedral scenes. Artif. Intell. **4,** 121–137 (1973)
D. Waltz: "Understanding Line Drawings of Scenes with Shadows", in *The Psychology of Computer Vision*, ed. by P. H. Winston (McGraw-Hill, New York 1975) pp. 19–91.
A. K. Mackworth: Model-driven interpretation in intelligent vision systems. Perception **5,** 349–370 (1976)
D. A. Huffman: "A Duality Concept for the Analysis of Polyhedral Scenes", in *Machine Intelligence* Vol. 8, ed. by E. Elcock, D. Michie (Ellis Horwood, Chichester, U.K. 1977) pp. 475–492
K. Sugihara: Picture language for skeletal polyhedra. Comput. Vision, Graphics Image Process. **8,** 382–405 (1978)
T. Kanade: A theory of origami world. Artif. Intell. **13,** 279–311 (1980)
T. Kanade: Recovery of the three-dimensional shape of an object from a single view. Artif. Intell. **17,** 409–460 (1981)
K. Sugihara: Classification of impossible objects. Perception **11,** 65–74 (1982a)
K. Sugihara: Mathematical structures of line drawings of polyhedrons — toward man-machine communication by means of line drawings. IEEE Trans. PAMI-4, 458–469 (1982b)
T. Kanade, J. R. Kender: "Mapping Image Properties into Shape Constraints: Skewed Symmetry, Affine Transformable Patterns, and the Shape-from-Texture Paradigm", in *Human and Machine Vision*, ed. by J. Beck, B. Hope, A. Rosenfeld (Academic, New York 1983) pp. 237–268
M. Brady, A. Yuille: An extremum principle for shape from contour. IEEE Trans. PAMI-6, 288–301 (1984)
K. Sugihara: A necessary and sufficient condition for a picture to represent a polyhedral scene. IEEE Trans. PAMI-6, 578–586 (1984)
S. T. Barnard: Choosing a basis for perceptual space. Comput. Vision, Graphics Image Process **29,** 87–99 (1985)
K. Kanatani: The constraints on images of rectangular polyhedra. IEEE Trans. PAMI-8, 456–463 (1986)
P. G. Mulgaonkar, L. G. Shapiro, R. M. Haralick: Shape from perspective: A rule-based approach. Comput. Vision, Graphics Image Process. **36,** 298–320 (1986)
K. Sugihara: *Machine Interpretation of Line Drawings* (MIT Press, Cambridge, MA 1986)
K. Kanatani: Constraints on length and angle. Comput. Vision, Graphics Image Process. **41,** 28–42 (1988)
L. M. Kirousis, C. H. Papadimitriou: The complexity of recognizing polyhedral scenes. J. Comput. System Sci. **37,** 14–38 (1988)
T. Shakunaga, H. Kaneko: Perspective angle transform: Principle of shape from angles. Int. J. Comput. Vision **3**(3), 239–254 (1989)

Chapter 9

The theory of distributions is found in Schwartz's own books:

L. Schwartz: *Théorie des Distributions*, Vols. 1, 2 (Hermann, Paris 1950, 1951)
L. Schwartz: *Méthodes Mathématiques pour les Sciences Physiques* (Hermann, Paris 1961)

The first may be too mathematical. The last one is a good textbook for non-mathematicians.

Gibson (1950, 1966, 1979) regarded converging texture as one of the most important clues to human perception of the outside world. This view is discussed in

J. J. Gibson: *The Perception of the Visual World* (Houghton Mifflin, Boston, MA 1950)
J. J. Gibson: *The Senses Considered as Perceptual Systems* (Houghton Mifflin, Boston, MA 1966)
J. J. Gibson: *The Ecological Approach to Visual Perception* (Houghton Mifflin, Boston, MA 1979)
H. A. Sedgwick: "Environment-Centered Representation of Spatial Layout: Available Visual Information from Texture and Perspective", in *Human and Machine Vision*, ed. by J. Beck, B. Hope, A. Rosenfeld (Academic, New York 1983) pp. 425–458

In computer vision, the first attempted methods took the "structure-based approach", computing such structures as regularity, periodicity, collinearity, parallelism, orthogonality, and symmetry. Works on this approach include:

K. Ikeuchi: Shape from Regular Patterns (An Example of Constraint Propagation in Vision). MIT AI Memo 567, March 1980.
H. Nakatani, S. Kimura, O. Saito, T. Kitahashi: "Extraction of Vanishing Point and Its Application to Scene Analysis Based on Image Sequence, in Proc. Intl. Conf. Pattern Recog., August 1980, Miami Beach, FL, pp. 370–372
T. Kanade: Recovery of three-dimensional shape of an object from a single view. Artif. Intell. **17,** 409–460 (1981)
Y. Ohta, K. Maenobu, T. Sakai: "Obtaining Surface Orientation from Texels Under Perspective Projection", in Proc. 7th Intl. Joint Conf. Artif. Intell., August 1981, Vancouver, BC, pp. 746–751
K. A. Stevens. The information content of texture gradients. Biol. Cybern. **42,** 183–195 (1981)
J. R. Kender: "Surface Constraints from Linear Extents", in Proc. 3rd Natl. Conf. Artif. Intell., August 1983, Washington, DC, pp. 187–190
K. A. Stevens: Slant-tilt: The visual encoding of surface orientation. Biol. Cybern. **46,** 183–195 (1983)
K. A. Stevens: Surface tilt (the direction of slant): A neglected psychological variable. Percept. Psychophys. **33,** 241–250 (1983)

The statistical method based on the assumption of isotropy was first proposed by Witkin (1981). The algorithm was later improved by Davis et al. (1983). Kanatani (1984) gave a rigorous mathematical description of the problem and explicit analytical formulae by invoking tensor calculus.

A. P. Witkin: Recovering surface shape and orientation from texture. Artif. Intell. **17,** 17–45 (1981)
L. S. Davis, L. Janos, S. M. Dunn: Efficient recovery of shape from texture. IEEE Trans. PAMI-5, 485–492 (1983)
K. Kanatani: Detection of surface orientation and motion from texture by a stereological technique. Artif. Intell. **23,** 213–237 (1984)

Statistical methods based on homogeneity have also been proposed and tried by many researchers. For example:

R. Bajcsy, L. Lieberman: Texture gradient as a depth cue. Comput. Vision, Graphics Image Process. **5,** 52–67 (1976)
A. Rosenfeld: A note on automatic detection of texture gradients. IEEE Trans. C-24, 988–991 (1975)
S. W. Zucker, A. Rosenfeld, L. S. Davis: Picture segmentation by texture discrimination. IEEE Trans. C-24, 1228–1233 (1975)
S. M. Dunn: Recovering the Orientation of Textured Surfaces, Ph.D. Thesis, Center for Automation Research, University of Maryland (1986)
J. Aloimonos, M. J. Swain: Shape from texture. Biol. Cybern. **58,** 345–360 (1988)
K. Kanatani T.-C. Chou: Shape from Texture: General Principle. Artif. Intell. **38,** 1–48 (1989)
D. Blostein, N. Ahuja: A multiscale region detector. Comput. Vision, Graphics Image Process. **45,** 22–41 (1989)

This chapter is based on Kanatani and Chou (1989). The Problem of resolution threshold and subtexture is treated in Blostein and Ahuja (1989).

Chapter 10

In this book, we do not discuss the *shape-from-shading* paradigm, which attempts to recover the surface gradient from the intensity of light reflectance from the object surface. Many papers and articles can be found on this subject. For example:

B. K. P. Horn: "Obtaining Shape from Shading Information", in *The Psychology of Computer Vision*, ed. by P. H. Winston (McGraw-Hill, New York 1975) pp. 115–155

R. J. Woodham: Analysing images of curved surfaces. Artif. Intell. **17,** 117–140 (1981)

K. Ikeuchi, B. K. P. Horn: Numerical shape from shading and occluding boundaries. Artif. Intell. **17,** 141–184 (1981)

B. K. P. Horn: *Robot Vision* (MIT Press, Cambridge, MA 1986)

The term *$2\frac{1}{2}$D sketch* was coined by David Marr. His philosophy is found in

D. Marr: *Vision: A Computational Investigation into the Human Representation and Processing of Visual Information* (W. H. Freeman, San Francisco 1982)

The argument on the constraints on polyhedron images in Sect. 10.2 — the degree of freedom and the criterion of regularity in particular — is based on

K. Sugihara: Classification of impossible objects. Perception **11,** 65–74 (1982)

K. Sugihara: Mathematical structures of line drawing of polyhedrons — toward man–machine communication by means of line drawings. IEEE Trans. PAMI-4, 458–469 (1982)

K. Sugihara: A necessary and sufficient condition for a picture to represent a polyhedral scene. IEEE Trans. PAMI-6, 578–586 (1984)

K. Sugihara: *Machine Interpretation of Line Drawings* (MIT Press, Cambridge, MA 1986)

The *regularization* concept in relation to computer vision has been endorsed by Tomas Poggio and his coworkers. Related works include:

A. N. Tikhonov, V. Y. Arsenin: *Solutions of Ill-posed Problems* (Winston, Washington, DC 1977)

V. A. Morozov: *Methods for Solving Incorrectly Posed Problems* (Springer, New York 1984)

T. Poggio, V. Torre, C. Koch: Computational vision and regularization theory. Nature **317,** 314–319 (1985)

T. Poggio, C. Koch: Ill-posed problems in early vision: from computational theory to analogue networks. Proc. R. Soc. Lond. B 226, 303–323 (1985)

For computational geometry, see

F. P. Preparata, M. I. Shamos: *Computational Geometry Texts and Monographs in Computer Science* (Springer, New York 1985)

G. Toussaint (ed.): *Computational Geometry* (North-Holland, Amsterdam 1985)

This chapter is based on:

K. Kanatani: Reconstruction of consistent shape from inconsistent data: Optimization of $2\frac{1}{2}$D sketches. Int. J. Comput. Vision **3**(4), 261–292 (1989)

Subject Index

Abelian group 27, 35, 402, 405
Abelian Lie algebra 427
aberration 5
absolute convergence 66, 99
absolute depth 36, 240, 356
absolute geometry 18
absolute invariant 26, 247
accumulation point, see limit point
active motion 193
active vision 357
addition 402, 405
adjacency condition 239, 260
adjacent flows, see linearly adjacent flows
adjoint (of a linear operator) 414
adjoint camera rotation transformation, see adjoint transformation
adjoint representation 216, 217, 430
adjoint transformation (of a Lie algebra) 216, 429
adjoint transformation (of an image feature) 181
admissible transformations 118–120, 155
affine connection 18
algebraic equation 136
algebraic expression 104, 136
algebraic invariant 136
alias 82
alibi 82, 213
almost everywhere 333
ambiguity 53, 255
analytical level 4
angle (of a half-line) 305
angle (of a rotation) 202
angular momentum 171
angular velocity 111
anharmonic ratio, see cross ration
anticommutative operation 65
anticommutative operator 426
anti-Hermitian matrix 212
anti-homomorphism 73
antipode 226
antisymmetric part 30, 76, 112, 129
antisymmetric tensor 157
antisymmetrization 157
aperture problem 46
apparent texture, see supertexture
approximation in the sense of a distribution 329
approximation in the weak sense 329
arcwise connected component 424
arcwise connected space 424
arrow 281, 299
artificial intelligence 5
aspect ratio 384
associated Legendre equation 434
associated Legendre function 170, 434
associated vector 311, 312, 386
associative law 399, 402, 406
asymptotic direction 55, 56
axis 113, 132
axis (of a rotation) 202
azimuthal quantum number 82, 171

backprojection 14
backward reasoning 5
base (of a loop) 424
basic operations 135
basis 407
Beltrami, Eugenio 155
Betti number 423
Bianchi, Luigi 155
bijection 400
bilinear operation 65, 426
binary relation 399
blocks world 278
boson 74
bottom-up approach 4
boundary 421
boundary point 421
boundary value problem 172
bounded subset 421
broken symmetry 85
buckling 85

camera calibration 5
camera constant, see focal length
camera rotation transformation (of images) 11, 25, 61, 92, 151
camera rotation transformation (of scenes) 190
canonical angle 305, 380
canonical basis (of an irreducible representation of $SO(3)$) 80
canonical form (of a matrix) 27, 41, 131, 413
canonical form (of infinitesimal generators of $SO(3)$) 80
canonical frame 131
canonical position 302
Cartesian product (of sets) 399
Cartesian tensor 155
Casimir operator (of a Lie group) 68
Casimir operator (of a representation of $SO(3)$) 71, 75, 76, 80, 93, 103, 165, 175, 432
Cauchy sequence 100
Cayley algebra 225
Cayley-Hamilton theorem 133, 144
Cayley-Klein parameters 209
Cayley numbers 225
center of projection, see viewpoint
central projection, see perspective projection
centroid (of a region) 187
centroid (of a texture) 342
character (of a representation) 420
characteristic equation 130, 202
characteristic function 180, 186, 330
Christoffel, Elwin B. 155
C^∞-differentiable manifold 423
C^∞-function 333
circular paraboloid 55
C^k-differentiable manifold 423
classical groups 424
classical theory of invariants 136
Clebsch-Gordon series 420
closed path, see loop
closed set 421
closure 421
Clowes, Maxwell B. 299
collinearity 13
collinearity test 390
collineation 13, 322
combinatorial algorithm 300
commutation relations (of a Lie algebra) 66, 427
commutation relations (of a Lie algebra of $SO(3)$) 70, 73, 75, 79, 83, 93, 167, 432
commutative group, see Abelian group
commutative law 406
commutative Lie algebra 66
commutator (of differential operators) 167
commutator (of linear operator) 426
commutator (of matrices) 65
compact (Lie group) 225
compact group 27, 418
compact space 422
compatibility test 380
completeness (of harmonics) 171
completeness (of spherical harmonics) 161, 434
complex linear space 406, 420
complex number plane 29, 209
complex vector space, see complex linear space
components (of a vector), see coordinates (of a vector)
composition, see multiplication
composition (of mappings) 399
computer tomography 13
concatenation (of paths), see product (of paths)
concurrency test 384
concurrent lines 310
conformal mapping 209
congruence 191
conic 321, 323
conjugate (of a quaternion) 221
conjugate class 404
conjugate lines 323
conjugate points 113, 323
conjugate representation 419
connected component 422, 428
connected Lie group 66, 429
connected set 422
connected space 422
connected topological group 201
connectivity 422
consistency condition 17
constraint (on an object) 4, 15, 278
constraint (on a polyhedron) 362
constraint propagation 301
continuous functional 333
continuous group 425
continuous mapping 421, 422
continuously deformed path 425
contraction (of a tensor) 156
contragradient representations 182
contravariant tensor 182
convergence (of a sequence) 422
convergence (of distributions) 334
convergence as distributions 334
convergence in norm 334, 434
coordinate change 423

coordinate system (of a rectangular polyhedron) 290, 294
coordinates (of a manifold) 423
coordinates (of a vector) 407
correlation (of a projective space) 323
correlation-based approach 46
correspondence detection 239
coset 404
cosine law 143
Coulomb potential field 169
covariance 355
covariant tensor 182
covering 422
covering group (of a Lie group) 67, 425
covering space (of a topological space) 425
covering space (of $SO(3)$) 74
cross ratio 13
CT, see computer tomography
curvature 18
curvilinear coordinate system 106, 177

deformable path, see continuously deformed path
degeneracy (of a polyhedron) 289
degeneracy (of invariants) 27, 35
degree (of a representation) 415
degree (of a symmetric polynomial) 135
degree (of a tensor) 155
degree of freedom (of a polyhedron) 360, 362
degree of homogeneity 330
delta function 333
density (of a texture) 49, 327, 328, 330
depth map 274, 357
derivation 219, 429, 430
derivative 244
derivative (of a distribution) 333
description 4
determinant 401, 413
deviator part 32, 78, 162
deviator tensor 94, 134, 161
diagonalization 27
diffeomorphic manifolds 423
diffeomorphism 423
difference 406
difference (of sets) 398
difference (of tensors) 155
differentiable manifold 423
differentiable mapping 423
differential 335
differential form 428
differential geometry 17, 327, 328
differential representation (of a Lie algebra) 219, 430
differential solid angle 157
dimension (of a Lie algebra) 65
dimension (of a Lie group) 62, 65
dimension (of a linear space) 406
dimension (of a manifold) 423
dipole moment field 169
Dirac delta function, see delta function
direct factors 403
direct product (of groups) 403
direct product (of topological spaces) 421
direct product topology 421
direct sum (of groups) 403
direct sum (of linear operators) 415
direct sum (of linear spaces) 411
direct sum (of representations) 24
directional cosine 48
disconnected set 422
disconnected space 422
discrete topology 421
discriminant (of a cubic equation) 145
disjoint sets 399
distribution 329, 333
divergence 39, 247
divergence theorem, see Gauss divergence theorem
divergent flow 39
domain (of knowledge) 4
domain (of a mapping) 399
dual (of a line) 108, 317
dual (of a point) 108, 317
dual basis, see reciprocal basis
dual representation, see transposed representation
duality (of points and lines) 108, 278, 311, 317, 319, 321

early vision 4, 359
edge detection 4, 260, 353
egomotion 47
eigenvalue 413
Einstein, Albert 155
Einstein's summation convention 155
element 398
elementary particle theory 85
elliptic surface 54, 57
elongation ratio 338
empty set 398
energy 85
entropy 85

epipolar constraint 273
epipolar line 273, 274
equality (of sets) 398
equation of continuity 180
equivalence (of images) 105, 137, 189
equivalence class 425
equivalence relation 425
equivalent representations 416
equivalent topologies 421
Euclidean inner product 411
Euclidean metric 411
Euclidean metric space, see Euclidean space
Euclidean norm 411
Euclidean space 411
Euclidean topology 421
Euler angles 205
Euler characteristic 423
Euler number, see Euler characteristic
Euler-Poincare characteristic, see Euler characteristic
Euler's formula (for complex exponent) 29, 223
Euler's formula (for symmetric polynomials) 136, 144
Euler's theorem (of 3D rotations) 62, 69, 202
even function 176
even parity 35
even permutation 157
exclusive or 290
expert vision 4
extended Gaussian sphere representation 5

factor group, see quotient group
faithful representation 28, 415
false image 18, 298
fanning (of a flow) 247, 255
feature (of a pattern) 3
feature point 46, 242
fermion 74
fiber bundle 428
field (of numbers) 224
field (of quaternions) 224
finite group 403
finite support 333
first fundamental form 58, 60, 178, 328, 335
first fundamental metric tensor 336
first homotopy group, see fundamental group
first variation 244
first-order predicate logic 5
fixed line 111
fixed point 111
flow equations 36, 86, 241, 257
flow parameters 37, 86, 239, 242
focal length 5, 7, 321
foreshortening (of a flow), see fanning
fork 281, 299
fork coordinate system 294
forward reasoning 5
4D spherical coordinates 232
Fourier transform 334
frame 5
free energy 85
Frobenius' theorem 368, 428
fully reducible representation 24, 417
fully reducible subspace 152
functional 333
functional basis 137
functional invariant 137
fundamental group 228, 425
fundamental sequence, see Cauchy sequence
fundamental symmetric polynomials 129, 134, 135
fundamental theorem on symmetric functions 134
fuzzy reasoning 5

gauge group 85
Gauss divergence theorem 195
Gaussian curvature 57
Gaussian plane, see complex number plane
general linear group, see $GL(n)$ or $GL(n, \mathbb{R})$
general position 280
generalized cone representation 5
generalized cylinder representation, see generalized cone representation
generalized inverse, see pseudo-inverse
generated subspace 411
Gibson, James Jerome 45
$GL(n)$ 136
$gl(n, \mathbb{C})$ 429
$GL(n, \mathbb{C})$ 405, 424, 428, 429
$gl(n, \mathbb{R})$ 429
$GL(n, \mathbb{R})$ 405, 424, 429
gradient (of a surface), see surface gradient
gradient space 341
gradient-based approach 46
Gram matrix, see Gram tensor
Gram tensor 410
Gram-Schmidt orthogonalization, see Schmidt orthogonalization
grand unified field theory 85
great circle 177, 322

Green's theorem 275, 367
group 402
group homomorphism, see homomorphism (of groups)
group of permutations 35
group of transformations 11, 17, 404
group of transformations (of corners) 290

Hamilton quaternion, see quaternion
Hamiltonian 74, 85, 171
hard problem 301
harmonic 168, 171, 432
harmonic function, see harmonic
Hausdorff space 422
Heaviside step function, see step function
Heine-Borel theorem 422
Hermitian conjugate (of a matrix) 211, 414
Hermitian matrix 410, 414
Hermitian operator 414
Hessian 57, 59
Hessian normal form 108
hidden edge 290, 294
hierarchical reasoning 5
high-level vision 4
Hilbert's theorem 137
homeomorphic spaces 422
homeomorphism 74, 225, 229, 422
homogeneity (of a state) 85
homogeneity (of a texture) 49, 328, 329, 330
homogeneous coordinates (of the complex number plane) 210
homogeneous coordinates (of the image plane) 12, 113, 311, 322, 394
homogeneous image plane 301
homogeneous space 232
homogeneous state 85
homogeneous texture 49, 327, 328, 329
homomorphism (of groups) 23, 64, 229, 404
homomorphism (of Lie algebra) 219, 427
homomorphism (of linear spaces) 219
homomorphism theorem (of groups) 227
homotopic class 425
homotopic loops 228
homotopic paths 425
homotopic to 0 425
homotopy 425
Hough transform 318
Huffman, David A. 299
Huffman-Clowes edge labeling 4, 278, 279, 299
Huffman-Clowes junction dictionary 299
hyperbolic surface 54, 55, 57

identity 399
identity element, see unit element
identity representation (of $SO(2)$) 28
identity representation (of $SO(3)$) 74, 153
ill-conditioned matrix 244
ill-conditioned problem 374
ill-defined problem 374
ill-posed problem 374
image 3, 148
image (of a mapping) 399
image characteristic space 374
image characteristics 14, 37, 50, 88, 103, 243
image coordinate system 6
image coordinates 6
image feature 3, 14, 148, 181, 240, 264, 339
image flow, see optical flow
image plane 5
image properties, see image characteristics
image segmentation 4
image space 148, 149
image sphere 126, 191, 322, 385
image understanding 3
image-centered coordinate system 7
imaginary circle 321
impossible object 278, 298
improper rotation 120
incidence (of a point and a line) 323
incidence pair 360
incidence structure (of a polyhedron) 360
incident vertex 360
inclusion (of sets), see set inclusion
inclusion mapping 424
incompatibility (of adjacency) 357, 358
incompatibility (of rectangular corners) 381
incompatible junctions 301
incompatible labels 300
incompatible rectangular corners 381
independence 407
induced topology 421
infinitesimal camera rotation 62
infinitesimal camera rotation transformation 88, 167
infinitesimal generators (of a Lie group) 66, 428
infinitesimal generators (of $SO(3)$) 75, 77, 79, 80, 92, 165, 167, 431
infinitesimal rotation 69, 72

infinitesimal transformation 62, 64, 67, 88, 103
infinitesimal transformation (of $SU(2)$) 213
inhomogeneous coordinate (of the complex number plane) 210
inhomogeneous coordinates (of the image plane) 12, 113
inhomogeneous image plane 301
initial point 424
injection, see one-to-one mapping
inner automorphism 430
inner point, see interior point
inner product 409
integrability condition (of a differential equation) 367, 368, 428
integrability condition (of a Lie algebra) 67, 70
integration in Lebesgue's sense 333
integrity basis 137
interior 421
interior point 421
intermediate description 3
intermediate vision 4
interpretation (of a line drawing) 295
intersection (of sets) 398
intersection line 257
intrinsic quantity 58
invariance property 17, 22
invariant 22, 26, 239
invariant angle 124, 127, 385
invariant area 127, 186
invariant basis 105, 115–125, 131
invariant centroid 187
invariant distance 122, 125, 126, 385
invariant inertia center 189
invariant inner product 158
invariant integration 157, 418
invariant measure (of a Lie group) 418
invariant measure (of $SO(3)$) 231–233
invariant measure (of the image plane) 177
invariant metric 418
invariant orthogonal decomposition 160
invariant principal axis 189
invariant principal value 189
invariant set (of image characteristics) 104
invariant subgroup 403, 404
invariant subspace (of a linear operator) 414
invariant subspace (of a representation) 152, 416
invariant vector field 428
inverse 402
inverse (of a homotopic class) 425
inverse (of a matrix) 401
inverse (of a path) 425
inverse (of a quaternion) 224
inverse (of a transformation) 400
inverse image (of a mapping) 400
inverse perspective projection operator, see inverse projection
inverse problem 374
inverse projection 149
irreducible polynomial 137
irreducible reduction (of a representation) 420
irreducible reduction (of a representation of $SO(2)$) 24, 61, 239
irreducible representation 416
irreducible representation (of $SO(2)$) 24, 25, 27, 247
irreducible representation (of $SO(3)$) 71, 74, 93, 98, 103, 160, 165, 170, 174, 175, 184, 185
irreducible representation (of $SO(n)$) 35
irreducible set (of image characteristics) 104
irreducible set (of invariants) 137
irreducible subspace 152
isomorphic groups, see isomorphism (of groups)
isomorphic Lie algebras 427
isomorphic linear spaces 408
isomorphism (of groups) 404
isomorphism (of Lie algebras) 68, 427, 429
isomorphism (of Lie groups) 213
isomorphism (of linear spaces) 172, 408
isotropic state 85
isotropic texture 328
isotropy (of a state) 85

Jacobi identity 65, 219, 426
Jacobi matrix 88
Jacobian 231, 332
junction 299

kernel 172, 227
knowledge (of the domain) 4, 359
Kronecker delta 113, 410
Kronecker product, see tensor product
Kronecker product representation, see product representation

L_1-norm 354
L_2-norm 354, 434

HL_∞-norm 354
Lagrangian 85
Laplace equation 168, 432
Laplace spherical harmonic 170, 434
Laplace-Beltrami equation 168, 175
Laplace-Beltrami operator 168, 175
Laplacian 168, 432
laser ranging 13
law of conservation of mass 180
least squares method 37, 88, 243
Lebesgue measure 333
left invariant tangent space 64
left inverse 402
left unit element 402
Legendre equation 434
Legendre polynomial 170, 434
lens equation 5
Levi-Civita, Tullio 155
lexicographical order 135
Lie, Sophus 62
Lie algebra 62, 65, 103, 213, 426, 428
Lie algebra (of $SU(2)$) 212, 216
Lie algebra homomorphism, see homomorphism (of lie algebras)
Lie algebra isomorphism, see isomorphism (of Lie algebras)
Lie bracket 65, 426
Lie group 62, 423
limit distribution 334
limit point 421, 422
line 107
line at infinity 12, 322
line drawing 260
line fitting 317
linear combination 406
linear fractional transformation group 209
linear functional 17, 148, 181, 329, 333
linear mapping 400, 412
linear operator 412
linear representation, see representation
linear space 406
linear space homomorphism, see homomorphism (of linear spaces)
linear space isomorphism, see isomorphism (of linear spaces)
linear transformation 401, 416
linearly adjacent flows 259
linearly dependent vectors 406
linearly independent vectors 406
linearly ordered set 421
Lipschitz, Rudolf O. S. 155
L-junction 282
local coordinate system 423
local homomorphism (of Lie groups) 429
local isomorphism (of Lie groups) 213, 429
locally arcwise connected space 424
locally connected space 422
locally isomorphic Lie goups 429
locally path connected space, see locally arcwise connected component
loop 424
Lorentz group 424
low-level vision 4

magnetic quantum number 82, 171
manifold 423
mapping 399
Marr, David 359
matching 18
matching condition 18
matrix 400, 416
matrix elements 198, 413
matroid 407
maximal compatible set 381
maximum contraction 41
maximum extension 41
maximum principle (of harmonics) 172
maximum shearing 44
mean curvature 55, 57
mean value theorem (of harmonics) 172
measure 0 365
medial axis, see skeleton
method of separation of variables 433
metric 18
metric linear space, see metric space
metric property (of a line drawing) 297
metric space 158, 409, 421
metric tensor 410
metric topology 421
middle image 378
minimal surface 57
mirror image 53
model (of an object) 274, 375
model (of the domain) 4
model-based object recognition 271
modeling 13
modulus (of a quaternion) 224
moment 187
moment of inertia 191
Monte Carlo method 330
motion field, see optical flow
motion parameters 241

multiplication 402
multipole moment field 169

n-ary relation 399
natural basis vectors 198
natural coordinates (of $su(2)$) 217
nD general linear group, see $GL(n, \mathbb{R})$
nD linear transformation 401
nD rotation group, see $SO(n)$
nD rotation matrix 201
nD special orthogonal group, see $SO(n)$
nD special unitary group, see $SU(n)$
nD symplectic group, see $Sp(n)$
nD unimodular group, see $SL(n)$
nD unitary group, see $U(n)$
Necker cube phenomenon 292
negative 402, 406
neighborhood 229, 421
Newton iterations 349
non-Euclidean metric 128
nonsingular matrix 401
norm 375, 409
normal curvature 57, 60
normal equations 58, 244, 313
normal subgroup, see invariant subgroup
normalized homogeneous coordinates 385
normed linear space, see normed space
normed space 409
NP-complete problem 301
null set, see empty set
nullity 413

$\boldsymbol{O}(3)$ 118, 120
$o(n)$ 429
$O(n)$ 136, 405, 424
object parameter space 374
object parameters 14, 36, 49, 50, 88, 243
observable 24, 264–267, 269, 337–339
octahedral group 133
octernions, see Cayley numbers
odd function 176
odd parity 35
odd permutation 157
one-parameter subgroup (of a Lie group) 64, 66
one-parameter subgroup (of $SO(3)$) 69
one-parameter subgroup (of $SU(2)$) 212
one-to-one mapping 400
onto mapping 400
open ball 421
open box 421
open neighborhood, see neighborhood
open set 229, 421
optical axis 6
optical flow 36, 45, 46, 239
optical flow detection 4
optimization 240, 263, 356, 358
orbiting motion 99, 111, 251
order (of a group) 403
order topology 421
orientation (of a corner) 292
orthogonal complement 159
orthogonal group, see $O(n)$
orthogonal matrix 199
orthogonality (of associated Legendre functions) 435
orthogonality (of irreducible representations) 420
orthogonality (of Laplace spherical harmonics) 435
orthogonality (of Legendre polynomials) 435
orthogonality (of spherical harmonics) 160, 170, 435
orthographic projection 7
orthonormal basis 410
orthonormal system 138
outer product, see vector product
overspecification 356, 357
overspecified problem 375

pan (of a camera) 304
parabolic surface 55, 57
parallel projection, see orthographic projection
parallel vectors 409
parallelism hypothesis 279, 384
parallelogram rule 409
parallelogram test 390
parity 35, 158, 198
passive motion 193
passive vision 356
path 424
path connected component, see arcwise connected component
path connected space, see arcwise connected space
pattern 3
pattern recognition 3
Pauli spin matrices 220
permutation 156, 403

perspective projection 6, 86
perspective projection operator, see projection
Pfaffian form, see differential form
photometric stereo 357
planar patch 243, 260, 357
planarity condition 243, 259, 260
point 105
point at infinity 12, 210, 322
point-to-point correspondence 18, 189, 193,
Poisson bracket, see Lie bracket
polar 317, 323
polar triangle 321, 323
polarity (of a projective space) 317, 321, 323
pole 317, 323
polyhedron 278
polynomial invariant 136, 142
positive-definite matrix 410
power set 399
principal axis 45, 142
principal curvature 57, 60
principal direction 55, 57, 60
principal value 142
product 402
product (of homotopic classes) 425
product (of mappings), see composition (of mapping)
product (of matrices) 402
product (of paths) 424
product (of tensors) 156
product representation 420
production rule 5
projection (of a scene) 15, 148, 149, 172
projection (of a topological space) 425
projection equations 6, 86
projection operator 148, 149, 172, also see projection (of a scene)
projective geometry 7, 113
projective space 13, 226
proper invariant subspace 152, 416
proper rotation 120
proper subgroup 403
proper subset 398
proper subspace 411
pseudo-inverse 244, 376
pseudo-orthographic approximation 240, 254, 255

quadratic convergence 349
quadric, see conic
quadrupole moment field 169
quantum mechanics 74, 82, 220
quantum state 74
quaternion 220–232
quotient group 227, 404, 415
quotient rule 156
quotient set 425

range (of a mapping) 399
ranging 13
rank 413
rational invariant 136
real linear space 406
real vector space, see real linear space
realization (of a Lie algebra) 68
realization (of spherical geometry) 128
reciprocal basis 412
reciprocal representation 11, 73
reciprocal system 139, 145
rectangular corner 280
rectangular polyhedron 278, 279
rectangularity hypothesis 278, 279, 379
rectangularity test 380
reduced surface parameters 362
reducible representation 24, 416
reducible set (of image characteristics) 104
reducible set (of invariants) 137
reducible subspace 152
reduction (of a representation) 24, 103, 417
reference point 36, 86
reflection 198
reflexive law 425
regular matrix, see nonsingular matrix
regularization 374, 375
regularization parameter 376
regularization with compromise 376
regularization without compromise 376
rejuvenation, see contraction (of tensors)
relation 360, 399
relative depth 260, 261
relative geometry 18
relative invariant 26, 247
relative rotation 287
relative topology 421
remainder theorem 258, 275
representation (of a group) 61, 62, 64, 415, 416
representation (of a Lie algebra) 430
representation (of $SO(2)$) 23
representation (of $SO(3)$) 11, 103
representation space 415
representative (of a homotopic class) 425
residual 58, 243, 245

resolution threshold 353
Ricci, Curbastro G. 155
Ricci calculus, see tensor calculus
Riemann, Georg F. B. 154
Riemann sphere 209
Riemannian geometry 18, 154
right inverse 402
right unit element 402
rigidity condition 44
Rodrigues formula (of 3D rotations) 204
Rodrigues formula (of Legendre polynomials) 170, 434
rotation 40, 431
rotation (of a flow) 247
rotation matrix 9, 198
rotation operator 431
rotation velocity 13, 36, 46, 72, 86, 241
rotational flow 40
rule of composition 205

scalar 28, 74, 103, 105, 155, 406
scalar invariant 105, 115, 119, 120
scalar part (of a quaternion) 225
scalar part (of a tensor) 32, 78
scalar product, see inner product
scalar triple product 117
scaling 245
scene 147, 148
scene coordinate 6
scene space 149
Schmidt orthogonalization 170, 410
Schrödinger equation 171
Schur's lemma 27, 69, 418
Schwartz, Laurent 333
Schwarz's inequality 117, 142, 409
second fundamental form 58, 60
segmentation 4
self-adjoint matrix, see Hermitian matrix
self-adjoint operator, see Hermitian operator
self-conjugate points 323
self-polar triangle 321, 323
self-reciprocal system 139, 145
semantic network 5
semi-simple Lie algebra 219
separable space, see Hausdorff space
set 398
set inclusion 398
shape from ... 4, 14, 21, 356, 357, 359
shape from motion 239
shape from shading 356
shape from texture 327
shear flow 41
shear strength 41
shearing (of a flow) 247
signature 122, 125, 157
similarity transformation 27, 413
simply connected Lie group 66, 67, 227, 229, 429
simply connected region 367
simply connected space 425
simply connected topological group 215
sine law 143
singular incidence structure 365
singularity (of a flow), see fixed point
singularity (of an incidence structure) 365
skeleton 353
skew-symmetric part, see antisymmetric part
skew-symmetric tensor, see antisymmetric tensor
skewed-symmetry hypothesis 279
$sl(n)$ 429
$SL(n)$ 405, 424, 429
slant 55
small rotation 121
smooth manifold 423
smooth function 333
$SO(2)$ 11, 23, 27, 28, 29, 30, 33, 35, 61, 62, 64, 83, 239
$SO(3)$ 11, 28, 29, 61–63, 69–84, 118, 119, 147, 148, 152–156, 165, 170, 181, 185, 197–235, 239, 418, 429, 431, 432
$SO(4)$ 232
$so(n)$ 429
$SO(n)$ 35, 136, 155, 201, 405, 424, 429
solid model 5
$Sp(n)$ 424
spanned subspace 411
spatio-temporal approach 46
special orthogonal group, see $SO(n)$
special unitary group, see $SU(n)$
spherical coordinates 432
spherical excess 128, 144
spherical function, see spherical harmonic
spherical geometry 126
spherical harmonic 147, 160, 171, 432
spherical harmonics expansion 161
spherical triangle 128, 143, 385
spherical trigonometry 127, 143
spin 74
spin angular momentum 74
spinor 74, 215
stability 374

stabilizer 376
stable solution 374
standard position 192
standard regularization 376
standard rotation 302, 304
standard transformation 302, 304
state (of a corner) 292
step function 354
stereo 13, 271, 357
stereo matching 4
stereographic projection 206
stress tensor 74
strong convergence 334, 434
structure (of a texture) 327, 328, 353
structure and motion parameters, see object parameters
structure constants 66, 103, 427
structure-based approach 328
su(2) 212, 213, 216–219
SU(2) 211–220, 227–230, 426, 429
su(*n*) 429
SU(*n*) 424, 429
subgroup 403
subpolyhedron 365
subset 398
subspace 411
subtexture 353
Sugihara, Kokichi 366
sum 402, 405
sum (of tensors) 155
superset 398
supertexture 353
support 151, 176
surface evolution equations 269
surface gradient 36, 240, 356, 357, 359, 377
surface model 5
surface parameters 240, 360
surjection, see onto mapping
swing (of a camera) 304
symmetric function 134
symmetric group 35
symmetric law 425
symmetric matrix 410
symmetric part 30, 76, 112, 129
symmetric polynomial 129
symmetric tensor 35, 129, 156
symmetrization 156
symmetry (of a state) 85
symplectic group, see *Sp*(*n*)
syzygy 118, 120, 137

tangent space 428
tangent space (of *Su*(2)) 212
tangent vector 428
tangential velocity 46
tensor 29, 35, 74, 76, 94, 103, 112, 155
tensor calculus 17, 154
tensor part (of an optical flow) 98, 251
tensor product 154
tensor representation (of *SO*(2)) 30, 35
tensor representation (of *SO*(3)) 76, 154
tensor representation (of *SO*(*n*)) 35
terminal point 424
test function 329, 333
tetrahedral group 133
texel 49
texture 49, 327
texture density, see density (of a texture)
texture element 49
theory of distributions 327, 328, 333
3D Euclidean approach 15
3D image 13
3D recovery equations 14, 21, 43, 52, 88, 239, 243, 249 340
3D rotation 198, 431
3D rotation group, see *SO*(3)
3D rotation matrix 9
3D scalar 28, 83
3D special orthogonal group, see *SO*(3)
3D tensor 29, 83
3D vector 28, 83, 109
3-sphere 225, 426
tilt (of a camera) 304
tilt (of a surface) 55
T-junction 282, 299
token 46
top-down approach 4
topological group, see continuous group
topological invariant 85, 228, 422, 425
topological space 421
topology 225, 421
total angular momentum 82, 171
totally ordered set, see linearly ordered set
trace (of a tensor) 161
transformation 400, 403
transitive law 425
translation (of a flow) 247
translation velocity 13, 36, 46, 86, 241
translational flow 39
transpose (of a matrix) 405
transposed representation 73
transversal velocity 46
triangular inequality 409
triangulation 271
trihedral corner 289
trihedral group 144

trivial topology 421
$2\frac{1}{2}$D sketch 4, 359
2D non-Euclidean approach 17, 264, 274
2D non-Euclidean metric 17
2D non-Euclidean space 17
2D projective geometry 17
2D projective space 12
2D projective transformation 12, 322
2D rotation group, see $SO(2)$
2D scalar 28, 83
2D special orthogonal group, see $SO(2)$
2D tensor 29, 83
2D vector 28, 83
2-sphere 226
type (of a corner) 290

$U(n)$ 424
ultrasonic ranging 13
underspecified problem 374
uniform convergence 333
unimodular group, see $SL(n)$
union (of sets) 398
unit element 402
unit vector 409
unitary group, see $U(n)$
unitary matrix 211, 414
unitary operator 414
unitary representation 417
universal covering 425
universal covering group 230, 425, 429
universal covering group (of a Lie group) 67

vanishing line 278, 279, 319, 341
vanishing point 278, 279, 310, 327
vanishing point heuristic 389, 391
variational principle 85, 244
vector 28, 75, 94, 103, 109, 155, 406
vector field (over a Lie group) 70, 428
vector field (over the image plane), see optical flow
vector part (of a quaternion) 225
vector part (of an optical flow) 98, 111
vector product 426
vector representation (of $SO(2)$) 28
vector representation (of $SO(3)$) 75, 153
vector space, see linear space
vertical velocity 46
viewer-centered coordinate system 7, 241
viewpoint 6
visibility (of edges) 290
visibility conditions 294
visible edge 290
vorticity 40

wall paradox 296
weak convergence 334
weaker topology 421
weight (of an irreducible representation of $SO(2)$) 25, 247
weight (of an irreducible representation of $SO(3)$) 71, 74
well-conditioned problem 374
well-defined problem 374
well-posed problem 374
Weyl, Hermann 24, 35
Weyl's theorem 35, 80
Weyl's thesis 22, 25, 29, 39, 42, 61, 98, 104, 105, 247
window 329
wire-frame model 5

Y-junction, see fork 28
Young diagram 35

zero 402, 406
zero-crossing 353